Electron Microscopy and Analysis Group Conference 2011

(EMAG 2011)

Journal of Physics: Conference Series Volume 371

Birmingham, United Kingdom
6-9 September 2011

ISBN: 978-1-62276-805-9
ISSN: 1742-6588

Printed from e-media with permission by:

Curran Associates, Inc.
57 Morehouse Lane
Red Hook, NY 12571

Some format issues inherent in the e-media version may also appear in this print version.

Printed by Curran Associates, Inc. (2013)

Institute of Physics
Dirac House, Temple Back
Bristol BS1 6BE UK

Phone: 44 1 17 929 7481
Fax: 44 1 17 920 0979

techtracking@iop.org

Additional copies of this publication are available from:

Curran Associates, Inc.
57 Morehouse Lane
Red Hook, NY 12571 USA
Phone: 845-758-0400
Fax: 845-758-2634
Email: curran@proceedings.com
Web: www.proceedings.com

Electron Microscopy and Analysis Group Conference 2011

(EMAG 2011)

Journal of Physics: Conference Series Volume 371

Birmingham, United Kingdom
6-9 September 2011

ISBN: 978-1-62276-805-9
ISSN: 1742-6588

IOP Institute of Physics

TABLE OF CONTENTS

BIOSCIENCES, BIOMATERIALS AND SOFT MATTER

CATALYSTS

EARTH AND ENVIRONMENTAL SCIENCES

FUNCTIONAL MATERIALS

IN-SITU ELECTRON MICROSCOPY

MODELLING AND QUANTIFICATION

NANOFABRICATION AND SCANNING ELECTRON MICROSCOPY

NANOMATERIALS

STRUCTURAL MATERIALS

Editorial

The biennial EMAG conference has established a strong reputation as a key event for the national and international electron microscopy community. In 2011 the meeting was held at The University of Birmingham, and I must first take this opportunity of thanking Birmingham for hosting the conference and for the excellent support we received from the local organisers. As a committee, we are delighted to see that enthusiasm for the EMAG conference series continues to be strong. We received more than 160 submitted abstracts, and 157 delegates attended the meeting.

The scientific programme organiser, Ian MacLaren, put together an exciting programme. Plenary lectures were presented by Professor Knut Urban, Dr Frances Ross and Dr Richard Henderson. There were a further 10 invited speakers, from the UK, Continental Europe, Australia, the USA and Japan. The quality of the contributed oral and poster presentations was also very high. EMAG is keen to encourage student participation, and a winner and two runners-up were presented with prizes for the best oral and poster presentations from a student.

I am always struck by the scientific quality of the oral and poster contributions and the vibrant discussions that occur both in the formal sessions and in the exhibition space at EMAG. I am convinced that a crucial part of maintaining that scientific quality is the opportunity that is offered of having a paper fully reviewed by two internationally selected referees and published in the *Journal of Physics: Conference Series*. For many students, this is the first fully reviewed paper they publish. I hope that you, like me, will be struck by the scientific quality of the 87 papers that follow, and that you will find them interesting and informative.

Finally I must thank the platinum sponsors for their support of the meeting. These were Gatan, Zeiss, FEI, JEOL and Hitachi. I must also thank the European Microscopy Society for their generous sponsorship and support for the travel costs of two invited speakers from Continental Europe.

Finally, keep an eye on www.emag-iop.org for details on EMAG 2013, which is to be held at the University of York.

P D Nellist
University of Oxford

Peer review statement

All papers published in this volume of *Journal of Physics: Conference Series* have been peer reviewed through processes administered by the proceedings Editors. Reviews were conducted by expert referees to the professional and scientific standards expected of a proceedings journal published by IOP Publishing.

STEM Image Post-processing for Instability and Aberration Correction for Transfer Function Extension

Lewys Jones and Peter D Nellist

Department of Materials, 13 Parks Road, Oxford, UK, OX1 3PH

Contact email: lewys.jones@materials.ox.ac.uk

Scanning transmission electron microscopy (STEM) images can be limited by either aberrations or, because of the technique's serial acquisition, by the effects of environmental instabilities. MatLab image-processing code was written which performs a two stage process to restore degraded high-resolution high-angle annular dark-field (HAADF) STEM images. Firstly, individual images were analysed to identify and correct for high-frequency scan noise; image resolution and signal-noise ratio (SNR) are used as performance metrics and were improved by up to 11.8% and 49% respectively. Secondly, a focal series was used to identify variations in information transfer as a function of defocus (aberrations) whereafter a single image was reconstructed yielding an increase in resolution and SNR of 9.88% and 205% respectively.

1. Introduction & Background

Many research facilities use both uncorrected and aberration-corrected scanning transmission electron microscopes (STEMs) to study a variety of imaging or spectroscopic signals [1,2]. Aberration-corrected STEM instruments are highly sensitive to environmental instabilities [1,3,4]; whilst aberrations can often limit the performance of uncorrected instruments [5,6].

Many images reported in the literature display, to varying extents, signs of image distortion arising from environmental or instrumental instabilities, which are often simply low-pass filtered, merely masking their effects and further deteriorating their resolution [7,8]. As an alternative, MatLab image-processing code has been written to examine individual images for signs of such distortion and correct for it. Subsequently, separate analysis code is used to analyse a focal-series of STEM images to identify variations in information transfer as a function of defocus. Such variation is the result of the form of the three-dimensional (aberrated) STEM probe and is here determined and corrected for.

2. Environment-Instability Image-Distortion Correction of Single Images

Microscope-suite instabilities range over several orders of frequency (0.1 Hz – 100 kHz) and can affect data recording in a variety of ways. The lowest frequency effects (over several scan-lines, perhaps <1 Hz), including drift, can cause images to appear warped [9]; whilst intermediate frequencies (across several dwell-times, say 10 Hz – 1000 Hz) can cause local distortion of atom column positions [10]. This work concentrates on the distortions arising from frequencies that correspond to both STEM image scan-line times and pixel-dwell-times. For a 50 Hz line-synced STEM image with a dwell time of 38.5 μs this corresponds to approximately 50 Hz – 25 kHz. Such distortions are examined both parallel and perpendicular to the scan direction.

Imaging distortions from probe instabilities parallel to the scan direction are often called scan-noise; this is described in detail by Muller et al. and is observed in images as atoms that appear 'cut' or 'torn', Figure 1a) [3].

a)
b)

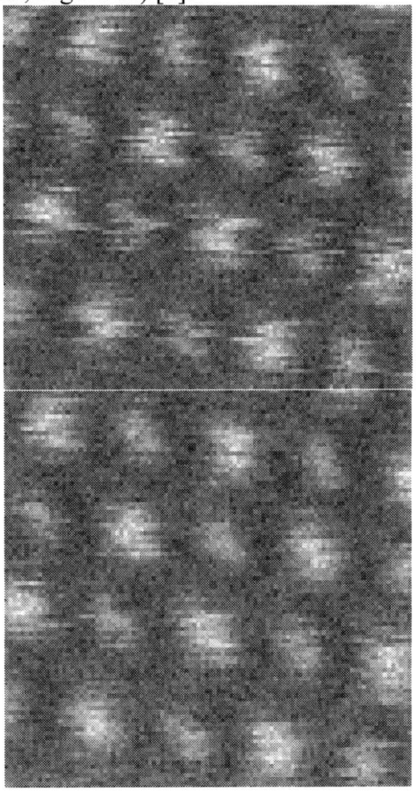

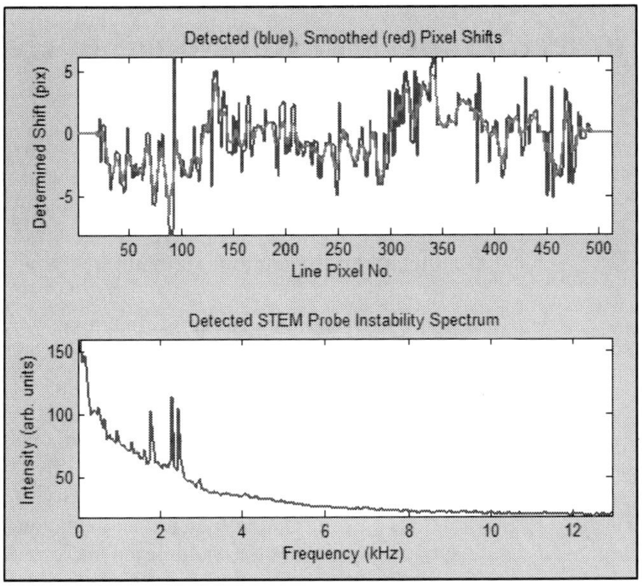

Figure 1 –STEM image enlargement; a) before, and b) after scan-distortion correction.

Figure 2 – Top; example pixel-offset profile showing detected offsets (blue) and applied corrections (red). Bottom; determined image-distortion power spectrum.

In the case of probe-instabilities perpendicular to the scan direction atom columns may display dissociated lines of features. This may be either a bright 'cap' of an atom separated from either its top or bottom or, equivalently, a dark background line introduced inside an atom image. Imaging probe instabilities both parallel and perpendicular to the scan direction can give rise to artefacts in diffractograms ; with spots becoming less sharp and additional 'copy-cat' spot patterns appear as weak duplicates of the genuine spots. These additional sets of spots may appear at any place in the diffractogram depending on the interaction of the line-scan time and the frequencies of distortions present. In the experiments conducted the satellite spot sets appeared half way to the top and bottom of the diffractogram similar to ref. [11]. This suggests the distortion may arise from a harmonic of the line-sync frequency of 50 Hz. Note, that these additional diffractogram spots are different and separable from those associated with incorrect black-level setting [12,13]. Such black-level artefact spots are aligned with the genuine crystallography of the diffractogram regardless of specimen rotation whereas the vertical probe instabilities are *always* duplicates of the first order spots shifted in the Fourier transforms. This introduces an advisable experimental practice where the specimen, or scan, is rotated so that the underlying crystallography deliberately does *not* align with the scan direction, the converse of what is most commonly reported in the literature.

Having identified the effects of distortions both parallel and perpendicular to the scan direction these are then corrected sequentially. To correct parallel to the scan direction a horizontal offset is determined for each pixel in the image by defining a 'test-band' about the pixel of interest, this is compared with the adjacent rows and an optimum fit determined; the strength of their cross-correlation suggests the confidence in this determined optimum result. A slight smoothing of these of these suggested offsets is used to avoid erroneous shifts being applied (Figure 2, top, red line). Using these determined offsets the image row is resampled corresponding to the pixel-dwell locations of an undisturbed probe.

To correct probe-offsets perpendicular to the scan direction the horizontally-integrated-intensity profile of features (atom columns) is examined and a sorting algorithm is performed such that within a local region-of-interest the image rows are allowed to swap such that after the swap the intensity is monotonically decaying away from the centre of the feature. This operation is not always necessary and is dependent on the nature of the instabilities present during the data acquisition. In addition, following the above restoration of the vertical probe wobble component an additional horizontal analysis may be merited.

Figure 1 then shows the results of such a restoration. Both the image resolution, as determined from the visibility of spots in diffractograms, and signal-noise-ratio (SNR), as determined from 15 real-space line profiles across each of 4 directions*, were calculated. Resolution was improved from the (400) at 0.976Å to the (420) at 0.873Å (11.8% better) and the SNR was improved from 2.77 to 4.14 (49% better). Figure 2 shows an example determined-distortion-profile and the image's distortion frequency spectrum which can be used to track down environmental instabilities. An underlying assumption of these methods is that the true sample is smoothly varying and should not contain excessive discontinuities or jumps in intensity between adjacent pixels. This is valid when the magnification allows for sampling of atoms to be at least a few times Nyquist.

3. Aberration Correction from Focal Series

It has been shown that focal series of images can access information at spatial frequencies beyond those of a Scherzer focus image [5,6]. By recording a focal series of STEM images all this data is within the data volume. By calculating the Fourier transform of each of these images separately an accompanying data volume exists that illustrates the variation in intensity of spatial frequencies over defocus. Analysis of this frequency space-defocus data volume yields a surface of maximum intensity which describes the defocuses at which each special frequency is most strongly transferred, Figure 3.

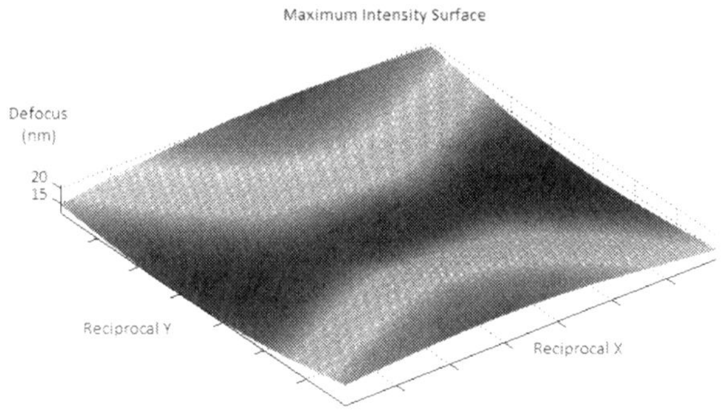

Figure 3 – Determined spatial frequency-defocus maximal-intensity surface, showing 2-fold astigmatism.

Figure 4 – Near-Scherzer focus STEM image of MoS_2 with reconstruction (overlaid).

* SNR was determined automatically using the authors Hawkeye v0.9 Matlab code (manuscript in preparation). Full algorithm description online at: www.lewysjones.com.

A Fourier image can then be reconstructed so that for every spatial frequency in the data volume the most strongly transferred, optimum focus, value is used. This transformed back to real-space then yields an image with improved resolution and SNR. Figure 4 shows a Scherzer-focus STEM image and reconstruction from a MoS_2 focal-series. The contrast variation seen from bottom-left to top-right is thought to be from thickness variation in this layered material. Here resolution was improved by 9.88% and SNR, as determined from 15 real-space line profiles across each of three directions*, from 6.70 to 20.44 (205% better). Were eight Scherzer focus images simply averaged, a SNR gain of $\sqrt{8}$ (+183%) would have been expected. The data shown here exceeded this SNR gain, demonstrating a real improvement in image quality. In addition the dumbbell inter-peak dip was improved from 36% to 63%, with the inter-peak dip also becoming more isotropic over the different directions of nearest neighbour pairs, suggesting an improvement it the homogeneity of the transfer function.

4. Conclusions

Environmental instabilities can and do affect the quality of STEM images that can be recorded. Whilst often image distortions are low-pass filtered to hide their effects, here a means of correcting for these has been presented. In the processing of single images correcting for such instabilities has allowed information to be recovered and resolution and SNR improved by 11.8% and 49% respectively.

Where aberrations are the limiting factor for imaging it has been shown that recording a focal-series can capture usually discarded image information. Frequency space analysis and subsequent reconstruction of a single image has expanded the contrast transferred and has again facilitated an increase in resolution and SNR of 9.88% and 205% respectively. This represents a SNR improvement better than $\sqrt{8}$ (+183%) that is attainable by image averaging.

Distortion correction of individual images does not preclude focal-series analysis pre-processing of each image in a focal-series is expected to yield still further gains.

References

[1] Browning N D, Chisholm M F & Pennycook S J 1994 *Advanced Materials* **6** 328-331
[2] Muller D A et al. 2008 *Science* **319** 1073-76
[3] Muller D A et al. 2006 *Ultramicroscopy* **106**, 1033-40
[4] Harrach H von 1995 *Ultramicroscopy* **58**, 1–5
[5] Loane R, Xu P & Silcox J 1992 *Ultramicroscopy* **40**, 121-38.
[6] Nellist P D & Pennycook S J 1998 *Physical Review Letters* **81**, 4156-59
[7] Voyles P M et al. 2002 *Nature* **416**, 826-29
[8] Voyles P M, Grazul J L & Muller D 2003 *Ultramicroscopy* **96**, 251-73
[9] Nakanishi N 2002 *Journal of Electron Microscopy* **51**, 383-390
[10] Recnik A, Möbus G & Sturm S 2005 *Ultramicroscopy* **103**, 285-301
[11] Klie R F & Zhu Y 2005 *Micron* **36**, 219-31
[12] Yu Z & Silcox J 2004 *Microscopy and Microanalysis* **10**, 570-71
[13] Nellist P D et al. 2004 *Science* **305**, 1741

Chromatic Confocal Electron Microscopy with a Finite Pinhole Size

P Wang, A I Kirkland and P D Nellist

Department of Materials, University of Oxford, Parks Road, Oxford OX1 3PH, UK

E-mail: peng.wang@materials.ox.ac.uk

Abstract. Scanning confocal electron microscopy (SCEM) is a new imaging mode in electron microscopy. Spherical aberration corrected electron microscope instruments fitted with two aberration correctors can be used in this mode which provides improved depth resolution and selectivity compared to optical sectioning in a conventional scanning transmission geometry. In this article, we consider the depth resolution and energy resolution in the confocal optical configuration for SCEM using inelastically scattered electrons with a finite pinhole size. We experimentally demonstrate energy-filtered optical sectioning in a double aberration-corrected instrument with uncorrected chromatic aberration without using a dedicated energy filter.

1. Introduction

In a state-of-art aberration corrected transmission electron microscope (TEM), the depth of field can be reduced to a few nanometres [1] due to an enlarged numerical aperture. This offers the capability of depth sectioning in the TEM in a similar fashion to confocal scanning optical microscopy [2]. In contrast to widefield microscopy, the confocal geometry provides controllable depth discrimination and reduction of background signal away from the focal plane, as the scattering from points away from the confocal point are detected less strongly than the in-focus scattering, as shown in Fig. 1 a).
The confocal mode detecting elastically scattered electrons, which is referred to as bright-field [3] SCEM (or coherent SCEM), was theoretically calculated by Cosgriff et al. [3] and experimentally demonstrated by Nellist et al. [4] on a double aberration corrected TEM/STEM instrument and further investigated by Mitsuishi et al. [5] and Wang et al.[6]. With an addition of an energy filter, Wang et al.[7, 8] have reported an incoherent confocal mode by filtering electrons that have been inelastically scattered with a specific energy-loss in the sample leading to energy-filtered (EF-) SCEM. They experimentally showed that this incoherent confocal geometry preserves depth resolution for a laterally extended object, unlike convention scanning TEM (STEM) imaging. This has been illustrated by considering the 3-D optical transfer function (OTF) [2] where the incoherent confocal mode does not have "missing cone".

The calculation by Wang et al.[7, 8] further showed that on an instrument with chromatic aberration (Cc) in the post-specimen lenses the confocal geometry with an infinitesimal point-like pinhole possesses a capability of energy selecting with a resolution of ~2 eV even without a physical energy filter present. This makes it possible to carry out EF-SCEM for 3D elemental distribution on instruments that are not fitted with a physical energy filter. In practice, a finite-size pinhole has to be used to obtain a sufficient signal to noise ratio, as the inelastically scattered electrons often suffer low counts. The extended pinhole size can affect the depth and energy discrimination.

In this paper, we will calculate the effect of the pinhole size on not only the depth resolution, but also the energy resolution in particular. We also experimentally illustrate the depth and energy discrimination in the confocal mode without an energy-filter using a through focal series of probe images recorded at the pinhole plane.

2. Energy filtered SCEM

To quantitatively examine the effective depth and energy resolution in EFSCEM, the plane-spread function or z–response of the SCEM for scattering by a planar film needs to be calculated. For a completely incoherent scattering object function $V(R)$ located at a height, z=0 and in the confocal condition, the three-dimensional SCEM image intensity [4] can be written as:

$$I(R_1, z) = \int V(\mathbf{R}) |P_1(\mathbf{R} - R_1, z)|^2 |P_2(R_1 - \mathbf{R}, -z)|^2 d\mathbf{R} \quad (1)$$

where we have used the notation given in Reference[4]. Therefore, the point spread function (*PSF*) for incoherent confocal imaging is:

$$PSF(\mathbf{R}, z) = |P_1(\mathbf{R}, z)|^2 |P_2(-\mathbf{R}, -z)|^2 \quad (2)$$

Using the reciprocity principle and the symmetry of the confocal configuration, the probe function P_2 can be considered as a virtual probe that would be formed by the post-specimen optics from a hypothetical source at the pinhole [4]. In this incoherent confocal mode, the pinhole can be treated as an incoherent virtual source in the same way as the finite source size. Therefore the *PSF* with the finite-size pinhole can be calculated as:

$$PSF(\mathbf{R}, z) = |P_1(\mathbf{R}, z)|^2 \left[|P_2(-\mathbf{R}, -z)|^2 \underset{\mathbf{R}}{\otimes} D(\mathbf{R}) \right] \quad (3)$$

where D is the circular pinhole function or top-hat function and $\underset{\mathbf{R}}{\otimes}$ is the notation for convolution of the coordinate $\mathbf{R}$. To setup a confocal condition for an energy loss of interest ΔE_{loss}, an increment of $k\Delta E_{loss}$ in the accelerating voltage E_0 needs to introduced, where $k = Cc_2/(Cc_1 + Cc_2)$ (Cc_1 and Cc_2 are the chromatic coefficients of the pre- and post-specimen optics, respectively) [7]. The SCEM *PSF* can be further written [8] as:

$$PSF(\mathbf{R}, \varepsilon) = |P_1(\mathbf{R}, z)|^2 \left[\left| P_2[-\mathbf{R}, -(z + \frac{\varepsilon}{E_0} Cc_2)] \right|^2 \underset{\mathbf{R}}{\otimes} D(\mathbf{R}) \right] \quad (4)$$

where a parameter, ε, which gives the deviation in the energy of scattered electrons from the desired energy loss ΔE_{loss}. The additional term of $\frac{\varepsilon}{E_0} Cc_2$ is included to take into account of focal spread at the pinhole plane due to the Cc_2 in the post-specimen lenses as shown in Fig. 1 b).

The EFSCEM z-response for a planar object for an energy loss with a deviation of ε from ΔE_{loss} can be written as:

$$Z_{\text{SCEM}}(z, \varepsilon) = \int |P_1(\mathbf{R}, z)|^2 \left[\left| P_2[-\mathbf{R}, -(z + \frac{\varepsilon}{E_0} Cc_2)] \right|^2 \underset{\mathbf{R}}{\otimes} D(\mathbf{R}) \right] d\mathbf{R} \quad (5)$$

By integrating over all possible energy deviations (ε) or all possible heights (z) of a scattering plane in Eq. 4, we obtain the *z*-responses or the energy filtering response, as shown in Fig. 1 c) and d), respectively, as a function of the pinhole size. Both the *z*-response and the energy-response are deteriorated with the increase of the pinhole size. Without a pinhole (or when the pinhole size is ∞), Eq. 4 is degraded to

$$Z_{\text{STEM}}(z) = \int |P_1(\mathbf{R}, z)|^2 d\mathbf{R} \quad (6)$$

which is the z-response for the STEM imaging [8]. Due to the total intensity of the beam propagating through the plane for each value of z being constant, Z is constant and hence the depth discrimination for a planar object completely disappears, as shown as the dotted black curve (···) in Fig. 1 c) . As a result, Eq. 6 is no longer a function of the energy deviation (ε), so that there is no energy response as shown as the dotted black curve (···) in Fig. 1 d).

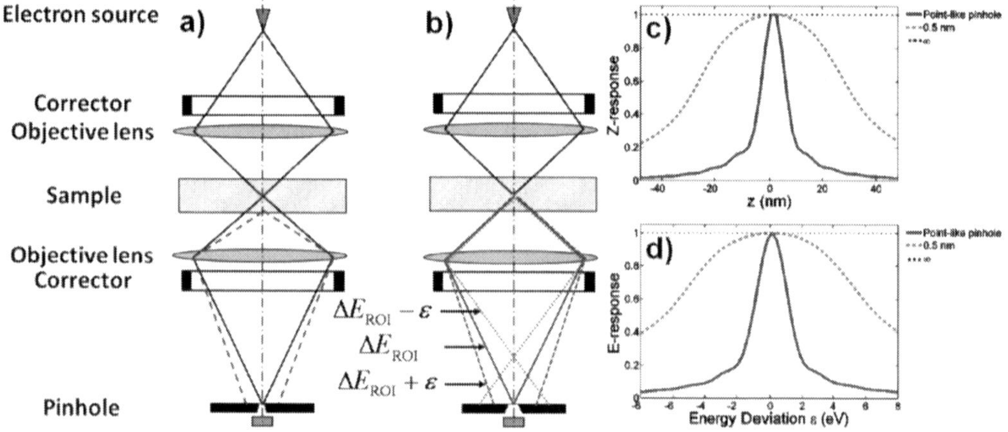

Figure 1 shows schematic diagrams of confocal trajectory illustrating the depth **a)** and energy loss **b)** discriminations given by a pinhole. It not only can reject the out-of-focus ray (----) in **a)** to improve depth resolution, but also can select the beam (—) with an energy loss ΔE of interest and excludes the beams (---) and (···) with an energy deviation of $\pm\varepsilon$ in **b)** due to Cc in the post-specimen optics. **c)** and **d)** show the z- and energy-response for different values of the pinhole size in diameter, respectively.

3. Experimental optical sectioning without an energy filter

To experimentally examine the z- and energy-response without an energy filter, we used a 2-D carbon thin film as a sample and setup a confocal condition for an energy loss of the carbon K-edge using the approach described in Ref [7, 8]. Fig. 1 b) shows that the post-specimen optics images the incident probe on the sample to form a probe image on the pinhole plane. Due to the absence of an energy filter and the residue Cc_2 in the post-specimen optics, the probe image was formed by all the electrons experiencing elastic and (or) inelastic scattering. By replacing the pinhole with a Gatan CCD camera, Fig. 2 show the probe images at the pinhole plane, with the 2-D carbon film at the confocal point a), 93nm b) and 106 nm away from the confocal point along the z direction, respectively. It is clear to see that the electrons which suffered the energy loss of carbon K-edge formed the bright and sharp central peak labeled with a red arrow, as shown in Fig 2 a), whereas the zero-loss beam was largely under-focused by $\Delta z = k\Delta E_{loss} \cdot Cc_2 / E_0$ and formed the majority intensity of the background disc. This experimentally illustrates that the electrons that undergo different energy loss with the energy deviations (ε) from ΔE_{loss} (here ΔE_{loss} is the carbon K-edge), became out-of-focus at the collection plane, as shown in Fig. 1 b). By inserting the pinhole, it is feasible to reject or detect less strongly the electrons with other energy loss to give energy discrimination. As predicted in Fig. 1 d), the pinhole size determines the energy resolution.

Fig. 2 d) shows line-profiles across the central peaks in the probe images as shown in Fig. 2 a), b) and c), respectively. As the specimen was moved away the confocal point, only the signal at the central peak in the probe image dropped off, whereas the intensities of the background disc remained relatively unchanged. This is because the electrons which contributed to the background had been largely defocused at the pinhole plane. Therefore, the variations of the sample heights only gave small and local fluctuations in the background, as shown in Fig. 2 a), b) and c). However, the sharp central peak, which was formed by the electrons with the energy loss in the confocal condition, was sensitive to the height variations. As the sample was moved away from the confocal point, the scattered electrons became out-of-focus at the collection plane, which gave the apparent signal drop-off, as shown in Fig. 2 d). Virtual pinholes with diameters of 0.03(single pixel of the CCD) and 0.5 nm were applied to a through focal series of the probe images by integrating the intensities within a circular region centred at each central peak. Fig. 2 e) shows that the larger the pinhole is used, the worse the depth resolution is. In an extreme case where the virtual pinhole is infinitely large, all the intensities of the probe images are integrated as shown as the dotted black curve (···) in Fig. 2 e). As predicted by Eq. 6, the system was reverts to the conventional STEM imaging case and no z-response exists, which consists with the dotted black curve (···) as shown in Fig. 1 c). Because of the integration of all the electrons, the energy response also disappears, as predicted in the dotted black curve (···) as shown in Fig. 1 d).

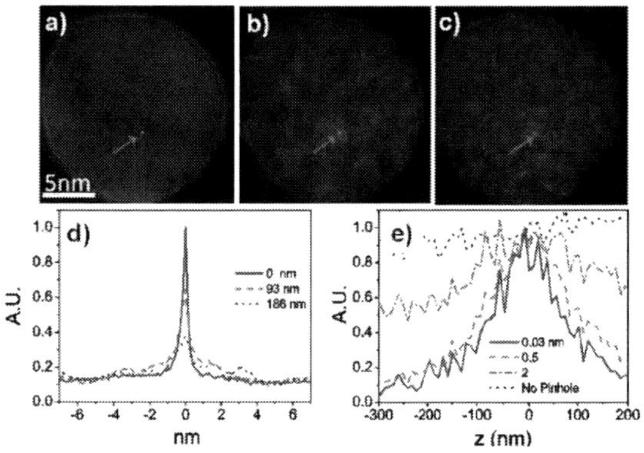

Figure 2 a) shows experimental probe images formed through a carbon film in the EFSCEM mode for the carbon K-edge without an energy filter, when the film is just at the confocal point **a)**, 93 **b)** and 106nm **c)** away from it, respectively. **d)** line-profiles across the centre peaks, labelled with a red arrow, in **a)**, **b)** and **c)**, respectively. **e)** shows an optical sectioning z line scan with the virtual pinhole with diameters of 0.03, 0.5, 2 and ∞ nm, respectively.

4. Conclusion

We showed the mechanisms of the depth and energy discriminations in EFSCEM on a chromatic electron microscope without an energy filter. We theoretically calculated that the depth and energy resolution can be controlled by the pinhole size, allowing the operator to trade image and energy resolution versus signal to noise ratio. which are further proved by the experimental optical sectioning using a through focal series of the probe images recorded on the CCD without an energy filter.

Acknowledgements

PW, AIK and PDN acknowledge financial support from the Leverhulme Trust (F/08 749/B) and the EPSRC (EP/F048009/1).

References

[1] van Benthem K, Lupini A R, Kim M, Baik H S, Doh S, Lee J-H, Oxley M P, Findlay S D, Allen L J, Luck J T, Pennycook S J, 2005 *Applied Physics Letters* **87** 034104-034103.

[2] Gu M, 1996 Principles of Three Dimensional Imaging in Confocal Microscopes, *World Scientific*.

[3] Cosgriff E C, D'Alfonso A J, Allen L J, Findlay S D, Kirkland A I, Nellist P D, 2008 *Ultramicroscopy* **108** 1558-1566.

[4] Nellist P D, Behan G, Kirkland A I, Hetherington C J D, 2006 *Applied Physics Letters* **89** 124105.

[5] Mitsuishi K, Hashimoto A, Takeguchi M, Shimojo M, Ishizuka K, 2010 *Ultramicroscopy* **111** 20-26.

[6] Wang P, Behan G, Kirkland A I, Nellist P D, Cosgriff E C, D'Alfonso A J, Morgan A J, Allen L J, Hashimoto A, Takeguchi M, Mitsuishi K, Shimojo M, 2011 *Ultramicroscopy* **111** 877-886.

[7] Wang P, et al., 2010 *Journal of Physics: Conference Series* **241** 012012.

[8] Wang P, Behan G, Takeguchi M, Hashimoto A, Mitsuishi K, Shimojo M, Kirkland A I, Nellist P D, 2010 *Physical Review Letters* **104** 200801.

Journal of Physics: Conference Series **371** (2012) 012003 doi:10.1088/1742-6596/371/1/012003

Three-Dimensional Crystal Structure Mapping by Diffractive Scanning Confocal Electron Microscopy (SCEM)

Lewys Jones, Peng Wang and Peter D Nellist

Department of Materials, 13 Parks Road, Oxford, UK, OX1 3UJ

Contact email: lewys.jones@materials.ox.ac.uk

Recently a diffracted-probe scanning confocal electron microscopy (SCEM) mode was demonstrated. Analysis of such diffracted-probe images yields the sample height (defocus), thickness and crystal orientation. An extension of this principle for three-dimensional (3D) mapping is described and preliminary bicrystal data presented. Additional experimental criteria are discussed and the 3D resolution of this new diffractive SCEM mapping is derived and was found to be ≈20 nm and is comparable with 3D FIB-EBSD in terms of both field-of-view, spatial resolution and acquisition time.

1. Introduction

Many material properties are highly dependent on the size or morphology of their grains. Examples of this include the yield strength or failure strain of metals and the conductivity of nano-scale films of copper or high-temperature super-conductors. The ability to accurately determine grain size, shape and orientation is essential for modern materials science.

Focused ion-beam milling with electron back-scattered diffraction (3D-FIB EBSD) offers large fields of view (10s of μm) and reasonable spatial resolution (≈ 25 nm) [1]; however this is very time consuming (tens of hours) and necessarily destructive, precluding any further analysis of the specimen.

Non-destructive techniques exist for characterising crystal structures at higher resolution including (scanning) transmission electron microscopy, HR-(S)TEM; offering both imaging and diffraction information for two-dimensional examination of thin samples [2]. Such wide-field STEM can be modified to set up a scanning-confocal geometry (SCEM) [3] ; rejecting out of focus rays and improving depth sensitivity [4].

Wang et al. has demonstrated for a single crystal that raising the specimen well above the confocal plane yields a family of diffracted probe images; yielding information about crystal structure, thickness and height (Δf, **Figure 1**) of the specimen [5]. The confocal plane is recorded by positioning the CCD in the real-space optically conjugate plane. The intensity and angular distribution of scattered spots will depend on several factors including: the size of the objective aperture, specimen thickness, specimen height and angular proximity to a zone axis.

Here this non-destructive alternative is presented, including example data from a bicrystal; showing the ability to discriminate between two grains across a boundary. Some experimental guidelines are also elaborated on, including deriving the 3D resolution, on-track to developing this as a practical grain mapping technique.

2. Optical Geometry

A geometric consideration shows that the radial distance of the diffracted probes, $\boldsymbol{r}$, is related to their reciprocal lattice vector, $\boldsymbol{g}$, the sample height, Δf, and the electron wavelength, λ [5].

$$\boldsymbol{r} = \lambda . \Delta f . \boldsymbol{g} \tag{1}$$

If the material and its orientation are known, then λ, $\boldsymbol{r}$ and $\boldsymbol{g}$ are all known and then defocus can be expressed as follows.

$$\Delta f = \frac{|r|}{\lambda . |g|} \tag{2}$$

Equation 2 is valid for each diffracted probe and the the gradient of a $|r|$ versus $|g|$ plot yields an accurate value for Δf, the sample height.

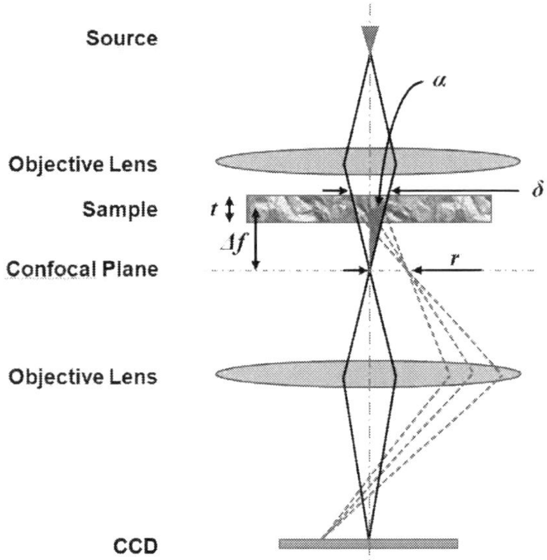

Figure 1. Diffractive SCEM optical geometry showing sample-probe intersection volume.

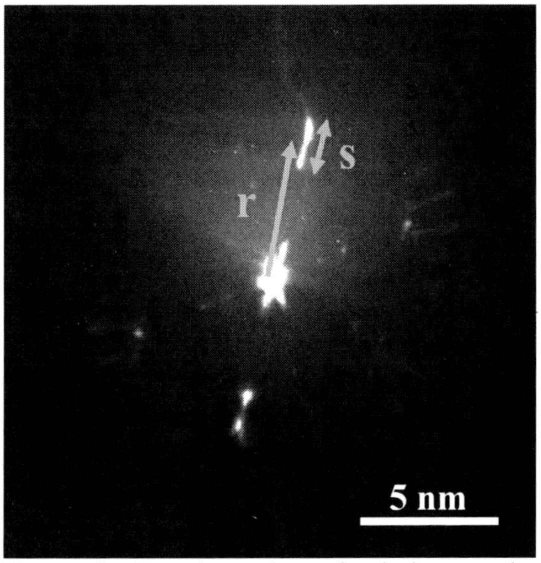

Figure 2. Experimental confocal-plane probe image from an YBCO film ($t \approx 200$ nm).

In **Figure 2Figure 1** the diffracted probe images are not sharp points, but have finite radial spread, s. Diffraction occurs throughout the thickness of the sample, t, and such spots are expected to be streaked between the solutions to Equation 1 from both the entrance and exit surfaces [5].

$$t = \Delta f_{ent} - \Delta f_{ext} = \frac{|r|_{max}}{\lambda . |g|} - \frac{|r|_{min}}{\lambda . |g|} \tag{3}$$

$$t = \frac{s}{\lambda \, . \, g_{hkl}}$$

There is occasionally observed some extra detail in the structure of the radial streaks and this is thought to be the effect of multiple scattering and is under investigation.

Equations 2 & 3 then allow a thin crystal's dimensions to be determined and Wang et al. has shown this thickness and depth resolution to be $\approx$5 nm [5]. Repeating a Δf versus t analysis over an XY raster means that the three dimensional profile of a grain boundary can be mapped.

3. Experimental Results & Discussion

The specimen studied in this work was an YBCO thin film prepared on a MgO substrate similar to those in reference [2]. Optical microscopy, **Figure 3**, shows many grains are visible sized 20-50 µm. The rough surface of these grains remain visible in the dual beam FIB-SEM and a Zeiss NVision was used to mill a specimen 10µm by 5µm with a thickness $\approx$200 nm.

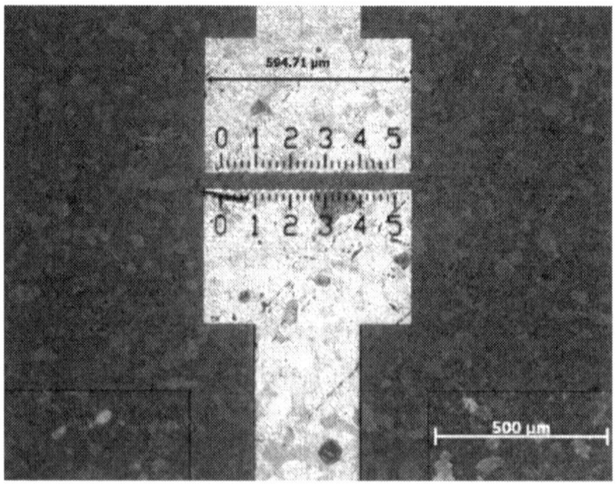

Figure 3. Optical micrograph of polycrystalline YBCO thin film between electrical contacts.

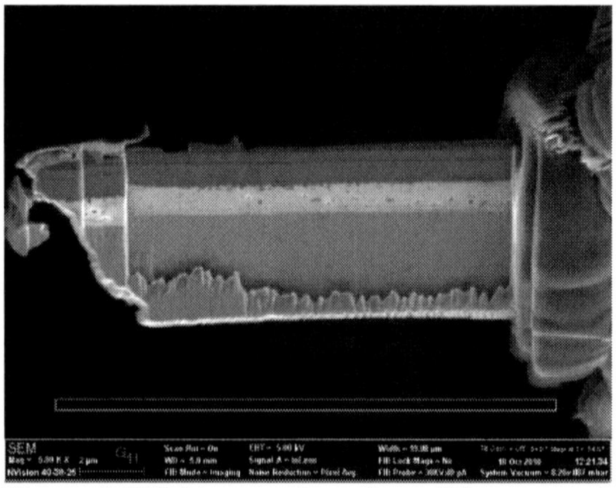

Figure 4. SEM image of one face of the prepared specimen (field of view ≈15x20 μm).

SEM imaging of both sample faces allowed for various types of grain interface to be identified (**Figure 5**). An edge-view bicrystal was investigated using the Oxford-JEOL 2200MCO with pre and post specimen spherical aberration correction. Confocal plane images were recorded along a 400 nm line at 20 nm intervals, **Figure 6**.

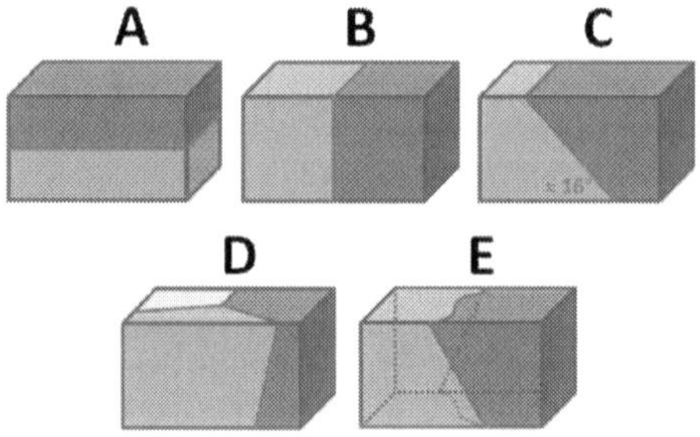

Figure 5. Various potential grain interfaces: a) plan-view bicrystal, b) edge-view bicrystal, c) inclined bicrystal, d) tricrystal and e) tortuous bicrystal boundary.

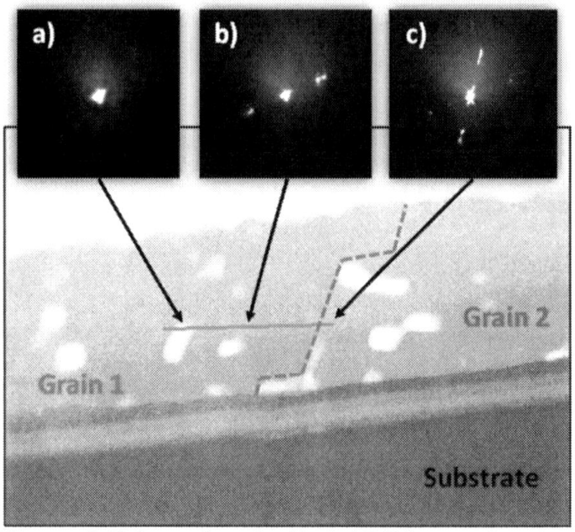

Figure 6. False-colour (online) ADF image showing two grains separated by a G.B. (dashed line). Solid line shows a 400 nm long measurement location path. Insets show confocal plane images corresponding to a) a hole, b) grain 1 and c) grain 2.

Inset a) shows no diffracted probes when the beam is at a hole in the specimen. Insets b) and c) show the patterns recorded either side of the grain boundary and how the two grains can be discriminated.

When the sample is placed well above the confocal plane the probe exhibits a finite width at the specimen. This probe spread limits the lateral resolution to much poorer than the atomic resolution of ADF STEM. To estimate the lateral resolution achievable we must first consider the types of specimens technologically relevant for such grain mapping where thicknesses up to 200 nm may be needed. At such thicknesses the first order diffraction spots will become heavily elongated. If the second order spots are also to be used for thickness/defocus determination then the defocus will need to be large enough to sufficiently separate the two orders. This criterion is comfortably met when the defocus is at least equal to the thickness. The lateral resolution then is limited by the width of the intersection volume, δ (**Figure 1**). Geometrically δ is found to be proportional to the defocus and the semi-angle of probe convergence.

$$\delta = 2\,.\,\Delta f_{ent}\,.\,\alpha \qquad (4)$$

For the imaging conditions used in reference [5] (Δf_{ent} = 186 nm, α = 22 mrad) this would correspond to a lateral resolution of 8.2 nm. However for the imaging of more realistic thicker specimens:

$$\Delta f_{ent} = 2\,.\,t \qquad (5)$$

$$\delta = 2\,.\,(2\,.\,t)\,.\,\alpha$$
$$\delta = 4\,.\,t\,.\,\alpha$$

For t = 200 nm and α = 22 mrad this gives a lateral resolution limit of ≈18 nm. In **Figure 6** inset c) a weak contribution from grain 1 is seen. This arises from the probe-specimen intersection volume containing more than one grain. As the radial distance of the spots remains unchanged from inset b), the originating crystals for the two patterns evident in inset c) must be at the same height (edge-view bicrystal, **Figure 5** B.). This is consistent with the predicted lateral resolution and an intersection volume being probed near the grain boundary.

This thickness dependence of the resolution requires the user to balance lateral resolution and mapable volume similar to EBSD. Additionally Wang et al. has shown that for specimens much thicker than 40 nm, and oriented to a strongly diffracting condition, dynamical diffraction imparts additional fine structure to the diffracted probe images making analysis of such patterns very difficult. This condition can be avoided by slightly tilting the sample to a more kinematical condition. In the case where this is not possible, data interoperability may impose a thickness limit.

4. Conclusions

A practical extension of the previously reported technique is described, with a bicrystal test sample prepared and explored. A line across the grain boundary was investigated at intervals using diffractive SCEM and the associated confocal plane images were used to demonstrate the ability to clearly discriminate between different grains, or holes, in the specimen. Some new practical considerations and criterion were discussed relating to sample thickness and a practical limit to 3D resolution of around 20 nm was derived. Extending such an analysis to an XY raster and incorporating determination of the specimen height and thickness is expected to yield a non-destructive three-dimensional grain mapping technique.

Acknowledgments

The author would like to thank Dr. G Hughes for assistance with FIB milling and Dr. S Speller for providing the specimen. The author is grateful to the EPSRC for supporting this work.

References

[1] Inkson B, Mulvihill M, Möbus G, 2001 *Scripta Materialia*, **45**, 753–758.
[2] Weigand M et al. 2010 *Physical Review B*, **81**, 174537.
[3] Nellist P D, Behan G, Kirkland A I, and Hetherington C J D, 2006 *Applied Physics Letters*, **89**.
[4] Jesson D and Pennycook S J, 1993 *Proc R Soc A*, **441**, 261-281.
[5] Wang P et al. 2010 *Ultramicroscopy*, **111**, 877-86.

Journal of Physics: Conference Series **371** (2012) 012004 doi:10.1088/1742-6596/371/1/012004

Development of Spin-Polarized Pulsed TEM

M Kuwahara[1], F. Ichihashi[2], S. Kusunoki[2], Y Takeda[2], K Saitoh[1,2], T Ujihara[2], H Asano[2], T. Nakanishi[3], N Tanaka[1,2]

[1] EcoTopia Science Institute, Nagoya University, Nagoya, Aichi 464-8603, Japan

[2] Graduate School of Engineering, Nagoya University, Nagoya, Aichi 464-8603, Japan

[3] Graduate School of Science, Nagoya University, Nagoya, Aichi 464-8602, Japan

E-mail: kuwahara@esi.nagoya-u.ac.jp

Abstract. Recent nanointegration of magnetic storage and spintronic devices is expected to lead to the development of systems that can analyze magnetic and spin states with a nanometer-order spatial resolution. Development of a polarized transmission electron microscope (SPTEM) that employs a spin-polarized electron source (PES) has commenced at Nagoya University. A spin-polarized electron beam is extracted from a photocathode that has a negative electron affinity surface. The temporal profile of the electron beam can be modulated by varying the temporal profile of the driving laser. The SPTEM can obtain temporal and spatial information about spin. The design and construction of the PES were optimized for the SPTEM. A beam energy of less than 40 keV (a low energy for a TEM) was used because the spin interaction with condensed matter is very small compared with the Coulomb interaction. A polarized electron gun was realized in an extreme high vacuum by applying a high field gradient of 4 MV/m to the photocathode surface. A 40-keV polarized electron beam with sub-millisecond pulses was generated using a backside-excited photocathode. The illumination system of the TEM was designed with the aim of producing a spin-polarized electron beam. The system contains two components that rotate spin: one performs azimuthal rotation and the other performs in-plane rotation in the condenser magnet system. The SPTEM has the potential to dynamically generate magnetic-field images with high contrast and high temporal resolution.

1. Introduction

Great advances have been made in magnetic recording technology and spintronic devices, which are promising for high-density storage devices [1–3]. Such devices are expected to lead to the development of systems that can analyze magnetic and spin states with a nanometer-order spatial resolution. Already an electron beam holography and a Lorentz electron microscopy have been used for observation of magnetization state of magnetic recording media as well as spin-detection scanning electron microscopy where image resolution is, however, restricted to larger than 5 nm [4]. A spin-resolved STM (SP-STM) realizes the performance in atomic scale [5]. Although the SP-STM can only observe the surface state of spin, and the measurable direction of magnetization is limited due to the probe tip. It is important that the inner spin-state can be observed directly in arbitrary direction with atomic scale.

Development of polarized electron sources (PES) has been advancing in the field of high-energy physics [6,7]. PES technology has been intensively developed from the mid-1980s to enable experiments using polarized electron beams to be performed at the International Linear Collider (ILC) [8]. PES technology has recently been applied to materials science instruments to observe surface magnetic phenomena [9,10]. Electron spin is very important for both fundamental physics and some key technologies such as spintronics and quantum-information technology [2,3].

We have been seeking to combine electron microscopy and accelerator technology to produce a spin-polarized transmission electron microscope (SPTEM), which consists of a PES and a conventional TEM. We have commenced developing a PES for a TEM [11]. Spin-polarized electrons can be generated using an optical orientation of appropriate III–V semiconductors and vacuum extraction that uses a negative electron affinity (NEA) surface. Circularly polarized laser light illuminates a semiconductor and spin-selectively excites valence electrons into the conduction band. These polarized conduction electrons diffuse and some of them drift to the surface region. They are then extracted through the NEA surface in a vacuum by applying a negative electrostatic field. Several beam parameters (including the electron spin polarization (ESP), the quantum efficiency (QE), and the beam emittance) of the PES are vastly superior to those of conventional electron beams. In addition, it has the ability to generate a sub-nanosecond multibunch beam. A high ESP of 92% and a high QE of 0.5% have been realized using a GaAs–GaAsP strained superlattice photocathode with a 100-nm-thick active layer that was excited with 780 nm wavelength light [12,13]. A normalized thermal beam emittance as low as 0.15π mm mrad can be achieved. This high-performance PES will enable dynamic observation of magnetic structures with high spatial and temporal resolutions in TEM. We have commenced developing a SPTEM. This paper presents an overview of the SPTEM and related systems.

2. Spin-polarized TEM system

Figures 1 and 2 respectively show photographs and a schematic diagram of the SPTEM system. The SPTEM consists of a PES, an illumination-lens system that includes a spin manipulator, an objective-lens system, a projector-lens system, and an image-detection system. The PES has been designed and constructed with optimization for SPTEM [11]. The beam energy is set to be below 40 keV, that is a lower energy type as a TEM, because the spin interaction with condensed matters is very small comparing with the Coulomb interaction [14]. The polarized electron beam emitted from the PES passes through a spin rotator and the condenser-lens (CL) system focuses the polarized electron beam onto the specimen. As the beam passes through the specimen, some electrons are scattered by

Figure 1. Photographs of SPTEM

interactions such as the Coulomb interaction, the spin–orbit interaction, and the spin–spin interaction. The modified beam then passes through the objective and projector lenses that focus it onto a fluorescent screen or the imaging plate of a CCD camera to obtain the final image. Since the electron beam passes through the specimen, the diffraction pattern obtained gives comprehensive information about the interior of the specimen, including its magnetic distribution. Except for the detection chamber, an ultrahigh vacuum (UHV) is maintained throughout the vacuum system in the SPTEM to prevent residual gas from destroying the NEA surface. The illumination-lens system was developed in this study. It consists of a spin rotator for polar rotation (SRB) and a CL system that contains in-plane spin-rotation magnetic lens (SRL). A 90° bending system was employed for the SRB; it generates both electric and magnetic fields to rotate spin by an arbitrary angle. The system can also correct for over-rotation due to relativistic effects such as the anomalous electron magnetic moment and Thomas precession [15]. This is important since if the spin rotator only used an electric field to bend the trajectory of a 40-keV beam by 90°, the actual spin-rotation angle relative to the beam axis will be 96.6° according to the Thomas–BMT (Bargmman-Michel-Telegdi) equation [15].

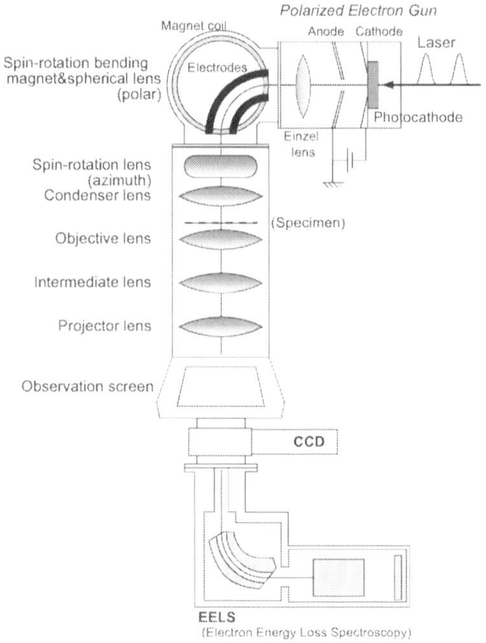

Figure 2. Schematic of SPTEM and components

3. Spin rotation system in illumination optics of TEM

Spin rotators are necessary to control the spin direction of the polarized electron beam. Electrons emitted from the photocathode have spins that either parallel or antiparallel to the beam axis due to the quantization axis of the semiconductor photocathode and the helicity of the excitation light. This system requires two spin rotators to rotate both the polar and azimuth angles.

A 90° bending system was employed as a spin rotator for the polar angle because the PES is installed on top of a TEM. In the SRB, spin precession due to an electric field rotates the spin vector relative to the injected beam axis, but bending of the beam trajectory by 90° is achieved by applying a magnetic field whose magnitude balances the Lorentz force [15]. The polar angle between the beam direction and the spin direction can be controlled to any value. The spin rotation system was carefully designed to maintain both the UHV and a high brightness. An electric field is generated by a spherical condenser with a radius of curvature of 100 mm. A transverse magnetic field is generated by a pair of magnetic poles that are separated by 30 mm. To prevent degradation of the UHV, the magnet coils that generate the magnetic field are located outside the vacuum chamber and the magnetic field is guided through the magnetic poles into the vacuum. On the other hand, the spin can be rotated azimuthally in the transverse plane by the magnetic field generated by the SRL magnet in the series of CL magnets. The CL system essentially collimates and concentrates the beam onto the specimen. In the present system, the CL system also acts as a spin rotator for in-plane rotation. The system consists of four magnetic lens: SCL1, SRL, SCL2, and SCL3. In the beam optics, the electron beam from the 90°-bending spin rotator is focused onto the center of the SRL by SCL1 since the beam trajectory will not be altered in SCL2 regardless of the magnetic field strength in the SRL. SCL2 and SCL3 play function as a CL in a TEM. This CL system can thus realize in-plane spin rotation and concentrate the beam onto the specimen.

4. Summary

We have constructed all the components of a SPTEM, namely a polarized electron source, spin rotators, double-gap CL magnets, an electron energy loss spectrometer, and a TEM. The PES can generate a high performance electron beam for the SPTEM. The spin rotation system can be used to completely control the spin of the electron beam since it accounts for relativistic effects. This SPTEM can rapidly image magnetic structures using spin-dependent contrast.

Acknowledgments

The authors would like to thank Dr. H. Shinada, Dr. M. Koguchi, M. Tomita, and M. Sugaya of Hitachi, Ltd. for fruitful discussions and encouragement. This research is supported by a Grant-in-Aid for Scientific Research (S) (No. 51996964) and a Grant-in-Aid for Young Scientists (B) (No. 22740157) from MEXT of Japan and the Kurata Memorial Hitachi Science and Technology Foundation.

References

[1] N. Gisin, G. Ribordy, W. Tittel, and H. Zbinden 2002 *Rev. Mod. Phys.* **74**, 145
[2] Loss D and DiVincenzo DP, 1998 *Phys. Rev.* A **57** 120
[3] Kuwahara M, Kutsuwa T, Ono K, Kosaka H 2001 *Appl. Phys. Lett.* **96** 163107
[4] Kohashi T, Konoto M, Koike K 2010 *J. Electron Microsc.* **59** 43
[5] Yang H and Smith R A 2002 *Phys. Rev. Lett.* **89** 226101
[6] Prescott CY, Atwood WB, Cottrell RLA, Destaebler H, Garwin EL, Gonidec A, Miller RH, Rochester LS, Sata T, Sherden DJ, Sinclair CK, Stein S, Taylor RE, Clendenin JE, Hughes VW, Sasao N, Schuler KP, Borghini MG, Lubelsmeyer K, Jentschke W 1978 *Phys. Lett.* B **77** 347
[7] SLD Collaboration 1995 *Phys. Rev. Lett.* **74** 2880
[8] ILC collaboration 2007, *ILC Global Design Effort and World Wide Study*
[9] Zdyb R and Bauer E 2002 *Phys.Rev. Lett.* **88** 166403
[10] Unguris J, Seiler A, Celotta JR and Pierce DT 1982 *Phys. Rev. Lett.* **49** 1047
[11] Kuwahara M, Takeda Y, Saitoh K, Ujihara T, Asano H, Nakanishi T and Tanaka N 2011 J. Phys.:Conf. Ser. 298 012016.
[12] Nishitani T, Nakanishi T, Yamamoto M, Okumi S, Furuta F, Miyamoto M, Kuwahara M, Yamamoto N, Naniwa K, Watanabe O, Takeda Y, Kobayakawa H, Takashima Y, Horinaka H, Matsuyama T, Togawa K, Saka T, Tawada M, Omori T, Kurihara Y, Yoshioka M, Kato K and Baba T 2005 *J. Appl. Phys.* **97** 094907
[13] Togawa K, Nakanishi T, Baba T, Furuta F, Horinaka H, Kurihara Y, Matsumoto H, Matsuyama T, Nishitani T, Okumi S, Omori T, Suzuki C, Takeuchi K, Wada K, Yamamoto M and Yoshioka M 2000 *Nucl. Instrum. Methods Phys. Res.* A **455** 118
[14] Vredevoe A L, and DeWames E R 1968 *Phys. Rev.* **176** 684
[15] Bargmman V, Michel L and Telegdi L V 1959 *Phys. Rev. Lett.* **2** 435

Journal of Physics: Conference Series **371** (2012) 012005 doi:10.1088/1742-6596/371/1/012005

Electron Beams with a Twist

L Clark, S Lloyd, M Babiker, J Yuan

Department of Physics, University of York, Heslington, York, YO10 5DD, UK

Email: lc586@york.ac.uk

Abstract: The properties and behaviour of electron beams possessing orbital angular momentum are considered. Methods of their creation are discussed, but we concentrate mainly on the method involving binarized holographic masks. It is shown how suitable masks can be modelled using Bessel functions as well as Laguerre-Gaussian functions. The theoretical far-field diffraction patterns of these holograms are simulated using Fourier transform techniques and compared with the recently reported experimental results of electron vortex beams. The limitations of the masks and the potential of the technique for future research are pointed out and discussed.

1. Introduction

1.1 Background Theory

The angular momentum of electrons is an intriguing subject as it draws together a wide array of key areas of physics within electromagnetism, optics and quantum mechanics in an unusual but elegant combination. From elementary quantum mechanics it is known that the total angular momentum of an electron can be separated into spin and orbital components. The spin angular momentum can be determined with established techniques such as the Stern-Gerlach experiment [1], whereas the orbital angular momentum (OAM) of electron beams has been significantly under studied until recently.

In the area of electron beams possessing OAM (known as electron vortex beams), the first paper of significance was published as recently as April 2010 [2]. It showed how, as for optical beams, an electron plane wave could be passed through a ramped phase plate within an electron microscope to impart OAM. However, there are a number of physical difficulties with this technique which led to results being approximate and limited to low orders of OAM. There is an alternative method, known as holographic reconstruction, the study and application of which forms the major focus of this paper. Both Laguerre-Gaussian (LG) and Bessel wavefunctions are considered as OAM-bearing entities, and some of their properties are discussed. This leads to the development of a series of hologram masks for creating electron vortex beams, the application and effectiveness of which are discussed with particular regard to their diffractive behaviour. They are compared with the theoretical and experimental results of electron vortex beam studies in published literature [3, 4].

1.2 Motivation and Objectives of the Study

It is imperative that the field of electron vortex beams is studied as there are many potential applications of this new development. An example that has already been demonstrated is electron energy-loss spectroscopy using such beams, allowing the properties of magnetic materials to be investigated on a very fine scale [3]. Additionally, there is potential for using such beams to rotate particles of matter as an analogue of the optical screwdriver [5], and there are possible applications in the field of quantum information transfer [6].

With these potential applications in mind, this article explores electron vortex beams made from holographic masks using the Laguerre-Gaussian and Bessel wavefunction forms and aims at a comparative study of the holographic masks produced. The theoretical diffraction patterns of the generated holographic masks are evaluated and compared with the published experimental results [3]. We conclude with a summary and brief comments.

2. Methodology

2.1. Mathematical Basis of OAM beams

A beam possessing OAM expresses this property through its azimuthal phase dependence, $e^{il\varphi}$ where $\varphi = \arctan(y/x)$ and l is the OAM quantum number, as seen in figure 1. This reveals a helical wavefront structure whereby, at the beam axis, the phase must be undefined – this is the phase vortex. The two most common beams with this property are LG beams and Bessel beams.

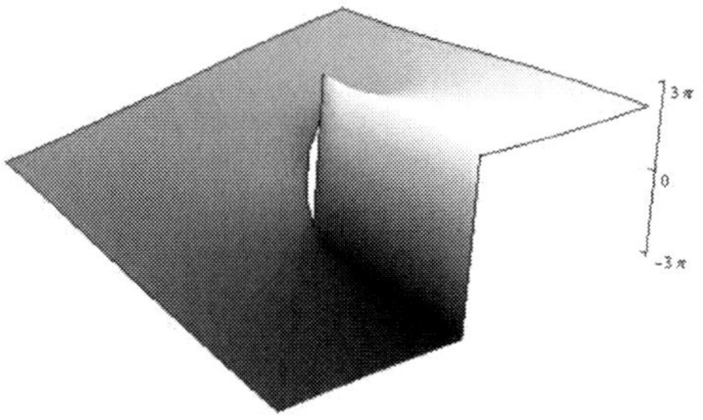

Figure 1: Wave front of a beam with $l = 3$, with the size of the phase ramp at 6π over the azimuth angular range of 2π

To describe beams possessing OAM, cylindrical polar coordinates may be conveniently used in solving the time-independent Schrödinger equation (TISE). Applying the paraxial approximation, i.e. when the diffraction and focussing occur over a distance much longer than the wavelength, one possible solution is the LG beam [7] in the long Rayleigh range limit:

$$\Psi_{LG} = C\left[\frac{\sqrt{2}.r}{W(0)}\right]^{|l|} L_p^{|l|}\left(\frac{2r^2}{W(0)^2}\right) e^{\frac{-r^2}{W(0)^2}} e^{ikz} e^{il\varphi} \qquad (1)$$

where C is an appropriate normalization constant, (r, φ, z) are the cylindrical polar coordinates and $W(0)$ is the beam waist. $L_p^{|l|}$ is the associated Laguerre function, with azimuthal index $l = \pm(0,1,2 \ldots)$ and radial index $p = 0,1,2 \ldots$. The wave vector is represented by k (considered here in the z direction).

Alternatively, Bessel beams are an exact solution of the TISE, as they are plane waves with a non-uniform intensity distribution. However, the infinite width of mathematical Bessel beams would require an infinite amount of energy, and thus they are not physically realisable in their true form. Good approximations are achievable, however, by applying a finite aperture to the function to produce a truncated Bessel function; a simpler alternative to the analytical Bessel-Gauss function [8]. This would allow the reduced diffraction properties of the Bessel beam to be employed [9] over a finite but still considerable distance. It is this non-diffractive property which makes electron Bessel beams an exciting prospect. The wavefunctions of a true Bessel beam take the form

$$\Psi_B = CJ_l(k_r r)e^{ik_z z}e^{il\varphi} \qquad (2)$$

where C is the normalization constant, J_l is a Bessel function of the first kind and is of the order l and a function of $(k_r r)$ where k_r is the radial component of the wave vector and k_z is the axial component of the wave vector. As for LG beams, (r, φ, z) are the cylindrical polar coordinates and l is the azimuthal index.

The similarities and differences between the Bessel and LG beams can be clearly seen in their radial distributions, displayed in figure 2.

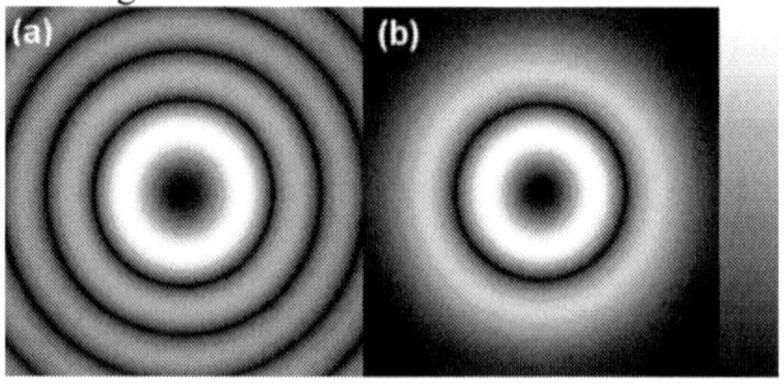

Figure 2: (Color Online) (a) the radial intensity distribution of an $l = 3$ Bessel beam, (b) the radial intensity distribution of an $l = 3, p = 1$ Laguerre-Gaussian beam. Maximum intensity represented as bright, minimum values as dark.

2.2. Holographic Reconstruction

Electron holography involves placing a diffraction grating in the path of a plane wave electron beam [10]. The OAM is imparted to the beam by the existence of a Y-shaped defect in the diffraction grating. Such holographic masks were first employed in the electron case by Verbeeck *et. al.*, and more recently by McMorran *et. al.* [3, 4]. Both groups produced electron vortex beams for a range of OAM.

Holographic masks represent the interference pattern of the intended beam (such as Ψ_{Bessel} or Ψ_{LG}) and a reference plane wave [10]. The interference of the desired beam with the reference wave allows the complex phase information of the desired beam to be combined with the amplitude, and retained in a real-valued array. This array is then binarized, by clipping away those areas below a threshold value, and setting those areas above the threshold to a maximum thickness. This array can be translated into the thickness of a foil to be placed in the beam path. This results in a mask resembling a typical diffraction grating, but with a Y-shaped defect appearing from the phase dependent term within the wavefunction. It is the passing of a beam around this defect which allows the transfer of OAM from the mask to the beam particles. This method was the technique first applied by Verbeeck *et. al.* to create electron vortex beams [3].

2.3 Fourier Transforms

The nature of the far-field diffraction pattern is one way to detect the form of the beam produced, and is described by a 2D Fourier transform of the aperture. Thus, theoretically expected results can be generated by assuming that the hologram is part of a periodic function, and performing a 2D Fourier transform on the mask. The modulus squared of the Fourier transform represents the expected intensity distribution created by the mask in the back focal plane of an electron microscope.

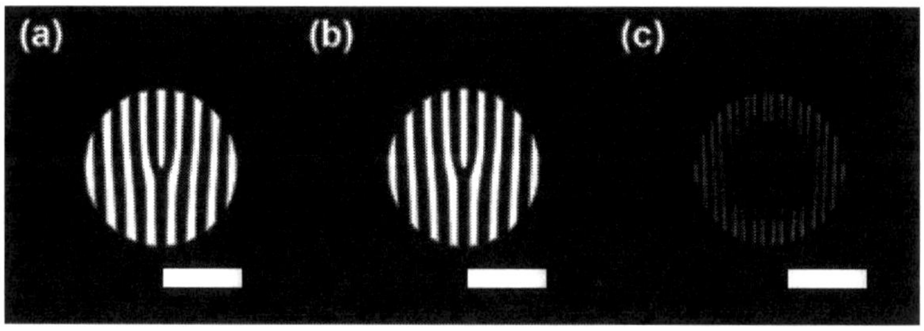

Figure 3: The binarized hologram masks made from $l = 1$ (a) a truncated Bessel wavefunction; (b) a truncated Laguerre Gaussian wavefunction; (c) the difference between the masks shown in (a) and (b). The scale bars represent the radial unit $r = j_1/k_r$, where $j_1 \approx 3.83$ is the first zero of the Bessel function.

3. Results and Discussion

The masks were simulated using Mathematica, following the process described above. The binarized masks produced using both the truncated Bessel and the LG wavefunctions are shown in figure 3.

3.1. Fourier Transforms

To analyse these masks, 2D Fourier transforms are applied to the arrays containing the mask data. This process allows for comparison with the experimental results in the literature. Figure 4 shows the Fourier transforms of the masks displayed in figure 3. It can be seen that both masks produce similar outputs. The binarized masks show the intended beams ($l = 0, \pm 1$) and, additionally, higher orders are generated as by-products of the binarization process.

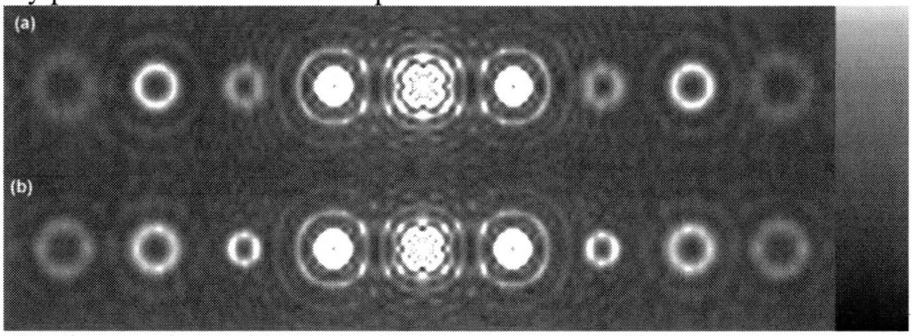

Figure 4: Fourier Transforms of the masks: (a) the Bessel case and (b) the LG case. Maximum intensity represented as bright, minimum values as dark.

These results agree closely with the experimental results for the beams displayed by Verbeeck *et. al.* [3]. However, the comparison of the masks (Figure 3, (c)) and the effect on the diffraction pattern (comparing Figure 4a and 4b), calls for a re-consideration of the effect of the chosen wavefunctions. The clearest difference is in the variation of the width of the fork tines as a function of the radius – as would be expected from a consideration of the beam profiles. However, this difference clearly has little effect on the far-field diffraction patterns. This suggests that to investigate the diffractive nature of electron vortex beams, a modification of this technique is required as the effect due to the approximation processes dominates over the differences between the these two wavefunctions.

4. Conclusions

We have considered the effect of differing radial profiles on electron vortex beams, by modelling holographic masks and the resulting diffraction patterns using Fourier transform techniques. We have shown that the radial differences of the Bessel and LG wavefunctions are small in the radially-truncated region from which holographic masks are produced. This implies that to investigate and control the spatial evolution of electron vortex beams, modifications to the current methods of producing the beams are required.

References

[1] Rae AIM, 2007, *Quantum Mechanics*, Taylor and Francis, London
[2] Uchida M, Tonomura A, 2010, *Nature*, **464**, 737-9
[3] Verbeeck J, Tian H, Schattschneider P, 2010, *Nature*, **467**, 301-4
[4] McMorran BJ et al, 2011, *Science*, **331**, 192-5
[5] Simpson NB et al, 1997, *Optics Letters*, **22**(1), 52-4
[6] Molina-Terriza G, Torres JP, Torner L, 2007, *Phys. Rev. Lett.*,**88**(1), 0136011-4
[7] Saleh BEA, Teich MC, 1991, *Fundamentals of Photonics*, Wiley, New York
[8] Gori F, Guattari G, 1987, *Optics Communications*, **64**(6), 491-5
[9] Durnin J, Miceli JJ, Eberly JH, 1987, *Phys. Rev. Lett.*, **58**(15), 1499-501
[10] Tonomura A, 1989, *Trends in Analytical Chemistry*, **8**(6), 226-9

Use of Aharonov-Bohm effect and chirality control in magnetic phase plates for transmission microscopy

C J Edgcombe[1] and J C Loudon[2]

[1]TFM Group, Department of Physics, University of Cambridge, JJ Thomson Avenue, Cambridge CB3 0HE, UK

[2]Department of Materials, University of Cambridge, Pembroke Street, Cambridge, CB2 3QZ, UK

E-mail: cje1@cam.ac.uk

Abstract. Initial holographic tests on thin rings of cobalt have demonstrated both onion (O) and vortex (V) states of magnetization, and show that the vortex state provides a uniform phase difference between inside and outside the ring as expected from the Aharonov-Bohm effect. Simple circular rings show a relatively small difference between the in-plane switching fields O→V and V→O and have unpredictable chirality (sense of flux rotation in the vortex mode). Simulations suggest that D-shaped rings provide both predictable chirality and a wider range between switching fields.

1. Introduction

In cryo- and other transmission electron microscopy of biological specimens, there is a need to minimise the electron dose by increasing the image contrast for given beam current. This can be achieved by use of a phase plate, designed to change by about 90 degrees the phase difference between scattered and direct electron waves. We propose to achieve the required phase change by use of the Aharonov-Bohm effect produced by a ring of magnetic flux set up in a thin magnetic ring inserted in the plane of the objective aperture. The ring dimensions are chosen so that scattered electrons pass outside the ring while direct electrons pass through it. For use in a microscope column, such rings would ideally be fabricated on a refractory support, but such fabrication is not discussed here.

2. Vortex and onion flux patterns observed in rings

Trial rings have been fabricated and tested by holography. Circular rings with ranges of diameters and widths were written by e-beam lithography on a silicon nitride membrane of thickness 50 nanometres, supported on a silicon frame [2]. The frame was approximately 2.9mm square so as to be suitable for insertion in an objective aperture holder. The rings were made by MBE deposition of 20 nanometres of cobalt on an adhesion layer of 10 nm of copper, followed by an anti-oxidation layer of 3 nanometres of gold, followed by lift-off. The sets of rings were magnetized by applying a field of –400 mT followed by one of +124 mT, both in-plane, and were then tested in a CM300 TEM equipped for holography. The first results shown here were obtained with no coating to minimize charging, and with no excitation of the objective lens. In further tests an anti-charging coating has been added.

Journal of Physics: Conference Series **371** (2012) 012006 doi:10.1088/1742-6596/371/1/012006

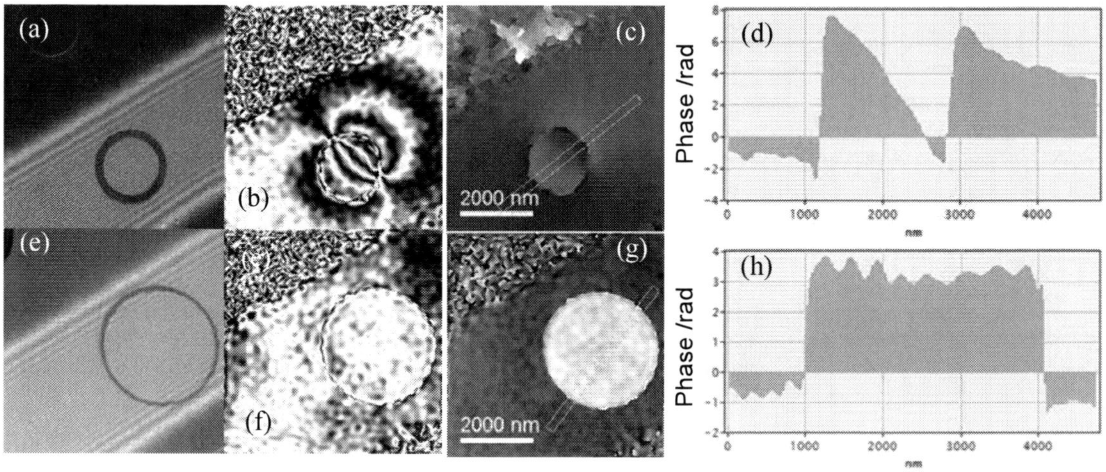

Figure 1. (a)-(d) Ring of outer diameter 2 micrometres, width 200nm; (e)-(h) Ring of outer diameter 3 micrometres, width 100nm. (a),(e) hologram; (b),(f) cos(2*phase); (c),(g) unwrapped phase; (d),(h) phase profile over path in (c) or (g).

Figure 1(a)-(d) shows results from a ring of diameter 2 micrometers and width 200nm, located at the specimen position. There is a nearly uniform phase gradient across the interior of the ring in the direction sampled, together with divergence of flux from two points on the circumference of the ring. This is the behaviour expected from a ring magnetized in the 'onion' mode. In contrast, figure 1(e)-(h) shows results obtained from a ring of diameter 3 micrometres and width 100nm. It is clear that the phase inside the second ring is approximately uniform and that there is a difference of phase between inside and outside the ring of about 3.7 radians. The uniformity of phase within the ring shows that it was magnetized in the vortex mode and that relative to a direct beam passing through the ring, scattered waves will be changed in phase uniformly at all azimuthal angles around the beam.

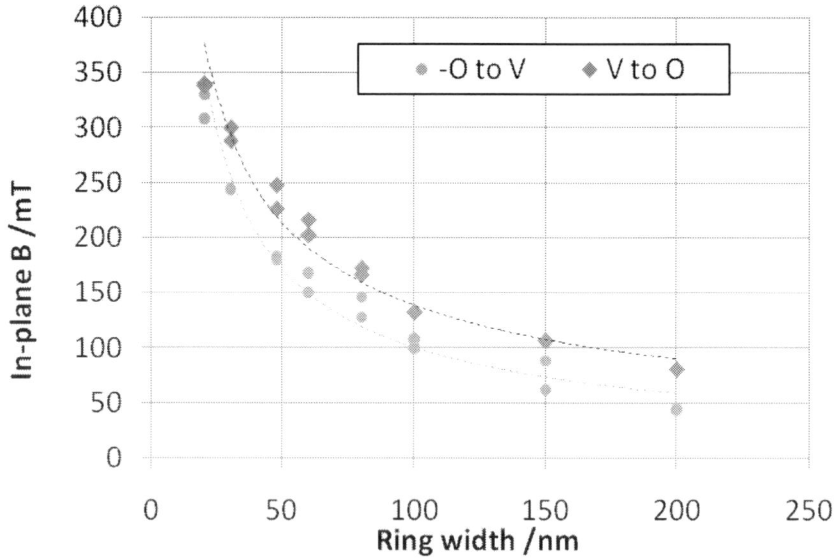

Figure 2. Switching fields as functions of ring width for circular cobalt ring of outer diameter 1 micrometre and thickness 20nm.

Simulations of the magnetic behavior using the OOMMF package [3] show why a single magnetizing field produced different magnetic states in the two rings. Figure 2 shows values for the switching fields, onion to vortex (O→V) and vortex to onion (V→O), found by simulation of circular cobalt rings. The field of 124mT used to magnetize the rings in Figure 1 was chosen to switch a ring of width 100 nm into the vortex state, while the same field applied to the ring of width 200 nm appears to have switched it through the vortex and into the onion state. The variation of the switching fields with ring diameter is small.

According to simulation, the vortex state in rings with width greater than their thickness should tolerate an out-of-plane field of 1T or more from the objective lens. Work is in progress to determine the experimental relation between ring dimensions, magnetization, M_s, resulting phase shift, and the out-of-plane field (from the objective lens) that these rings can tolerate.

3. Disadvantages of circular rings

Figure 2 shows that for a given width of ring, the switching fields differ by about 40 mT over the range of width simulated. In the simulations, a fixed anisotropy file was used with cells set to random directions. Crystalline anisotropy then had little effect in the simulations, but the polycrystalline deposition used in practice may produce larger grains and so may cause some variability of switching fields. Thus it is desirable to extend the range between switching fields if possible. Also, when circular rings are magnetized in vortex mode, the chirality, or sense of flux circulation, is liable to be set at random by thermal effects [4]. This happens because switching from onion to vortex state proceeds by movement of one domain wall around half the ring. Motion in either direction is possible, and in each instance behaviour depends on any slight asymmetry of crystal anisotropy or the initial location of the domain wall in the ring, relative to the applied field. Thus when circular rings are magnetized, the resulting chirality is unpredictable.

4. Behaviour of modified rings

The behaviour of circular rings suggested that if their shapes were modified so that the two possible paths for a domain wall differed geometrically, this might help to ensure that the wall followed one particular path on switching from onion to vortex. Simulations were therefore carried out on rings with the general form of a semicircle joined to its diameter, as shown in figure 3. The tests showed that a variation in width at these junctions could cause the domain walls to be pinned there more

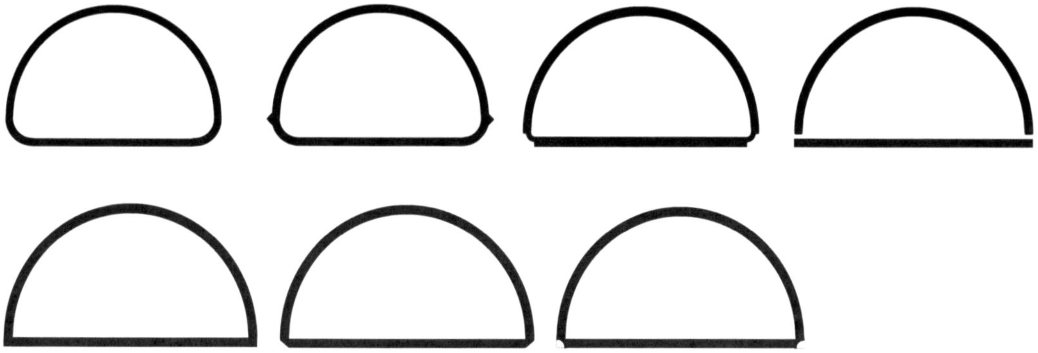

Figure 3. Some non-circular rings used for simulations.

precisely than in purely circular rings, and that the difference in paths ensures that the domain wall always moves along the same path. Then, with a switching field of given sense, the vortex mode in a particular ring is always created with the same chirality. In addition, the size of the O→V switching field is strongly influenced by the detailed shape of the junctions. Unexpectedly, a restriction in width there can cause the domain wall to start moving at a lower field than for a uniform width of wall. The upper switching field (V→O) varies much less with the shape of the ring.

The switching fields found for the shape with a 45 degree chamfer in figure 3 are shown in figure 4. It can be seen that the range between the two switching fields is very much greater than that found for circular rings in figure 2, thereby giving more tolerance of effects of crystal anisotropy and of possible variations in fabrication.

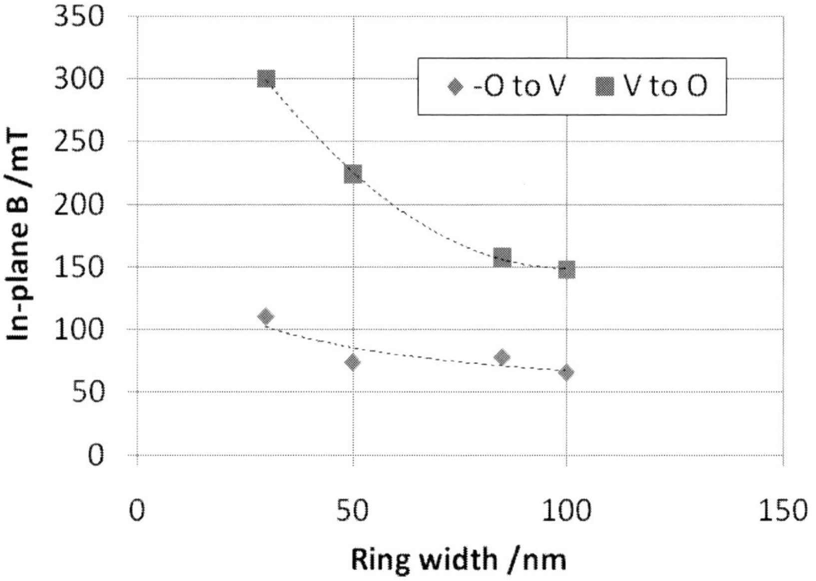

Figure 4. Switching fields as functions of ring width for chamfered semi-circular Co ring of outside radius 0.5 micrometre and thickness 20nm.

5. Summary

Initial holographic tests on thin rings of cobalt have demonstrated both onion and vortex states of magnetization, and show that the vortex state provides a uniform phase difference between inside and outside the ring as expected from the Aharonov-Bohm effect. Simple circular rings show a limited range between the in-plane switching fields O→V and V→O and are unpredictable as to chirality. Simulation suggests that D-shaped rings can provide both control of chirality and a wider range between switching fields.

Acknowledgments

The authors thank Arthur Blackburn (Hitachi Lab, Cambridge), Adrian Ionescu and Theo Trypiniotis for e-beam lithography, MBE deposition and magnetization.

References

[1] Edgcombe C J 2009 *Electron Microscopy and Analysis Group Conf. (Sheffield)* Journal of Physics: Conf. Series **241** ed R T Baker (Bristol: IoP Publishing) p 012005
[2] Silson Ltd, http://www.silson.com
[3] Donahue M J and Porter D G 1999 *OOMMF User's Guide, Version 1.0* National Institute of Standards and Technology Interagency Report NISTIR 6376 (Gaithersburg, MD: NIST); http://math.nist.gov/oommf
[4] Hayward T J, Moore T A, Tse D H Y, Bland J A C, Castano F J and Ross C A 2005 *Phys. Rev.* B **72**, 184430.

Atomic motion on various surfaces of ceria nanoparticles in comparison

U M Bhatta[1*], I M Ross[2], Z Saghi[1], A Stringfellow[1], D Sayle[3], T X T Sayle[3], A Karakoti[4], D Reid[4], S Seal[4] and G Möbus[1]

[1] Dept of Materials Science and Engineering, University of Sheffield, Sheffield, S1 3JD, UK
[2] Dept of Electronic and Electr. Engineering, University of Sheffield, Sheffield, S1 3JD, UK
[3]Department of Engineering and Applied Science, University of Cranfield, UK.
[4]Mechanical Materials and Aerospace Engineering, University of Central Florida, Florida, 32816, USA

[*]Email: u.m.bhatta@sheffield.ac.uk

Abstract: Aberration corrected HRTEM has been successfully applied to image the atomic motion at the edges and surfaces of nanoparticles of ceria of various types under electron irradiation. Here we identify movements not only on {100} facets, but also on {110} and even {111} facets, previously considered stable. However, the degree of movement varies strongly and HRTEM is evidently the preferred technique to measure relative stability at high spatial resolution as it does not require extended surfaces as in scanning probe microscopy (SPM) or chemical methods. The advantage of aberration correction shows in suppression of contrast from the carbon support films and the absence of delocalisation fringes at particle edges, apart from improving point resolution.

1. Introduction

The efficiency of a nanoparticle as a catalyst is strongly influenced by its surface reactivity, which in turn depends on its surface irregularities, atomic steps, majority of crystal orientations exposed etc. Ceria is one such example, utility of which as a catalyst [1,2] originates from surface oxygen content. The catalytic activity and mobility of Ce and O ions on the surface of ceria nanoparticles is linked with how strongly each atom is bound to the surface with the order of binding energy being: {111} > {110} > {100} [3]. So it would be highly beneficial to study relative activities of ceria low index surfaces experimentally.

High resolution transmission electron microscopy (HRTEM) has been used effectively in nanoparticle studies as it gives direct information on surface instabilities and reconstruction of nanocrystals [4]. Recent developments (e.g. aberration corrected TEMs and modern CCD cameras) have made *in-situ* observation of atomic hopping possible including structural transformations in gold at elevated temperatures, e.g. [5, 6]. Here we report the use of aberration-corrected bright-field phase-contrast TEM in monitoring atomic movements in oxide crystals. In extension of our earlier work studying motion exclusively on small facets of {100} type [7], a comparison of {111}, {110}, and large faces of {100} is presented, and some motion is now identified on each of them.

2. Experimental details

Both commercial CeO_2 (Sigma Aldrich) nanopowder and specifically hydrothermally prepared CeO_2 nanocrystals [8] were used for measurements in this work. The former are predominantly of octahedral shape with only small caps [9] (see Figures 1, 2 and 4), while the latter are tuned to show a maximum fraction of cube-faces (see Figure 3).

Aberration corrected TEM with spherical aberration close to zero and near Gaussian focus has been used in standard imaging conditions. As Ce provides mixed amplitude/phase contrast even at the smallest thicknesses due to its high atomic number, images with vanishing weak-phase contrast were found to provide the best single-atom visibility and minimized carbon background. Image series were automatically acquired in 2 s intervals using a GatanUltrascan 2kx2k camera. Gatan Digital Micrograph and ImageJ softwares were used for post-processing of the images. Experiments for figures 1 and 2 were obtained on the Sheffield JEOL JEM-2200FS-AC at 200kV, while experiments for figures 3 and 4 used the Sheffield JEOL JEM 3100 R005 at 300kV.

3. Results and Discussion

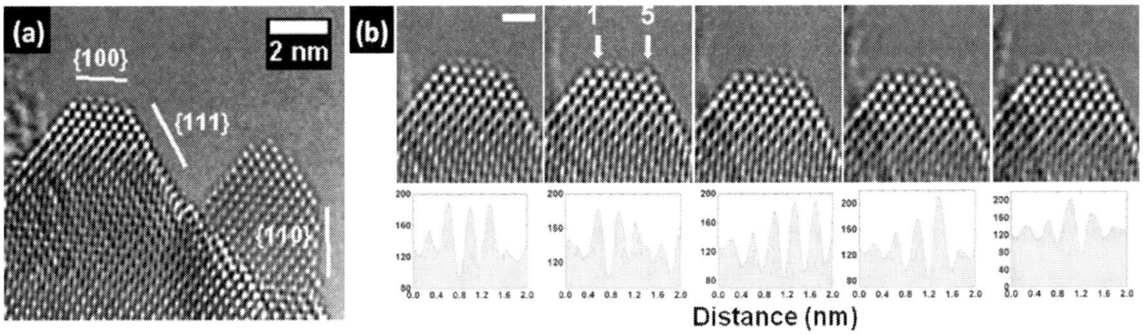

Figure1: (a) Ceria nanocrystals with {100}, {110} and {111} facets, (b) Time series (at 0, 20, 38, 40 and 44 s) of bigger particle (left) and the corresponding profile plots of the 1st monolayer. (scale bar = 1 nm); C-film ablated by e-beam.

Figure 1(a) shows the bright field overview of two epitaxially adjoining faceted ceria octahedral nanocrystals. While the tiny {100} facet of the right hand side particle has been analysed in detail in [7], we demonstrate here that the hopping phenomena observed can be generalised to the much larger {100} facet of the bigger particle (left hand side), and are therefore not corner-effects.

Figure 1(b) shows some representative images of the time series taken for a length of 2 min (60 images). During the length of recording {111} facets remained stable while the {100} facet underwent major atomic hopping processes as indicated by the accompanying profile plots. This kind of high instability of {100} facets compared to adjoining stable {111} facets is directly dependent on variable surface oxygen contents/vacancies on these surfaces which strongly influences the ceria surface chemistry. In particular we assume {100} surfaces semi-occupied with oxygen to suppress surface dipoles [7]. The variable surface occupancy of the top monolayer due to Ce-hopping, should lead to variations of lattice constant in the 2nd monolayer: A similar line profile was taken on the second monolayer, in order to measure interatomic distance. Difference between two consecutive Gaussian peak centres would give corresponding inter atomic distance. Using the first two frames in figure 1(b), corresponding to 0 and 20 s, we find atomic positions 1, 4 and 5 on the top monolayer undergoing major changes in terms of atomic visibility and corresponding plot intensity.. As a consequence, the interatomic distance below Pos. 1 increases from 0.39 ± 0.005 nm to 0.405 ± 0.005 nm. Pos. 4 also loses atoms from frame 1 to 2 and again the corresponding interatomic distance below increases from 0.382 ± 0.005 nm to 0.41 ± 0.005 nm. Pos. 5 seemingly gains a large number of atoms and the projected Ce-Ce distance below decreases from 0.438 ± 0.005 nm to 0.426 ± 0.005 nm. In summary, when there is no atoms/less atoms present in the first monolayer the corresponding position in the second monolayer becomes a virtual exposed 1st monolayer which reconstructs by expansion, also accounting for the larger Ce^{3+} ionic radius in these oxygen depleted (reduced) top layer positions.

Figure 2(a) shows an overview of another particle of the same system, this time with two {111} facets, enclosing a small {110} facet now. A part of the time series (Fig 2(b)) clearly shows that the {110} surface reconstructs itself into {111} facets under the e-beam. Atomically flat and smooth facets are not always the most stable surfaces, and it is established, that through formation of dislocations a facetted or a stepped surface can be preferred by such a system [10]. Surface oxygen content (or depletion of oxygen

Journal of Physics: Conference Series **371** (2012) 012007 doi:10.1088/1742-6596/371/1/012007

depending on the system and the environment) plays an important role in such reconstruction of the surfaces [11]. So it is evident that in this case the relatively unstable {110} surface transforms into a faceted surface covered by more stable {111} faces, as predicted in [3]. Similar reconstruction, albeit on larger scale, has also been reported by Nie *et a*l [12], as below a critical thickness ceria forms {111} faceted islands when ceria was deposited on sapphire as a thin film.

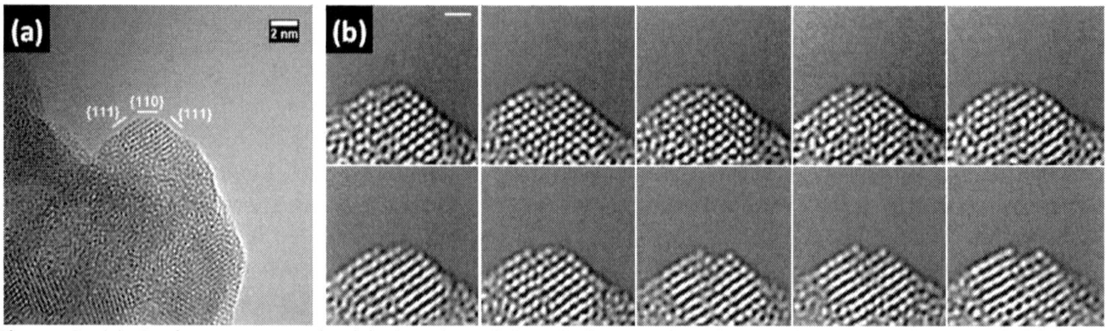

Figure 2: (a) Ceria nanocrystal with {110} and {111} facets, (b) Time series (at 0, 8, 18, 24, 26, 30, 32, 40, 46 and 50 s) (scale bar = 1 nm). C-film ablated by e-beam. Contrast inversion between (a) and (b) to facilitate comparison with experiment I (Figure 1).

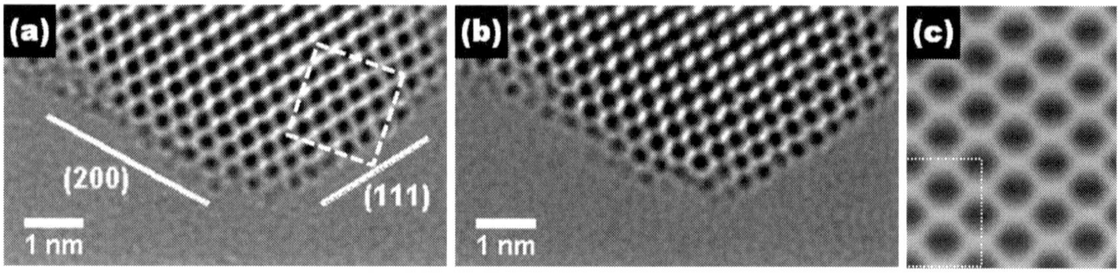

Figure 3: (a, b) Two time steps from a time series of mainly cube-shaped nanoparticles with dominant {200} facets; amorphous C contrast suppressed via Cs = 0, JEOL-JEM-R005. (c) Java EMS simulation corresponding to ROI, dashed on (a): Defocus +14 nm, thickness 1.5 nm. Unit cell (white box) with all Ce atom columns appearing as black spots.

It is important to clarify whether cuboid ceria surfaces with {100}-type faces larger than {111} behave in the same way. Fig 3(a, b) confirms extensive atomic hopping on {100} on a hydrothermal specimen with mainly cubes, while for the first time (minor) hopping on the adjacent {111} face, probably due to overspill of atoms from the very active corner region, is observed (see [13] for further details).

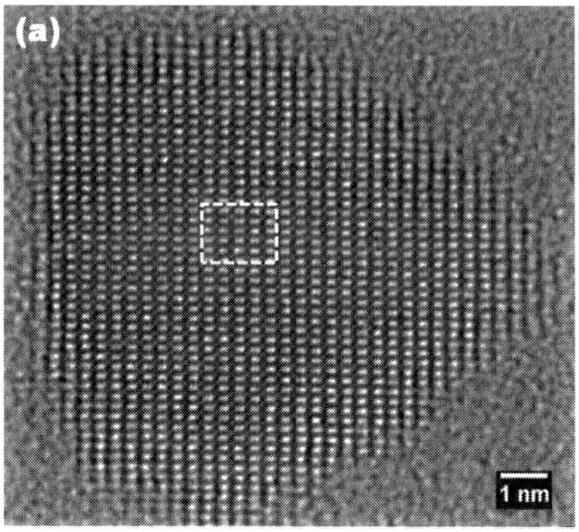

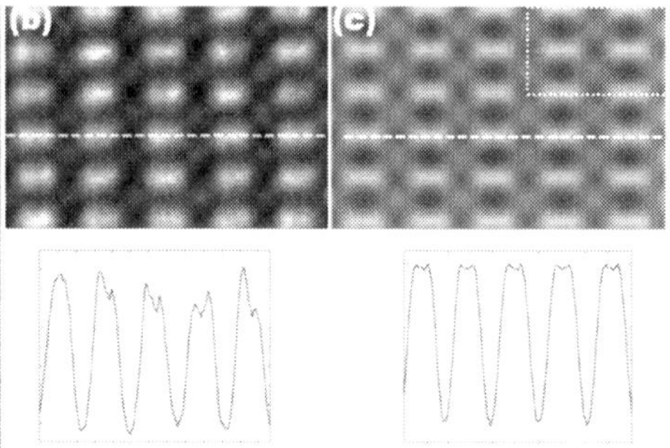

Figure 4: (a) A ceria nanocrystal imaged along [112] zone axis, (b, c) experimental and JEMS simulated section (dashed region on (a)) with corresponding profile plots.

4. JEMS simulation

To confirm the relative orientations of unit cells, the positions of Ce columns, and the local thickness of various regions of interest, High Resolution TEM (HRTEM) image simulations were employed using Java EMS software [14,15]. For the [110] ceria images of Figure 3, simulations were done for a range of defocus values starting from -25 to +25 nm in steps of 5 nm. For the center of the dashed region on Figure 3a, Figure 3c shows a match with defocus value of +14 nm and thickness of 1.5 nm. The other simulation parameters were used as follows: energy spread 0.3 eV/ defocus spread 3.0 nm, incoherent envelope (MTF/vibrations) fitted to 0.05-0.06 (preliminary guess only), while the beam convergence envelope is of negligible influence for any setting of 0.5mrad or smaller. To confirm visibility of even finer details in the ceria structure, simulations of [112] zone axis are added. Figure 4(a) represents the HRTEM images of a ceria nanocrystal imaged along [112] direction. Simulation parameters used were similar except for the defocus value which was +25 nm and the thickness of the ROI was found to be 8.0 nm. Profile plots corresponding to the experimental image (Fig 4(b)) shows apparent dumbbell contrast of the white spots along the dotted line and same is confirmed by the simulation as well. Ce positions are in corners and edge centres of the unit cell, oxygen would coincide with the "dumbbells", however this is coincident and not atomic contrast, as it rapidly fluctuated with focus.

5. Conclusions

Electron beam irradiation appears to be a promising probe in measuring or establishing the relative functional activity of different nanoparticle surfaces in comparison, as it probes one individual face of one nanoparticle rather than a wide average as with chemical, SPM or X-ray methods. Here we found exceptionally high Ce mobility, linked to oxygen deficit, for extended {100} ceria facets, but also mobility on {110} and to a minor degree on {111}. The importance of aberration correction to suppress Fresnel fringing and support film contrast and the ability to tune to maximum contrast for Ce-columns is confirmed.

Acknowledgements
This work has been supported by the EPSRC, under grant number EP/H001298/1.

References

[1] Kummer J T 1980 *Proc. Energy Combust. Sci.* **6** 177
[2] Trovarelli A 2002 *Catalysis by Ceria and Related Materials*, 2nd ed (Dunod: London)
[3] Sayle T X T, Parker S C and Catlow C R A 1994 *Surf. Sci.* **316** 329
[4] Cochrane H D, Hutchison J L, White D 1989 *Ultramicroscopy* **31** 138
[5] Young N P, van Huis M A, Zandbergen H W, Xu H, Kirkland A I 2010 *Ultramicroscopy* **110** 506
[6] Barnard A S, Young N P, Kirkland A I, Marijn A. van Huis and Huifang Xu 2009 *ACS Nano* **3** 1431
[7] Möbus G, Saghi Z, Sayle D C, Bhatta U M, Stringfellow A and Sayle T X T 2011 *Adv. Func. Mater.* **21** 1971
[8] Hao-Xin Mai, Ling-Dong Sun, Ya-Wen Zhang, Rui Si, Wei Feng, Hong-Peng Zhang, Hai-Chao Liu, and Chun-Hua Yan 2005 *J. Phys. Chem. B* **109** 24380
[9] Saghi Z, Ross I M and Moebus G 2009 *J. Phys. Conf Ser.* **241** 012071
[10] Alerhand O L, Vanderbilt D, Meade R D and Joannopoulos J D 1988 *Phys. Rev. Lett.* **61** 1973
[11] Guillemot L and Bobrov K 2011 *Phys. Rev. B* **83** 075409
[12] Nie J C, Yamasaki H, and Mawatari Y 2004 *Phys Rev. B* **70** 195421
[13] Bhatta U M, Ross I M, Sayle T X T, Sayle D, Parker S C, Reid D, Seal S, Kumar A and Möbus G *ACS Nano* DOI: 10.1021/nn2037576.
[14] Stadelmann P 1995 *Ultramicroscopy* **60** 103
[15] *http://cimewww.epfl.ch/people/stadelmann/jemsWebSite/jems.html*

Analysing interface reactions using EELS

A J Craven[1,2], B Schaffer[2,1] and M C Sarahan[2,1]

[1]SUPA, School of Physics and Astronomy, University of Glasgow, Glasgow G12 8QQ Scotland UK
[2]SuperSTEM, STFC Daresbury Laboratories, Keckwick Lane, Warrington, WA4 4AD, UK

E-mail: alan.craven@glasgow.ac.uk

Abstract. Using EELS spectrum imaging, an HfN or Hf(O,N) reaction layer has been identified at the TiN/HfO_2 interface in a metal inserted high-k gate stack. The reaction layer has a mean thickness of 0.45nm over an 18nm length of the interface. This reaction layer formed in the original HfO_2. By binning the data ×4 along the interface, a variation of the width from 0.35nm to 0.65nm along the interface can be seen. The 10%-90% widths of the elemental profiles can also be found but the binning required is greater (×10). The profile widths are approximately constant along the interface but the values differ from element to element. Thus the reaction has not formed a conformal layer of uniform thickness. An efficient way of processing the data at different levels of binning is described.

1. Introduction

As structures and devices become smaller, interfaces and surfaces play an increasingly important role in determining their properties. Since the layers are frequently polycrystalline, the interface may not be atomically flat. This is typical of many of the interfaces found in technologically important systems and so the ability to characterise such "rough" interfaces on the sub-nanometre scale is crucial.

Even if the average plane of such an interface can be made accurately parallel to the electron probe axis, which is often not the case, the electron probe will sample several layers as it traverses the interface region. This is the result of effects such as interface roughness and beam spreading. Thus interpretation of the image and spectroscopic data obtained is a major challenge. Here we discuss how the data from a standard EELS spectrum image, obtained using a NION UltraSTEM equipped with a Gatan ENFINA spectrometer, can be analysed to provide key information on a reaction at the HfO_2/TiN interface in a TiN inserted HfO_2 high-k gate stack on Si. This reaction was first suspected and commented on in references [1,2].

A key problem is to ensure that the specimen is not modified by the electron fluence. This condition sets a limit to the spatial resolution that can be achieved. This limit depends on the signal-to-noise ratio required to extract the information and this, in turn, depends on the type of information sought. Once acquired, the signal-to-noise ratio can be improved by binning the data at the expense of loss of spatial resolution. If the mean interface position can be aligned parallel to one edge of the data set, the data can be binned in the direction parallel to the interface. In this way, the spatial resolution is preserved across the interface and lost along it. What is required is an efficient method of processing the data with different levels of binning so that the optimum binning can be chosen for the extraction of a given type of information. An approach to this is discussed below.

2. Experimental details

The details of the deposition and processing of the high-k gate stack are given in references [1,2]. A sample for STEM was prepared using an FEI Nova 200 Dualbeam focused ion beam/SEM with a Sidewinder ion column and in-situ lift-out. The final thinning was carried out at 2kV, resulting in a lamella with t/λ in the range 0.3-0.4 at 100kV accelerating voltage. This lamella was examined in a Nion UltraSTEM at 100kV with a probe half-angle of 36mrad and an EELS collection half-angle of 36mrad. Figure 1 is a bright field image of the interface.

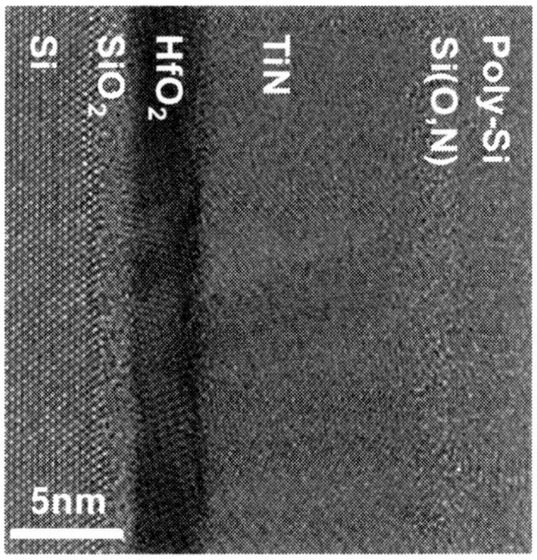

Figure 1. BF image of the gate stack showing the structure and the non-flat interfaces

Figure 2a shows the survey image on which the regions used for a spectrum image and for drift correction are shown. The spectrum image had 36×188×1340 voxels with a pixel step of 0.1nm, an energy step of 0.3eV and a dwell time per step of 50msec. The voltage on the flight tube of the spectrometer was changed by a random amount after every spectrum so that the fixed pattern noise would be reduced when spectra were added together [3]. The spectra were corrected for the change in the flight tube voltage after acquisition, requiring the number of channels per spectrum to be reduced from 1340 to 1182 to remove end effects. During acquisition, spatial drift was corrected after every 10 lines. Figure 2b shows the HAADF image recorded as part of the dataset, showing the steps in position introduced by the drift correction. The dataset was aligned line by line using the contrast change at the interface. Figure 2c shows the resulting HAADF image. At this stage, the dataset has been reduced to 30×180×1182 voxels in size. The reduction to 30 in the horizontal direction was required to remove edge effects resulting from the alignment while the reduction to 180 in the vertical direction was chosen to give a value with a large number of factors so that the data could be binned by different amounts. Principal component analysis was applied to this dataset. The resulting components proved difficult to interpret physically. However, by reconstructing the data using only 10 components, the noise in the background was reduced significantly [4].

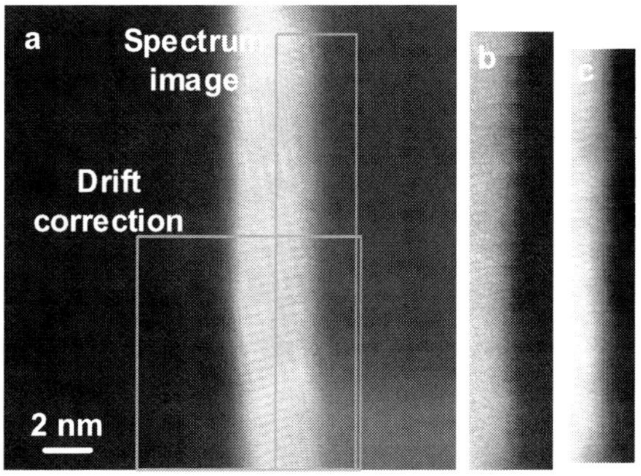

Figure 2. a) HAADF survey image showing the regions for the spectrum image and drift correction; b) the HAADF image recorded as part of the spectrum image showing the discontinuities left by drift correction; c) the HAADF image after alignment of the interface; d) a schematic of the extended data set composed of the unbinned data followed by blocks with successively greater degrees of binning, as described in the text.

To avoid having to repeat the processing for each level of binning, extended datasets were constructed from the original data and the reconstructed data. These each contained the unbinned data and data binned by 2, 3, 4, 5, 6, 10, 15, 20, 30, 90 and 180. For the 20× binning, four starting positions (lines 0, 4, 9 and 14) were selected to allow the investigation of the positioning of the binning boundaries. A schematic of the resulting extended dataset, 30×513×1182 voxels in size, is shown schematically in Figure 2d.

The weak Hf $N_{2,3}$-edge at 380eV lies just before the N K-edge (*ca* 400eV). A power law background model was fitted in a region prior to the Hf edge on the extended PCA reconstructed dataset. The backgrounds obtained were subtracted from the extended original dataset so that no information in the edges was lost due to the limited number of components in the reconstruction. The Hf signal was integrated up to the N edge onset. Using the same background subtraction, the N K-edge intensity was integrated over a 54 eV window from its onset. No correction was made for the small contribution from the Hf $N_{2,3}$-edge. A check was made to confirm that correction was always small. This was done by subtracting a suitably scaled Hf $N_{2,3}$-edge from a region of HfO_2. A power law background was fitted before the Ti $L_{2,3}$-edge and its intensity integrated over 70 eV window from its onset. The same procedure was used for the O K-edge but the integration window was over 22eV.

For the intensity and fit coefficient profiles, it is useful to define a mid-point (the 50% height) and a profile width (the separation of the 10% and 90% heights). The 0% and 100% heights are defined as the value in the bulk material on either side of the interface. The experimental profiles have noise and so, to give a consistent measure of the heights, a function was fitted to each profile. The function chosen consisted of the sum of two sigmoids with independent mid-points and widths but constrained to change from 0 to 100%.

3. Results and discussion

Figure 3 shows the elemental profiles across the interface for the fully binned data. Small errors in the background subtraction caused the Hf and O signals to go slightly negative in the bulk TiN on the right hand side. Offsets were added to bring them to zero. The profiles were then scaled to have the same value at the left (Hf and O) and right (Ti and N) sides. It is clear that the N profile is to the left of the Ti profile and the O profile is to the left of the Hf profile. The Hf and Ti profiles cross at around the 50% level as do the O and N profiles. The relative shift of elemental profiles identifies the presence of a reaction layer [1,2]. The separation of the profiles at the 50% heights is ~0.45nm and so there is an HfN or possibly Hf (N,O) reaction layer of this width at the interface. Since the light element 50% heights are to the left of the metal ones, the reaction has been a "nitridation" of the original HfO_2 rather than an "oxidation" of the subsequent TiN.

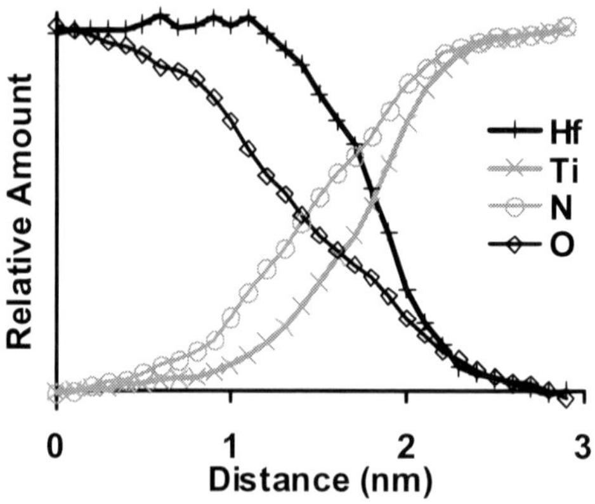

Figure 3. Elemental profiles across the interface.

The separation of the 50% heights is a measure of the average width of the reaction layer along the interface. Since double sigmoid fitting has been carried out on the extended dataset, the loci of the 50% heights as a function of position along the interface can then be examined as a function of the degree of binning. Figure 4a shows the loci for ×4 binning along the interface. Even with this modest level of binning, it can be seen clearly that the positions of the Ti and Hf profiles are always close together, as are those of the O and N profiles. It is also clear that the interface layer is wider in the top part of the interface (~0.65nm) than it is in the bottom part (0.35nm).

Figure 4b shows the separation of the 10% and 90% heights of each profile, which gives a measure of the profile width. The positions of these heights, and thus the profile widths, are much more susceptible to noise in the original data. Even with the dual sigmoid fits and ×10 binning, there is still considerable scatter in the values of the separations. Nonetheless, it is clear that the widths of the profiles are relatively constant along the interface. The O profile (~1.8nm) is wider than those of the Hf and Ti profiles (~1.2nm) while that of the N profile is intermediate (~1.4nm). These are all considerably wider than the width of the interface layer, showing clearly that such interface layers can be detected even if they are much narrower than the widths of the corresponding elemental profiles.

The width of the profiles is the result of a number of effects e.g. beam spreading and interface roughness. Since beam spreading would affect all the profiles in a similar way, the differing values of the profile widths is providing information on the nature of the HfO_2-"HfN" and the "HfN"-TiN interfaces. The fact that the widths of the O and N profiles differ is providing information on the nature of the reaction that has occurred.

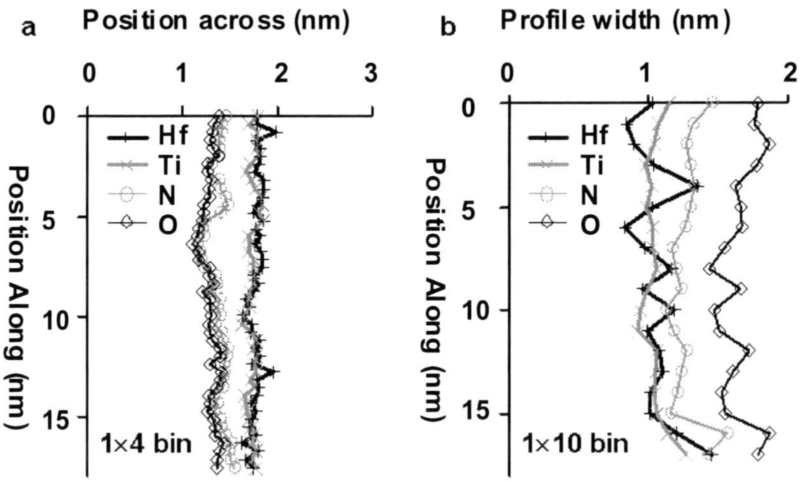

Figure 4. a) The positions of the 50% heights of the elemental profiles, determined by fitting a double sigmoid function to the profiles, as a function of position across and down the interface. Binning ×4 reduces the noise in the data allowing the change in the width of the reaction layer along the interface to be seen. b) The separation of the 10% and 90% intensities as a function of the position along the interface. Even at ×10 binning, there is still considerable scatter in the values.

4. Conclusions

Using EELS spectrum imaging, an HfN or Hf(O,N) reaction layer, has been identified at the TiN/HfO_2 interface in a metal inserted high-k gate stack. The layer has a mean thickness of 0.45nm over an 18nm length of the interface. This reaction layer has formed in the original HfO_2. By binning the data ×4 along the interface, a variation of the width from 0.35nm to 0.65nm along the interface can be seen. The 10%-90% widths of the elemental profile can also be found but the binning required to overcome noise is greater (×10). The profile widths are approximately constant along the interface but the values differ from element to element. Thus the reaction has not formed a conformal layer of uniform thickness. To allow binning along the interface, the interface was aligned parallel to one edge of the dataset. To provide an efficient way of processing the data with different degrees of binning, extended datasets containing the original data and data binned to the different levels were constructed. In this way, the processing can be carried on all levels of binning at the same time and the appropriate level of binning to allow extraction of a particular type of information can be easily identified.

Acknowledgements
The authors wish to thank Professor S de Gendt of IMEC for supplying the gate stack material and EPSRC for funding under grant EP/D040205/1.

References

[1] MacKenzie M, Craven A J, McComb D W and de Gendt S 2006 *App. Phys. Lett.* **88** 192112
[2] MacKenzie M, Craven A J, McComb D W and de Gendt S 2006 S *J. Electrochem Soc.* **153** F215
[3] Bosman M, Keast V J, Garcia-Munoz J L, D'Alfonso A J, Findlay S D and Allen L J 2007 *Phys. Rev. Lett.* **99** 086102
[4] Borglund M, Åstrand P-G and Csillag S *Microscopy and Microanalysis* 2005 **11** 88
[5] Craven A J, Schaffer B and Sarahan M C 2011 *Microelectronic Eng.* **88** 1488

Quantitative transmission electron microscopy at atomic resolution

L J Allen[1], A J D'Alfonso[1], B D Forbes[1], S D Findlay[2], J M LeBeau[3] and S Stemmer[4]

[1] School of Physics, University of Melbourne, Victoria 3010, Australia.
[2] School of Physics, Monash University, Victoria 3800, Australia.
[3] Department of Materials Science and Engineering, North Carolina State University, Raleigh, NC 27606, U.S.A.
[4] Materials Department, University of California Santa Barbara, Santa Barbara, CA 93106, U.S.A.

E-mail: lja@unimelb.edu.au

Abstract. In scanning transmission electron microscopy (STEM) it is possible to operate the microscope in bright-field mode under conditions which, by the quantum mechanical principle of reciprocity, are equivalent to those in conventional transmission electron microscopy (CTEM). The results of such an experiment will be presented which are in excellent quantitative agreement with theory for specimens up to 25 nm thick. This is at variance with the large contrast mismatch (typically between two and five) noted in equivalent CTEM experiments. The implications of this will be discussed.

1. Introduction

High-resolution conventional transmission electron microscopy (CTEM) and scanning transmission electron microscopy (STEM) are both powerful techniques capable of investigating local atomic structure. STEM uses a finely focused probe and CTEM uses plane wave illumination conditions. Interpretation of images in both techniques relies critically on image simulations [1]. For example, the intensity of columns in high-angle annular dark-field (HAADF or Z-contrast) imaging in STEM depends on the physics of thermal diffuse scattering (TDS) and Debye-Waller factors. Phase contrast and associated contrast reversals can dominate CTEM images. Quantitative agreement has recently been shown between experiments and theory in Z-contrast STEM [2,3] to within a few percent. Those results demonstrate that the current understanding of image formation is adequate in the modeling of TDS, which dominates Z-contrast images. However, quantitative comparisons in CTEM have often been plagued by a large contrast mismatch (typically between two and five) between theory and experiment, a phenomenon often referred to as the Stobbs factor [4]. Recently, it has been shown that careful consideration of the image recording process may account for the Stobbs factor in CTEM [5].
Bright-field STEM, in which lattice images are formed by interference of overlapping convergent beam discs, is related to CTEM via the principle of reciprocity [6]. As in CTEM, these images are dominated by phase contrast. Bright-field STEM images can be acquired simultaneously with Z-contrast images without the need to change the electron optical conditions of the microscope (except for slight adjustments in the defocus), providing near-simultaneous complementary information.

Similarly to CTEM, bright-field STEM images are more sensitive than Z-contrast images to light columns of atoms.

Here we report on quantitative comparisons between experiments and theory in bright-field STEM [7]. Experimental bright-field STEM images are placed on an absolute scale [8] and can thus be directly and quantitatively compared with image simulations. We show near perfect agreement between simulations and experiment and discuss the implications for CTEM.

2. Results and discussion

An FEI Titan 80–300 with a super-twin lens (Cs = 1.2 mm) was operated at 300 kV with a probe convergence semi-angle of 9.4 mrad. Experimental bright-field STEM images of a $SrTiO_3$ single crystal (2.8 mrad collection semi-angle) are shown in the left-hand column of each column pair in figure 1 for a range of thickness and defocus values. Contrast reversals and changes in the pattern occur for different defocus and thickness values, characteristic of coherent interference (phase

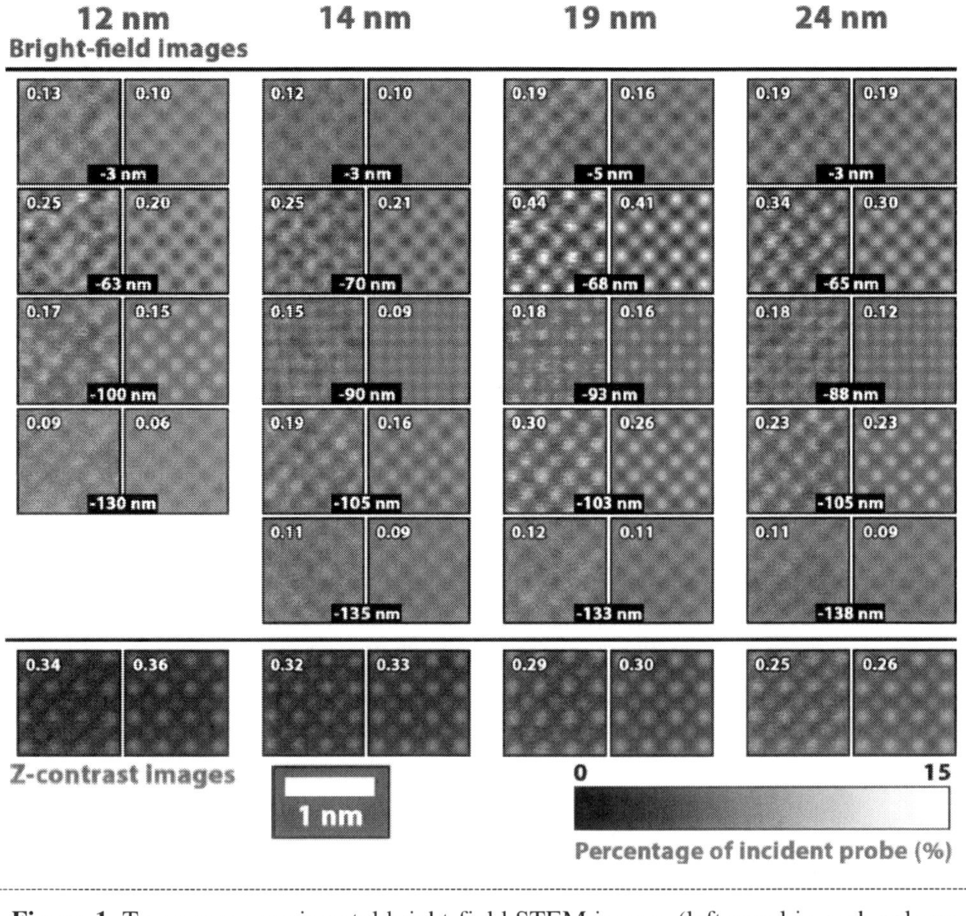

Figure 1. Top rows: experimental bright-field STEM images (left panel in each column, unfiltered) compared to multislice absorptive model calculations (right panel in each column). The upper labels in each image show their contrast values. The lower (black background) labels state the defocus, with underfocus being negative. Bottom row: experimental and simulated Z-contrast images (54 nm underfocus). Note that all images are on an absolute intensity scale relative to the incident probe and reported as a fraction of the incident probe intensity. The simulations have been convolved with a Gaussian of 0.11 nm FWHM to account for the effects of a finite source size.

contrast) images. The labels on each image state its measured contrast value, defined here as the ratio of the standard deviation of the image intensities to the mean image intensity.

Results from calculations using the multislice absorptive model [9] are shown in the right-hand column of the column pairs in figure 1. Spatial incoherence is the combined effect of a finite illumination source size, mechanical instabilities, sample drift, etc. and it is taken into account by convolving the simulated images with an effective source distribution function, a Gaussian with a full-width at half-maximum (FWHM) of 0.11 nm. Because spatial incoherence is difficult to measure directly in non-aberration-corrected STEM, several control experiments were performed to ensure the effective source size used here provides a realistic estimate for the spatial incoherence in this instrument. In particular, the bottom row in figure 1 shows Z-contrast images and comparisons with frozen phonon simulations [10] accounting for spatial incoherence. The Z-contrast images (detector inner angle of 65 mrad) were recorded at the same locations as the bright-field STEM images. A Gaussian effective source function with a FWHM of 0.11 nm yielded excellent agreement between simulation and experiment in *both* Z-contrast imaging (see bottom row in figure 1) *and* in bright-field STEM. The function is thus independent of the scattering processes, which are very different in bright-field STEM and Z-contrast imaging. The function is furthermore independent of the material and sample thickness, as has been shown previously using materials with a wide range of different scattering cross-sections, and thus it is *not* caused by a scattering mechanism not accounted for in the simulations. Furthermore, the *mean* image intensities, which are independent of spatial incoherence, agree quantitatively in both experiments and simulations in both bright-field STEM and Z-contrast imaging (not shown). Discrepancy in the mean intensities would indicate a redistribution of intensities by a scattering process not accounted for in the simulations, which is clearly not the case.

Figure 1 shows near perfect agreement between simulations and experiments across the entire range of thickness and defocus values. The average contrast mismatch factor between experimental and simulated bright-field STEM images is 1.15. This small residual mismatch is entirely within the experimental uncertainty caused by the noise of the detector, residual astigmatism, drift and, most importantly, the thickness determination (±1 nm).

3. Quantum mechanical model for TDS

The frozen phonon model has been used for calculating the Z-contrast images shown in Figure 1. That approach [10] models elastic scattering from atoms displaced from their equilibrium positions. It is motivated by the idea that the time taken for the fast electron to traverse the crystal is much smaller than the oscillation period of an atom. Within this semi-classical model the electron sees a snapshot of the atom frozen mid-vibration. Each electron "sees" a different configuration, and the contributions of different electrons are summed incoherently in the detector plane. In practice this is implemented by a Monte Carlo integration. The frozen phonon model approach has been shown to agree well with an absorptive model of thermal scattering and has produced simulations that compare well with experiment. However, the frozen phonon model does not contain within its conceptual framework the momentum or energy transfer one would normally associate with inelastic scattering, in this case phonon excitation. This shortcoming has contributed to the mistaken assumption

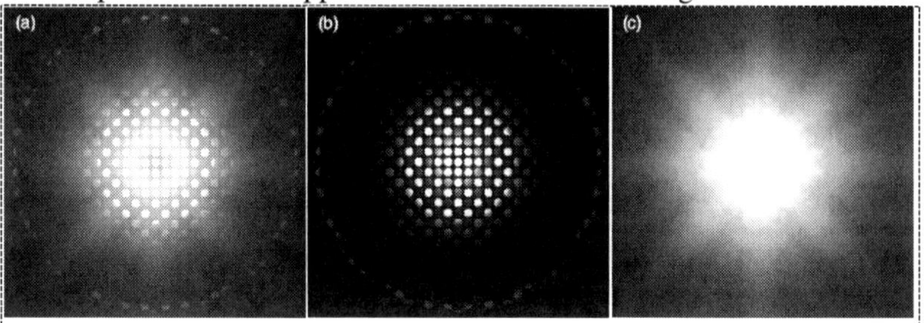

Figure 2. Diffraction pattern components for a 60-Å-thick $SrTiO_3$ specimen formed with a convergent probe of aperture 0.112 $Å^{-1}$ positioned over a Sr column and displayed on a log scale; (a) the full intensity, (b) elastically electrons only, and (c) inelastically (thermally) scattered electrons only.

that scattering from a moving lattice and the excitation of inelastic transitions are distinct effects.

We have introduced a fully quantum mechanical model for TDS by solving the many body Schrödinger equation using a Born-Oppenheimer approximation [11]. The Born-Oppenheimer (BO) model predicts the scattered intensity for a single electron, including multiple elastic and inelastic phonon scattering to all orders. This is an advantage over other approaches based on the Yoshioka formalism [12] in which the single inelastic scattering approximation is usually made to allow for tractable calculations. In this model, we can explicitly calculate the elastic component and inelastic components of the scattered electron wave. This is illustrated in figure 2. Whilst the semi-classical conceptual underpinnings of the frozen phonon model are quite different from the quantum mechanical model it is important to note that the frozen phonon model yields results which are numerically equivalent to the BO model [11]. The bright-field and Z-contrast images in the previous section can be calculated in the BO model and agree with those simulations. This confirms that the match between theory and experiment demonstrated there is on a secure footing.

4. Summary

In summary, we have demonstrated excellent agreement between simulation and experiment in bright-field STEM. The excellent agreement in this and in prior studies of Z-contrast images [2,3] shows that current models of image formation adequately model both low and high-angle scattering, including thermal diffuse scattering. Other inelastic scattering processes do not play a significant role in contrast formation for samples that are sufficiently thin. Furthermore, inelastic scattering, which by reciprocity is present in bright-field STEM as much as it would be in CTEM, is not the origin of the contrast mismatch observed in the CTEM literature for similar specimen thicknesses.

Acknowledgements

L.J. Allen and S.D. Findlay acknowledge support from the Australian Research Council. S. Stemmer was supported by the U.S. National Science Foundation under Grant No. DMR-0804631.

5. References

[1] Spence J C H 2003 *High-Resolution Electron Microscopy* (Oxford: Oxford Science Publications)
[2] LeBeau J M, Findlay S D, Allen L J and Stemmer S 2008 *Phys. Rev. Lett.* **100** 206101
[3] LeBeau J M, Findlay S D, Wang X, Jacobson A J, Allen L J and Stemmer S 2009 *Phys. Rev. B* **79** 214110
[4] Howie A 2004 *Ultramicroscopy* **98** 73
[5] Thust A 2009 *Phys. Rev. Lett.* **102** 220801
[6] Cowley J M 1969 *Appl. Phys. Lett.* **15** 58
[7] LeBeau J M, D'Alfonso A J, Findlay S D, Stemmer S and Allen L J 2009 *Phys. Rev. B.* **80** 174106
[8] LeBeau J M and Stemmer S 2008 *Ultramicroscopy* **108** 1653
[9] Allen L J, Findlay S D, Oxley M P and Rossouw C J 2003 *Ultramicroscopy* **96** 47
[10] Kirkland E J 1998 *Advanced Computing in Electron Microscopy* (New York: Plenum Press)
[11] Forbes B D, Martin A V, Findlay S D, D'Alfonso A J and Allen L J 2010 *Phys. Rev. B* **82** 104103
[12] Yoshioka H 1957 *J. Phys. Soc. Jpn.* **12** 618

Intensity calibration and atomic imaging of size-selected Au and Pd clusters in aberration-corrected HAADF-STEM

Z W Wang and R E Palmer*

Nanoscale Physics Research Laboratory, School of Physics and Astronomy, University of Birmingham, *Birmingham*, B15 2TT, U.K.

* E-mail: r.e.palmer@bham.ac.uk

Abstract. We report an experimental calibration of the exponent α, which relates intensity (I) and atomic number (Z) in HAADF-STEM imaging. The method employs size-selected Au_{923} and Pd_{923} clusters in an aberration-corrected 200 kV microscope. We find $\alpha = 1.46 \pm 0.18$ for an inner collection angle of 62 mrad. The value of α is needed in the quantitative compositional analysis of, e.g., bimetallic nanostructures. The atomic structures of these two examples of size-selected magic-number clusters are also reported.

1. Introduction

High Angle Annular Dark Field (HAADF) imaging [1-4] in the aberration-corrected Scanning Transmission Electron Microscope (STEM) has established itself as a powerful method to obtain structural and chemical information with atomic resolution [5-8]. An appealing feature of the technique is that the measured image intensity (I) follows a simple relationship with specimen thickness (t) and atomic number (Z), $I \sim tZ^{\alpha}$, in the limit of sufficiently thin samples. The exponent α would be 2 for pure Rutherford scattering, although this value is reduced by the electronic screening of the nuclear charge [9]. In practice, the exponent α obtained in the STEM will vary with the experimental parameters employed (mainly the inner collection angle). An exact value of α is required for quantitative HAADF imaging, e.g. the determination of chemical composition in bimetallic clusters. A precise experimentally calibration of α is however not trivial, due to the difficulties of accurately measuring the sample thickness inside the microscope. For example, a difference of 25% is found when convergent beam electron diffraction and low-loss electron energy spectra are used to measure the film thickness for an SrTiO3 sample [10]. In this study, we report an experimental calibration of the exponent α based on statistical HAADF intensity analysis of size-selected Au and Pd nanoclusters, specifically Au_{923} and Pd_{923} clusters that were generated in a cluster beam source with a mass-selection accuracy of 5%. The atomic imaging and structural comparison of these two examples of magic number clusters is also presented, with an aim to provide a reference for understanding some morphology/structure-relevant catalytic phenomena of nanoclusters.

2. Experimental method

Size-selected Au_{923} and Pd_{923} clusters were prepared in a magnetron-sputtering, gas-condensation cluster-beam source coupled to a lateral time-of-flight mass filter ($M/\Delta M$=20 nominally) [11-13]. An impact energy (500 eV per cluster) in the "soft-landing" regime (<1 eV per atom) was used when depositing the clusters onto amorphous carbon films supported by Cu TEM grids. STEM imaging was performed in a 200 kV JEM2100F STEM (JEOL) with a spherical aberration corrector (CEOS). The signal was collected by a high-angle annular dark-field detector with inner and outer collection angles of 62 and 164 mrad at a probe convergence angle of 19 mrad.

3. Results and discussion

HAADF-STEM imaging of the Au_{923} and Pd_{923} clusters was performed under the same experimental conditions to allow for direct intensity comparison. Figure 1 presents examples of HAADF images of the two kinds of clusters. The intensity difference between the two clusters can be clearly seen and is attributed to their widely different atomic numbers (Au, 79; Pd, 47).

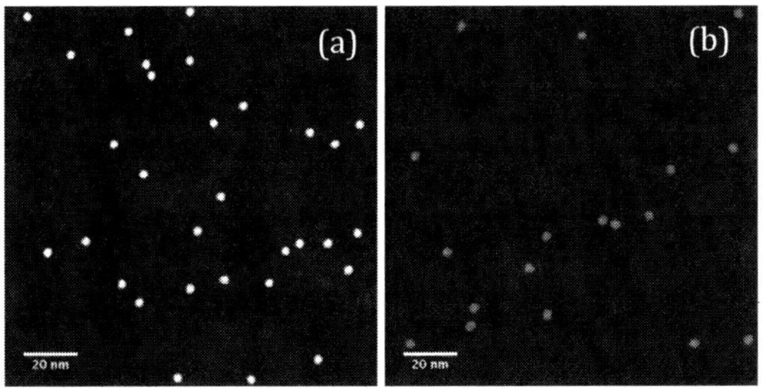

Figure 1. Typical HAADF-STEM images of (a) Au_{923} and (b) Pd_{923} clusters on amorphous carbon films acquired under the same imaging conditions (see text). The same greyscale settings (brightness/contrast) were also used in both cases to allow direct visual comparison of the relative intensities.

The quantitative integrated intensity analysis for the two kinds of cluster is performed using "as – recorded" HAADF images. The integrated intensity from a circular annulus around the cluster is taken as the background intensity. The annulus has the same normalized area as the projected cluster area. After the background subtraction, the intensity measurement of the two cluster ensembles is obtained, figure 2. It can be seen that the peak intensity values of the two clusters are well separated. The calculated mean intensities and corresponding standard deviations are 2.42 ± 0.13 for the Au_{923} and 1.10 ± 0.09 for the Pd_{923} clusters, in arbitrary unit. Since the two reference clusters contain the same number of atoms, and identical experimental conditions were used for recording HAADF images, the exponent α is then simply derived from an intensity comparison using the relationship $I_{Au}/I_{Pd} = (Z_{Au}/Z_{Pd})^{\alpha}$. By this method, we obtained the value $\alpha = 1.46 \pm 0.18$.

Obviously, the experimental value of the exponent α is much smaller than the value for true Rutherford scattering (α=2). We found similar behavior in our uncorrected HAADF-STEM relative intensity measurements of Au_{923} and Pd_{923} clusters [14]. We identified the inner collection angle of the detector as chief culprit – a value of α as large as ~ 1.8 was obtained when the inner collection angle was increased to 103 mrad. Measurements of α as a function of scattering angle in the new aberration-corrected instrument are in progress. In general, we find that very large collection angles are problematic for the quantitative measurements, especially for thin samples, due to poor signal to noise ratios (SNR). The significant contribution from the background noise will undermine the measurement precision. Thus, collection angles such as the one used in the present experiment, where the value of α may be far from 2, are perhaps best suited to quantitative analysis [15]. In consequence, an accurate calibration of the value of α will be critical for some kinds of compositional analysis.

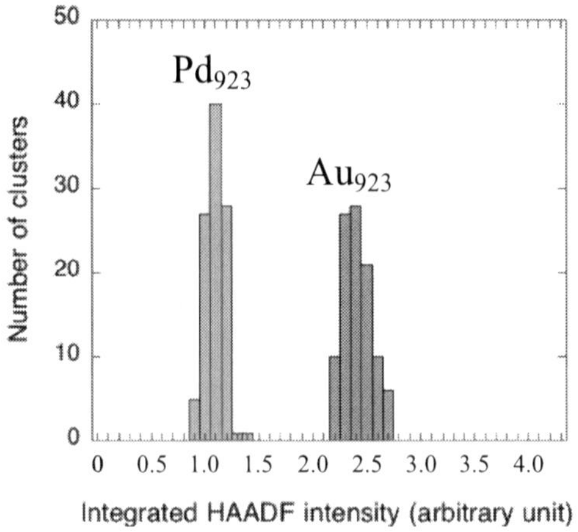

Figure 2. Histogram of the HAADF intensities for Au_{923} and Pd_{923} clusters obtained under the same imaging conditions.

One question that might arise is – what is the effect of diffraction contrast on the value of α obtained? Since the measurement depends on the integrated cluster intensities this is a fair question. However, considering that in this study the exponent α is calibrated by using the statistical HAADF intensities, which are obtained by averaging over a large number of clusters that locate on the amorphous carbon films with random orientation, the effect of orientation and thus diffraction contrast should be minimal. This is also reflected in the previous experimental calibration of mass-intensity relationship for Au clusters [5], which demonstrated a linear relationship between the integrated HAADF intensity from the clusters and the number of constituent atoms up to at least 1000, indicating that the influence of the diffraction contrast is trivial for the small clusters.

It is believed that the catalytic activity (and selectivity) of nanoparticles depends on their size and geometric structure [16-17]. Thus, while the atomic structure of the size-selected Au and Pd clusters is not our focus here, we show in figure 3 examples of the measured high-resolution STEM images of these particles. The Au_{923} and Pd_{923} clusters have significantly different internal structures, even though they have the same nuclearity (size). In general, the Au_{923} clusters show more ordered atomic arrangements than Pd_{923}. Three typical structures are observed for Au_{923} – decahedron, FCC polyhedron and icosahedron, while Au adatoms on the cluster facets are also observed. The Marks-like decahedron and FCC polyhedron are the dominant structural forms, and are illustrated in figure 3(a)-(c). By contrast, Pd_{923} often show more disordered structures; an example image and corresponding FFT transform are shown in figure 3(d). In some cases, icosahedral-type motifs can be observed, as illustrated in figure 3(e)-(f). The cluster presents rather like a fivefold-axis orientated icosahedron according to the ring features observed, but it is difficult to match the data precisely with a STEM simulation obtained using a standard icosahedral atomic model.

4. Conclusion

We have calibrated the exponent α in the formula $I \sim Z^{\alpha}$, applicable to HAADF imaging in the (aberration-corrected) STEM, using size-selected Au_{923} and Pd_{923} clusters. We find α = 1.46 ± 0.18. This value is far from that defined by ideal Rutherford scattering (α=2), which is highly relevant to compositional analysis. The atomic structures of the Au_{923} clusters show more ordered atomic arrangements than the Pd_{923} clusters.

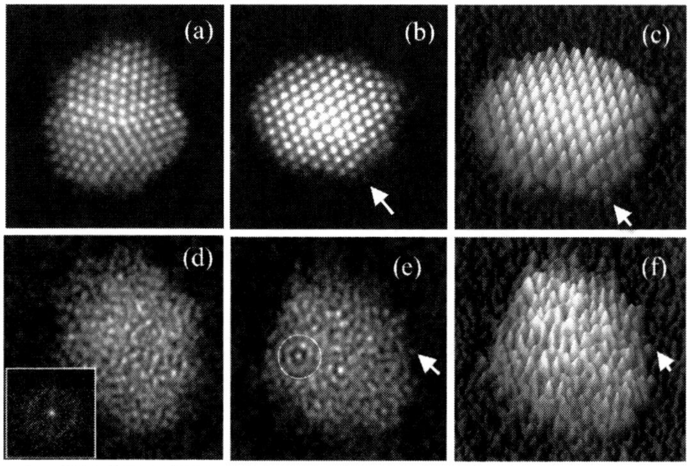

Figure 3. Representative HAADF-STEM images (5 × 5 nm) of (a)-(b) Au_{923} clusters and (d)-(e) Pd_{923} clusters deposited on amorphous carbon films. The inset in (d) is an FFT power spectrum. (c) and (f) are 3D colour plots of (b) and (e), respectively. The surface Au or Pd adatoms marked by arrows show the relationship between the orientations of the corresponding 2D and 3D images. The circle in (e) marks a ring feature observed in the icosahedra-like Pd_{923} cluster.

Acknowledgements
We acknowledge financial support from the EPSRC and TSB. The STEM instrument used in this research was obtained through the Birmingham Science City project "Creating and Characterising Next Generation Advanced Materials", supported by Advantage West Midlands (AWM) and partly funded by the European Regional Development Fund (ERDF).

References

[1] Crewe A V, Wall J and Langmore J 1970 *science* **168** 1338
[2] Howie A 1979 *J. Microscopy* **117** 11
[3] Treacy M M J, Howie A and Wilson S J 1978 *Phil. Mag. A* **38** 569
[4] Loane R F, Xu P and Silcox J 1992 *Ultramicroscopy* **40** 121
[5] Li Z Y, Young N P, Di Vece M, Palomba S, Palmer R E, Bleloch A L, Curley B C, Johnston R L, Jiang J and Yuan J 2008 *Nature* **451** 46
[6] Wang Z W, Toikkanen O, Quinn B M and Palmer R E 2011 *Small* **7** 1542
[7] LeBeau J M, Findlay S D, Allen L J and Stemmer S 2008 *Phys. Rev. Lett.* **100** 206101
[8] Yin F, Wang Z W and Palmer R E 2011 *J. Am. Chem. Soc.***133** 10325
[9] Hartel P, Rose H and Dinges C 1996 *Ultramicroscopy* **63** 93
[10] Klenov D O and Stemmer S 2006 *Ultramicroscopy* **106** 889
[11] Pratontep S, Preece P, Xirouchaki C, Palmer R E, Sanz-Navarro C F, Kenny S D and Smith R 2003 *Phys. Rev. Lett.* **90** 055503
[12] Pratontep S, Carroll S J, Xirouchaki C, Streun C M and Palmer R E 2005 *Rev. Sci. Instrum.* **76** 045103
[13] Di Vece M, Young N P, Li Z Y, Chen Y and Plamer R E 2006 *Small* **2** 1270
[14] Wang Z W, Li Z Y, Park S J, Abdela A and Palmer R E 2011 submitted to *Phys. Rev. B*
[15] Wang Z W, Toikkanen O, Yin F, Li Z Y, Quinn B M and Palmer R E 2010 *J. Am. Chem. Soc.* **132** 2854
[16] Kaden W E, Kunkel W A, Kane M D, Roberts F S and Anderson S L 2010 *J. Am. Chem. Soc.* **132** 13097
[17] Kesavan L et al 2011 *Science* **331** 195

Journal of Physics: Conference Series **371** (2012) 012011 doi:10.1088/1742-6596/371/1/012011

High-resolution SEM observation at the atomic level using a dedicated STEM with aberration correction.

M Konno [1], Y Suzuki[1], H Inada[2] and K Nakamura[2]

* Global Application Center, Naka Division, Nanotechnology Products Business Group, Hitachi High-Technologies, 11-1, Ishikawa, Hitachinaka, Ibaraki, 312-0057, Japan

** Advanced Microscope System Design 2nd Department, Naka Division, Nanotechnology Products Business Group, Hitachi High-Technologies, 882, Ichige, Hitachinaka, Ibaraki, 312-8504, Japan

E-mail: konno-mitsuru@naka.hitachi-hitec.com

Abstract. We have developed a high-resolution secondary electron (SE) imaging technique using a dedicated scanning transmission electron microscope (STEM) instrument with a high efficiency SE detector (Hitachi HD-2700) in order to investigate the structural characterization at the atomic level without thin film processing. The HD-2700 is equipped with a CEOS GmbH spherical aberration (Cs) corrector and can routinely achieve 1Å spatial resolution in SE and STEM imaging. We applied this technique for high accuracy critical dimension measurements of semiconductor materials using aberration corrected SE imaging. We have observed the cross sectional SE image of a semiconductor transistor along the Si[011] zone axis. The specimen thickness was 1μm and an accelerating voltage of 200 kV was used to observe the semiconductor structures, such as a metal gate. Under these conditions the high-k gate insulating layer, source, and drain structures can be clearly seen. The higher magnification SE images show that it is possible to resolve the Si111 plane, which has a spacing of 0.314 nm, and these images can be used for magnification calibration. These results indicate the ability to perform high precision measurements of crystal lattices for the structural characterization of semiconductor materials.

1. Introduction

The manufacturing of semiconductor devices requires highly precise measurement of film thickness, such as a gate oxide layer and barrier metal layers, with atomic resolution using transmission electron microscopy. However, recent improvements in semiconductor transistors have led to a reduction of gate dimensions bellow 20 nm. The decreased gate dimensions require precise site-specific and time intensive sample preparation that leads to decreased throughput. In order to reduce the demands placed on sample preparation we have developed a high-resolution secondary electron (SE) imaging technique using the Hitachi HD-2700 dedicated scanning transmission electron microscope (STEM) instrument [1]. The incorporation of the high efficiency SE detector makes high resolution imaging of semiconductor specimens possible while avoiding the time consuming sample preparation required to

[1] E mail: konno-mitsuru@naka.hitachi-hitec.com

produce thin sections for transmission electron microscope (TEM). Focused ion beam (FIB) systems are commonly used in sample preparation of semiconductor devices for electron microscopy. FIB prepared samples are covered with a Ga-ion damage-layer, which is reduces the resolution in SE imaging. In this study, we have investigated the possibility of applying this new technology for highly accurate critical dimension measurements of semiconductor materials.

2. Experiment

The HD-2700 is equipped with a CEOS GmbH aberration corrector and a high efficiency secondary electron detector making it possible to routinely achieve 1Å spatial resolution in SE imaging and STEM imaging [2]. The beam current for high resolution SE imaging was measured using a Faraday cup in the specimen holder and the beam current was found to be 50 pA using a probe size of 0.1 nm diameter and an accelerating voltage of 200 kV. The samples used in this study have a pillar geometry (Figure.1) and are covered with a gallium (Ga) ion damage-layer induced by FIB milling. A NB5000 FIB-SEM (Hitachi) that is equipped with a micro-sampling system was used for the FIB milling [3]. The unique capabilities of the FIB-STEM compatible sample holder that allows a 360° sample rotation inside the FIB system and the microscope were taken advantage of for this work [4]. The damage layer caused by FIB milling at accelerating voltages ranging from 2-40 kV were measured along the Si [110] axis by using the HD-2700 in the STEM mode at 200 kV. The HD-2700 is equipped with a cold field emission gun (CFE), which has a high brightness and narrow energy distribution, and a CCD camera (Gatan, ORIUS, 14bit, 2.6k×2.6k) for acquiring electron diffraction and Ronchigram patterns [5]. Figure 2 indicates a clear relationship between the accelerating voltage of the FIB and the thickness of the damage-layer, which consists amorphous silicon (Si) on top of the bulk Si crystalline material. Measurements of the of amorphous Si layer determined that the thickness of the layer was 28 nm at 40 kV, 8 nm at 10 kV and 3 nm at 2 kV.

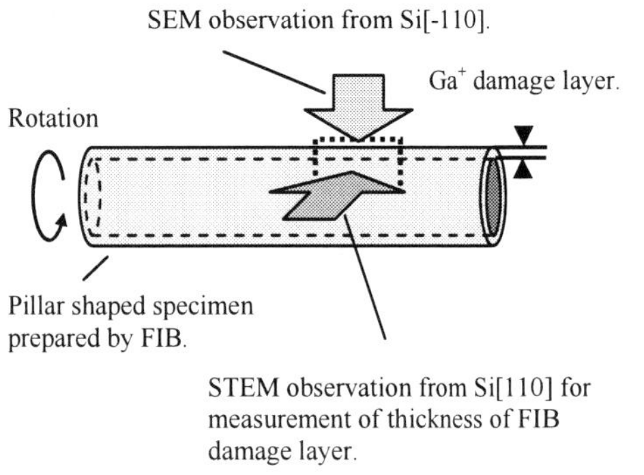

Figure 1. The specimen geometry and the observation direction of the SEM/STEM .

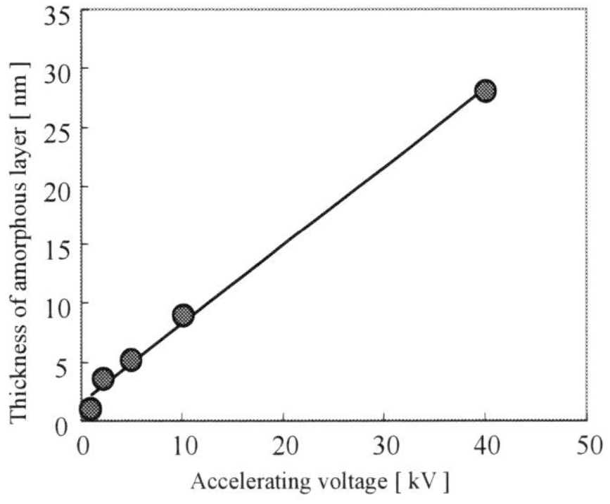

Figure 2. Measured results of the thickness of the amorphous layer formed on top of the Si crystalline layer as a function of the FIB accelerating voltage.

3. Results and discussion

3.1. SE image resolution

Figure 3 shows the SE images of the specimen prepared using a FIB beam at accelerating voltages of 40 kV (a), 10 kV (b) and 2 kV (c) and the Fourier transform (d, e, f) of each corresponding SE image. The images were obtained by observation of the Si[011] zone axis at an accelerating voltage of 200 kV. Not surprisingly, the SE image of the sample prepared at 2 kV had the thinnest amorphous layer and that the sample prepared at 40 kV yielded a very poor signal noise ratio. The Fourier transform images show the achievable frequencies of the 004 reflection spot (silicon dumbbell) and the 333 and 115 reflection spot (105 pm). These results demonstrate the ability to use SE images, obtained at high accelerating voltages, for determining the surface characteristics of materials. The results show that it is possible to achieve atomic resolution imaging light elements such as Si. What is even more exciting is the fact that this imaging technique can reveal structures underneath of a 2-3 nm amorphous Si layer. Further studies are needed to fully understand the underlying effects of the Si amorphous layer on SE imaging.

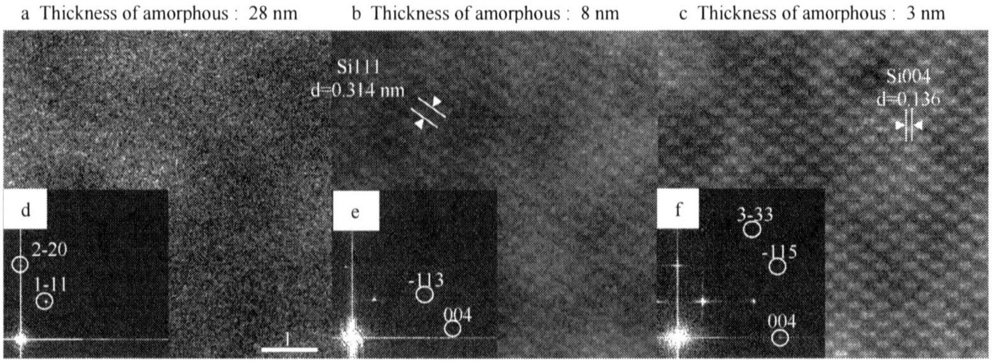

Figure 3. SE images taken with HD-2700 at 200 kV along the Si[011] zone axis with the corresponding FFT of a Si samples prepared by a FIB beam at accelerating voltages of (a) 40 kV , (b) 10 kV and (c) 2 kV.

Image magnification: ×8,000,000

3.2. Focus depth

Samples prepared by FIB at 40kv had an amorphous layer thickness of 28nm and it was still possible to detect the frequency of the Si111 reflection spot. In this study we have investigated the impact of the focus depth for SE imaging and the results will be discussed in this section. Figure 4 shows a series of through-focus SE and ADF-STEM images for strontium titanate ($SrTiO_3$) samples prepared by ion-milling where the through-focus was changed from -50 nm to +50 nm. The images were observed along the strontium titanate [001] zone axis using an accelerating voltage of 200 kV. Electron energy loss spectroscopy (EELS) was used to measure the thickness of the sample, which was found to be around 45 nm. For the STEM imaging, the point at which the resolution of the atomic column was the sharpest was used to determine the just focus condition (defocus amount Δf is 0 nm). With SE imaging the white spots that correspond to the atomic column of strontium (Sr) and Titanium (Ti) are visible for defocus settings from -30 nm to +10 nm. In contrast, the ADF-STEM images show that the atomic

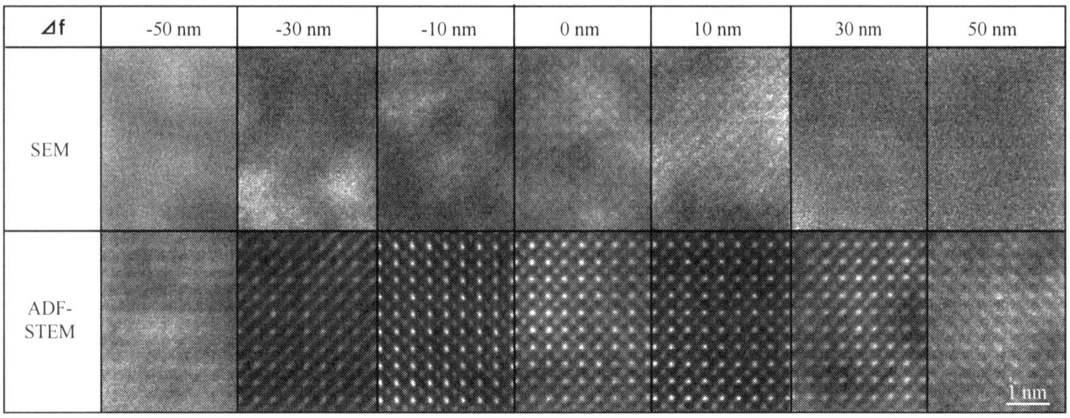

Figure 4. Series of through-focus SE and ADF-STEM images observed along the strontium titanate ($SrTiO_3$)[001] zone axis at accelerating voltage of 200 kV prepared using an ion-milling.

Image magnification: ×8,000,000

column is visible at a broader range of defocus settings and demonstrate a better focus depth compared to SE imaging. Conversely, SE images are less affected by the overlap of the atomic structures in the in the thickness of the sample. This result expands the possibility for overlap free structural characterization at atomic level without needing to produce a thin section.

4. Application for semiconductor

Figure 5 shows the cross sectional SE image of semiconductor transistor (a, b) and the magnified SE image (c) observed along the Si[011] zone axis for a sample prepared by using a FIB beam. The sample thickness was 1μm. Figure 5a and 5b clearly show the semiconductor structures such as a metal gate, high-k gate insulating layer, source and drain with great clarity. In Figure 5c, the magnified SE image shows that it is possible to resolve the Si111 plane, which has a spacing of 0.314 nm. This capability may also be used as a technique for precisely calibrating the magnification settings for STEM instruments. These results demonstrate the ability to perform high precision measurements of crystal lattices for the structural characterization of semiconductor materials.

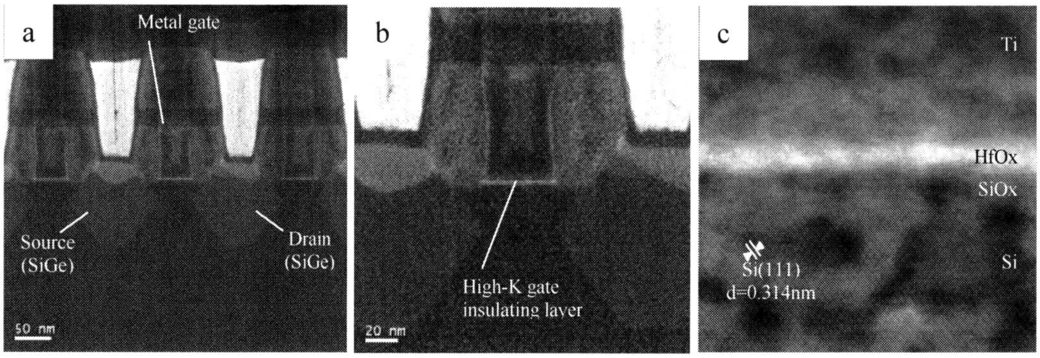

Figure 5. (a,b) The cross sectional SE image of semiconductor transistor and (c) the magnified SE image observed along the Si[011] zone axis.

Image magnification: ×180,000(a), ×5,000,000(b)

5. Summary

We have shown high-resolution SE images semiconductor structures using an aberration corrected dedicated STEM at 200kV and an SE detector for imaging. The Si dumbbell structure found in the samples had a size of 136 pm and while the atomic column of Sr and Ti structure had a size of 400 pm. We have also shown cross sectional SE images of a semiconductor transistor with a Si111 plane, corresponding to 314 pm which will be used for calibration of the magnification. This technique demonstrates high accuracy critical dimension measurements of semiconductor materials without the need for thin sectioning of the sample.

References

[1] Inada H, Wu L, Wall J, Su D and Zhu Y 2009 J. of Electron Microsc. 58, 111-122
[2] Zhu Y, Inada H, Nakamura K and Wall J 2009 Nature Mater. 8, 808-812
[3] Ohnishi T and Ishitani T, JP Patent 2774884 and US patent 5270552
[4] Kamino T, Yaguchi T, Konno M, Ohnishi T and Ishitani T 2004 J.Electron Microsc. 53, 583-588T
[5] Konno M, Suzuki Y, Inada H and Nakamura K 2010 Journal of Physics: Conference Series 241
[6] Konno M, Suzuki Y, Inada H, Nakamura K, Kimoto K and Zhu Y 2010 Microsc. And Microanal. 16, 128-129
[7] Rose H, 1990 Optics, 85, 19-24
[8] Haider M, Rose H, Uhlemann S, Shwan E, Kabius B and Urban K 1998 Ultramicroscopy, 75 53-60

Configuring a 300kV cold field-emission gun for optimum analytical performance

I M Ross and T Walther

Kroto Centre for High Resolution Imaging & Analysis, Department of Electronic and Electrical Engineering, University of Sheffield, Mappin Street, Sheffield S1 3JD, UK

i.ross@shef.ac.uk; t.walther@shef.ac.uk

Abstract: The performance of a new 300kV scanning transmission electron microscope (STEM) equipped with cold field-emission gun (c-FEG) is characterised with respect to the stability for and the energy width in electron energy-loss spectroscopy (EELS) and the probe size and beam current obtainable for convergent beam and scanning probe operation. These are compared to values from a 200kV FEG-(S)TEM with standard Schottky field-emission.

1. Introduction

The recent development of stable aberration corrected electron optics has seen an increase in the commercialisation of a number of analytical transmission (TEM) and scanning transmission electron microscopes (STEM). Different analytical problems often require an appropriate electron source and adjustments to the electron gun settings are therefore needed to provide the most appropriate conditions for the required analysis, e.g. high energy resolution for energy-loss near edge fine-structure (ELNES) analysis, small stable probes for high resolution (HR-STEM) and medium size probes with high intensity for energy-dispersive X-ray (EDX) mapping [1]. For field-emission TEM/STEM instruments, these conditions are usually a function of the settings of the extraction and accelerating anodes in the electron gun combined with factors such as beam convergence and detector collection angles. Here, we investigate the operation conditions of a newly developed W<111> cold field-emission gun fitted to a double aberration corrected TEM/STEM developed by JEOL (JEM Z3100F-R005) [2]. We have studied the influence of the extraction anode (A1), accelerating anode (A2) and illumination conditions (beam convergence, spot size) to establish the optimum electron source parameters for a range of analytical conditions. A comparison of the 300kV cold-FEG data to that obtained for a 200kV thermally assisted Schottky FEG (JEM 2010F) is also made. For a 200kV Schottky FEG probe currents approaching 1nA have been reported for a probe ~1nm in diameter [3].

2. Experimental methods

Measurements from the JEOL double aberration cold-FEG TEM/STEM were obtained at an accelerating voltage of 300kV. Those obtained from the JEOL 2010F Schottky FEG were obtained at 197kV. On both instruments, the normal means of indicating the beam current is via measurement of the apparent current density on either the large or small phosphorescent screen. However, this method is restricted by the sensitivity of the built-in current density meter to a range from ~10pA/cm^2 to ~511pA/cm^2. Consequently, in this study we have measured the beam current by directly coupling the output of the phosphor screen to a high sensitivity pico-ammeter (Keithley Model 2635). To assess the likely effect of variations in the "small" 25mm diameter phosphor screen coating, measurements were initially taken systematically at the small screen centre and at the very edge of the screen. The location at the edge was determined as the point which gave the maximum current reading. For each instrument, the small phosphor screens where calibrated against the beam current values obtained using a Faraday cage in the specimen plane (Gatan Model 641).

The nature of the 300kV c-FEG gun requires "*flashing*" after several hours of operation to remove gas absorbed on the tip. Beam current measurements were obtained prior to and after periodic *flashing* (*low-flash*: 1.62A for 20 seconds) and also after tip re-build (*high-flash*: 2.3A for 3 seconds followed by *Heat-1A*: A1 -2.70kV, 2.2A for 10 seconds). Values of the beam current were obtained as a function of nominal spot size in both convergent beam (CBED-α9) in which only the imaging corrector is optimized and STEM mode in which both imaging and probe correctors are optimized. In CBED mode a 150μm diameter condenser aperture was used and A2 was fixed at 4.6kV with A1 varied between 1.2 and 1.7kV for each nominal spot size. In STEM mode the value of A1 was maintained at 1.4-1.5kV while A2 was varied from 4.60kV to 4.80kV. A nominal camera length setting of 15cm was utilized with the condenser aperture setting varied from 30 and 40μm, corresponding to convergence semi-angles of 21 and 28mrad respectively. After fine adjustment of the corrector optics the probe image was focused onto the small phosphor screen edge by adjustment of the intermediate lens IL3 using free lens control. The probe diameter was estimated by consideration of the smallest resolvable lattice fringes observed in a corresponding annular dark-field (ADF) image for a given nominal spot size.

Similar investigations were performed for comparison with a Schottky FEG (JEOL 2010F at 197kV) in both CBED mode (α9) and STEM with an A1 setting of 2.75kV; A2 of 7.16kV; bias of 300V and a filament current of 2.24A, corresponding to an emission current of 147μA.

The energy spread within the primary beam was assessed in diffraction coupled mode using a 1mm GIF entrance aperture, 0.01-0.02eV/pixel dispersion and exposure times of 0.1 and 1.0s without specimen in the beam path (R005). The extraction anode (A1) of the c-FEG (R005) was adjusted from 0.9kV to 1.7kV with the accelerating anode (A2) set at 4.60kV in each case. Comparative data for the energy spread obtained for the Schottky FEG was obtained using an exposure time of 0.05-0.1s, a dispersion of 0.02eV/pixel and either 0.6mm or 3mm GIF entrance aperture. In all cases the energy resolution is stated as the full width half maximum (FWHM) of the resulting zero loss energy peak.

3. Results and Discussion

Fig. 1 shows that the zero loss peak (ZLP) in the JEOL R005 depends strongly on the setting of the extraction voltage, A1, and will typically be ~0.5eV wide (FWHM) at 300kV (~0.4eV at 80kV) for A1~1.5kV and A2 of 4.6kV. The form of the curves indicate that, for both acceleration voltages, the ZLP can be as narrow as ~0.33eV for A=1.25kV but below this the energy resolution seems to be limited by combination of the Fowler-Nordheim distribution of the field emitter and the spectrometer resolution. In the case of the 200kV Schottky emitter, however, the energy resolution is only ~0.6eV and, over the range of extraction voltages examined, depends linearly on A1 (i.e. here, A1 is limiting the performance). It is worth noting that the c-FEG produces a highly asymmetric ZLP (Fig. 2), with a tail on the low-energy side (higher energy losses) due to the tunneling of the electrons through the surface potential of the tip [4]. This will make band-gap investigations somewhat awkward, compared to the use of a monochromator [5].

The probe current depends exponentially on the settings on A2 and in particular A1, as can be seen from fig. 3. This means that for each application the A1 setting should be set to the current needed to yield a particular dose. In particular, lattice images should be recorded with the maximum current and probe that agree with the spatial resolution just necessary to resolve atomic columns in the lattice.

Fig. 4 compares probe currents recorded in CBED and STEM modes and shows, firstly, that the beam current depends strongly on the tip flashing; the tip is ~9 times brighter after a *high flash* at elevated temperature (which requires a subsequent "rebuild" of the tip apex and may ultimately limit its lifetime) and ~2 times brighter after a periodic *low flash* is performed. Secondly, our findings suggest that the setting of A2 is more important than the choice of a slightly larger condenser aperture.

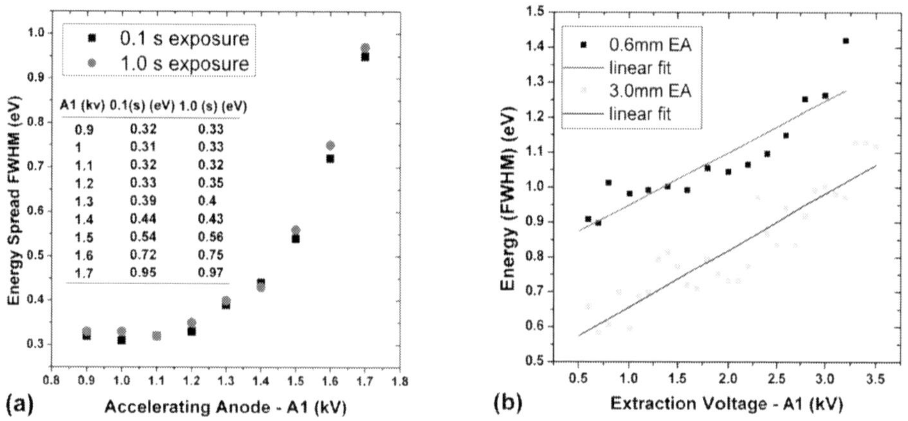

Fig. 1: plots of full width at half maximum (FWHM) of the zero loss peak distribution without specimen; (a): JEOL R005 at 300kV, 0.01eV/channel dispersion, 1mm entrance aperture (EA); (b): JEOL 2010F at 0.02088eV/channel, 0.05s (0.6mm EA) and 0.1s (3mm EA).

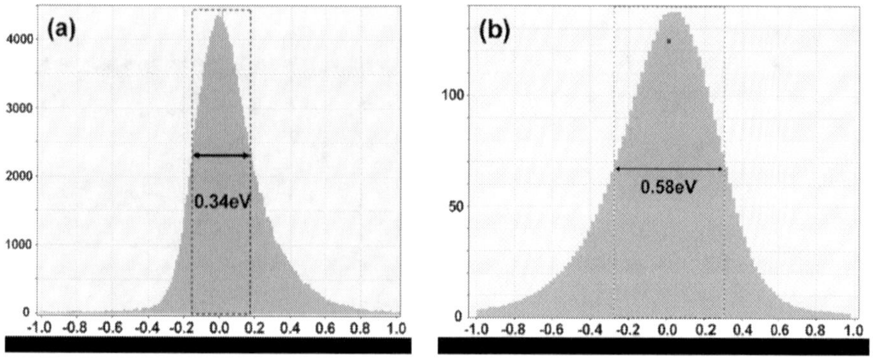

Fig. 2: comparison of the zero electron energy-loss spectra from (a) JEOL R005 at 300kV (0.34eV FWHM) and (b) JEOL 2010F at 197kV (0.58eV FWHM) at the same energy scale

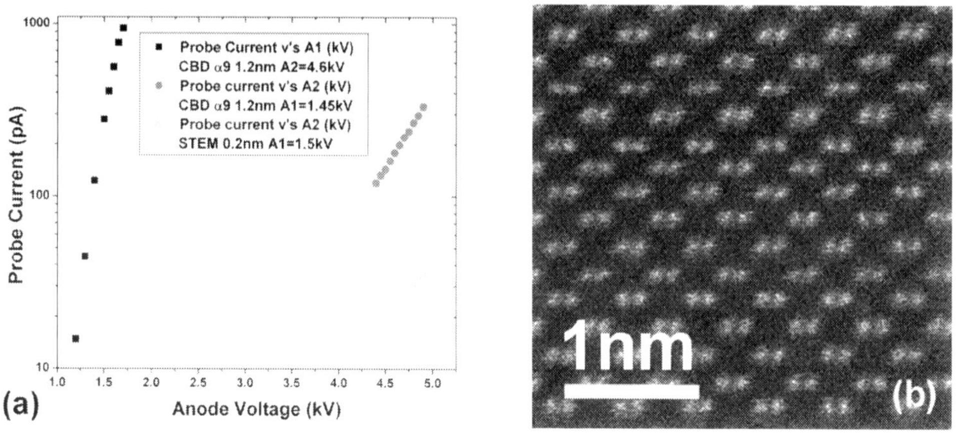

Fig. 3: Performance of the JEOL R005 at 300kV: (a) plot of the dependence of probe current on anode voltage settings for CBED and STEM and (b) a typical lattice image of silicon [110] recorded in STEM mode with ~0.08nm spot size (probe current ~20pA).

The advance offered by Cs correction is clear if one considers the beam current versus probe size data for the 2010F, Fig 5a, where it can be seen that for the smallest probe (~0.3nm diameter) a beam current ~17pA is obtained resulting in very poor signal to noise in the subsequent ADF STEM image as illustrated in Fig 5b.

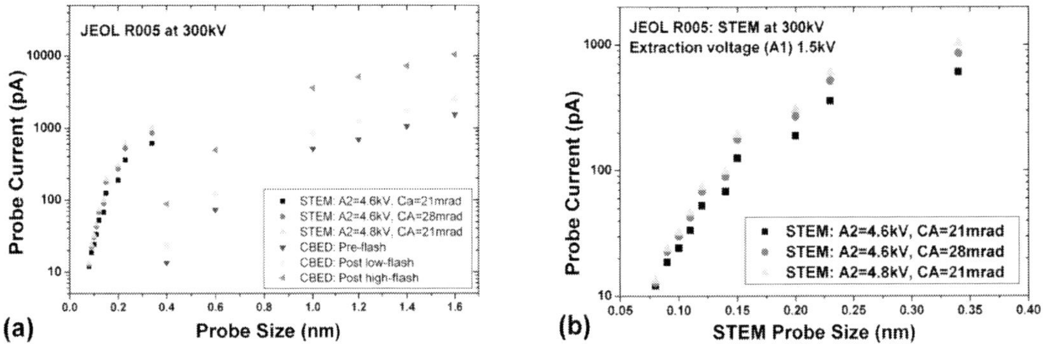

Fig. 4: plot of probe current for JEOL R005 in CBED and STEM mode. CBED conditions: α9, 150μm diameter condenser aperture (CA), 2μA emission, A1=1.44 kV, A2=4.6kV; STEM conditions: A1=1.5kV, 30/40μm CA. The STEM data are re-plotted on an expanded scale on the right.

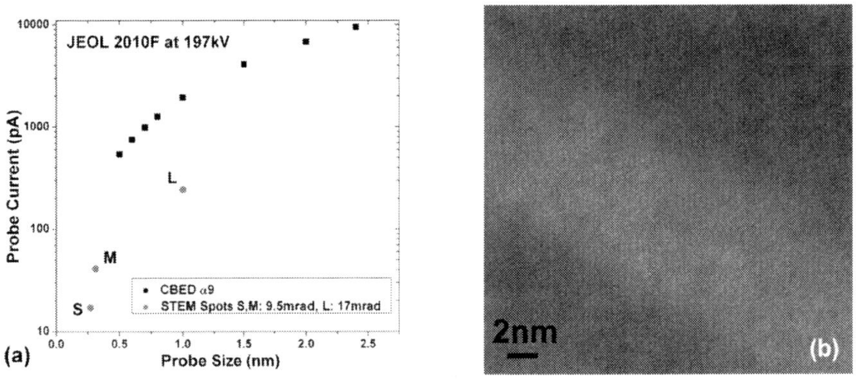

Fig. 5: (a) plot of probe current vs. probe size for the JEOL 2010F (A1=2.75kV, A2=7.16kV, 147 μA emission) and (b) typical lattice image, showing noisy 0.32nm lattice fringes in an InGaAs quantum well [6].

4. Summary

The performance of a c-FEG on a newly developed 300kV FEG-STEM has been evaluated and compared to a Schottky FEG installed on a 200kV instrument. We can attain 0.3-0.5eV energy resolution and ~0.1nA current sufficient for EELS in an electron beam ~0.15nm in diameter. The authors gratefully acknowledge the assistance of Dr Mike Fay (Univ. Nottingham) for loan of the Faraday cup holder.

References

[1] Walther T and Stegmann H 2006 Microscopy & Microanal. **12**, 498
[2] Sawada H, Tanishiro Y, Ohashi N, Tomita T, Hosokawa F, Kaneyama T, Kondo Y, Takayanagi K 2009 J. Electron Microsc. **58**, 357
[3] Honda T, Tomita T, Kaneyama T, Ishida Y 1994 Ultramicroscopy **54**, 132
[4] Walther T 2007 *Lecture Manuscripts of the 38th Spring School of the Institute of Solid State Research, Research Centre Jülich, Germany* (eds. K. Urban et al.) Matter & Materials **34**, E1.1-30
[5] Walther T, Quandt E, Stegmann H, Thesen A and Benner G 2006 Ultramicroscopy **106**, 963
[6] Qiu Q, Zhang ZY, Hogg RA, Cullis AG and Walther T 2010 J. Phys.: Conf. Ser. **209**, 012036

Performance of a cold-field emission gun double aberration corrected TEM/STEM at 80kV

I M Ross[1], M Kawazu[2], H Sawada[2], A Papworth[3] and T Walther[1]
[1]Kroto Centre for High Resolution Imaging and Analysis, Department of Electronic and Electrical Engineering, University of Sheffield, Sheffield S1 3JD, United Kingdom
[2]JEOL Ltd, 3-1-2 Musashino, Akishima, Tokyo 196-8558, Japan
[3]Gatan UK, 25 Nuffield Way, Abingdon, Oxon, OX14 1RL, United Kingdom

i.ross@shef.ac.uk

Abstract: In this report we demonstrate the successful 80kV alignment of a double aberration corrected TEM/STEM (JEM-Z3100F-R005) equipped with a cold field emission gun designed to operate at 300kV. In TEM mode we demonstrate information transfer out to 0.078nm in the corresponding power spectrum. For high angle annular dark-field STEM, direct resolution of the 0.082nm Ge (444) dumbbells was achieved. The energy spread of the primary electron beam was studied using energy-loss spectroscopy. Reduction of the extraction anode voltage (A1) to 1.2kV results in an energy resolution in the primary electron beam, taken as the full width half maximum (FWHM) of the zero loss peak, of 0.28eV with a corresponding beam current of 12pA whereas operation of A1 at 1.5kV results in a FWHM at the zero-loss peak of ~0.4eV for a beam current of 1.1nA.

1. Introduction

Lower operating voltages in the transmission and scanning transmission electron microscope (TEM/STEM) are particularly desirable to reduce the effects of knock-on damage during the study of beam sensitive materials such as those based on carbon and in particular the study of graphene [1-4]. Lower operating voltages also offer the potential for improved sensitivity energy-loss spectroscopy and X-ray spectroscopy due to the increased ionization cross-section. However, at lower accelerating voltages the resolution is reduced and chromatic aberrations are increased due to the longer wavelength of the incident primary electrons. Recent attempts to overcome these limitations have applied a combination of a cold field-emission gun (c-FEG) source and aberration correction of the probe and image forming lenses [5,6].

In this report we present the first results obtained at 80kV from a JEOL Z3100F-R005 (R005) TEM/STEM nominally designed to operate at 300kV. This instrument is equipped with a cold field emission gun consisting of a W <111> nano-tip emitter [7] and JEOL probe (STEM) and imaging (TEM) lens correctors. An asymmetric corrector configuration is utilized consisting of two dodecapole assemblies of different thicknesses coupled via transfer lenses of different focal lengths. Such an asymmetric corrector design has the advantage of suppressing chromatic aberrations otherwise introduced by symmetrical corrector designs and also reduces the effects of higher order parasitic aberrations [8].

2. Experimental methods

The c-FEG was operated at 80kV by utilisation of appropriate gun shorting electrodes to ensure the optimum position of the first cross over of the electron beam with respect to the subsequent condenser optics. Residual aberrations in the imaging and probe forming lens were minimised using the diffractogram tableau (Zemlin tilt-tableau) [9] and segmental Ronchigram auto correlation function matrix method respectively [10]. STEM high angle annular dark-field (HAADF) images of Si (110) and Ge (112) were obtained using a convergence semi-angle of ~28mrad and an inner and outer annular dark field detector angle of 86mrad and 229mrad respectively (nominal 15cm camera length). The energy resolution of the 80kV primary electrons was assessed as the FWHM of the zero energy-loss peak in vacuum as a function of the accelerating anode (A1) between 1.15 and 1.70kV (A2 held constant at 4.6kV) using image coupled mode (TEM spot size 1, α2). The defocus of the illumination was adjusted using the C3 lens (brightness) to give a maximum ZLP intensity of ~60000 counts in each case. Spectra were obtained using a post column GATAN Tridium GIF™ with a 1mm GIF entrance aperture, 0.02eV/pixel dispersion and an exposure of 0.1 and 1.0 seconds respectively.

3. Results and Discussion

Typical residual aberrations for both the probe (STEM) and imaging (TEM) correctors at 80kV, after routine alignment using the automated corrector routine, are given in Table 1. Final adjustment of the 2-fold astigmatism A_2 and (for STEM) the axial coma P_3 is normally performed manually using the appropriate corrector deflector coils and the condenser lens 2nd beam deflector (bright tilt) respectively. The residual aberrations at 80kV are of a similar magnitude to those under optimised conditions at 300kV (also listed in Table 1 for comparison) with the exception of 80kV probe correction which shows a general increase of the fifth order Cs and six-fold astigmatism values. This is perhaps to be expected since the influence of fifth-order 6-fold aberrations (A_6) can become significant at lower operating voltages and the asymmetrical dodecapole-type aberration correctors of the R005 cannot compensate for 6-fold astigmatism [5].

Aberration Coefficient	**80kV**		**300kV**	
	TEM	**STEM**	**TEM**	**STEM**
Defocus: O_2	-223.18nm	1.60nm	-226nm	3.81nm
2-fold astigmatism: A_2	6.03nm	21.40nm	6.3nm	0.68nm
Axial coma: P_3	190.01nm	213.61nm	87.4nm	27.97nm
3-Fold astigmatism: A_3	33.50nm	105.27nm	60.3nm	7.70nm
Spherical aberration: O_4	-0.67μm	3.28μm	0.516μm	0.219μm
Star aberration: Q_4	1.22μm	2.03μm	0.298μm	0.36μm
4-fold astigmatism: A_4	1.45μm	1.48μm	3.05μm	0.80μm
Fourth-order axial coma: P_5	111.31μm	123.68μm	6.01μm	32.24μm
Three-lobe aberration: R_5	4.25μm	99.19μm	30.24μm	6.07μm
5-fold astigmatism: A_5	37.15μm	95.03μm	49.84μm	9.81μm
Fifth-order spherical aberration: O_6	-0.37mm	-2.05mm	-0.614mm	-0.54mm
6-fold astigmatism: A_6	0.22mm	3.20mm	0.78mm	0.41mm

Table 1: Typical coefficients of the residual aberrations measured at 80kV and 300kV for TEM and STEM after aberration correction using the automated correction routine (note: the large TEM defocus value is a necessity of the correction routine).

Fig. 1(a) shows a high resolution TEM image from a Si [110] single crystal while the corresponding fast Fourier transform (FFT) indicates information transfer out to the Si (444) reflection, corresponding to 0.078nm. The Zemlin tilt-tableau generated to calculate the aberrations within the image forming optics is shown in Fig.1(b) with a maximum tilt angle of 45.25mrad. Fig. 1(c) shows a typical Ronchigram after probe correction from an amorphous film recorded by a 2k x 2k CCD camera (Gatan Ultra-Scan) mounted within the camera chamber. The Ronchigram shows a flat contrast region in the central circle with a half-angle of 38.8mrad compared to the typical flat region in the

Ronchigram at 300kV of 45-50mrad. STEM resolution is demonstrated in Fig. 2, (a) showing a lattice resolved HAADF image from Si [110], (b) the corresponding FFT showing transfer out to the (2-2 4) reflection and (c) the resolved 0.136nm dumbbell spacing. By careful adjustment of the corrector alignment to suppress the fifth order spherical aberrations to a minimum it has been possible to just resolve the (444) dumbbell spacing in Ge [112] with a contrast minima ~27±2% of the maximum contrast intensity (after background subtraction) as illustrated in Fig. 3.

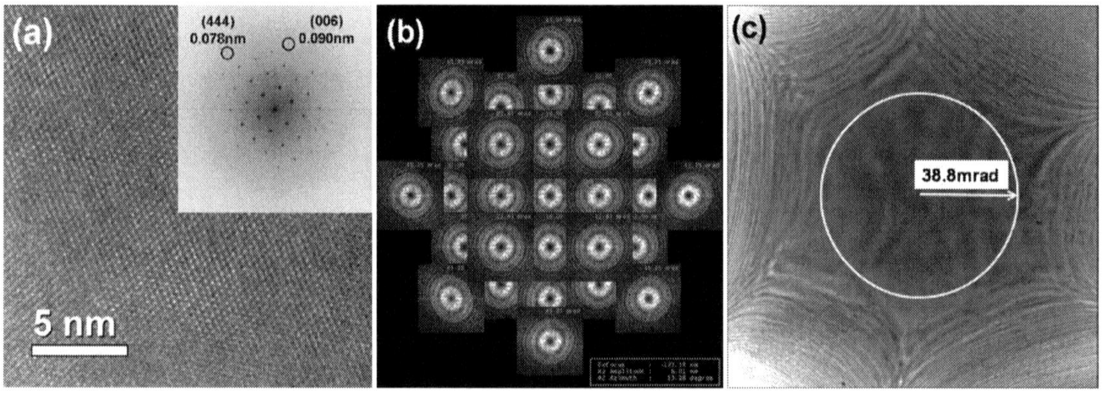

Fig. 1: (a) TEM image of Si [110] single crystal with corresponding FFT (insert), illustrating information transfer beyond 0.1nm, (b) 80kV TEM Zemlin tilt tableau with outer beam-tilt angle of 45.25mrad, (c) 80kV STEM correction Ronchigram showing a flat contrast region out to 38.8mrad.

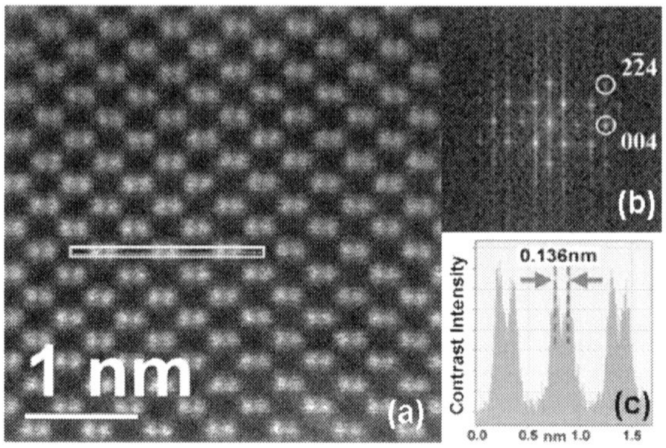

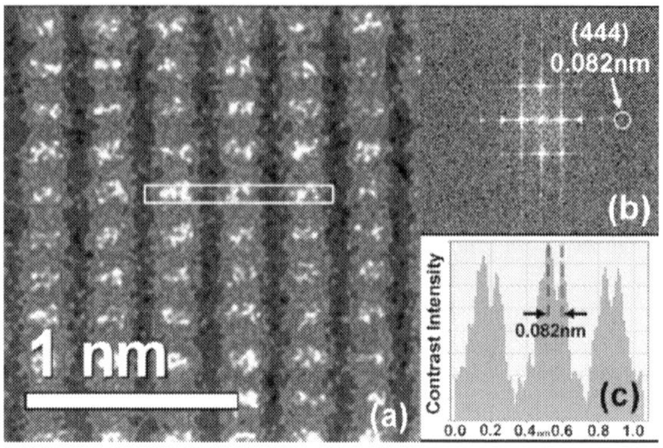

Fig. 2: (a) Raw HAADF STEM image from Si [110], (b) corresponding FFT indicating transfer out to 0.111nm (2-2 4) spot and (c) intensity profile across the Si (004) dumbbells with a peak minima 50.5±2.5% of the maximum contrast.

Fig. 3: (a) Raw HAADF STEM image from Ge [112] (shown in false colour), (b) corresponding FFT indicating transfer out to 0.082nm and (c) intensity profile from 3(a) illustrating the splitting of the Ge (444) dumb-bells.

A plot of the energy spread of the electron source at 80kV, measured as the FWHM of the zero energy-loss peak, as a function of the extraction anode voltage (A1) is presented in Fig. 4(a). A minimum energy spread of 0.28eV was achieved with an A1 setting of 1.20kV (shown in Fig 4(b)). Further reduction of the extraction anode resulted in no further improvement in energy resolution until below an A1 of 0.9kV a rapid reduction in emission occurs with insufficient beam current to record a meaningful energy-loss spectrum. Under the conditions used to generate the zero-loss peak shown in Fig. 4b the emission current was significantly lower than 1μA to minimize the Boersch effect. The peak shape has a distinctive asymmetry with a broader tail on the higher energy-loss side. This effect was also observed in the minimum energy-spread ZLP at an operating voltage of 300kV (A1 at 0.9eV, emission current <1μA, FWHM 0.32eV). Increasing A1 to 1.70kV results at 80kV a corresponding increase in the energy spread to 0.6eV FWHM as shown in Fig. 4c. The equivalent energy spread at 300kV for an A1 of 1.7kV is 0.95eV FWHM.

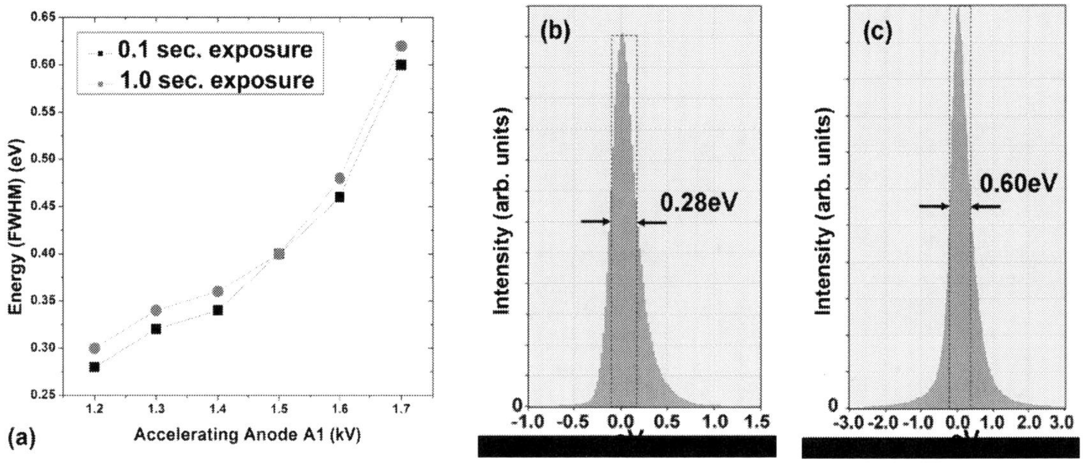

Fig 4: (a) Plot of primary electron beam energy spread as a function of extraction voltage (A1), (b) zero-loss peak (ZLP) for A1=1.20kV with a FWHM of 0.28eV and (c) corresponding ZLP for an A1 of 1.7kV with a FWHM of 0.6eV. The corresponding emission current for an A1 of 1.20kV and 1.70kV was <1μA and 4μA with beam currents of 12pA and 10.7nA respectively.

4. Summary

Sub-Ångstrøm resolution has been successfully demonstrated at 80kV in a double aberration corrected c-FEG TEM/STEM designed to operate at 300kV. For TEM imaging, information transfer out to 0.078nm has been observed in the FFT of Si [110] while direct evidence of HAADF STEM resolution of the Ge (444) dumbbells of 0.82nm has been recorded. A best energy spread of the primary electron beam of 0.28eV was given for an extraction voltage of 1.20kV with an emission current of <1μA.

Acknowledgements

The authors gratefully acknowledge the contribution of JEOL ltd (Japan), JEOL UK and Gatan UK towards this project.

References

[1] Warner J H, Rümmeli M H, Bachmatiuk A, Wilson M, Büchner B 2010 *ACS Nano* **4** (1), 470
[2] Krivanek O L, Dellby N, Murfitt M F, Chisholm M F, Pennycook T J , Suenaga K, Nicolosi V 2010 *Ultramicroscopy* **110**, 935
[3] Meyer J C, Kisielowski C, Erni R, Rossell M D, Crommie M F 2008 *Nano Letts.* **8**, 3582
[4] Jinschek J R, Yucelen E, Calderon H A, Freitag B 2011 *Carbon* **49** 556
[5] Sasaki T, Sawada H, Hosokawa F, Kohno Y, Tomita T, Kaneyama T, Kondo Y, Kimoto K, Sato Y, Suenaga K (2010) *J. Electron Micros.* **59**, S7
[6] Konno M, Suzuki Y, Inada H, Nakamura K 2010 *J. Phys. Conf. Ser.* **241** 012011
[7] Kurata H, Isoda S, Tomita T 2006 *Proc. IMC* **16**, 583
[8] Sawada H, Tanishiro Y, Ohashi N, Tomita T, Hosokawa F, Kaneyama T, Kondo Y, Takayanagi K 2009 *J. Electron Micros.* **58,** 357
[9] Uhlemann S, Haider M 1998 *Ultramicroscopy* **72** 109
[10] Sawada H, Sannomiya T, Hosokawa F, Nakamichi T, Kaneyama T, Tomita T, Kondo Y, Tanaka T, Oshima Y, Tanishiro Y, Takayanagi K 2008 *Ultramicroscopy* **108**, 1467

Characterization of thickness, elemental distribution and band-gap properties in AlGaN/GaN quantum wells by aberration-corrected TEM/STEM

H Amari[1, 2], I M Ross[1, 2], T Wang[2] and T Walther[1, 2]

[1] Kroto Centre for High-Resolution Imaging and Analysis
[2] Semiconductor Materials and Devices Group
Department of Electronic & Electrical Engineering, University of Sheffield, Sir Frederick Mappin Building, Mappin Street, Sheffield S1 3JD, UK

E-mail: h.amari@sheffield.ac.uk, t.walther@sheffield.ac.uk

Abstract. This work presents an investigation of the thickness, the chemical composition and the band-gap properties of AlGaN/GaN quantum wells grown by Metal-Organic Chemical Vapour Deposition (MOCVD) on sapphire (0001) substrates. Various methods of analysis by transmission electron microscopy (TEM) were used for the characterization of these layers, such as scanning transmission electron microscopy STEM, energy-dispersive X-ray spectroscopy (EDXS) and electron energy loss spectroscopy (EELS). Annular dark-field STEM provides more accurate layer thicknesses and Al concentrations than EDXS profiling.

1. Introduction

Annular dark-field STEM (ADF-STEM) and X-ray point analysis and line scans were used to investigate the elemental distribution of AlGaN/GaN quantum wells with atomic resolution using a new spherical aberration-corrected TEM/STEM microscope equipped with both probe and imaging correctors and a cold field-emission gun operating at 300 kV (JEOL-R005). Also, plasmon loss spectroscopy and valence EELS (VEELS) were used to investigate the band-gap properties of thin AlGaN and GaN layers, and to compare them to measurements.

2. Sample description and experimental setup

GaN quantum wells (QW) with nominal thicknesses decreasing from 8 nm to 0.5 nm were embedded between 10 nm AlGaN barriers. The AlGaN/GaN layers were grown on the top of c-plane sapphire substrates. First, an AlGaN buffer layer was grown at 1210 °C, followed by the active region consisting of a multiple quantum well (MQW) structure with 5 GaN layers seperated by 10 nm $Al_{0.15}Ga_{0.85}N$ barrier layers grown at 1125 °C. Cross-sectional TEM samples were prepared by mechanical polishing and Ar^+ ion milling using both Gatan precision ion polishing system (PIPS) at 5 kV, and Ar^+ polishing at 3 kV in a Technoorg Linda IV3 system. TEM analysis was carried out in a JEOL R005 which can achieve a sub-50 pm electron probe with 28 mrad semi-angle and is equipped with an energy dispersive X-ray spectrometer (JEOL 50 mm^2 Si:Li detector with ultra thin window and related acquisition software), annular dark field (ADF) and bright field (BF) STEM detectors coupled with a Gatan Digiscan system, and a Gatan Imaging Filter model 865 GIF Tridium.

3. Results and Discussion

3.1. STEM analyses

The annular-dark field (ADF) image in Fig. 1a shows the structure investigated. The thick layer preceding the active region represents the AlGaN buffer. The active region with darker and brighter bands, perpendicular to the growth direction, as shown enlarged in Fig. 1b, represents the AlGaN barrier layers and GaN QWs respectively. The GaN QWs appear brighter because of their higher mean atomic number compared to the AlGaN layers.

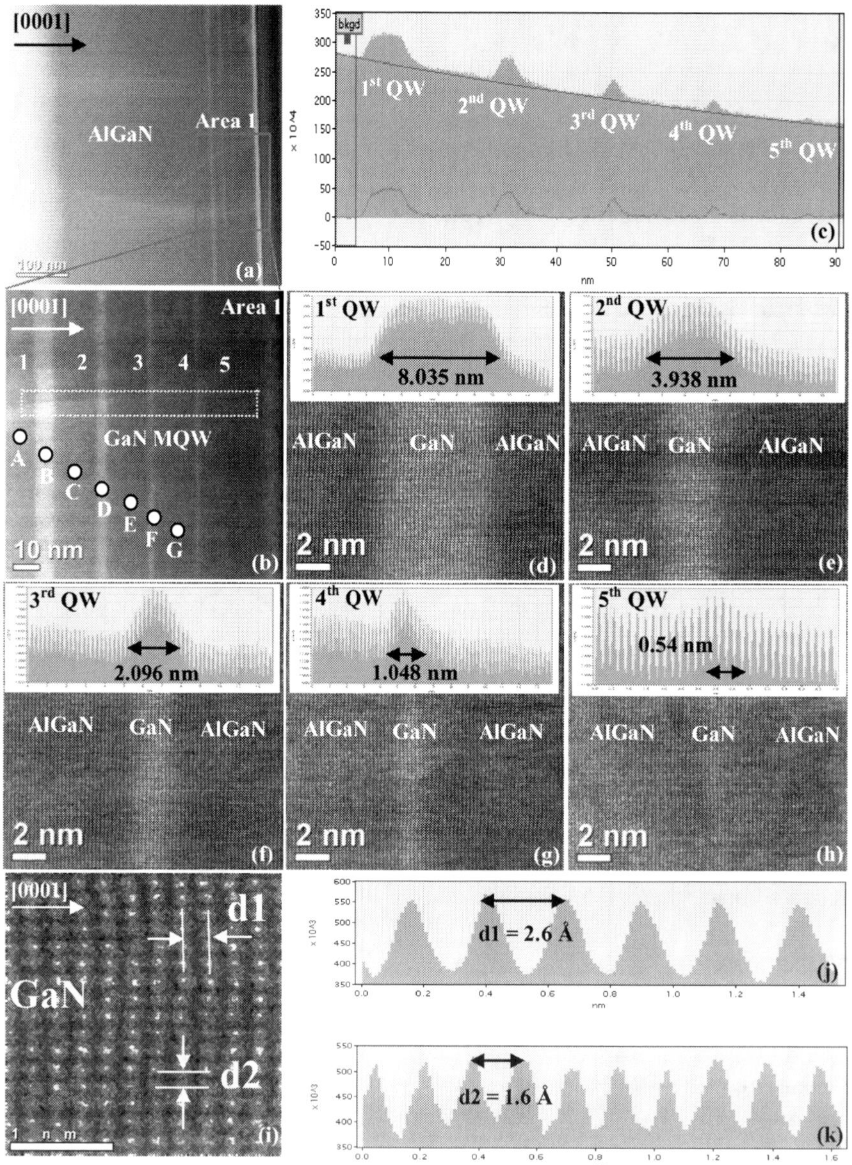

Fig. 1 ADF images showing (a) overview of the structure, (b) 5 GaN QWs separated by AlGaN barrier layers, (c) ADF STEM intensity profile of the 5 GaN QWs, high magnification image of (d) 1st QW, (e) 2nd QW, (f) 3rd QW, (g) 4th QW, (h) 5th QW, and (i) high resolution STEM image of GaN lattice. A-G indicate positions of EDXS and EELS point analyses reported below.

The possibility to obtain images using large illumination angles in aberration corrected microscopy [1] allows us to visualize the Ga concentration in QWs with high spatial resolution as shown in Fig. 1b.

Also, an intensity line profile across the active region was extracted from Fig. 1b and is shown in Fig. 1c.

The intensity profile and high resolution ADF-STEM images (Figs. 1d, 1e, 1f, 1g, and 1h) show the existence of all 5 GaN QWs and the decrease of the thickness of the layers from 8 to 0.5 nm. The 5th QW may be damaged by ion beam milling as it is located topmost and appears at lower contrast.

Fig. 1i shows a high resolution STEM image of the centre of a GaN QW taken near the [1-100] zone-axis. The lattice planes labelled d_1 (~2.6 Å) and d_2 (~1.6 Å) correspond to the basal (0001) and a-plane (11-20) distances, respectively.

The higher Ga concentration of the GaN QWs relative to the AlGaN barrier layers is reflected in the intensity profile in Fig.1 c. The Al content in AlGaN can be directly measured from the ADF intensity ratio (R) of the centre of the GaN QW relative to the AlGaN [2], as presented in Table 1(a).

QW #	*t (nm)*	*R*
1	8.03	1.170 ± 0.040
2	3.94	1.174 ± 0.047
3	2.09	1.150 ± 0.053
4	1.05	1.111 ± 0.058
5	0.54	1.062 ± 0.064

		$\varepsilon = 1.6$	$\varepsilon = 2$
x_{Al}	1st QW	0.214 ± 0.061	0.186 ± 0.052
	2nd QW	0.218 ± 0.071	0.189 ± 0.060
	3rd QW	0.192 ± 0.080	0.167 ± 0.068
	4th QW	0.147 ± 0.088	0.128 ± 0.075
	5th QW	0.086 ± 0.096	0.075 ± 0.082

Table 1 (a) Comparison of peak GaN/AlGaN intensity ratios, (b) evaluation of Al concentrations for for $\varepsilon = 1.6$ and $\varepsilon = 2$.

Assuming the quantum wells were all pure GaN, the Al content in $Al_xGa_{1-x}N$ can be obtained using the following approach [2]: $I_{GaN} / I_{AlGaN} = R = \dfrac{Z_{Ga}^{\varepsilon} + Z_{N}^{\varepsilon}}{xZ_{Al}^{\varepsilon} + (1-x)Z_{Ga}^{\varepsilon} + Z_{N}^{\varepsilon}}$ (1)

where R is the ratio obtained from the intensity profile, x the elemental Al concentration in atomic percent (at %), Z the atomic number, and ε lies somewhere between 1.6 and 2 [2]. Table 2(b) lists the Al concentrations obtained from equation (1) for $\varepsilon = 1.6$ and $\varepsilon = 2$. For the two widest QWs the assumption is probably valid, which suggests $x_{Al} \approx 0.19$ in the AlGaN, and $x_{Al} \approx 0.10$ in the thinnest QW, demonstrating interdiffusion.

3.2. EDXS analyses

Using a probe size of less than 1 nm along the beam direction, EDX spectra were acquired at points A, B, C, D, E, F, and G as shown in Fig. 1b, for acquisition times of 30 s, and an energy resolution of ~130 eV. Figs. 2a and 2b depict typical EDX spectra acquired at points A and B in Fig. 1b.

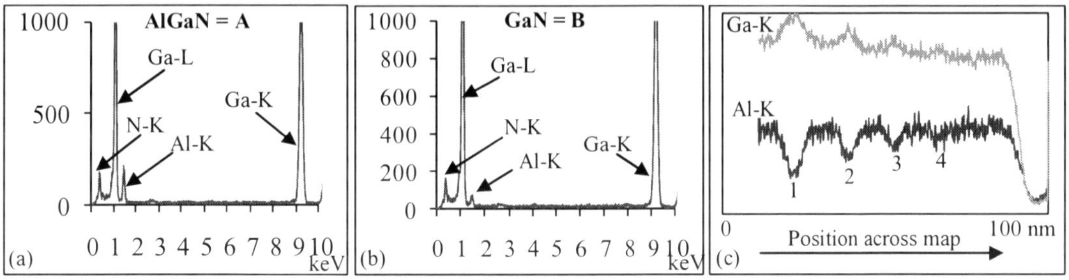

Fig. 2 (a) Typical EDX spectra acquired from (a) AlGaN barrier layer, (b) GaN QW, and (c) EDX line scan across the GaN QWs.

The concentration of Al was obtained using the JEOL X-ray quantification routine, considering only Al-K, Ga-K and Ga-L peaks. The sample thickness t was estimated using low-loss energy spectra acquired close to the positions A, B, C, D, E, F and G in Fig. 1b using a convergence angle of 21mrad and a collection angle of 1.75 mrad. The corresponding thickness was obtained via the compute thickness rule using the EELS routines of Gatan Digital Micrograph [3]. The quantitative comparison of Al concentrations from X-ray spectra of different thicknesses are shown in Table 2. The presence of a weak Al peak in the GaN spectrum (points B, D, and F) is due to beam broadening so that probe tails extend from the thin GaN QW into the AlGaN barrier layers. This can also affect calculations of Al:Ga ratios in AlGaN barriers. As the layers are very thin EDX point analyses were performed slightly off the zone-axis but with the layers edge-on, resulting in fewer counts. To estimate the beam broadening effect (b) for a thickness $t \sim 40$ nm, we used the following equation [4]:

$$b = 7.21\text{x}10^5\text{x}\ (Z/E_0)\ \text{x}\ (\rho/A)^{0.5}\text{x}t^{1.5} \qquad (2)$$

position	t (nm)	x_{Al} from Al-K/Ga-K	x_{Al} from Al-K/Ga-L
A	50.33	0.116	0.094
B	38.05	0.024	0.021
C	43.97	0.101	0.083
D	38.49	0.042	0.035
E	37.49	0.110	0.087
F	38.64	0.067	0.054
G	33.87	0.120	0.096

Table 2 Al concentrations from EDXS.

where Z is the atomic number, E_0 the beam energy in kV, ρ the density in gcm^{-3}, and A the atomic weight [4]. The beam broadening is estimated as ~ 1.35 nm. In addition, an X-ray line scan has been carried out, as shown in Fig. 2c. Since Al substitutes Ga atoms in III-V nitrides, the Al content in the AlGaN and GaN layers can be calculated from the corresponding depletion of the Ga-K signal, which is ~ 14 at% for the 1st QW.

3.3. EELS analyses

Low-loss EEL spectra were acquired in STEM mode with an energy resolution of 0.5 eV (measured from full width at half maximum (FWHM) of the zero loss peak (ZLP)), exposure times of 1s, and dispersion of 0.05 eV/channel. Typical low-loss spectra from AlGaN and GaN layers are presented in Fig. 3 which shows, first, the main plasmon excitations are positioned at different energies (E_p = 19.36 eV for AlGaN and E_p = 19.28 for GaN), and second, the strong Ga $M_{4,5}$ edge in the GaN spectrum positioned at ~ 23 eV compared to the AlGaN spectrum.

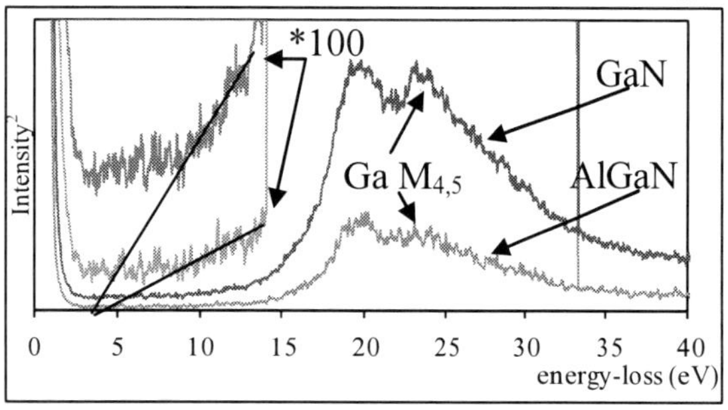

Fig.3 Square of experimental low-loss intensity from the GaN QW (blue) and the $Al_{0.15}Ga_{0.85}N$ barrier layer (grey).

For direct transitions in semiconductors, the intensity (I) has a square-root dependence on the band gap (E_g). If we plot I^2 vs. energy-loss (E) and extrapolate the linear fitting of the experimental spectra over energy range from 6.25 to 14eV to zero, we can determine the band-gap as the crossing of the abscissa [5]. We obtain E_g = 3.80 ± 0.50 eV with a linear regression coefficient (R^2) of ~ 78% for $Al_{0.15}Ga_{0.85}N$, and E_g = 3.16 ± 0.50 eV with the R^2 of ~ 77% for GaN, which is in good agreement with previous work [6], and suggests $x_{Al} \approx 20\%$ in the AlGaN barrier, in agreement with the above ADF data.

4. Conclusion

Thicknesses, elemental distribution and band-gap properties were obtained for AlGaN/GaN quantum wells using different analytical techniques in an aberration corrected electron microscope. Thickness measurements of the multilayers were found in agreement with the nominal values. The Al concentrations of AlGaN extracted from Z-contrast imaging were found to be larger and the values from EDX to be lower than the nominal Al concentration of 15%. It can also be concluded that VEELS shows changes in the plasmon excitations and band-gap properties from different layers.

References

[1] Walther T and Stegmann H, 2006, *Microscopy and Microanalysis* **12**, 498
[2] Walther T, 2006, *J. Microsc.* **221**, 137
[3] Ahn C.C., *et al.*, *EELS Atlas*, Gatan, Warrendale, 1983
[4] Williams DB and Carter CB, 1996, *Transmission Electron Microscopy*: Springer, New York
[5] Schubert EF, *Light-Emitting Diodes*, 2006, 2nd edition: Cambridge University Press, Cambridge
[6] Amari H et al. 2011, Proc. MSM-XVII, Cambridge, *J. Phys. Conf. Ser.* **326**, 012039

Quantitative HAADF STEM study of β'-like precipitates in an Al-Mg-Ge alloy

R Bjørge[1], C Dwyer[2,3,4], M Weyland[2,3], P N H Nakashima[3,4], J Etheridge[2,3] and R Holmestad[1]

[1] Department of Physics, Norwegian University of Science and Technology (NTNU), 7491 Trondheim, Norway
[2] Monash Centre for Electron Microscopy, Monash University, Clayton 3800, Australia
[3] Department of Materials Engineering, Monash University, Clayton 3800, Australia
[4] ARC Centre of Excellence for Design in Light Metals, Monash University, Clayton 3800, Australia

E-mail: randi.holmestad@ntnu.no

Abstract. Precipitates in an Al-Mg-Ge alloy similar to the β' phase in Al-Mg-Si alloys were investigated using quantitative aberration-corrected high-angle annular dark-field scanning transmission electron microscopy (HAADF STEM). Quantitative investigation into the Ge-rich columns of the precipitates found the intensity of these columns to be consistent with an Al concentration of 30±10 %, or a vacancy concentration of 20±10 %. Possible errors involved in the determination of the column composition are discussed.

1. Introduction

Al-Mg-Si alloys are an industrially important group of materials due to their high strength-to-weight ratio, formability, and resistance to corrosion. These alloys get their strength from the formation of nanometre-sized, metastable precipitates during heat treatment. One of the precipitate phases that form is known as β' [1]. A phase similar to β' also forms in Al-Mg-Ge alloys [2-3].

In the present study, the β'-like precipitates (henceforth referred to as β'-Ge) that occur in the Mg-rich Al-Mg-Ge alloy in [2], have been investigated using quantitative high-angle annular dark-field scanning transmission electron microscopy (HAADF STEM). The hexagonal β'-Ge phase precipitates as needles with the hexagonal axis parallel to the needle direction and a $<001>_{Al}$ direction. The precipitate is fully coherent with the matrix along this direction.

β'-Ge precipitates are well-suited for HAADF STEM because of the considerable atomic number difference between Ge (Z=32), and Mg (Z=12) and Al (Z=13). Here, we use quantitative HAADF STEM to estimate the composition of Ge-rich columns. This technique has recently been used to determine the structure of the T_1 precipitate phase in Al-Li-Cu alloys [4].

2. Methods

An Al-0.87Mg-0.43Ge (at. %) alloy was cast and homogenized for 4 hours at 550 °C, before extrusion. The alloy was then solution heat treated for 2 hours at 600 °C, water quenched to room temperature and held for 4 hours at this temperature. The alloy was subsequently aged at 250 °C for 5 hours before water quenching to room temperature. TEM samples were prepared by twin-jet electropolishing with a 33 % nitric acid / 67 % methanol solution at -25 °C with a voltage of 14 V. The data was acquired on a double aberration-corrected FEI Titan^3 80-300 FEG-TEM operating at 300 kV.

The response of the detector was measured by scanning the focused STEM probe across the detector. The acquired HAADF STEM images were normalized to obtain the intensity I_{norm}, from the raw intensity, I_{raw}, using the equation $I_{norm} = (I_{raw} - I_{vac})/(I_{det} - I_{vac})$, where I_{vac} is the intensity measured when there is no scattering to the detector, and I_{det} is the average intensity when the probe is placed on the detector. The thickness of the specimen in the vicinity of each precipitate was determined using convergent-beam electron diffraction (CBED) patterns of the Al matrix. Using CBED means that the determined thickness does not include the thickness of any amorphous layers, unlike when using electron energy-loss spectroscopy. The probe was scanned over an area corresponding to several unit cells in order to remove coherent effects, resulting in a position-averaged CBED pattern [5]. A smaller condenser aperture was used for CBED in order to reduce the amount of overlap between discs. The thickness was determined by comparing the acquired CBED pattern with CBED patterns simulated using JEMS [6]. The accuracy of the method was estimated to be ±2 nm.

STEM images were simulated using the multislice program *stemimg* [7], using the following parameters: convergence semi-angle: 18 mrad, HAADF detector: 59-200 mrad. The supercell used for aluminium was 3.24 x 3.24 nm^2, with a slice thickness of 0.2025 nm. The supercell used for β'-Ge was 2.86 x 2.48 nm^2, with the same slice thickness. A STEM image was calculated every two slices. 8 frozen phonon configurations were averaged over for each simulation, which should be more than enough for the thicknesses considered here since a different phonon configuration is calculated for each slice. Partial occupancy was implemented by using a random-number generator when creating each slice so that each atomic site was either fully occupied or empty. Temperature factors for the elemental phases at 300 K were used [8].

Due to amorphous layers on the surface of the specimen, the average intensity measured in the Al matrix was consistently greater than in simulations of the Al matrix for the thickness determined from CBED. To reduce the effect of the amorphous layers, this intensity difference was subtracted from the normalized experimental image [4].

The effective source size, which can have a significant effect on peak intensities, was determined for each image by matching simulated images of aluminium for different Gaussian source sizes to the acquired images. Line profiles were used in order to fit both the peaks and the background. The effective source size was found to lie between 0.10 nm and 0.13 nm full-width at half maximum (FWHM).

Once the effective source size had been determined using the Al matrix, the intensities of many Ge-rich columns and Al matrix columns in each acquired image were determined by measuring the average intensity within a circle of radius 76 pm, 68 pm, and 59 pm, for corner Ge-rich columns, interior Ge-rich columns, and Al matrix columns, respectively. The circle was shifted around in order to find the maximum average intensity for each column. The radius of the circle was chosen large enough to reduce the effect of a finite source size, but small enough to limit the intensity contribution from neighbouring columns. Columns close to the precipitate/matrix interface were not used. The same method was applied to simulated images of β'-Ge for different compositions of the Ge-rich columns and of Al.

3. Results

A HAADF STEM image of a β'-Ge precipitate viewed in a <001>$_{Al}$ zone axis is shown in figure 1. The structural similarity with the model of β'-Si shown in the figure is clear. In the *c*-direction, the repeat distance is 1.215 nm. In the corner Si column in β'-Si, there are 4 Si sites per repeat distance, while in the interior Si columns, there are only 3. This difference in occupancy between the corner and the interior columns is clear in the β'-Ge precipitate as well.

A position-averaged CBED pattern was acquired near the precipitate in figure 1. Comparison with simulated CBED patterns for aluminium gave a thickness of 50±2 nm. A line profile through a part of

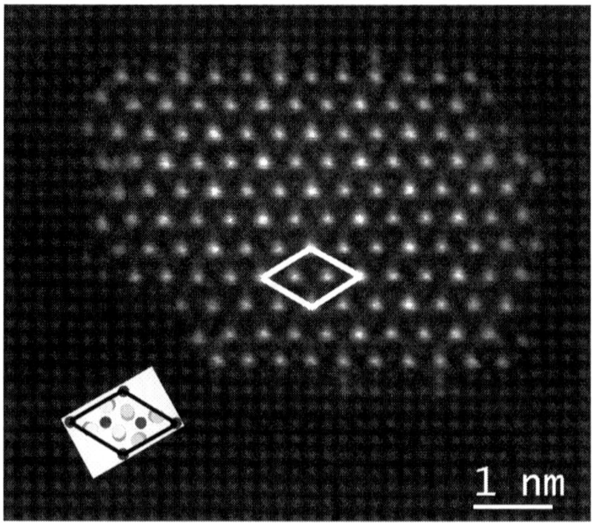

Figure 1. Quantitative HAADF STEM image of a β'-Ge precipitate. The hexagonal unit cell is indicated with white lines. The dimmer Mg atoms are arranged hexagonally around the corner Ge-rich columns, and triangularly around the interior Ge-rich columns, just as in the model of β'-Si shown in the lower left corner. The Si atoms are black and the Mg atoms are white in the model.

the Al matrix was used to determine the effective source size. The best match was found to be a source size of 0.11 nm FWHM.

Measurement of the column intensities of corner and interior Ge-rich columns in this precipitate as well as 20 other precipitates showed that these were considerably lower than the expected intensity if the Ge-rich columns were fully occupied. Two types of modification of the Ge sites were therefore considered in order to get a match between simulation and experiment. The first kind consisted of replacing some of the Ge with Al. In the second kind, Ge was replaced by vacancies. These replacements were done by making all the Ge sites partially occupied, that is, the Al atoms or vacancies were distributed randomly on Ge sites throughout the crystal.

In order to reduce the effect of the source size, we plot the ratio of the Ge-column to matrix-column intensities as a function of thickness for the 21 precipitates considered (figure 2). Comparing the experimental data points with simulations, it can be seen that for both corner and interior Ge columns, the experimental data corresponds to an Al concentration of roughly 30±10 %. Simulations where Ge was replaced by vacancies (not shown), gave a vacancy concentration of 20±10 %.

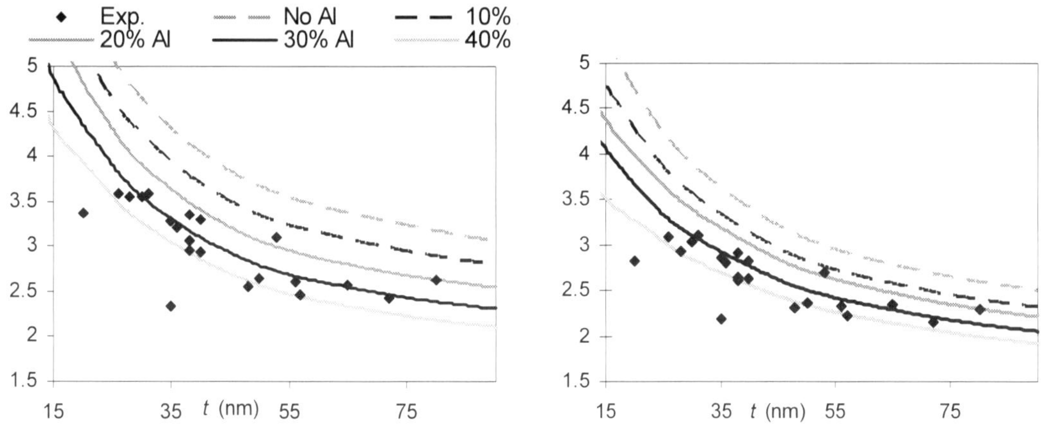

Figure 2. Plot of Ge peak/Al peak ratio as a function of thickness, t, for the corner Ge-rich columns (left) and interior Ge-rich columns (right). Each experimental data point corresponds to one image. The simulated peak ratios are for a source size of 0.10 nm FWHM.

4. Discussion

The measured Ge composition of the Ge-rich columns in β'-Ge is significantly lower than the Si composition of corresponding Si columns of β' in Al-Mg-Si alloys. It was not possible in this work to distinguish between a replacement of Ge atoms with Al, or a smaller amount of vacancies. It is possible that the Ge columns contain both Al and vacancies.

The fact that the precipitates are embedded in the matrix presents a challenge as it is crucial that the precipitates and the surrounding matrix are equally thick. It is simple to check that the precipitate and matrix are at the same height at the entrance surface, but this is not possible at the exit surface. The composition for several precipitates and for different specimen thicknesses was therefore determined in order to verify the results. The results show a fair amount of variation in the composition. This is probably due to measurement errors, but it could also be partly due to compositional variations among the precipitates.

Other possible sources of error include errors in the determination of the CBED thickness and the detector inner angle, non-uniformities in the detector response, and also the presence of an oxide layer and contamination on the entrance surface of the specimen. Simulations of the first two errors suggest that their effect would be small. Non-uniformities in the detector response could affect the results. However, the uniformity as determined from the detector image was very good except for a dip at one side for low angles. The non-uniformity should therefore only have a small effect on the result, also because ratios of column intensities were considered. An amorphous layer on the specimen surface might modify the column intensity, but the effect of this is partly taken into account by determining the effective source size from the matrix surrounding the precipitate.

The precision of this result was estimated by investigating the sensitivity of the composition to errors in thickness measurement in the following two ways. The effect of an error in the thickness measurement of the matrix, with the precipitate having the same thickness, can be assessed from the column intensity ratio plot in figure 2: Because the slope of the curves is not very steep, an error in the thickness measurement of a few nanometres would have only a small effect. A precipitate that is shorter than the matrix thickness would have a more serious effect on the calculated composition. This could happen if the specimen preparation preferentially attacked the precipitates. This was rarely observed at the entrance surface since the precipitate and surrounding matrix could in most cases be focused at the same time, and such an effect is therefore unlikely to have caused a systematic error.

5. Conclusions

β'-Ge precipitates in an Al-Mg-Ge alloy were investigated using aberration-corrected HAADF STEM. The composition of columns in β'-Ge precipitates thought to consist of Ge was investigated using quantitative aberration-corrected HAADF STEM. The intensity of the Ge columns was found to be consistent with an Al content of 30±10 %, or a vacancy concentration of 20±10 %.

Acknowledgements

RB and RH were supported by Hydro, Steertec Raufoss and the Research Council of Norway through Project no. 176816/I40. The FEI Titan^3 80-300 FEG-TEM was funded via the Australian Research Council infrastructure grant LE0454166.

References

[1] Vissers R et al. 2007 *Acta Mater.* **55** 3815–23
[2] Bjørge R et al. 2010 *Metall. Mater. Trans.* A **41** 1907–16
[3] Bjørge R et al. 2011 *Acta Mater.* **59** 6103
[4] Dwyer C, Weyland M, Chang L Y and Muddle B C 2011 *Appl. Phys. Lett.* **98** 201909
[5] LeBeau J M, Findlay S D, Allen L J and Stemmer S 2010 *Ultramicroscopy* **110** 118–25
[6] Stadelmann P A 1987 *Ultramicroscopy* **21** 131–45
[7] Dwyer C 2010 *Ultramicroscopy* **110** 195–8
[8] Gao H X and Peng L M 1999 *Acta Cryst.* A **55** 926–32

Journal of Physics: Conference Series **371** (2012) 012016 doi:10.1088/1742-6596/371/1/012016

A Study of the relationship between brown colour and extended defects in diamond using core-loss electron energy loss spectroscopy

I S Godfrey[1,2] and U Bangert[2]

[1]SuperSTEM Laboratory, STFC Daresbury Laboratory, Warrington, WA4 4AD, UK
[2]School of Materials, The University of Manchester, Manchester M1 7HS, UK

E-mail: iain.godfrey@postgrad.manchester.ac.uk

Abstract. There is strong evidence to support the theory that the vacancy cluster, a by-product of dislocation interaction, is the extended defect responsible for brown colour in natural type IIa diamonds. A characteristic of these open volume structures is the hybridised sp^2 bonding of the 3-fold coordinated carbon atoms that line their cage-like surfaces. When present in sufficiently large concentrations the sp^2 bonds alter the electronic and optical properties of type IIa diamond dramatically. Similar sp^2 bond concentrations are found in dark brown CVD diamond, although the defects in this material appear much larger in size than in the natural diamond. In this study we employ core-loss electron energy loss spectroscopy (EELS) and aberration corrected scanning transmission electron microscopy (AC-STEM) to compare the relative sp^2 content of brown coloured natural type IIa diamonds and dark brown CVD diamonds, using 100% sp^2 bonded graphite as a reference. The effect of high-pressure high-temperature (HPHT) treatment on sp^2 content is also investigated.

1. Introduction

Both high resolution electron microscopy (HREM) [1] and positron annihilation spectroscopy (PAS) [2] [3] studies have provided tangible evidence to support the theory that vacancy defects contribute to brown colouration in both naturally occurring type IIa and CVD diamonds. In the case of type IIa diamonds it is probable that large clusters of 50 or so vacancies are the main contributors to the brown colour, as indicated by optical absorption spectroscopy measurements [4]. Annealing studies have shown that the brown colour in type IIa diamond can be removed by heating to >2500°C [5]. However, in the case of CVD diamond it is found that the brown colour is significantly reduced at a comparatively low annealing temperature of 1400°C, presumably due to a reduction in the concentration of smaller nitrogen-vacancy defects through clustering [6] [7].

The exact mechanism for the production of vacancies through plastic deformation of type IIa diamond is not completely understood, however one popular theory is that they are formed by the dissociation of shuffle type dislocations [8]. It is probable that the increased mobility of these dislocations at the elevated temperatures experienced in the earth's upper mantle provides suitable conditions for the emitted vacancies to cluster together. Density functional theory (DFT) calculations show that both the formation of clusters and the hybridisation of the surrounding carbon bonds lower the overall energy of the crystal [9]. In the case of CVD diamond the dislocation density is very low compared with type IIa diamond, which suggests that the vacancy defects are introduced during the deposition process.

To quantify the sp^2 content of diamond crystals, a technique that is sensitive to electronic structure is required. One such technique is core-loss EELS, which produces inner shell loss edges that are dependent on the local bonding arrangement of atoms. EELS studies of diamond materials have shown that the π^* peak of the carbon k-edge provides a quantitative measure of the amount of sp^2 bonding in a material [10] [11]. The intensity of the π^* peak relative to that of the σ^* peak (sp^3 bonded carbon) can therefore be used to approximate the sp^2 content in brown coloured diamonds. In order to provide a means of studying the relationship between brown colour and sp^2 content, core-loss EEL spectra were acquired for untreated and HPHT treated brown type IIa diamonds together with dark brown CVD diamonds. The HPHT treatment temperatures were 1900, 2200 and 2500°C.

2. Experimental

Diamond specimens for AC-STEM and EELS analysis were provided by DeBeers research centre. Thinning of these materials to electron transparency was carried out by conventional tripod polishing followed by argon ion milling. A 100kV Vacuum Generators HB501 STEM with Nion aberration corrector was used to acquire high resolution high angle annular dark field (HAADF) and bright field (BF) images. Electron imaging was carried out under UHV conditions ($\sim 1\times10^{-9}$ Torr) and the specimens were free from carbon contamination. EELS measurements were performed with a Gatan Enfina spectrometer. The experimental conditions were: convergence angle 24 mrad, nominal probe size 1.3 Å, collection angle ~19mrad and energy resolution 0.3 eV (measured at FWHM of zero loss peak). Spectrum images were recorded over the energy range 230-364 eV using a dispersion setting of 0.1 eV/channel. The background intensity was removed from the spectra by applying a power-law model. Thickness calculations were performed using the formula given by Egerton [12]. For the purpose of comparing the integrated edge intensity, the acquired spectra were normalised to the maximum intensity at 295 eV. The average integrated intensities corresponding to the π^* and σ^* excitations were extracted over the energy ranges 282-287 eV and 287-292 eV respectively and used to calculate the sp^2 content relative to 100% sp^2 bonded polycrystalline graphite.

3. Results and discussion

The core-loss spectra for type IIa brown and colourless diamonds are shown in figure 1, along with spectra for type IIa brown diamonds that have been HPHT treated to 1900°C, 2200°C and 2500°C. A characteristic k-edge is visible at 290 eV corresponding to the 1s to σ^* transition that is associated with sp^3 bonded tetrahedral carbon. In addition, a prominent pre-edge peak is observed at ~286eV for the untreated brown diamond signifying the presence of sp^2 bonds in the crystal. This feature is absent from the spectrum for the colourless diamond. This characteristic feature correlates with low-loss EELS measurements that have identified a π-π^* transition at 6-7eV in brown type IIa diamond, which is indicative of a graphitic type structure and hence sp^2 bonding [1]. Employing the integrated intensity comparison method, the sp^2 content of the untreated brown diamond was found to be 16 %.

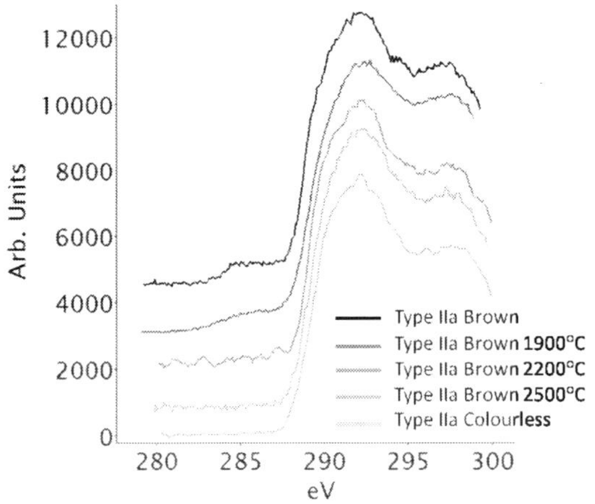

Figure 1. K-edge spectra for type IIa brown and colourless diamonds together with spectra for brown type IIa diamonds heat treated at 1900°C, 2200°C and 2500°C.

The spectra for the HPHT treated type IIa brown diamonds reveal that the effect of annealing on sp^2 content does not become apparent until temperatures in excess of 1900°C are reached. A significant pre-edge peak remains after annealing at this lower temperature, with the calculated sp^2 content ranging from 10 % - 20 % depending on the region sampled. One possible explanation for this non-uniformity is that the mono-vacancy population starts to decrease through the growth of larger vacancy clusters and / or the elimination of interstitial carbon atoms, thereby creating a dispersed sp^2 network. Annealing studies show that there is no apparent change in the brown colour at this temperature. By comparison, the colour is observed to lighten considerably when the annealing temperature is raised to 2200°C and this coincides with a corresponding drop in sp^2 content to between 0 % and 6 % depending on the region sampled. It is possible that the larger vacancy clusters become unstable when their energy of formation reaches a critical level, at which point they collapse to form inactive dislocation loops. This might explain the marked reduction in the pre-edge peak intensity at 2200°C. Increasing the annealing temperature still further to 2500°C has the effect of completely eliminating the sp^2 absorption peak from the core-loss spectrum which corresponds with the removal of the brown colour. Additional studies have shown that there is a correlation between the disappearance of contrasting features from STEM images, the removal of the sp^2 peak and the elimination of the brown colour at 2500°C [13].

Furthermore, there is evidence to suggest that the dislocations found in both brown and colourless type IIa diamonds contribute additional electronic states to the band structure of diamond. This is demonstrated by the spectra shown in Figure 2 which show a separate absorption peak at ~284 eV near to dislocations in both brown and colourless natural diamonds.

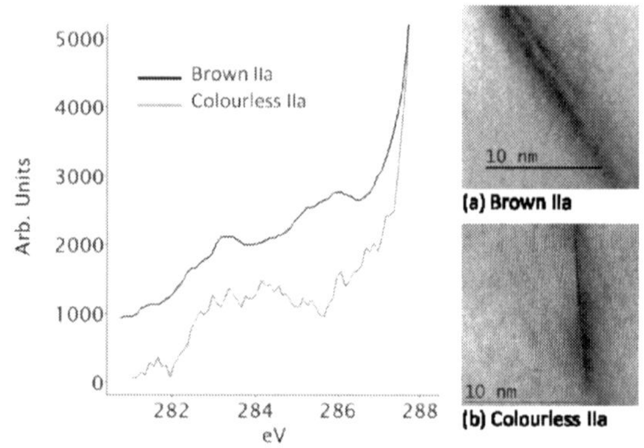

Figure 2. Additional pre-edge feature at ~284eV at dislocations in brown and colourless type IIa diamonds.

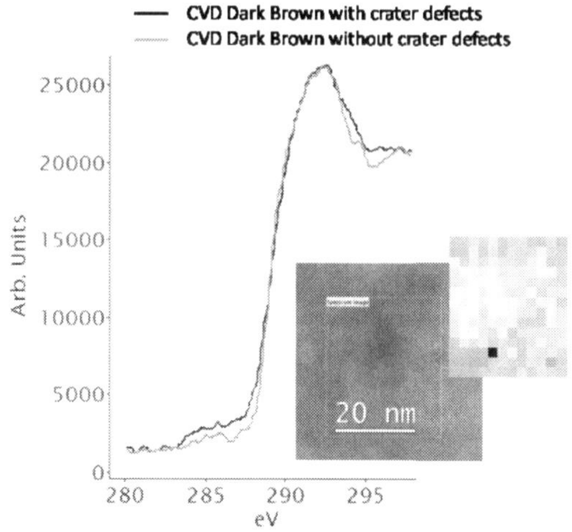

Figure 3. EEL spectra showing increased sp^2 content at crater-like defect in dark brown CVD diamond.

An investigation of large (10-15 nm) crater-like defects in dark brown CVD diamond has revealed that significant quantities of sp^2 bonds exist around the perimeter of the defects, figure 3. The exact origin of these open volume structures is unknown, however it is likely that they have been introduced during the deposition process and are in fact much larger versions of the vacancy clusters observed in the natural brown diamonds. This hypothesis would be consistent with the PAS measurements which show increased positron lifetime components for brown CVD diamond [7]. It is important to note that these crater type defects are not found in all dark brown CVD material and that the EEL spectra exhibit a background sp^2 contribution regardless of their presence.

4. Conclusion

Core-loss EELS shows a double pre-edge peak for brown diamond that is annealed out by heat treating to 2500 °C. The peak at 286 eV originates from sp^2 states associated with the π-bonded lining of vacancy clusters. However, the 284 eV peak can be attributed to electronic states at dislocations in both brown and colourless natural diamonds. Large open volume defects of the order of 10nm have been identified as sites of sp^2 bonding in dark brown CVD diamond.

Acknowledgements

The authors are grateful to the DTC and EPSRC for funding the experimental work and to the staff at the SuperSTEM Laboratory for their assistance with STEM and EELS work.

References

[1] Barnes R, Bangert U and Scott A 2007 Investigating large vacancy clusters in type IIa diamond with electron energy loss spectroscopy (EELS) *Phys. Stat. Sol.* A **204** 9 3065-3071

[2] Avalos V and Dannefaer S 2003 Vacancy-type defects in brown diamonds investigated by positron annihilation *Physica* B **340-342** 76-79

[3] Mäki J-M, Tuomisto F, Kelly C J, Fisher D and Martineau P M 2009 Properties of optically active vacancy clusters in type IIa diamond *J. Phys. Condens. Matter* **21** 364216 10pp

[4] Hounsome L, Jones R, Martineau PM, Fisher D, Shaw MJ, Briddon PR and Oberg S 2006 Origin of brown colouration in diamond *Phys. Rev.* B **73** 125203

[5] Willems B 2006 PhD Thesis Antwerp

[6] Wotherspoon A, Steeds J W, Coleman P, Wolverson D, Davies J, Lawson S, and Butler J 2002 Photoluminescence studies of type IIa and nitrogen doped CVD diamond *Diamond and Rel. Mat.* **11** 692-696

[7] Maki J, Tuomisto F, Kelly C, Fisher D and Martineau P M 2007 Effects of thermal treatment on optically active vacancy defects in CVD diamonds *Physica B Condensed Matter* **401-402** 613-616

[8] Hull D and Bacon D J 2001 *Introduction to Dislocations* 5th Ed. Elsevier Ltd.

[9] Fujita N, Jones R, Oberg S and Briddon P R 2009 Large spherical vacancy clusters in diamond – origin of the brown colouration *Diamond and Rel. Mat.* **18** 843-45

[10] Okada K, Kimoto K, Komatsu S and Matsumoto S 2002 Sp^2 bonding distributions in nanocrystalline diamond particles by electron energy loss spectroscopy *J. Appl. Phys.* **93** 3120

[11] Gilkes K W R, Gaskell P H, and Yuan J 1994 The detailed atomic structure of tetrahedral amorphous carbon *Diamond and Rel. Mat.* **3** 369-372

[12] Egerton R F 1989 *Electron Energy Loss Spectroscopy in the Electron Microscope* 2nd Ed. Plenum Press

[13] Godfrey I and Bangert U 2011 Imaging vacancy clusters in natural diamond using aberration corrected high angle annular dark field and bright field scanning transmission electron microscopy *J. Phys. Conf. Series* **281** 012024

Structure of different grades of nuclear graphite

B E Mironov[1*], A V K Westwood[1], A J Scott[1], R Brydson[1], A N Jones[2]

[1]Institute for Materials Research, University of Leeds, Leeds, LS2 9JT, UK
[2]Nuclear Graphite Research Group, Materials Performance Centre, University of Manchester, Manchester, M60 1QD, UK

[*]E-mail: pmbem@leeds.ac.uk

Abstract. Owing to its low neutron absorption cross-section, large scattering cross section and thermal and chemical stability, graphite is a key component of operational nuclear reactors where it is used as a moderator, reflector and as major structural component for 90% of current UK nuclear plants. It is also of interest for use in developing the future high temperature gas-cooled reactors. The properties of the nuclear graphite are influenced by its structural characteristics, which change as a function of neutron irradiation, temperature and oxidation. The principal structural changes during neutron irradiation that affect the integrity and dimensions of nuclear graphite components, thereby affecting service lifetime, are that the a-axis contracts and the c-axis expands in the crystallites. Characterization of virgin graphite structure and of the damage evolution after irradiation of nuclear graphite has an important role to play in the understanding and development of materials used in current and future nuclear reactors, respectively.

1. Introduction

Graphite is a key component of operational nuclear reactors where it is used as a moderator, reflector and for other structural components in 90% of current UK nuclear plants and is proposed for use in future high temperature gas-cooled reactors.

Inside a reactor, fast neutron irradiation significantly changes the crystallite size and influences bulk property changes. The solidity and working life of reactor components is influenced by dimensional change, by creep and weight loss, amongst other properties [1]. These effects however are not completely understood and the body of current data regarding irradiation of nuclear graphite is incomplete.

Graphite is used as a moderator due to its low absorption cross section and high scattering cross section compared to other materials. Neutron impact leads to carbon atom displacements which damage the graphite lattice and change the crystallite size and spacing [1,2]. Characterization of the virgin nuclear graphite structure and of the evolution of damage after irradiation has an important role to play in developing a comprehensive understanding of the working behaviour of current nuclear graphite. This will also have a major role in the development of material for use in future nuclear reactors.

This work aims to characterise two types of graphite in terms of differing crystallinity using techniques such as Raman spectroscopy, X-ray diffraction (XRD) and electron microscopy. The graphite employed are from two Gen I and Gen II UK nuclear plants materials, namely; "Pile Grade A" (PGA) and "Gilsocarbon" (GIL). Both of these materials are based on a filler and a binder, petroleum coke and asphalt coke as a filler in PGA and GIL respectively, mixed and formed with pitch binder, then baked and finally graphitised at high temperature (~3000° C).

2. Experimental

The samples of PGA and GIL were machined with a Struers Accutome-2 cutting machine at a speed of 1200rpm/sec for Raman spectroscopy, XRD and Scanning Electron Microscopy (SEM) analyses, while for Transmission Electron Microscopy (TEM), samples were either ground, dispersed in acetone and drop cast onto a carbon grid or using Focused Ion Beam (FIB) sections made in an FEI Nova200.

For SEM studies an FEI XL30 was used operating at 3keV while TEM analysis was conducted on an FEI Tecnai F20 FEG-TEM operating at 200kV and equipped with a CCD camera. Raman spectra were collected from 1cm thick sample slices using a commercial Renishaw Invia micro-Raman spectrometer equipped with a CCD detector and a modified optical microscope. The spectra were recorded for 10 sec/acquisition with a number of 10 acquisitions/sample. We used a 514.5nm Ar/Kr ion laser with P=25mW and a 50X objective lens for focusing the laser light onto the samples. X-ray diffraction patterns were recorded on a Philips X'Pert diffractometer, making use of a copper X-ray source ($\lambda K_{\alpha 1}$= 0.15405 nm) and working in θ-2θ mode with X'pertCollector software.

3. Results and discussion

The characterization of the two nuclear graphites that we studied (PGA and GIL) began with the morphology characterization. It was observed by SEM, that the porosity is very high (~20% after impregnation with pitch and 25% before that) [1], with GIL having a slightly lower porosity (with a density ~1.81g/cm^3) compared with PGA (density ~1.74/cm^3). Also the coke particles look different in shape and size with PGA showing needle shaped filler particles relative to GIL (Fig.1), this aspect is due to differences in manufacturing procedure and materials used. The differences between PGA and GIL are that PGA has anisotropic properties as it is made using extrusion which imparts preferred orientation to the coke particles, while GIL is semi-isotropic, made by moulding of particles which have an "onion" like structure with random orientation

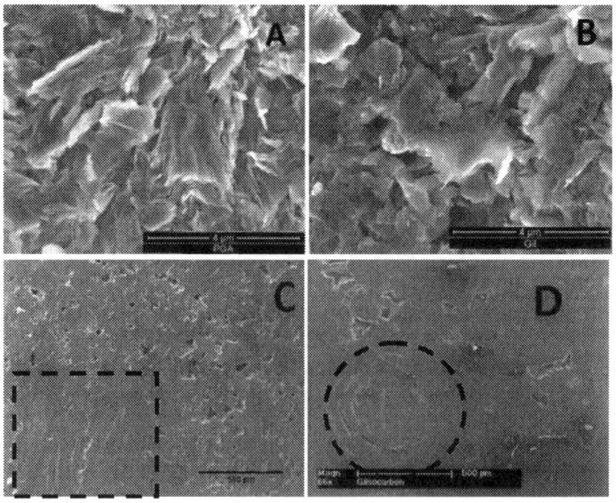

Figure 1. SE image taken in SEM of: typical needle like platelets of PGA(A) and GIL(B) in random orientation; Surface of PGA(C) and GIL(D) - dotted lines denote the filler particles specific porosity, longitudinal cracks for PGA and circular cracks for GIL

The pores observed were distinctive for both the filler and binder. For the filler part of the graphite, highlighted with dotted circles (Fig.1 C-D), elongated pores follow the structure of the filler particles, needle-shape particles for PGA (Fig.1 C) and onion like-circular structures for GIL (Fig.1 D).

Raman analysis of both graphites was used in order to determinate the disorder present inside the structure. Two peaks could be observed that are characteristic of (nuclear) graphite [4]: the D peak (disorder ~1354cm^{-1}) and the G peak (order ~1583cm^{-1}). A second order feature could also be detected (~2708cm^{-1}) for both samples. The ratio of the intensities of these peaks, I_D/I_G, was lower for PGA than for GIL (Fig.2) reflecting a larger crystallite size which can be attributed to a textured product. From this ratio it is also possible to calculate the value for the coherence length in the **a**-direction, L_a, using a corrected version of Tuinstra-Koening [5] formula:

$$L_a(\text{nm}) = C(\lambda)\left(\frac{I_D}{I_G}\right)^{-1}, \text{ where } C(514.5\text{nm}) \sim 4.4.$$

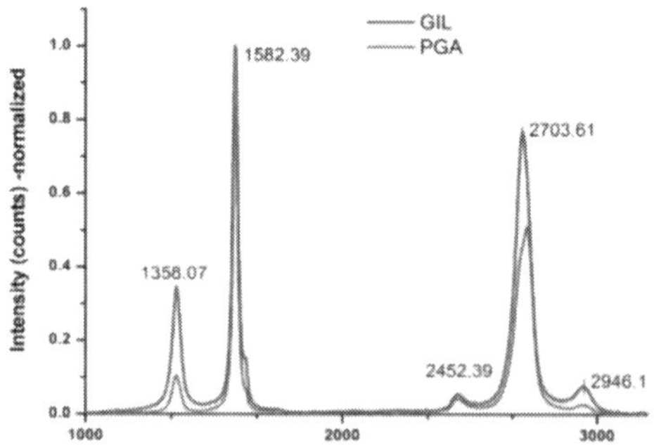

Figure 2. Raman spectra of PGA and GIL

The calculated values of L_a for PGA and GIL were 21±4nm and 20±11nm, respectively. The difference in standard deviation values may be due to variation in coherence lengths within the samples (presence of filler or binder in the acquiring spot) coupled with the local rather than global nature of this analysis.

XRD was used in order to confirm the Raman results and also to obtain global information regarding crystallite size (coherence length) in the **c**- and **a**-directions [6]. The coherence lengths in the **a** direction were 44.8nm for PGA and 47.2nm for GIL, whereas in the **c** direction the calculated values were 23.7nm for PGA and 35.6nm for GIL. These were calculated using Scherrer's equation ($L=K\lambda/\beta\cos\theta$), where the K constants used were: K_a=1.84(for **a** direction) and K_c=0.9(for **c** direction); β represents broadening at half the maximum intensity in radians and θ the angle that corresponds to the measured peak.

Though the L_a values obtained from XRD and Raman have the same order of magnitude, there are clear differences between the values derived from the two techniques. The smaller values obtained from the Raman measurement could be due to the fact that the sample was cut and hence a bigger disorder was created which increased the intensity of the D peak and influenced the calculated Raman values for L_a.We have used Bright Field (BF) TEM to confirm these values.

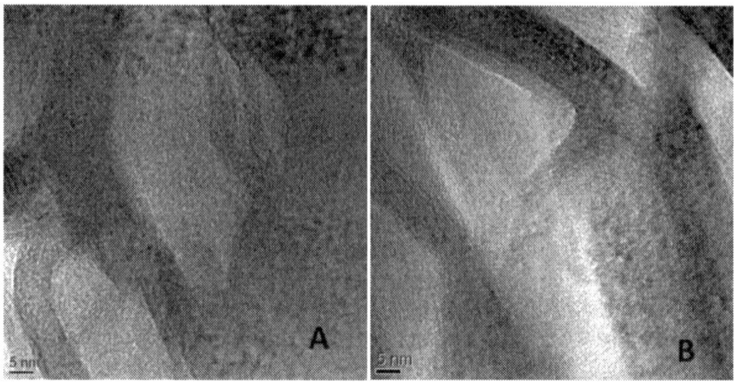

Figure 3. BF TEM images of FIB samples of PGA(A) and GIL (B) in random orientation

TEM of the FIB samples of both PGA and GIL gave information about the internal structure of the materials, and showed curved crystallites surrounding voids (Fig.3) which make measurement of the coherence lengths uncertain. For simplicity we report here measurements taken on the crushed and ground materials shown in Fig. 4.

The coherence lengths in the **c** direction (L_c) exhibited a large range: between 2-26 nm for PGA and 2-36nm for GIL. The average of the dimensions of L_c measured from TEM (for PGA 10.3±7.4nm and for GIL 11.2±9.1nm) were slightly lower than the values from XRD but they showed the same trend with the value for PGA being larger and showing a larger standard deviation than that for GIL.

A histogram of the crystallite size data (L_a) from TEM is shown in Figure 5. Similar to the coherence length in the **c** direction, the two graphites show a large range in L_a. PGA exhibits a range of L_a values for the crystallites between 5-80nm and for GIL the range was bigger and lay between 5-200nm or higher. This could explain the discrepancy between Raman, a more local analysing technique, and XRD, a more global analysing technique giving an average value.

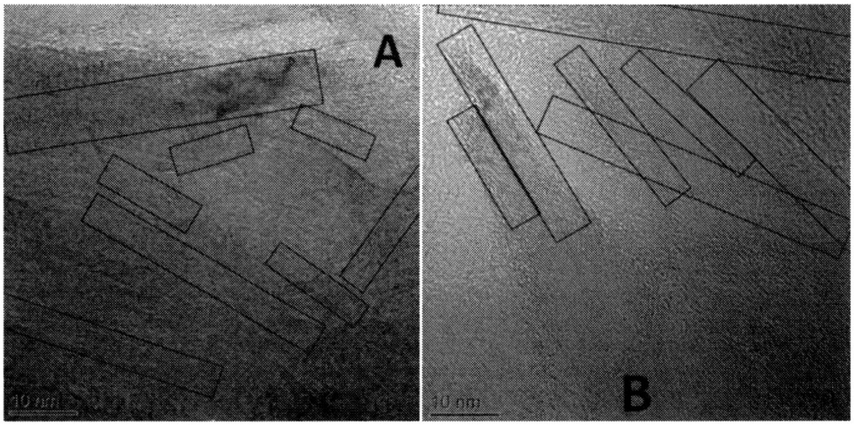

Figure 4. BF TEM images of: powder PGA(A) and GIL (B) with highlighted different size crystallites for both of graphites

From the histogram (Fig.5) of the L_a dimensions measured by TEM, average L_a values of 48±20nm and 53±38nm can be calculated for PGA and GIL, respectively. These are close to the ones calculated from the XRD.

As expected from the comparison between XRD and Raman, the L_a values calculated from Raman (PGA L_a =21±4nm and GIL L_a=20±11nm) were different compared with the ones found by TEM, which could mean that the Raman analysis can only give information regarding the degree of order/disorder in the sample and does not allow precise evaluation of L_a as has been suggested.

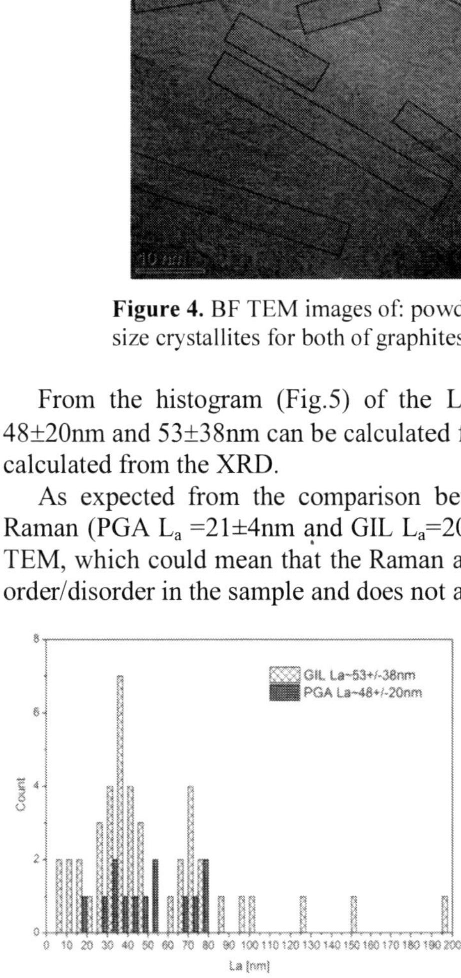

Figure 5. Histogram of dimensions occurrence in both PGA and GIL

4. Conclusions

XRD and Raman investigations of graphite for nuclear applications can provide both global and local information respectively with regards to the order, disorder or crystallite orientation size and spacing. However it is desirable to support these studies with SEM and TEM to understand the details of the microstructure and to minimize ambiguities.

Further investigations of how the structure changes in graphite are influencing the working life of nuclear reactors are needed. Parameters like irradiation duration, dose, gas environment and temperature will be taken in consideration in predicting the failures and operating time of nuclear reactors for further development.

Acknowledgments

The authors would like to acknowledge the financial support of EPSRC (EP/I003312/1) for this project entitled "Fundamentals of current and future uses of Nuclear Graphite".

References

[1] Nightingale R 1962 *Nuclear Graphite* Academic Press 87-117
[2] Simmons JHW 1965 *Radiation Damage in Graphite*
[3] Thomsen C and Reich R 2004 *Phil. Trans. R. Soc. Lond.* **362** 2271-2288
[4] Wang Y et.all 1990 *Chem. Mater* **2**, 557-563
[5] Ferrari AC and Robertson J 2000 *J.Physics Review B* **61** 14095-14107
[6] Seehra MS and Pavlovic AS 1993 *Carbon* **31** 557-564

Aberration corrected imaging of a carbon nanotube encapsulated Lindqvist Ion and correlation with Density Functional Theory

J Sloan*, E Bichoutskaia[1], Z Liu[2], N Kuganathan[1], E Faulques[1,3], K Suenaga[2], I J Shannon[4]
Department of Physics, University of Warwick, Coventry, CV4 7AL UK.
[1]School of Chemistry, University of Nottingham, Nottingham, NG7 2RD UK
[2]Nanotube Research Center, National Institute of Advanced Industrial Science and Technology (AIST), Higashi 1-1-1, Tsukuba, 305-8565, Japan
[3]Institut des Matériaux Jean Rouxel, CNRS, UMR6502, Université de Nantes, 2 rue de la Houssinière, 44322 Nantes, France
[4]Department of Chemistry, University of Birmingham, Edgbaston, B15 2TT UK

E-mail: j.sloan@warwick.ac.uk

Abstract. 80 kV aberration-corrected transmission electron microscopy (AC-TEM) of discrete $[W_6O_{19}]^{2-}$ polyoxometalate ions mounted within double walled carbon nanotubes (DWNTs) allow high precision structural studies to be performed. W atom column separations within the octahedral W_6 tungsten template can be visualized with sufficient clarity that correlation with full-scale density functional theory (DFT) can be achieved. Calculations performed on the gas phase and DWNT-mounted $[W_6O_{19}]^{2-}$ anions show good agreement, in the latter case, with measured separations between pairs of W_2 atom columns imaged within equatorial WO_6 octahedra and single W atoms within axial WO_6 octahedra. Structural data from the tilted chiral encapsulating DWNT was also determined simultaneously with the anion structural measurements, allowing the nanotube conformation to be addressed in the DFT calculations.

1. Introduction

In tandem with the burgeoning interest in a wide diversity of nanoscale structures, associated studies into single molecules are becoming more prevalent. Advances in atomic probe-type (i.e. AFM, STM or similar) [1] and electron microscopy [2,3] atomic and molecular imaging techniques are making efficient comparison between experimentally determined discrete structures and *ab initio* or *a posteriori theory* more routine. A key aspect of these investigations is the correlation of intra-molecular perturbations with effects induced by the local environment. Polyoxometalate ions (POMs) are of interest in this respect as they have an extensive structure-synthetic chemistry enabling a wide variety of ordered molecular-scale cluster anions to be produced with variable charge-ordering arrangements [4]. In the present context, they provide molecular-scale heavy atom motifs with observable symmetry than can be imaged by conventional high resolution transmission electron microscopy (HRTEM) when mounted in single or double walled carbon nanotubes (i.e. SWNTs or DWNTs) or graphene oxide [5,6]. We have previously been able to image the octahedral O_h symmetry Linqvist $[W_6O_{19}]^{2-}$ ion in DWNTs [5] (Fig. 1(a)) by HR-TEM and have imaged the W_2-W_2 atom column separations in the equatorial plane of the anion. A limitation of this study was that information from the corresponding axial W atom columns was blurred however a minimum requirement for an effective comparison with theory is that at least two dimensions of the anion be resolvable. We demonstrate here that it is possible to do this for individual ions by aberration corrected transmission electron microscopy (AC-TEM) and also that we are able to obtain conformational information from the nanotube simultaneously [7]. This allows us to produce trial composite anions and nanotube conformations for investigating both structural and electronic interactions between the encapsulated anion and the nanotube using DFT.

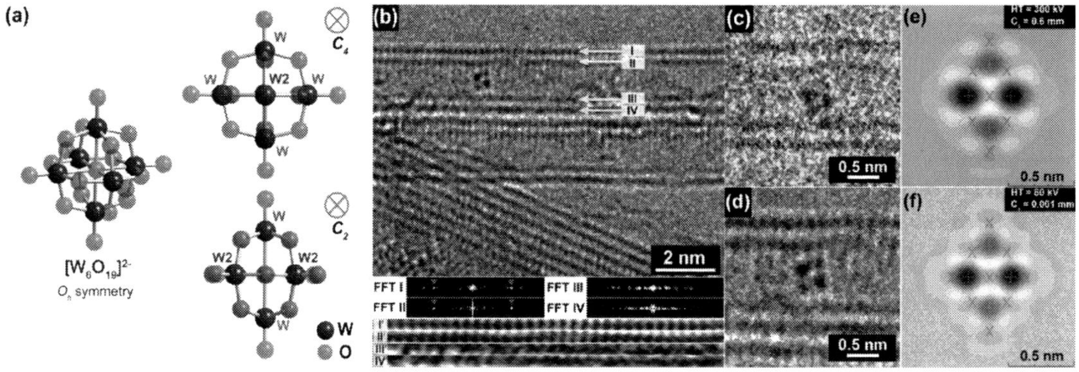

Fig. 1 a) Perspective view, C_4 projection and C_2 projection of the $[W_6O_{19}]^{2-}$ ion; b) 80 kV AC-TEM image of $[W_6O_{19}]^{2-}$ ion in a DWNT with fast fourier transforms of the two pairs of opposing walls **I-IV** (i.e. **FFT I-FFT IV**) and corresponding filtered images (i.e. **I'-IV'**) at bottom; c) & d) Details from 300 kV (C_s = 0.6 mm) image and 80 kV (C_s = 0.001 mm) image of $[W_6O_{19}]^{2-}$ ions in a DWNT. e) & f) corresponding image simulation of the $[W_6O_{19}]^{2-}$ ion, using the same imaging conditions as c) and d).

2. Experimental Details

The DWNTs used in this experiment were supplied by Carbon Nanotechnology Inc. $[n\text{Bu}_4\text{N}]_2[W_6O_{19}]$ was prepared and inserted into DWNTs from an enthanolic solution. The previous 300 kV HR-TEM image of the $[n\text{Bu}_4\text{N}]_2[W_6O_{19}]$@DWNT composite was obtained on a JEOL 3000F instrument (Oxford) with an objective C_s of 0.6 mm and a 300 kV accelerating voltage [5]. 80 kV images were obtained from the freshly prepared $[n\text{Bu}_4\text{N}]_2[W_6O_{19}]$@DWNT composite in a JEOL 2010F AC-TEM (Tsukuba) equipped with a CEOS corrector for which C_s was tuned to 0.001 mm. Image simulations were performed using the multislice package SimulaTEM. Nanotube models were generated using the program Nanotube Modeler from JCrystalSoft (©2004-2011). Full scale CASTEP DFT calculations were performed using High Performance Computing (HPC) facility at the University of Nottingham.

3. Results

The previous imaging study of $[W_6O_{19}]^{2-}$ ions mounted within DWNTs [5] revealed that within sterically matched nanotubes with an innermost ~1.2 nm diameter SWNT the ions lie with a preferred orientation within the encapsulating cylindrical volume, specified by the van der Waals radii of the wall carbons (i.e. 0.17 nm), of diameter ca. 0.9 nm. This orientation (which is projected along the C_2 rotation axis of the anion – see Fig. 1(a)) conveniently enables W_2-W_2 atom column pairs within the equatorial plane of the $[W_6O_{19}]^{2-}$ superoctahedron, to be directly visualised by a conventional HR-TEM with an objective lens coefficient of spherical aberration, C_s (alternatively, C_3), of 0.6 mm. Multislice image simulation studies showed that this level of aberration blurs the image contrast from the W_2-W_2 atom column pairs once the contribution of the equatorial terminal oxygen atoms is taken into account. Additionally, image contrast from the single atom W-W pair was blurred or missing. Fig. 1(b) shows a wider field of view image of two parallel DWNTs, one empty, one containing a discrete $[W_6O_{19}]^{2-}$ anion. Comparison of a detail of the same structure anion as imaged using the 300 kV HR-TEM (C_s = 0.6 mm) in Fig. 1(c) with the same structure anion using the lower voltage (i.e. 80 kV) AC-TEM (C_s = 0.001 mm) imaging conditions in Fig. 1(d) shows the improvement in detail.

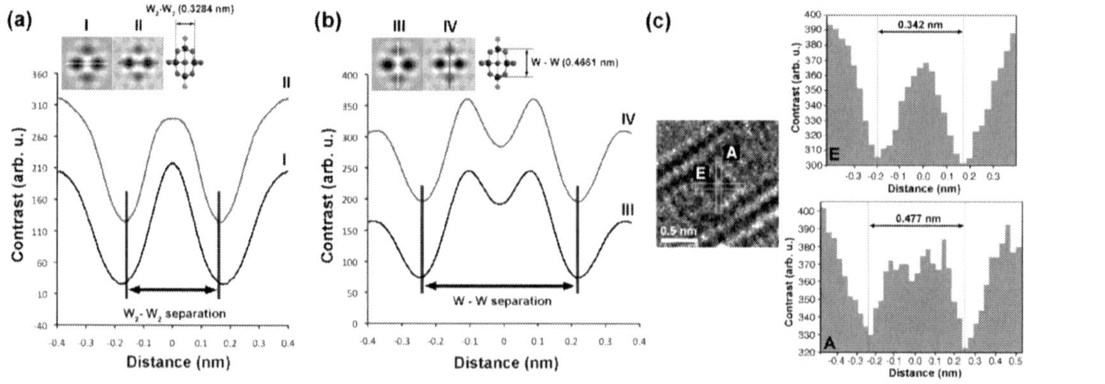

Fig. 2 a) Line profiles through simulated equatorial W_2-W_2 atom column pairs (**I** 300 kV, C_s = 0.6 mm; **II** 80 kV, C_s = 0.001 mm); b) as for a) but for line profiles through W-W atom column pair (**I** 300 kV, C_s = 0.6 mm; **II** 80 kV, C_s = 0.001 mm). c) (left) 80 kV image (C_s = 0.001 mm) of the $[W_6O_{19}]^{2-}$ ion from Figs. 1(b) and (d). (right) Experimental line profiles **E** and **A** through W_2-W_2 atom column pair and W-W atom column pair respectively.

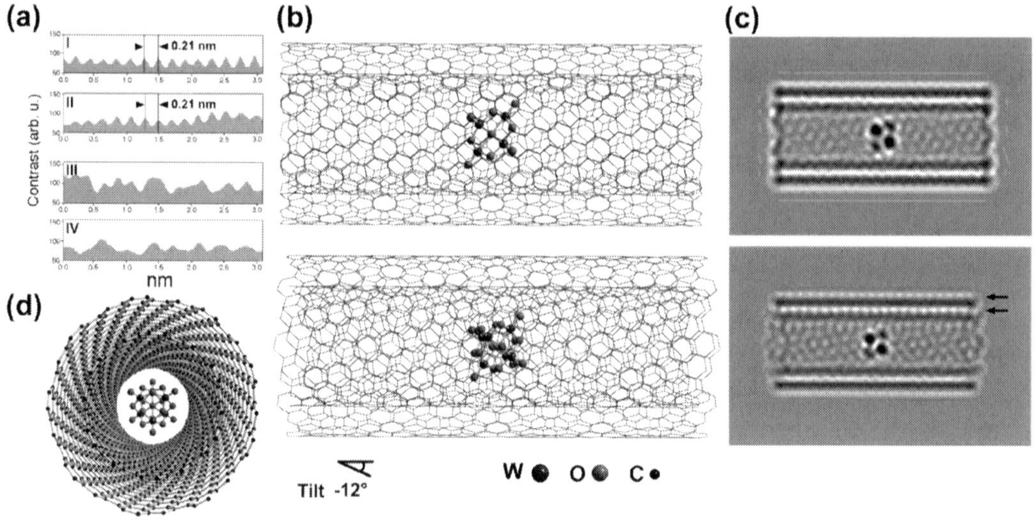

Fig. 3 a) Line profiles **I-IV** obtained through the four filtered images **I′-IV′** in the bottom part of Fig. 1b). The upper two walls have a periodicity of ca. 0.21 nm whereas the lower walls are aperiodic. b) Models corresponding to a $[W_6O_{19}]^{2-}$@(12,6)@(18,9) DWNT. The top model is untilted and the bottom model tilted 12° out of plane. c) Multislice image simulations from the two configurations in Fig. 4(b). Only the bottom simulation exhibits wall carbon spacings on the uppermost two walls. d) Perspective end-on model of the $[W_6O_{19}]^{2-}$@(12,6)@(18,9) DWNT in Fig. 2(b).

An indication of the improvement in information in the AC-TEM images can be made using line profiles through W_2-W_2 and W-W atom column pairs in multislice image simulations for both cases (i.e. Figs. 1(e) and (f)) as shown by the extracted equatorial and axial profiles (i.e. Figs. 2(a) and (b)). As noted previously, there is a ~5% distortion in the W_2-W_2 column separation for the more aberrated case partly due to higher C_s but also to the smearing contribution of the O_3 columns on the anion periphery [5]. This distortion is reproduced in Fig. 2(a) in which the troughs of profile **I** corresponding to the W_2 columns are displaced ~5% with respect their expected positions. Profile **II** in Fig. 2(a) corresponds to the simulated W_2-W_2 atom column pairs for the AC-TEM but with C_s = 0.001 mm. In this case, the troughs are displaced ~1%, reflecting a greater confidence in the new measurements. According to profiles **III** and **IV in** Fig. 2(b), there is little distortion of the axial W-W columns in

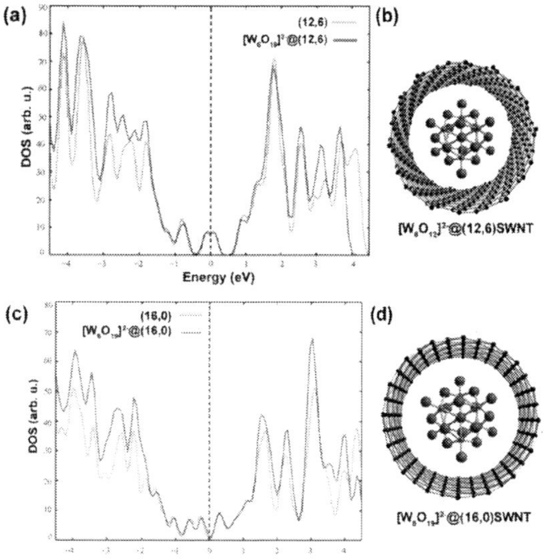

Fig. 4 a) DOS plots for metallic $[W_6O_{19}]^{2-}$@(12,6) composite. b) Optimised structure of $[W_6O_{19}]^{2-}$@(12,6)SWNT. c) and d) DOS plot and optimized structure for semiconducting $[W_6O_{19}]^{2-}$@(16,0)SWNT.

either the 300 kV or 80 kV images but the image contrast from the 300 kV image (Fig. 1(c)) is too noisy to produce a reliable measurement. In Fig. 2(c) we see that it is possible to measure the atom column spacings for both equatorial W_2-W_2 and axial W-W separations for the 80 kV case (i.e. 0.342 and 0.477 nm respectively, precision ± 0.026 nm). Additionally we have also been able to resolve the wall carbon spacings on one side of the DWNT (Figs. 1(b) and (d)). Line profiles through all four walls (Fig. 3(a)) show that the top walls have an average periodicity of 0.21 nm whereas the bottom walls are aperiodic. Observation of wall carbon spacings on one side the DWNT indicates that it is chiral and tilted (Figs. 3(b) and (c)).

Reliable comparisons can now made between the observed measurements and optimised structure models of $[W_6O_{19}]^{2-}$@DWNT composites produced by DFT (in practice the innermost SWNT is only considered). In the first instance, we can differentiate optimised models produced for metallic $[W_6O_{19}]^{2-}$@(12,6) (Fig. 4(a) and (b)) and semiconducting $[W_6O_{19}]^{2-}$@(16,0) composites (Figs. 4(c) and (d)) The densities of states (DOS) plot in Fig. 4(a) is evidently more appropriate for comparison with the experimental case. We can also now compare both experimentally determined and DFT-optimised equatorial W_2-W_2 and axial W-W separations. The experimental measurements for the equatorial and axial separations (i.e. 0.342 and 0.477 nm, see above) compare reasonably with the corresponding DFT optimised gas phase separations (i.e. 0.333 and 0.471 nm, respectively) and optimised $[W_6O_{19}]^{2-}$@(12,6) SWNT separations (i.e. 0.331 and 0.471 nm, respectively) [7].

4. Conclusions

High quality structural detail can be obtained by AC-TEM at 80 kV from DWNT-encapsulated $[W_6O_{19}]^{2-}$ ions. The obtained structural information compares favourably that obtained from DFT and the enhanced resolution also allows for conformational information from the DWNT to be included.

References

[1] Eigler D M and Schweitzer E K 1990 *Nature* **344** 524.
[2] Smith B W, Monthioux M and Luzzi D E 1998 *Nature* **396** 323.
[3] Sato Y, Suenaga K, Okubo S, Okazaki T and Iijima S 2007 *Nano Lett.* **7** 3704.
[4] Volkmer D, Bredenkötter B, Tellenbröker J, Kögerler P, Kurth D G, Lehmann P, Schnablegger H, Schwahn D, Pipenbrink M and Krebs B 2002 *J. Am. Chem. Soc.* **124** 10489.
[5] Sloan J, Matthewman G, Dyer-Smith C, Sung A-Y, Liu Z, Suenaga K, Kirkland A I and Flahaut E 2008 *ACS Nano* **2** 966.
[6] Sloan J, Liu Z, Suenaga K, Wilson N R, Pandey P, Perkins L M, Rourke J P and Shannon I J 2010 *Nano Lett.* **10** 4600.
[7] Bichoutskaia E, Liu Z, Kuganathan N, Faulques E, Suenaga K, Shannon I J and Sloan J, 2012 *Nanoscale*, in press.

Acknowledgements

We are indebted to the Warwick Centre for Analytical Science (EPSRC Grant EP/F034210/1). E. B. acknowledges a Career Acceleration Fellowship, NanoTP "Cost" action and HPC. Z. L. and K. S. acknowledge support from JST-CREST and Grant-in-aid from MEXT (19054017).

Quantitative biological measurement in Transmission Electron Tomography

Judith M Mantell[1], Paul Verkade[1,2] and Kenton P Arkill[3]

[1] Wolfson Bioimaging Facility and School of Biochemistry, University of Bristol, UK
[2] School of Physiology and Pharmacology, University of Bristol, UK
[3] MVRL, School of Physiology and Pharmacology, University of Bristol, UK

Email:j.mantell@bristol.ac.uk

Abstract. It has been known for some time that biological sections shrink in the transmission electron microscope from exposure to the electron beam. This phenomenon is especially important in Electron Tomography (ET). The effect on shrinkage of parameters such as embedding medium or sample type is less well understood. In addition anisotropic area shrinkage has largely been ignored. The intention of this study is to explore the shrinkage on a number of samples ranging in thickness from 200 nm to 500 nm. A protocol was developed to determine the shrinkage in area and thickness using the gold fiducials used in electron tomography. In brief: Using low dose philosophy on the section, a focus area was used prior to a separate virgin study area for a series of known exposures on a tilted sample. The shrinkage was determined by measurements on the gold beads from both sides of the section as determined by a confirmatory tomogram. It was found that the shrinkage in area (approximately to 90-95% of the original) and the thickness (approximately 65% of the original at most) agreed with pervious authors, but that almost all the shrinkage was in the first minute and that although the direction of the in-plane shrinkage (in x and y) was sometimes uneven the end result was consistent. It was observed, in general, that thinner samples showed more percentage shrinkage than thicker ones. In conclusion, if direct quantitative measurements are required then the protocol described should be used for all areas studied.

1. Introduction

The often-observed effect of "bleaching" of biological material in the electron microscope (where the observed area appears lighter than the surrounding area) can be attributed to local shrinkage of the sample. Luther et al, [1] used the relative measurement between beads on the sample surface in tilted samples, also containing tropomyosin paracrystals of "known" spacing, to measure this collapse of plastic sections. However, images were taken on photographic plates and due to initial sample "drift", the first images used for measurements were after some initial sample irradiation. Under their fluence rate (<50 $e/nm^2/s$), they recorded an initial period of fairly rapid collapse followed by a longer period (40 minutes) of slower change. It has since become accepted practice to pre-irradiate the area of interest for up to an hour, before collecting a tomography tilt series [2]. However, if one would like to make a quantitative size or volume measurement of a certain feature within cellular material, then it is vital to know the initial sample volume, preferably at the region of interest. In this study, modern CCD camera technology and microscope Low Dose software procedures have been applied to investigate these first moments of irradiation. A protocol was developed that can be applied to any sample prepared for tomography using fiducial markers. This has then been used to study the shrinkage in z, as well as anisotropic movement in the x,y plane, and to investigate the possible protective effect of carbon coating [2].

2. Materials and Methods

2.1 *Materials*
High pressure frozen, freeze substituted, Epon embedded samples of isolated rat islets of Langerhans [3] and standard resin embedded sample of equine metocarplophalangeal calcified cartilage, were cut at nominal 200, 300, 400, and 500 nm thickness on a Reichert Ultracut S microtome and collected on pioloform coated, copper slot grids. 15 nm fiducial markers (Aurion, NL) were applied to both sides. The sections were examined in a Tecnai 20 LaB6 TEM (FEI Company), operating at 200 kV with images captured on a 4k x 4k Eagle CCD camera (FEI Company). One of the 300 nm sections was subsequently carbon coated in an Edwards Auto 306 Evaporation system and re-examined. The electron fluence rate was measured via the Low Dose software.

2.2 *Microscope protocol*
The specimen was initially imaged at low magnification and beam settings to find an area of cellular material, as in a typical Low Dose procedure, with an electron fluence rate of less than 10 $e/nm^2/s$. The sample was then tilted to 45°. The Low Dose Focus mode was used to align and focus an adjacent area 10 μm further along the tilt axis from the area to be exposed. The electron fluence rate was now increased to around 300 $e/nm^2/s$, typical of our beam settings for tomography. In the Exposure mode, a series of 60 exposures was then recorded at 1 second intervals, using the pre-specimen shutter between exposures. Subsequently a quick tomography series between 45° and -45° at 5° increments was recorded to identify those fiducials above/below each other on the top and bottom surfaces. This tilt series was used to calculate the "final" tomogram thickness, using IMOD tomography software [4], following what was a further 15 minutes of electron irradiation and a total electron fluence of approximately $3x10^5$ e/nm^2. Finally a reference image at zero tilt was taken.

2.3 *Measurement protocol*
Changes in thickness in the z direction were calculated using the geometry shown in Figure 1. At each sample position on a grid (3 positions analysed per grid, or sometimes on separate grids), measurements were made through the image stack between at least three pairs of fiducials on top and bottom surfaces. Fiji software, a version of ImageJ,[5]was used for this. This is then one dataset.

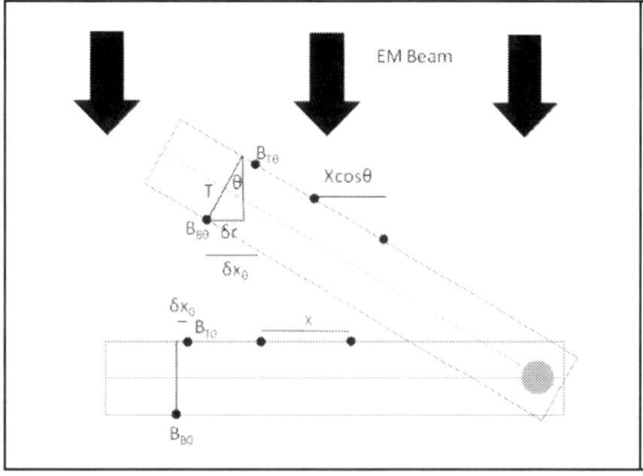

Figure 1: Schematic of geometry of pairs of gold fiducials on both surfaces of section, showing thickness calculation where,

$\delta x_0 = B_{T0} - B_{B0}$
$\delta x_\theta = \delta r + \delta x_0 \cos\theta$
hence
$B_{T\theta} - B_{B\theta} = T\sin\theta + (B_{T0} - B_{B0})\cos\theta$
$T = ((B_{T\theta} - B_{B\theta}) - (B_{T0} - B_{B0})\cos\theta)/\sin\theta$

These were averaged and normalised to the thickness calculated from the image taken at zero tilt, immediately after the exposure series. Changes in shape in the x, y plane were also investigated using relative shifts between images of fiducials on one surface, taken from each quadrant of all images. These were again averaged and normalised.

3. Results

Initially measurements on 200 and 300 nm sections, which are our standard thicknesses for ET, were used to test our protocol extensively. As data appeared consistent, the protocol was then applied to samples of increased thickness. Averaged results for all the different sample thicknesses, with number of data sets analysed from different positions given in brackets, are shown in Figures 2 and 3. The initial collapse of the sections is extremely rapid at this fluence rate. It was also observed that in three out of four cases, the thinner samples shrink proportionately more than thicker samples for the same exposure. There was no intuitive reason why the 400 nm sections behaved differently. Measurements of thickness at the end of the first minute were then compared with the "final" thickness after the tomogram collection and presented in Table 1. Data from the 300 nm section which was subsequently carbon coated is also presented here as well as provisional data from sample resin, in which no biological material was contained. A complete data set could be collected and analysed in a few hours.

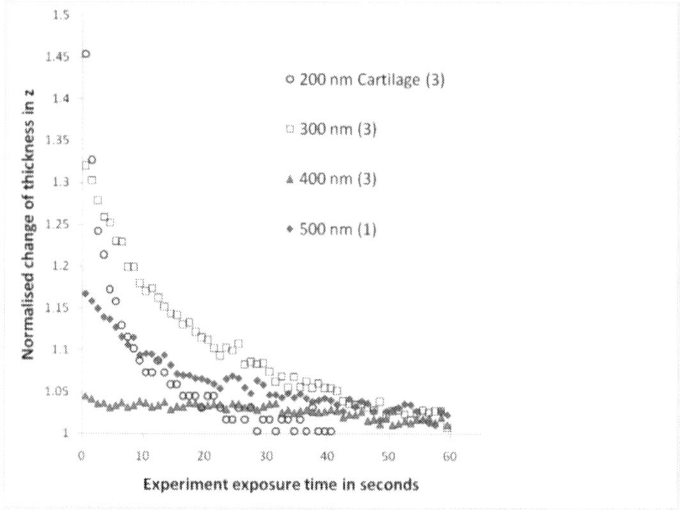

Figure 2. Proportional change in thickness in z- direction over time. The figure in brackets gives the number of different sample data sets analysed.

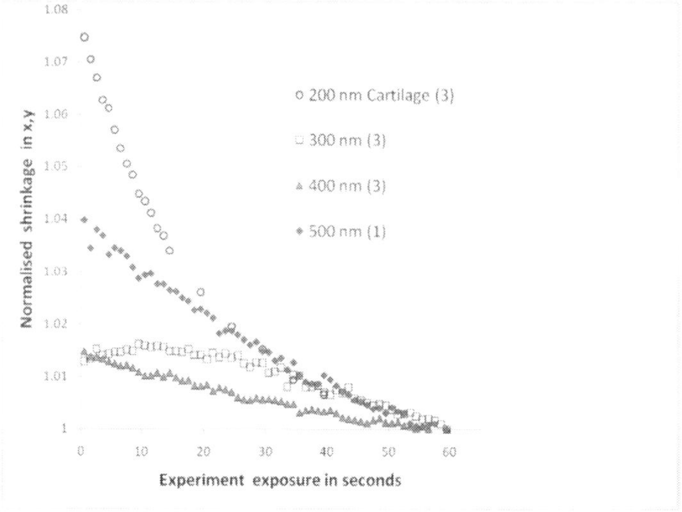

Figure 3. Proportional in-plane shrinkage in x,y direction over time. The figure in brackets gives the number of different sample data sets analysed.

Table 1. Thicknesses at start, end and after completed tomogram for different samples

Sample and thickness	Range of thickness in nm ± 10 nm at 1 s. Fluence: 300e/nm^2	Range of thickness in nm ± 10 nm at 60 s. Fluence: 1.8 x10^{3}e/nm^2	Thickness from tomogram ±10 nm. Fluence : 3x10^{5}e/nm^2
200 nm (3)	184 - 219	139 - 179	116 - 142
200 nm resin	169	159	132
300nm (3)	333 - 375	243 - 293	246 - 264
300nm C Coated (1)	267	266	248
400nm (3)	322 - 349	304 - 340	335 - 340
400nm resin (1)	360	329	318
500nm (1)	476	417	414

4. Discussion and conclusions

A review of shrinkage data and possible mechanisms involved, as well as inhibiting factors, is available in Luther's contributed chapter to the Electron Tomography Methods book [2] and is beyond the scope of this paper. The x,y movement, although displaying initial variations, showed a consistent change of about 1.5–3 %. The greatest z shrinkage (to 65% of the original) was measured in the thinnest samples: the 300 nm sections of rat islets and the 200 nm calcified cartilage material. The maximum z shrinkages were similar to the results from Luther et al for 80 nm sections, though as their first measurements correspond to our data after 3-4 s, we would expect that some shrinkage had already taken place on both the paracrystals and the section. The greatest change happens in the first few seconds and appears to follow a first order exponential decay model. Although further small shrinkage was measured after collection of the tomogram (see Table 1), we would suggest that in this type of biological sample, even when quantitative measurements are not required, further pre-irradiation of samples beyond setting up the tomography experiment, is not necessary. Our initial data on carbon coating supports that it does indeed reduce shrinkage. However results also suggest some pre-shrinkage occurring either in the microscope vacuum or during the coating process. Interferometry might be useful here to measure sample thickness at different stages in the analysis and is planned for future studies. It was also found that resins alone can behave differently to areas where the resin contains the biological material. As sample behavior is individual, it is our conclusion that this quick and simple protocol should be applied to any sample if volume measurements of cellular components within cellular material are desired. The authors would be happy to supply a template spreadsheet to this end. The protocol could also be used to investigate further the influence of other parameters on shrinkage: parameters such as specimen cooling, electron microscope operating voltage, and other differences due to sample preparation methods.

References

[1] Luther, P. K., M. C. Lawrence, et al. (1988) A method for monitoring the collapse of plastic sections as a function of electron dose. *Ultramicroscopy* **24**(1) 7-18.

[2] Luther, P. K. (2006). Sample shrinkage and radiation damage of plastic sections. *Electron Tomography: Methods for Three-Dimensional of Structures in the Cell* . Publisher: Springer edited by J. Frank. Pp.17 – 48.

[3] Verkade P. (2008). Moving EM: The Rapid Transfer System as a new tool for Correlative Light and Electron Microscopy and high throughput for High-Pressure Freezing. *J Microsc* **230** 317-328.

[4] Mastronarde, D. N. (1997). Dual-axis tomography: An approach with alignment methods that preserve resolution _*J Struct Biol* **120**(3) 343-352.

[5] Rasband, W. S. (1997-2009. ImageJ, U. S. NIH, Bethesda, Maryland, USA,.

Journal of Physics: Conference Series **371** (2012) 012020 doi:10.1088/1742-6596/371/1/012020

TEM analysis of nanoparticle dispersions with application towards the quantification of *in vitro* cellular uptake

N Hondow[1], P Wang[2], R Brydson[1], M D Holton[3], P Rees[3], H D Summers[3] and A Brown[1]

[1]Institute for Materials Research, School of Process, Environmental and Materials Engineering, University of Leeds, Leeds, LS2 9JT, UK.
[2]Institute of Molecular and Cellular Biology, University of Leeds, Leeds, LS2 9JT, UK.
[3]Centre for Nanohealth, College of Engineering, Swansea University, Swansea, SA2 8PP, UK.

N.Hondow@leeds.ac.uk

Abstract. TEM analysis of nanoparticle dispersions in solution has previously been limited, in the main, to the measurement of primary particle size due to the drying effects that occur during sample preparation. We show that solutions prepared for TEM by plunge freezing specimen grids coated with a blotted solution is a sensitive and potentially representative route to measuring a nanoparticle dispersion. We have started to benchmark the technique against the more commonly used dynamic light scattering method and show that the TEM route is potentially more sensitive route to counting individual nanoparticles in a dispersion that also contains nanoparticle agglomerates. Accurate measurement of nanoparticle dispersion in biological solutions could be a key step in the application of nanoparticles in medicine.

1. Introduction

Nanoparticles are finding increasing applications in the field of medicine [1], with extensive research being conducted on the design of nanoparticles to enhance such applications [2]. Preliminary investigations of nanoparticles with potential applications in medicine involve *in vitro* testing to explore cellular uptake and action. Transmission electron microscopy (TEM) is a valuable tool for the intracellular localization of nanoparticle uptake and for the development of mechanistic modes of action. TEM studies are often conducted on the nanoparticles prior to *in vitro* delivery and subsequent to it on ultrathin sections of cells.

The use of nanoparticles in medicine is far more complicated than the use of soluble chemicals [3]. Characterisation of nanoparticles prior to *in vitro* delivery is affected by two factors. Firstly, nanoparticles are rarely used in an as-received or as-synthesised state; the nanoparticles are commonly dispersed in biological media which affects the surface chemistry and the agglomeration state. Secondly, the most commonly used dispersion characterization technique of dynamic light scattering (DLS), is not able to distinguish between different types of particles in the solution and more importantly becomes significantly less reliable for the measurement of wide particle size distributions (because of the sixth power dependence of light scattering intensity on the scattering particles' sizes). DLS accuracy is also reliant on knowing the refractive index of the nanoparticles and the viscosity and refractive index of the dispersant [4]. On the other hand TEM imaging of samples prepared by drop-casting is not representative of the dispersed state of the nanoparticles owing to drying effects and can only really be used to examine the primary nanoparticles [5].

We have investigated an alternative TEM sample preparation route to produce solution derived specimens with more representative dispersions of nanoparticles. We are primarily interested in the dispersion of cadmium selenide/zinc sulphide quantum dots, on which studies regarding statistical analysis of the cellular uptake have been reported and in which DLS results have shown a hetro-dispersion of the dots [6]. Using sample preparation, which involves the plunge freezing of TEM grids coated with blotted solutions and a combination of cryo- and room temperature-TEM we demonstrate that this preparation route produces a specimen with a range of nanoparticle agglomerate sizes. We also discuss correlation with dynamic light scattering size profiles of the same solutions.

2. Experimental

CdTe/ZnS core/shell quantum dots with specific peptide molecules on their surface to facilitate cell entry (Qtracker 705, ~10 nm in length, Invitrogen Ltd) were dispersed in McCoys Medium 5A Modified Medium (Sigma Aldrich) at a concentration of 10 nM. Ludox SM-30 silica nanoparticles (Ludox Colloidal Silica, ~15 nm in diameter) were dispersed in Dulbecco's Modified Eagle's Medium (Lonza, Slough, UK) at a concentration of 100 μg/mL.

Samples for cryo-TEM were prepared by placing a 3.5 μL droplet on a glow discharge treated carbon film (R1.2/1.3 Quantifoil MicroTools GmBH), blotted and plunge frozen in liquid ethane [7].

Cryo-TEM was conducted at liquid nitrogen temperatures (77 K) on an FEI Tecnai F20 FEG-TEM operated at 200 kV and equipped with a Gatan Ultrascan 4000 CCD camera. Standard TEM was conducted on an FEI Tecnai F20 FEG-TEM operated at 200 kV and equipped with a Gatan Orius SC600A CCD camera. DLS was conducted on a Malvern Zetasizer Nano ZS with subsequent data manipulation conducted using the Zetasizer software (Version 6.20).

3. Results and Discussion

A sample of the CdSe/ZnS quantum dots (QDs) dispersed in water was prepared by plunge freezing a blotted TEM grid into liquid ethane [7] and examined by cryo-TEM. This allowed imaging of the QD dispersion through a thin film of vitrified ice, such that a range of agglomerates in the solution are visible (figure 1). A lack of QD signal to background is evident in the low magnification images and this is due to two major factors: the low electron fluence required to avoid devitrification of the ice and the variable thickness of the ice. These combine to make identification of individual or small groups of QDs difficult, and consequently skew any data collection towards the identification and counting of larger agglomerates.

To overcome these problems we heated the sample in the TEM from cryo conditions (~77 K) to a temperature (~160 K) at which the ice devitrifies and then sublimes. It is known that during devitrification the water and any macromolecules present exhibit little movement [8]. TEM imaging of the same agglomerates after sublimation of the ice show that the QDs do not move and that the same size agglomerates could be measured (figure 1 (b) and (c)). This allows for the sample to be prepared via the plunge freezing route, but then analyzed in a TEM at room temperature after freeze drying. Without the ice, a higher electron fluence can be used (the specimen is far less beam sensitive) and there is a weaker more consistent background level (from the carbon support film) and so it is much more straightforward to also identify small groups and individual dots across the grid.

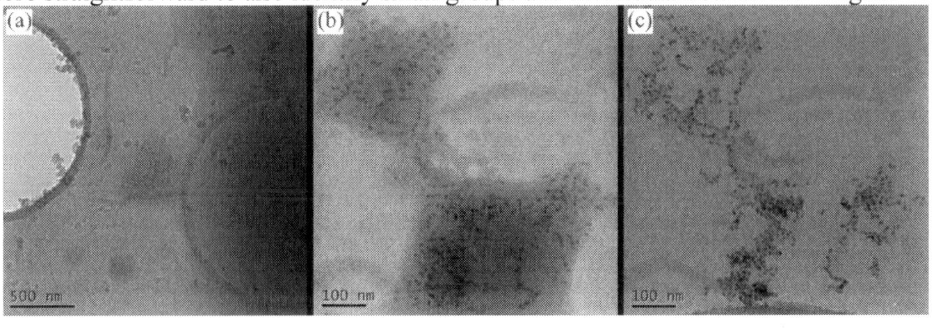

Figure 1. (a) Low magnification cryo-TEM image of the vitrified solution on the support film. Agglomerates of QDs are clearly visible however it is difficult to identify individual dots. TEM images of an agglomerate (b) before and (c) after devitrification and sublimation of the ice.

In order to qualitatively validate this preparation route for TEM sizing against DLS sizing, a simple system of uncoated silica nanoparticles was investigated. The silica nanoparticles, as observed by DLS, are monodisperse in water and agglomerate in biological media (figure 2 (a)). Samples of the water and biological media dispersions were prepared for TEM via the plunge freezing method. Room temperature TEM imaging of the two solutions are consistent with the DLS size profiles, monodispersed nanoparticles were visible in the water dispersion (figure 2 (b)) and agglomerates of several hundred nanometers were visible in the water plus biological media dispersion (figure 2 (c)).

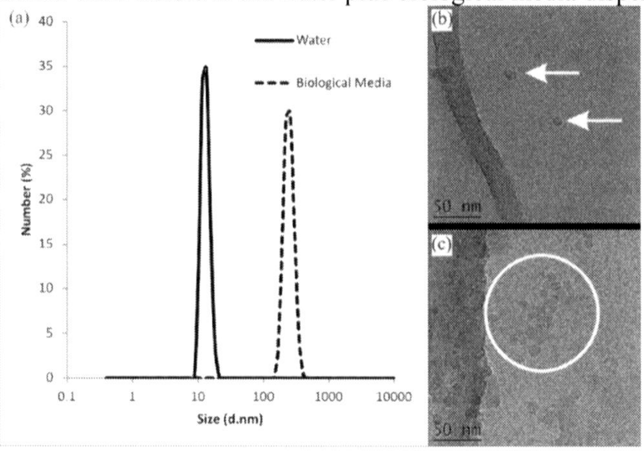

Figure 2. Analysis of silica nanoparticle dispersions in water and water plus media (a) DLS number plot derived from the original intensity-based light scattering data, (b) TEM image of monodisperse silica nanoparticles in the water dispersion and (c) agglomerates in the water plus media dispersion.

For a more quantitative comparison with DLS we measured a dispersion of the QDs in water plus biological media, however this is now complicated by the peptide coating on the nanoparticle surface. Coating is common for nanoparticles with medical applications in order to target specific interactions, in this case cellular uptake by endocytosis. The coating however will also affect the dispersion of the dots in the biological media prior to uptake. The inorganic salts, amino acids, vitamins and other components of the media will affect the agglomeration of the QDs.

A TEM grid containing a dispersion of QDs in biological media was prepared by plunge freezing and was imaged by TEM at room temperature. In total, 480 individual or agglomerated groups of QDs were imaged from 216 images, which were recorded by examining 17 grid squares from two separate grids. The size of each agglomerate was given by the distance between the two most spaced QDs in the cluster. To provide a comparison to the DLS results (figure 3 (a)) the TEM derived measurements were then grouped to the nearest 10 nm, and presented on a frequency plot (figure 3 (b)). The TEM results, as shown by both the size distribution plot (figure 3 (b)) and several images (figure 3 (c)-(e)), suggest that there is a fraction of agglomerates on the scale of 100 nm in size (as suggested by DLS) but that there is also a more significant number of mono-dispersed/isolated dots. This difference may be due to an artifact of the TEM preparation route or to the relative insensitivity of light scattering to the smallest fraction of the QD dispersion. The latter would be significant if individual dots were also taken up by cells during *in vitro* exposure. To explore the difference between the two techniques in more detail we are currently measuring simpler dispersions of QDs in water.

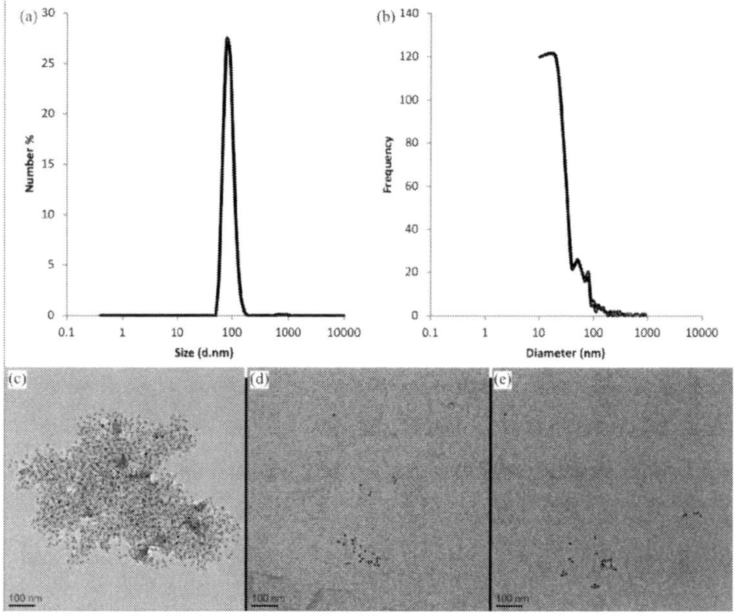

Figure 3. Analysis of QDs in biological media (a) DLS number plot (b) frequency plot from TEM measurements (c)-(e) TEM images of agglomerated and individual QDs.

4. Conclusions

The preparation of samples via the plunge freeing route provides a potential avenue for the analysis of the dispersion state of nanoparticles in biological fluids. It is known that when nanoparticles are dispersed in biological fluids the sizes formed are variable and dependent upon both the nanoparticles and dispersants. Other characterization techniques such as DLS are potentially insensitive to the finer fraction of a nanoparticle dispersion if the size range is wide, whereas samples prepared via plunge freezing and then analyzed by TEM do enable counting of agglomerates and individual particles.

Acknowledgements

This work has been supported by the EPSRC in Leeds (EP/E059678/1 and EP/H008578/1) and Swansea (EP/H008683/1).

References

[1] Shi J, Votruba A R, Farokhzad O C and Langer R 2010 *Nano Lett.* **10** 3223
[2] Petros R A and DeSimone J M 2010 *Nat. Rev. Drug Discov.* **9** 615
[3] Teeguarden J G, Hinderliter P M, Orr G, Thrall B D and Pounds J G 2007 *Toxicol. Sci.* **95** 300
[4] Berne B J and Pecora R 2000 *Dynamic Light Scattering with Applications to Chemistry, Biology, and Physics* (New York: Dover Publications Inc.)
[5] Dutta D, Sundaram S K, Teeguarden J G, Riley B J, Fifield L S, Jacobs J M, Addleman S R, Kaysen G A, Moudgil B M and Weber T J 2007 *Toxicol. Sci.* **100** 303
[6] Summers H D, Rees P, Holton M D, Brown M R, Chappell S C, Smith P J and Errington R J 2011 *Nature Nanotechnol.* **6** 170
[7] White H D, Walker M L and Trinick J 1998 *J. Struct. Biol.* **121** 306
[8] Dubochet J 2007 *Cellular Electron Microscopy* ed J R McIntosh (London: Elsevier) p 7

STEM mode in the SEM for the analysis of cellular sections prepared by ultramicrotome sectioning

N Hondow, J Harrington, R Brydson and A Brown

Institute for Materials Research, School of Process, Environmental and Materials Engineering, University of Leeds, Leeds, LS2 9JT, UK.

N.Hondow@leeds.ac.uk

Abstract. The use of the dual imaging capabilities of a scanning electron microscope fitted with a transmitted electron detector is highlighted in the analysis of samples with importance in the field of nanotoxicology. Cellular uptake of nanomaterials is often examined by transmission electron microscopy of thin sections prepared by ultramicrotome sectioning. Examination by SEM allows for the detection of artefacts caused by sample preparation (eg. nanomaterial pull-out) and the complementary STEM mode permits study of the interaction between nanomaterials and cells. Thin sections of two nanomaterials of importance in nanotoxicology (cadmium selenide quantum dots and single walled carbon nanotubes) are examined using STEM mode in the SEM.

1. Introduction

The increasing presence of nanoparticles in various aspects of our lives (healthcare, energy, consumer products, etc) has led to studies examining their potential health risks and the development of the field of nanotoxicology [1]. *In vitro* investigations in this field comprise cyto- and geno-toxicity testing of nanomaterials, with electron microscopy proving to be an excellent technique to study uptake through a combination of imaging and spectroscopy. However, to enable imaging in the TEM extensive sample preparation is required, involving resin embedding fixed cells with thin sections being cut using an ultramicrotome. As with all sample preparation methods, valid concerns have been raised regarding the potential for the sectioning process to cause the formation of artefacts [2].

For nanotoxicology studies it is vital to be able to confidently state the location of a nanomaterial in relation to an exposed cell. Nanomaterials that have been pulled out of the resin and dragged along the surface of the thin section by the action of the microtome knife could easily lead to misidentified nanomaterial–cellular interactions. One method to determine the three-dimensional location of a nanomaterial is through electron tomography, which has been used extensively in the chemical, biological and materials sciences [3]. In the nanotoxicology field, electron tomography and consequent reconstructions have been used to locate single walled carbon nanotubes (SWCNTs) within cell organelles after *in vitro* exposure [4]. Electron tomography can provide definitive confirmation as to the location of nanomaterials, however it is an involved technique, requiring much time for data collection and post-acquisition analysis, and it not necessarily suited towards the rapid screening of samples.

We have suggested an alternative technique in which we have used the combination of transmitted and secondary electron imaging in the SEM to determine whether nanomaterials are present *in* or *on* a thin section [5]. The dual imaging capabilities of a SEM fitted with a transmitted electron detector allow for projection images to be produced using the transmitted electrons (similar to those in STEM conducted in a TEM) while the same area of sample can be imaged using the secondary electrons at a low accelerating voltage giving a surface sensitive signal. STEM mode in the SEM has been used since the 1970s [6], however there is not many reports in the literature which have recorded images in both secondary and transmitted electron modes, and until recently no examples of using both of these modes of images for the study of thin sections [5]. The inclusion of a transmission detector has become a common addition to a SEM, and in the study of samples of nanotoxicological importance it has numerous benefits of wide field of view, high throughput of samples and ease of analysis, in addition to the capability to identify surface artefacts using the secondary electron signal.

2. Experimental

Scanning electron microscopy (SEM) was conducted on a Leo 1530 Gemini FEG-SEM fitted with a KE Developments scanning transmission electron microscopy (STEM) detector. SEM imaging using the in-lens detector was conducted at an accelerating voltage of 3 kV and at a working distance of 3 mm. STEM imaging was conducted at an accelerating voltage of 20 kV and at a working distance of 8 mm (resolution ~ 13 nm [5]). Using this specimen-detector configuration we estimate the acceptance angle of the bright-field detector to be about 17 mrad and the dark-field detectors to range from ~ 120 mrad up to ~ 1000 mrad. The Bragg angle for 20 keV electrons diffracted by 0.35 nm lattice planes is 12.3 mrad, indicating that imaging with the dark field (DF) STEM mode in the SEM corresponds to a medium angle annular dark field configuration. For further details please see [5].

Results from two nanotoxicology related samples are discussed in this paper. The first consists of MCL-5 human lymphoblastoid B (suspension) cells that were exposed to polyethylene glycol coated cadmium selenide-zinc sulphide quantum dots (Invitrogen); the second of adherent BEAS-2B normal human bronchial epithelial cells that were exposed to single-walled carbon nanotubes (SWCNT; 1-3 μm length, diameter 1-2 nm, kindly supplied by Prof. A. Barron, Rice University). The samples were prepared using standard techniques for electron microscopy analysis and from the resulting resin block silver sections (~ 70 nm thick) were cut using an ultramicrotome (Leica Ultra-cut E) and placed on a copper grid (Agar Scientific Ltd). No conventional heavy metal stain (uranyl acetate or lead citrate) was employed on the thin sections. For further experimental details regarding the cell culture and nanoparticle dosing, and sample preparation for electron microscopy please see [5].

3. Results and Discussion

A large field of view can be imaged using the STEM mode in the SEM, as shown by the nanotoxicological sample of interest, lymphoblastoid B-cells exposed to CdSe/ZnS quantum dots (figure 1(a)). It is very easy to increase the magnification to image an individual cell (figure 1 (b)), and then to identify collections of quantum dots that are present at higher magnifications (figure 1 (c)), without the changes in lens settings (and therefore focus) that is experienced in the conventional TEM. We have previously shown that it is possible to identify the elemental composition of nanoparticles using energy dispersive X-ray spectroscopy in this mode of imaging [5].

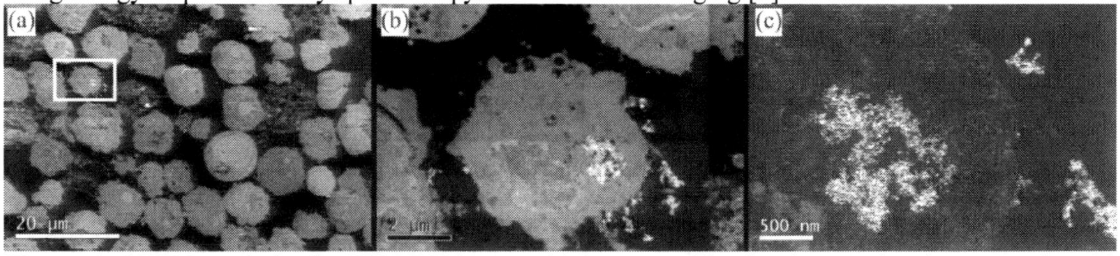

Figure 1. STEM mode in the SEM DF images (a) low magnification image showing more than 30 lymphoblastoid B-cells exposed to CdSe/ZnS quantum dots, (b) image of a cell that is highlighted in the previous image by a white box, (c) higher magnification image of quantum dots present in and surrounding the cell in the previous image.

In addition to being able to easily increase the magnification from imaging individual cells to collections of quantum dots, it is also possible to image the surface using the secondary electrons. A change in conditions is required (in our instrument this is primarily a change in voltage from 20 kV to

3 kV and a decrease in working distance) however it is relatively simple to image the same area of the sample (figure 2). The corresponding secondary electron image in general has very little contrast, and although general features of the cell can be identified, the collections of quantum dots cannot be distinguished from the surrounding cell. What is obvious in the secondary electron image is the artefact that is present in the centre of the cell (figure 2 (b)), as it is on the surface of the section. The lack of contrast in the higher magnification secondary electron image (figure 2 (d)) corresponding to the quantum dots located within the cell in the transmitted electron image (figure 2 (c)) confirms that they are within the thin section and that any interpreted nanomaterial–cell uptake is therefore genuine.

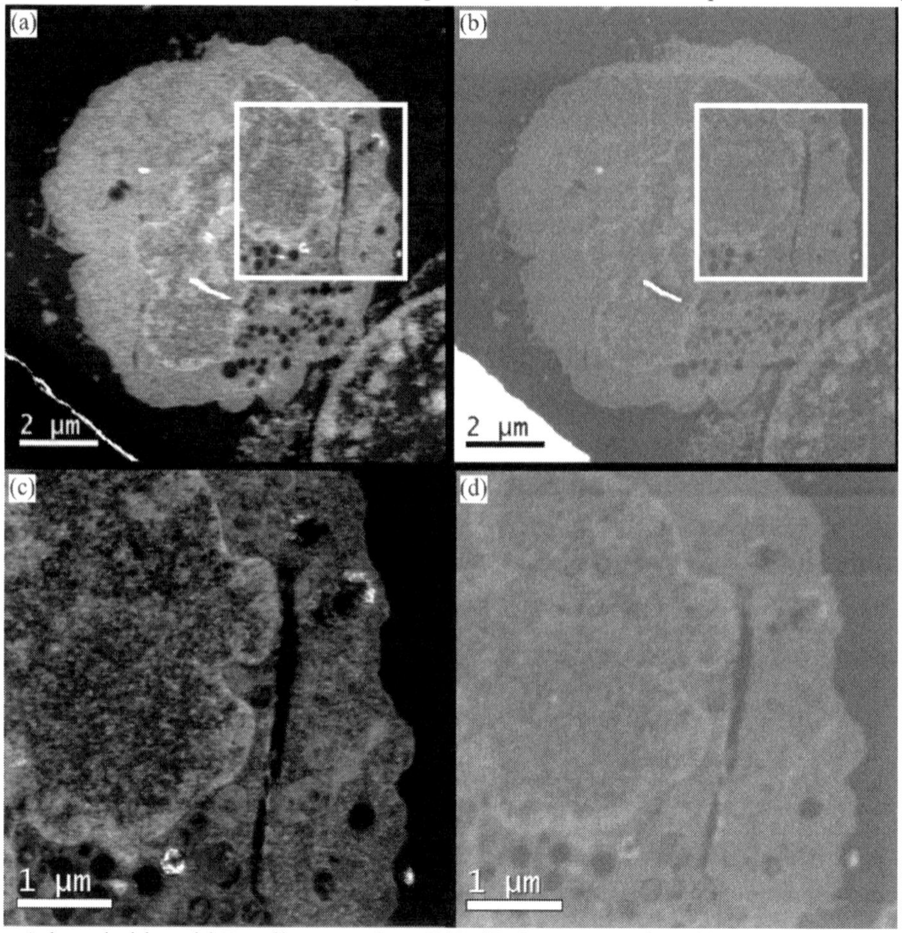

Figure 2. A lymphoblastoid B-cell exposed to CdSe/ZnS quantum dots (a) STEM mode in the SEM image of the whole cell, (b) corresponding SEM image, (c) higher magnification STEM mode in the SEM image showing quantum dots within and surrounding the cell from the area indicated in the previous images, (d) corresponding SEM image.

We have also examined the uptake of SWCNTs by human bronchial epithelial cells by TEM and STEM mode in the SEM [5]. The SWCNTs, which tend to form bundles, are dispersed through the resin section and can be seen in the DF STEM mode in the SEM image (figure 3 (a)). However in this case, extra information can be gathered from the corresponding bright field (BF) STEM mode in the SEM image; along with the SWCNT bundles the knife marks from the sample preparation can be seen especially in the BF STEM image. Of most interest however is the corresponding secondary electron image (figure 3 (c)) in which the bundles of SWCNTs are clearly seen to have been pulled out and dragged along the surface of the section. In some cases the bundles of SWCNTs may be dragged across the surface of the section near a cell (figure 3 (d) (e)). This type of analysis allows identification of artefacts and prevention of discussion of false nanomaterial uptake.

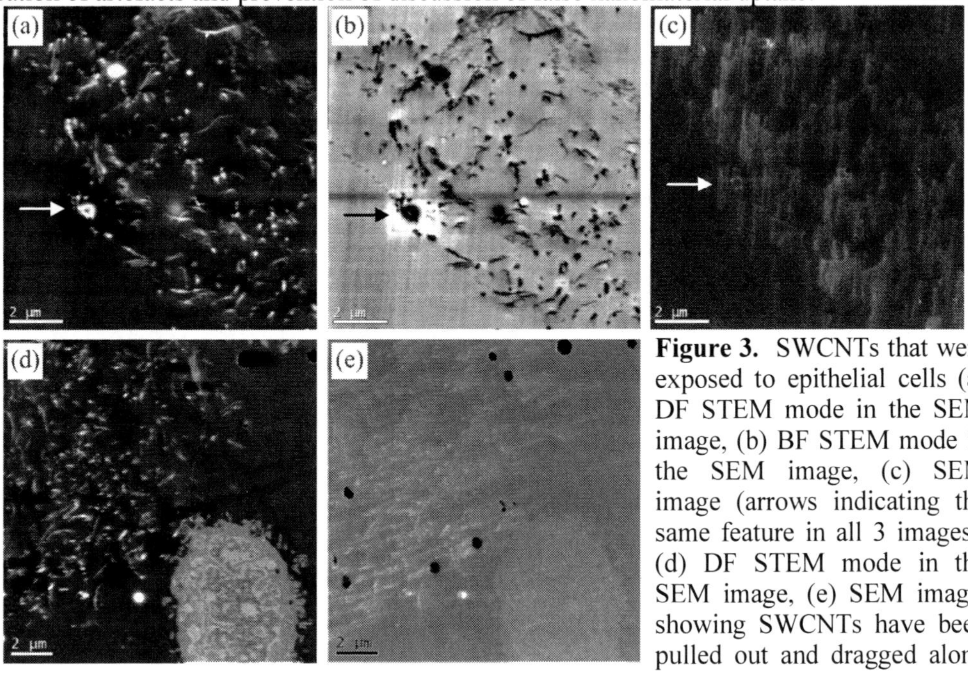

Figure 3. SWCNTs that were exposed to epithelial cells (a) DF STEM mode in the SEM image, (b) BF STEM mode in the SEM image, (c) SEM image (arrows indicating the same feature in all 3 images), (d) DF STEM mode in the SEM image, (e) SEM image, showing SWCNTs have been pulled out and dragged along the section during preparation.

4. Conclusion

The combined use of STEM mode in the SEM with secondary electron imaging allows for the examination of thin sections and identification of genuine nanomaterial-cellular interactions whilst avoiding the misinterpretation of any sample preparation derived artefacts. The location of CdSe/ZnS quantum dots was easily identified in dark field STEM imaging, and surface particulate contamination was noted by examining the corresponding in-lens SEM image. STEM mode in the SEM in conjunction with in-lens SEM imaging identified locations where carbon nanotubes had been pulled out and dragged across the surface of a thin section. This technique can be used ensure that cellular uptake of nanomaterials is correctly identified and is ideally suited to the rapidly screening of specimens due to the large field of view that can be imaged.

Acknowledgments

This work has been supported by the EPSRC (EP/E059678/1). The authors wish to thank Dr Shareen Doak and Dr Bella Manshian (Swansea University) for samples and Dr Alice Warley, Dr Anthony Brain and Mr Ken Brady (King's College London) for expert sample preparation.

References

[1] Maynard A D, Warheit D B and Philbert M A 2011 *Toxicol. Sci.* **120(S1)** S109-29
[2] McIntosh J R 2007 *Cellular Electron Microscopy* (San Diego: Elsevier Inc)
[3] Midgely P A, Ward E P W, Hungria A B and Thomas J M 2007 *Chem. Soc. Rev.* **36** 1477-94
[4] Porter A E, Gass M, Muller K, Skepper J N, Midgely P A and Welland M 2007 *Nature Nanotechnol.* **2** 713-7
[5] Hondow N, Harrington J, Brydson R, Doak S H, Singh N, Manshian B and Brown A 2011 *Nanotoxicology* **5** 215-27
[6] Joy D C 1976 *Scanning Electron Microscopy* **9** 361-8

Scanning electron microscopy as a tool for the analysis of colony architecture produced by phenotypic switching of a human pathogenic yeast *Candida tropicalis*

M C Furlaneto[1], C G T J Andrade[2], P H A Aragão[2], E J G França[1], A T P Moralez[1] and L C S Ferreira[1]

[1]Department of Microbiology, Paraná State University at Londrina, Brazil
[2]Electronic Microscopy and Microanalysis Laboratory, Paraná State University at Londrina, Brazil

Abstract. *Candida tropicalis* has been identified as one of the most prevalent pathogenic yeast species of the *Candida*-non-*albicans* group. Phenotypic switching is a biological phenomenon related to the occurrence of spontaneous emergence of colonies with different morphologies that provides variability within colonizing populations in order to adapt to different environments. Currently, studies of the microstructure of switching variant colonies are not subject of extensive research. SEM analysis was used to verify the architecture of whole *Candida* colonies. The strain 49/07 exhibited a hemispherical shape character, while the strain 335/07 showed a volcano shape with mycelated-edge colony. The ring switch variant is characterized by a highly wrinkled centre and an irregular periphery. The rough phenotype exhibited a three-dimensional architecture and was characterized by the presence of deep central and peripheral depressions areas. The ultrastructural analysis also allowed the observation of the arrangement of individual cells within the colonies. The whole smooth colony consisted entirely of yeast cells. Differently, aerial filaments were found all around the colony periphery of the volcano shape colony. For this colony type the mycelated-edge consisted mainly of hyphae, although yeast cells are also seen. The ring and rough colonies phenotypes comprised mainly yeast cells with the presence of extracellular material connecting neighbouring cells. This study has shown that SEM can be used effectively to examine the microarchitecture of colonies morphotypes of the yeast *C. tropicalis* and further our understanding of switching event in this pathogen.

1. Introduction

Candida tropicalis is an opportunistic yeast pathogen that causes superficial and systemic mycoses [1-3]. In Latin America, *C. tropicalis* accounts for the majority of non-*albicans Candida* species associated with candidemia episodes [4].

Among several virulence factors that contribute to pathogenesis of *Candida* it has been suggested that phenotypic switching provides variability within colonizing populations in order to adapt to challenges at different environments, including various anatomical sites in the human body [5].

Phenotypic switching represents an epigenetic state that occurs in a small fraction of the population, is random and reversible. For fungi, phenotypic switching is defined as the spontaneous emergence of colonies with altered colony morphology at rates higher than the somatic mutations rates that enables the microorganism to undergo rapid microevolution [6]. For *Candida albicans*, a relationship between colony shape of switching variants and the constituent cells (blastospores, true and psedo-hyphae) has been demonstrated by scanning electron microscopy analysis [7]. According to these authors the relationship between switched variant colonies and microstructure may help elucidate the relationship between pathogenicity *in vivo* and colonial morphology *in vitro* [7].

In contrast to the species *C. albicans*, the study of switching in *C. tropicalis* has not been the subject of extensive research. Our work focuses on using scanning electron microscopy for the analysis of switched variant colonies of clinical isolates of *C. tropicalis*. An understanding of this virulence determinant would provide insight into *C. tropicalis* pathogenic mechanisms.

2. Experimental techniques

2.1 Fungal strains

C. tropicalis strains 49/07 and 335/07 included in this study were recovered from tracheal secretion and belong to the *Candida* culture collection of the Fungal Genetics Laboratory, The University of Londrina-Brazil. The switched variants ring and rough were obtained as previously described [8].

2.2. Scanning electron microscopy of intact colonies

To verify the architecture of *C. tropicalis* colonies morphotypes, whole yeast colonies were removed from YPD (1% yeast extract, 2% peptone, 2% dextrose) agar plates using a scalpel blade. Colonies were fixed for 18 h at 4°C in 3% glutaraldehyde (Electron Microscopy Sciences) in 0.1 M phosphate buffer, pH 7.2. They were then immersed in liquid nitrogen for 30 sec and freeze-dried for 90 min at $2x10^{-3}$ MPa (Juan LP3). Then, colonies were coated with gold (BALTEC SDC 050 Sputter Coater) and viewed in a FEI Quanta 200 Scanning Electron Microscope at 30kV.

3. Results and Discussion

3.1. Microarchitecture of whole Candida colony

Despite the pioneer study on colony variants from *C. tropicalis* clinical strains [9] little information is available concerning the microstructure of individual colonies. Thus, we employed SEM to verify the architecture of *C. tropicalis* morphotypes. The preparation of colonies by a freeze-drying technique allowed their architecture preservation (Fig. 1) with maintenance of the phenotypes observed at lower magnitude (data not shown). The strain 49/07 exhibited a hemispherical shape character (Fig. 1A), while the strain 335/07 showed a volcano shape with mycelated-edge colony (Fig. 1B). The ring switch variant is characterized by a highly wrinkled centre and an irregular periphery (Fig. 1C). The rough phenotype exhibited more complex architecture and was characterized by the presence of deep central and peripheral depressions areas (Fig. 1D). França et al. [8] were the first to describe the architecture of whole *Candida* colonies at ultrastructural level. Here, we extend these observations of both non-variant colonies and switch variants produced by phenotypic switching of *C. tropicalis*.

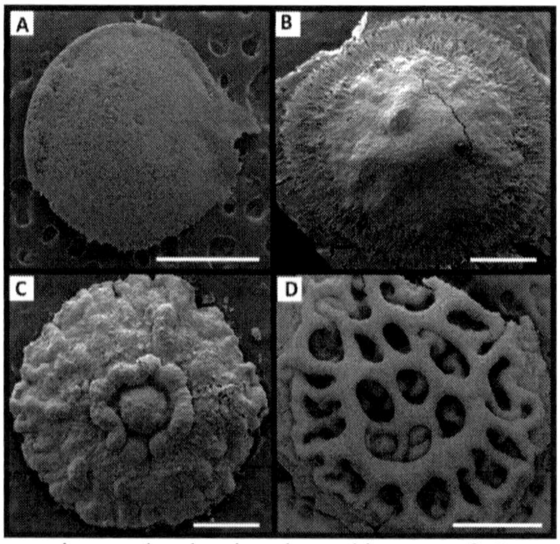

Figure 1 Scanning electron micrographs showing the architecture of *C. tropicalis* colonies morphotypes following 96 h incubation on YPD agar. (A) smooth, (B) volcano shape with myceliated-edge, (C) ring and (D) rough. Scalebar = 1mm.

3.2. Ultrastructural analysis of morphotypes

The ultrastructural analysis allowed the observation of the arrangement of individual cells within the colonies. After 4 days of colony development, the whole smooth colony consisted entirely of yeast cells (not shown). Aerial filaments were found all around the colony periphery of the volcano shape colony (Fig. 2A). Although the central area of this colony morphotype also comprises blastoconidia, the mycelated-edge consisted mainly of hyphae (Fig. 2B), although yeast cells are also seen.

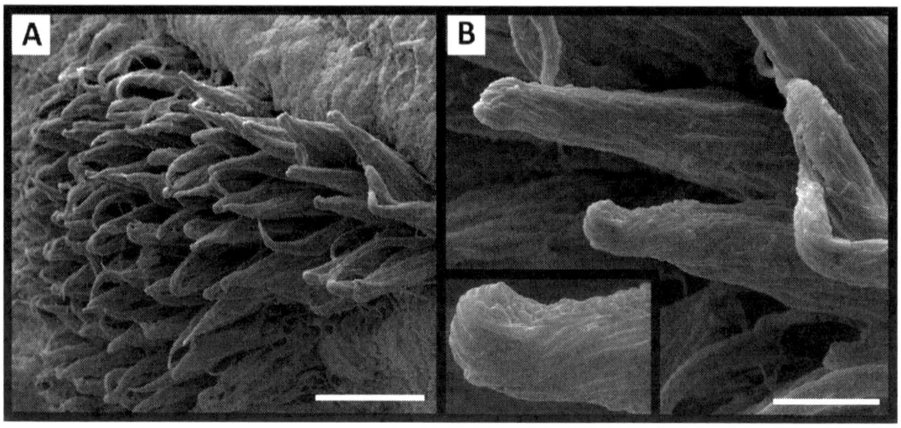

Figure 2 Electron micrographs of the *C. tropicalis* volcano shape with myceliated-edge morphotype following 96 h incubation on YPD agar. Scalebar = 200μm (A); 50μm (B). (B) Insert shows detail of myceliated edge that is composed of hyphae and blastoconidia at higher magnification (5000x)

The ring and rough colonies phenotypes also comprised mainly yeast cells as observed at the deep peripheral depressions areas (Fig. 3). Most interesting is the presence of extracellular material exclusively at these areas (Fig. 3 A, B), where many of the cells are almost hidden by this material. It was observed as fibrils, with enlarged structures, connecting neighbouring cells. In a recent study we report the presence of extracellular material, resembling a biofilm-like colony, throughout the development of *C. tropicalis* switch colonies, suggesting that its presence is correlated with the complex architecture of colonies in *C. tropicalis* [8].

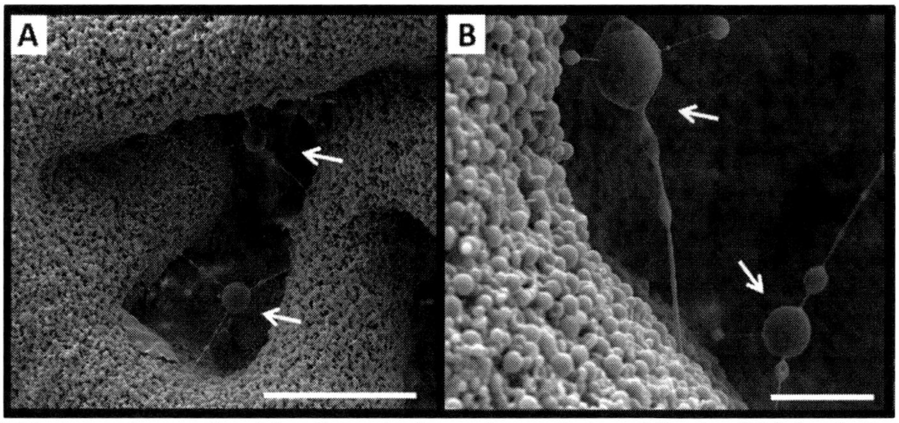

Figure 3 Electron micrographs of the *C. tropicalis* rough morphotype following 96 h incubation on YPD agar. Extracellular material (arrows) is seeing forming a biofilm-like colony.
Scalebar = 200μm (A); 50μm (B).

4. Conclusions

This study has shown that scanning electron microscopy can be used effectively to examine the microarchitecture of colonies morphotypes of the yeast pathogen *C. tropicalis*. Thus, SEM seems to be a useful tool to analyse the relationship between switched variant colonies and microstructure. The information derived from this study will further our understanding of the switching event in *C. tropicalis*.

5. Acknowledgments

The authors would like to acknowledge the financial support of the Conselho Nacional de Desenvolvimento Científico e Tecnológico (CNPq) – Brazil and PROPPG-UEL-Brazil.

6. References

[1] Marol S and Yücesoy M. 2008. *Mycoses* **51.** 40-49.
[2] Kliemann D A, Pasqualotto A C, Falavigna M, Giaretta T and Severo L C. 2008. *Revista do Instituto de Medicina Tropical São Paulo* **5.** 261-263.
[3] Kremery V and Barnes A. 2002. *Journal of Hospital Infection* **50.** 243-260.
[4] Colombo A L, Nucci M, Park B J, Nouer A S, Arthington-Skaggs B, Matta D A, Warnock D and Morgan J. 2006. *Journal of Clinical Microbiology* **44.** 2816-2823.
[5] Soll D R. 2009. *FEMS Yeast Research* **9**. 973-989.
[6] Soll D R. 1992. *Clinical Microbiology Review* **5**. 183-203.
[7] Radford D R, Challacombe S J and Walter J D. 1994. *Journal of Medical Microbiology* **40**. 416-423.
[8] França E J G, Andrade C G T J, Furlaneto-Maia L, Serpa R, Oliveira M T, Quesada R M B and Furlaneto M C. 2011. *Micron* **42**. 726-732.
[9] Soll D R, Staebell M, Langtimm C, Pfaller M, Hicks J and Gopala Rao T V. 1988. *Journal of Clinical Microbiology* **26.** 1448-1459.

Journal of Physics: Conference Series **371** (2012) 012023 doi:10.1088/1742-6596/371/1/012023

Scanning Ion Conductance Microscopy of Live Keratinocytes

V Hegde, A Mason, T Saliev, FJD Smith, WHI McLean & PA Campbell

Division of Molecular Medicine, College of Life Sciences, University of Dundee, Wellcome-MSI Complex, Dow Street, Dundee DD1 5EH. Scotland.

ABSTRACT: Scanning ion conductance microscopy (SICM) is perhaps the least well known technique from the scanning probe microscopy (SPM) family of instruments. As with its more familiar counterpart, atomic force microscopy (AFM), the technique provides high-resolution topographic imaging, with the caveat that target structures must be immersed in a conducting solution so that a controllable ion current may be utilised as the basis for feedback. In operation, this non-contact characteristic of SICM makes it ideal for the study of delicate structures, such as live cells. Moreover, the intrinsic architecture of the instrument, incorporating as it does, a scanned micropipette, lends itself to combination approaches with complementary techniques such as patch-clamp electrophysiology: SICM therefore boasts the capability for both structural *and* functional imaging. For the present observations, an ICnano S system (Ionscope Ltd., Melbourn, UK) operating in 'hopping mode' was used, with the objective of assessing the instrument's utility for imaging live keratinocytes under physiological buffers. In scans employing cultured HaCaT cells (spontaneously immortalised, human keratinocytes), we compared the qualitative differences of live cells imaged with SICM and AFM, and also with their respective counterparts after chemical fixation in 4% paraformaldehyde. Characteristic surface microvilli were particularly prominent in live cell imaging by SICM. Moreover, time lapse SICM imaging on live cells revealed that changes in the pattern of microvilli could be tracked over time. By comparison, AFM imaging on live cells, even at very low contact forces (<nN), could not routinely image microvilli: rather, an apparently convolved image of the underlying cytoskeleton was instead prevalent. We note that the present incarnation of the commercial instrument falls some way behind the market leading SPMs in terms of technical prowess and scanning speed, however, the intrinsic non-obtrusive nature of SICM imaging leads us to advocate its use for monitoring the most delicate living structures with attendant high spatial resolutions.

1. INTRODUCTION

Investigating the structural and functional character of living cell membranes is a highly challenging process. Scanning probe microscopy (SPM) represents one approach for imaging live cells with nanoscale resolution (Driscoll et al., 1990; Arkawa et al., 1992, Campbell *et al.* 2001). However, due to the necessary contact the probe makes with the fragile surface of the cell, this invariably leads to some deformation, sometimes damaging the cell, and almost certainly affecting the local reaction of the cell to the probe environment. The risk of damage in such circumstances is greatly reduced by the use of atomic force microscopy (AFM) utilizing appropriately low contact forces, however even then, understanding the image generation mechanism, and its interpretation, remains complicated (Dufrene, 2008). One emerging solution to this challenge is scanning ion conductance microscopy (SICM), which facilitates cell surface imaging under physiological conditions, without physical contact (Hansma et al., 1989, Korchev et al., 1997) but importantly retaining a capability for nanometer scale resolution under optimal conditions.

SICM uses a glass micropipette as the sensitive distil probe, and relies on a [constant] ion current for feedback controlled imaging of even insulating surfaces, submerged within an electrolyte bath. Briefly, as topography sweeps below the scanning pipette, the active ion flow through the aperture will be partially occluded, so that by direct comparison with a set-point ion current, electronic feedback can then control the displacement of the micropipette relative to the target surface and this adjustment is interpreted as target topography. Previously, this technique was restricted to imaging flat surfaces, as the probe had a propensity to collide with the specimen upon encountering vertical surfaces (Novak et al., 2009). Nowadays, a countermeasure has been developed that dispenses with the continuous feedback and raster scan methodology, and where instead the pipette undergoes approach to the surface in order to measure its relative height difference, whereupon it is quickly withdrawn to a position well above the sample, scanned laterally, and the process repeated. This protocol, which is now known as 'hopping mode' in SICM parlance (Novak et al., 2009), has proved to be a useful tool for the study of the structures on live cell membranes, and also for studying the pattern evolution of microvilli on cell membranes by time lapse imaging (Gorelik et al., 2003, Novak et al., 2009). The study presented here highlights some of the preliminary data that we have achieved on live HaCat cells by both SICM and AFM, where the focus for our attention was directed to the appearance of surface microvilli under both imaging regimes. For comparison, chemically fixed cells were also analysed by both methods, as described below.

2. METHOD & RESULTS

A keratinocyte cell line (HaCaT cells), as well as Xenopus laevis kidney epithelial cell line, A6 cells, were cultured separately in Dulbecco's Modified Eagle's Medium (DMEM), this solution having been supplemented with 10% (v/v) fetal bovine serum at 37 ^{0}C and 5% CO_2 in a humidified atmosphere. These cells were cultured in 35 mm dishes and used for scanning when they were 90 – 95% confluent. For live cell imaging, HaCaT cells were bathed with an iso-osmotic solution similar to the growth medium with pH of 7.4 and an osmolarity of 300 milliosmolar/kg containing (in mM): 137 NaCl, 5.4 KCl, 0.5 $MgCl_2$, 1.8 $CaCl_2$, 11.8 HEPES (4-(2-hydroxyethyl)-1-piperazineethanesulfonic acid) and 5.56 glucose before the imaging. To image the fixed cells, 90 – 95% confluent cells in 35 mm dishes were washed with Dulbecco's Phosphate Buffered Saline (DPBS) solution, followed by fixing with 4% paraformaldehyde in DPBS for 20mins before rinsing and storing in DPBS. The glass micropipette probes were pulled from borosilicate glass on a laser puller (P-2000, Sutter Instrument Co., Novato, CA, USA). Micropipettes had resistances in the range 100 - 200 MOhms when filled with phosphate buffered saline (PBS). Initial studies employed silicon gratings (normally used as calibration objects for atomic force microscopy) to gain familiarization with the technology by optimising scan parameters on a somewhat less challenging target object, but also to validate absolute topography measurements downstream.

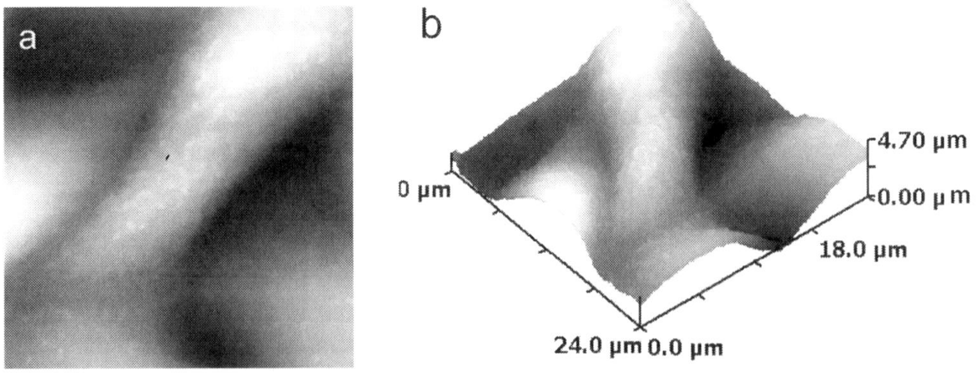

Figure 1: Initial SICM images in (a) birdseye, and (b) 3D perspective topography of A6 cells fixed using 4% paraformaldehyde in DPBS. Note the appearance of microvilli across the cell membrane.

In operation, the user-declared reference current is continually measured while the pipette is retracted, and upon extension towards the target, the current is then reduced by some predefined amount based on the user experience with the subject under study, but which might typically be 0.25–1%. The position of the *z*-dimension actuator during this process is then recorded and interpreted as the topographic height differences across the sample. Notably, even a 1% reduction of the current leads to a situation where the pipette is still at a distance of about one inner pipette radius from the surface. The software then forces the *z*-actuator to withdraw the pipette away from the surface and the sample is moved laterally to the next imaging point. By continuously updating the reference current while the pipette is away from the surface, this procedure automatically adjusts for any slow drifts in the pipette current.

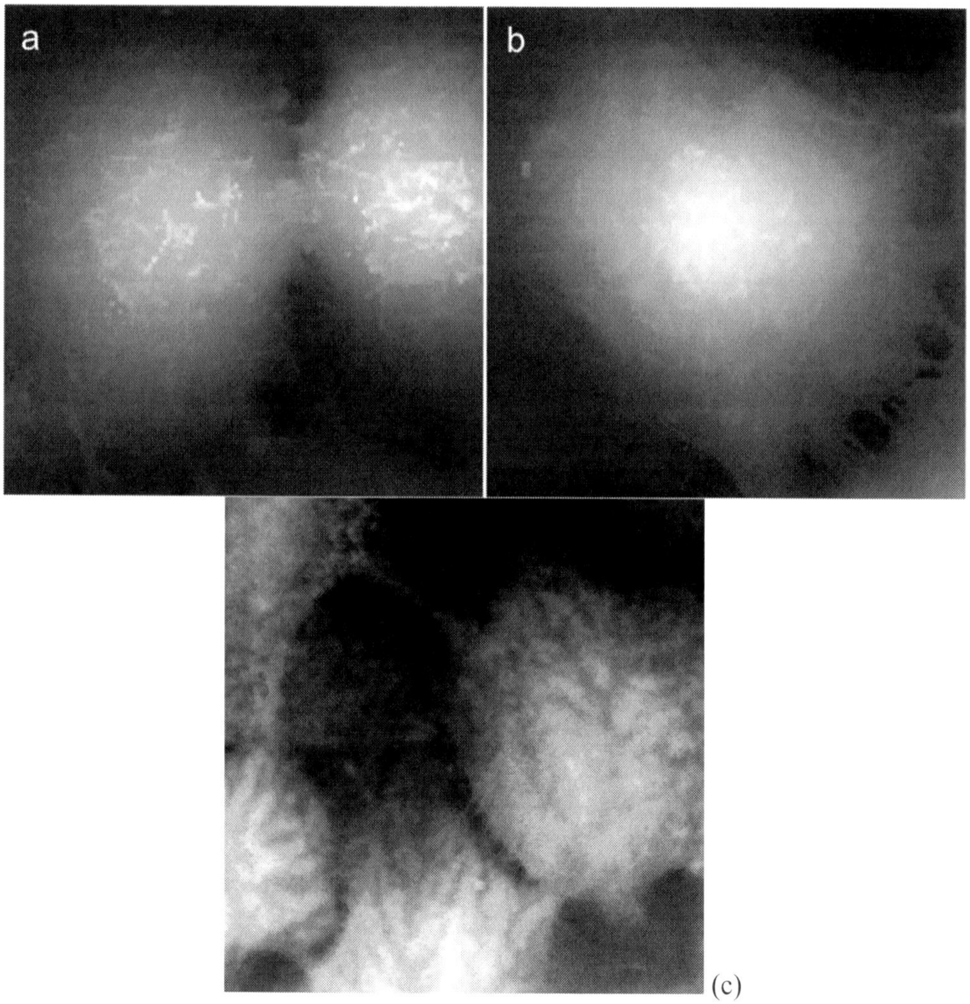

(c)

Figure 2: (a) SICM observations on live HaCaT cells, imaged under physiological buffer and showing characteristic microvilli on the surface. (b) SICM image of a paraformaldehyde-fixed HaCaT cell - the delicate microvilli structures are preserved somewhat, but evidently not with the clarity of (a) [Note that both images (a) and (b) are presented with the same z-scale.] (c) AFM of a live monolayer of HaCat cells imaged under physiological buffer solution. The intercell junctions are comparable to those observed using SICM however microvilli are not obvious at all. Image windows in (a) and (b) are 30μm x 30μm, whereas (c) has exhibits an image field of 50μm x 50μm.

Figure 2 above illustrates the strength of the SICM approach for live cell imaging: to the left in (a) two live HaCaT cells are imaged under physiological buffer whereas in (b) the same monolayer has been reimaged after chemical fixing according to the prescription detailed earlier. The extraordinary sensitivity of SICM to capture the native structure on live cells, in terms of, for example, the microvilli which stand proud of the membrane surface, is clearly evident. On fixed cells (b) and where the z projection is identical to that in 2(a), it is clear that the fixing process, whilst 'freezing' in the macroscopic cell topography, appears to alter [flatten] the microvilli from their natural state.

Most notably however, is the difficulty in obtaining accurate membrane surface structure on such biological targets when regular AFM is employed. Figure 2(c) shows a typical AFM image on live HaCat cells under the same physiological conditions used to capture the image in Fig 2(a) – notably, microvilli can not be distinguished at all, and the topography arising appears to represent a convolved image of the underlying cytoskeletal architecture, this despite the use of sub-nN set-point forces. On fixed cells however (data not shown, but for example highlighted in the paper by Ali et al (2010)) the enhanced rigidity of the chemically modified cells does allow for high resolution imaging of microvilli by AFM. Here, we note that spatial resolutions are typically higher on fixed cells, and that AFM with nominal tips, when compared with SICM, will lead to superior resolution for that circumstance. Moreover, with alternative imaging schemes, such as environmental SEM, it must be noted that the attractiveness of achieving a much wider areal coverage is somewhat undermined by the diminished z-resolution, so that detection of subtle sub-micron topography alteration, as we intend to measure, is unlikely. It would therefore seem that SICM has much to offer those seeking to quantitative measurements on such delicate, sub-micron sized target structures. Previously, the development of distinct topographical signatures on HaCat cells in response to biochemical stimulus has been observed, for example, PAR-2 was seen to upregulate phagocytosis in HaCat cells, resulting in a specific morphological change in podia (Sharlow et al. (2000)) that was characterized by fixing and reimaging in SEM. We anticipate that with SICM's enhanced facility to image delicate natural topography on live cells, then similar expectations might be achievable in response to [potentially therapeutic] physical stimuli, but with the significant advantages that surface alteration could be monitored in real time, with quantitative and high spatial resolution (in z as well as x-y), and on live target cells.

3. CONCLUSIONS

In summary, we presented SICM images of HaCaT keratinocytes in the live state. We note that intrinsic native structure is captured without interference, and without the need for chemical fixing, which itself alters the appearance of the true topography. Furthermore, we demonstrate that this sensitivity outperforms AFM for routine live cell imaging.

ACKNOWLEDGEMENTS

We are grateful to the MRC for a Milstein Award to IMcL, FS & PC which funded the team on this project. VH and TS are partially supported by Northern Research Partnership studentships. The instrument was provided by an EPSRC Basic Technology 5 Award (EP/D048958) and PC is presently working directly with Ionscope to develop the technology under a Royal Society Industry Fellowship.

REFERENCES

Ali M, Campbell P, Kondo T. *Drug Discovery Today* **15** (Issue 21-22), p892 (2010)
Arkawa H, Umemura K, Ikai A. *Nature* **358** 171 (1992)
Campbell PA et al. *Inst. Phys. Conf. Series* No 168, Section 11, 513 (2001)
Driscoll RJ, Youngquist MG, Baldeschwieler JD. *Nature* **346** 294 (1990)
Dufrene YF. *Nature Reviews Microbiology* **6** 674 (2008)
Gorelik J, Shevchuk AI, Frolenkov GI, Diakonov IA, Lab MJ, Kros CJ, Richardson GP, Vodyanoy I, Edwards CRW, Klenerman D, Korchev YE. *PNAS* **100** 5819 (2003)
Hansma PK, Drake B, Marti O, Gould SA, Prater CB. *Science* **243** 641 (1989)
Korchev YE, et al. *Biophysical Journal* **73** 653 (1997)
Novak P, Li C, , et al.. *Nature Methods* **6** 279 (2009)
Sharlow R, Paine CS et al. *J. Cell Science* **113**, 3093 (2000)

Quantitative HAADF-STEM tomography of unsupported intermetallic Ga-Pd catalysts

Rowan Leary[1], Zineb Saghi[1], Marc Armbrüster[2], Robert Schlögl[3], John Meurig Thomas[1] and Paul Midgley[1]

[1]Department of Materials Science and Metallurgy, University of Cambridge, Pembroke Street, Cambridge, CB2 3QZ, UK.
[2]Max-Planck-Institut für Chemische Physik fester Stoffe, Nöthnitzer Strasse 40, 01187 Dresden, Germany.
[3]Department of Inorganic Chemistry, Fritz Haber Institute of the Max Planck Society, Faradayweg 4-6, 14195 Berlin, Germany.

E-mail: rkl26@cam.ac.uk

Abstract. HAADF-STEM tomography has been used for characterisation of novel unsupported intermetallic Ga-Pd catalysts, with accompanying analysis by HRTEM and EDXS. Image processing techniques applied to the tomogram have facilitated segmentation and the subsequent extraction of size and shape parameters. The fidelity of the analysis has been critically examined, enabling identification of reconstruction artefacts and thereby more reliable determination of catalytically relevant properties. Further steps towards robust and accurate metrology by electron tomography are discussed.

1. Introduction

Ga-Pd intermetallic compounds have recently been shown to be highly stable and selective catalysts for the semihydrogenation of acetylene, where they outperform established systems [1]. This new class of catalyst has been realized by exploiting the covalent interactions in the well-ordered crystal structures of the intermetallic compounds, to stabilize isolated Pd sites on the catalyst surface. The stability of the Ga-Pd compounds under reaction conditions allows the direct relation of size, composition and morphology with catalytic properties and a knowledge-based development of new intermetallic catalysts. HAADF-STEM tomography has been undertaken for 3D characterization of unsupported Ga-Pd nanoparticle clusters (expected phase from XRD: $GaPd_2$) with particular interest in the size and shape of the particles, as well as their 3D arrangement with respect to one another. The 3D characterization has been complemented by HRTEM and EDXS for verification of phase and compositional homogeneity.

2. Experimental

HAADF-STEM tomography was performed on an FEI Tecnai F20 TEM operated at 200 kV, with a detector inner angle of ~45 mrad. The tilt-series was acquired in a semi-automated manner over a range of ±76° with a 2° tilt increment using FEI Xplore3D. The image series was aligned by cross-correlation and reconstruction performed using 40 iterations of the SIRT algorithm in FEI Inspect3D. Post-processing and visualisation were carried out using ImageJ and Avizo Fire. EDXS was also performed on the Tecnai F20 using an EDAX r-TEM ultra-thin window X-ray detector. For HRTEM a

JEOL 4000EX-II dedicated high-resolution microscope was operated at 400 kV. Phase identification was based on local Fourier transform (FT) analysis of the HRTEM images, with comparison to simulated diffraction patterns generated using CrystalMaker and SingleCrystal.

3. Results and discussion

3.1. HRTEM and EDXS

EDX point spectra shown in figure 1 suggest consistent composition, in keeping with the expected phase $GaPd_2$ [2]. The HRTEM data also supports the phase $GaPd_2$, and has been considered in some detail. The Ga-Pd phase diagram [2] shows a number of possible phases, which are known to have very similar lattice spacings and are therefore not easy to distinguish by FT analysis. Measured FTs from isolated particles have been carefully compared to simulated diffraction patterns for $GaPd_2$, using the recently refined structure parameters from Kovnir *et al.* [3] (orthorhombic, *Pnma* (no. 62), a = 5.4829(8) Å, b = 4.0560(4) Å, c = 7.7863(8) Å). An example supporting the suggested phase $GaPd_2$ is shown in figure 2a, and the crystallinity of smaller particles is confirmed by the diffractogram in figure 2b. While alternative phases could not be unequivocally ruled out in every case, simulated diffraction patterns for $GaPd_2$ were found to consistently fit all particles giving clear FTs with enough spots for reliable identification. The HRTEM images also show an amorphous layer around the particles, which is thought to be remnant organic from the synthesis process [4]. This may account for the oxygen peak in the EDX spectra.

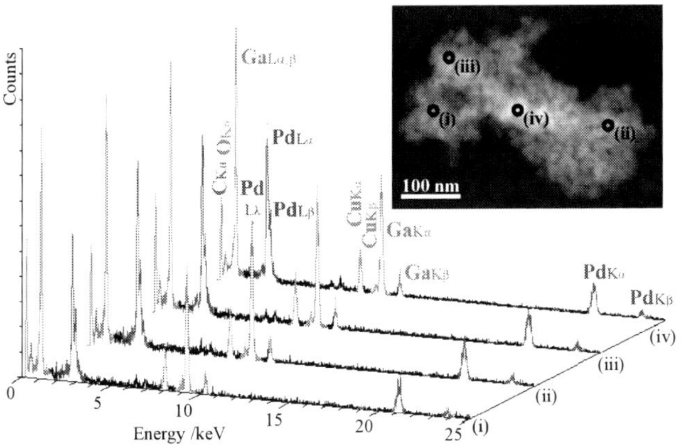

Figure 1. EDX point spectra. All spectra have been normalised to their GaL_α peak. (Colour online)

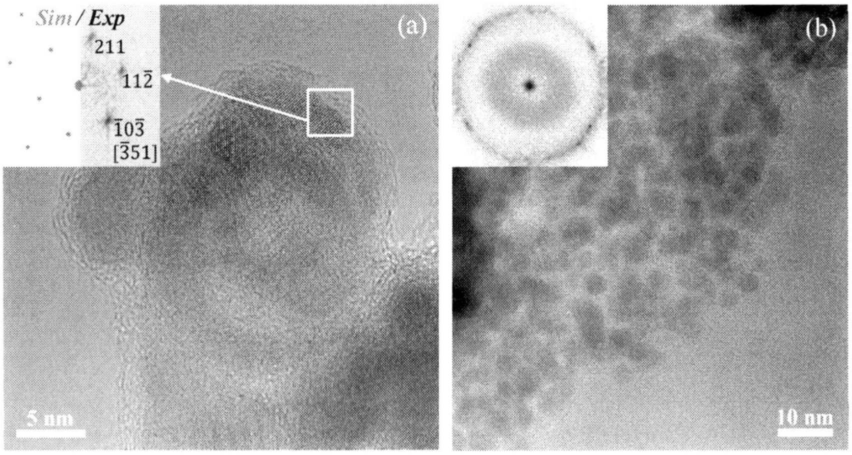

Figure 2. HRTEM images of (a) large and (b) smaller Ga-Pd particles. In (a) the matching simulated diffraction pattern for the phase $GaPd_2$ is overlaid on the left of the local FT taken from the area indicated. The inset in (b) is the diffractogram for the entire image.

3.2. HAADF-STEM tomography

Figure 3a shows a typical image from the tilt-series, where it can be seen that the cluster analysed contains significant numbers of both small (<10 nm) and large (~20 nm) particles. A central 2D slice from the tomogram (figure 3b) displays this varied size distribution with greater clarity, and reveals a diverse range of particle morphologies. Some particles show indications of faceting, but the majority appear to be rounded and irregular in shape.

The variation in size and shape and the dense packing of the particles, as well as the noise in the tomogram, prevents immediate thresholding and binarisation to enable quantification. These factors also mean that manual segmentation would be highly subjective and prohibitively time consuming. A semi-automated segmentation routine was therefore developed, which consisted of the following steps. A Difference of Gaussians (DoG) operation was first applied in the *x-y* planes of the tomogram to highlight the large particle boundaries as well as the smaller particles. The DoG filtered and original tomograms were then averaged to yield both enhanced boundaries and uniform intensity in the interiors of the particles, facilitating thresholding where the individual binarised particles could be more readily distinguished. A 3D watershed transform was then applied to separate particles that were incorrectly joined after thresholding, acting on the contrast inverted Euclidean distance map (EDM) of the binary tomogram. The severity of the watershed separation was tuned by a controlled degree of merging of the intensity minima of the contrast inverted EDM.

Figure 3c shows voxel projections of the segmented tomogram, where individual particles have been assigned a colour that differs from those of their nearest neighbours, confirming that the 3D structure of many of the larger and smaller particles has been successfully captured by the segmentation routine. Visualization in this way was used to judge the appropriate settings of the watershed transform, identifying over- and under-segmented particles.

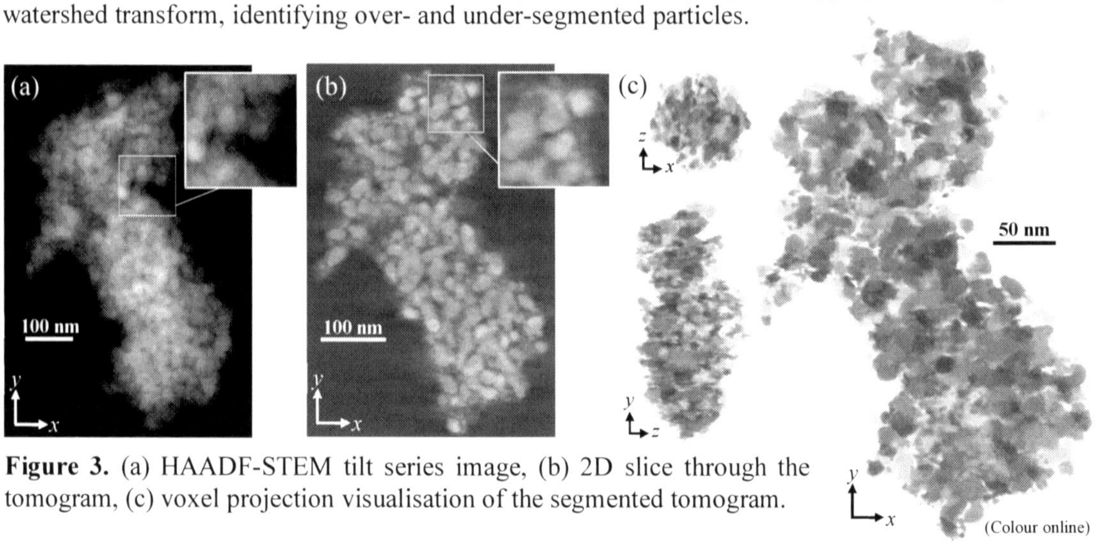

Figure 3. (a) HAADF-STEM tilt series image, (b) 2D slice through the tomogram, (c) voxel projection visualisation of the segmented tomogram.

A variety of catalytically relevant properties can be readily determined from the segmented data. Two examples are briefly presented here. The particle sizes can be gauged by considering their equivalent diameter, *d*, defined as the diameter of a sphere having the same volume as the particle. A useful indicator of the particle shape is their 3D Feret ratio (FR), defined as the maximum divided by the minimum Feret diameter. The statistical distributions of these measures are shown in figure 4.

Further insight into the size, shape and spatial distribution of the particles, and assessment of the fidelity of the segmentation and quantitative analyses, is aided by filtering of the tomogram based on the obtained measures. In figure 4a the tomogram has been split according to particle sizes, showing that particles in each size range are relatively uniformly distributed throughout the cluster. From careful inspection of each size-filtered tomogram, the particles predominantly making up each of the classes indicated in figures 4a(i)-(v) can be described as: (i) fine particles, (ii) small particles, (iii) medium particles and agglomerates of small particles, (iv) large particles and medium sized agglomerates, and (v) large agglomerates. The presence of agglomerated particles in the segmented tomogram reflects true agglomeration visible in the unprocessed data, which may be contributed to by the (suspected) remnant organic layer (cf. figure 2). However, some degree of false agglomeration may have been introduced into the tomogram by elongation effects and blurring of particle boundaries caused by the limited angular sampling of Fourier space in the original tilt-series (the so-called 'missing wedge' of unsampled information), as well as under-segmentation during the binarisation or watershed transform.

The distribution of FRs (figure 4b) suggests that the majority of particles and agglomerates are mildly anisotropic in their overall shape. Filtered tomograms showing particles with FRs lying in different ranges are shown in figures 4b(i)-(iii). This analysis will be affected to some degree by the false elongation (by a factor of ~1.2) in the missing wedge (z) direction. In particular, it is visible from figure 4b that the measured distribution is extended to the right of the mean, and the filtered tomogram for FR > 3 (figure 4b(iii)) shows that the highly anisotropic particles are mostly of the small to fine size range. Viewing this filtered tomogram in the x-z plane (not shown here) reveals that many of these particles are actually agglomerates of small/fine particles that are likely to have been falsely merged by missing wedge artefacts, as their longest dimension is almost exclusively in the z-direction.

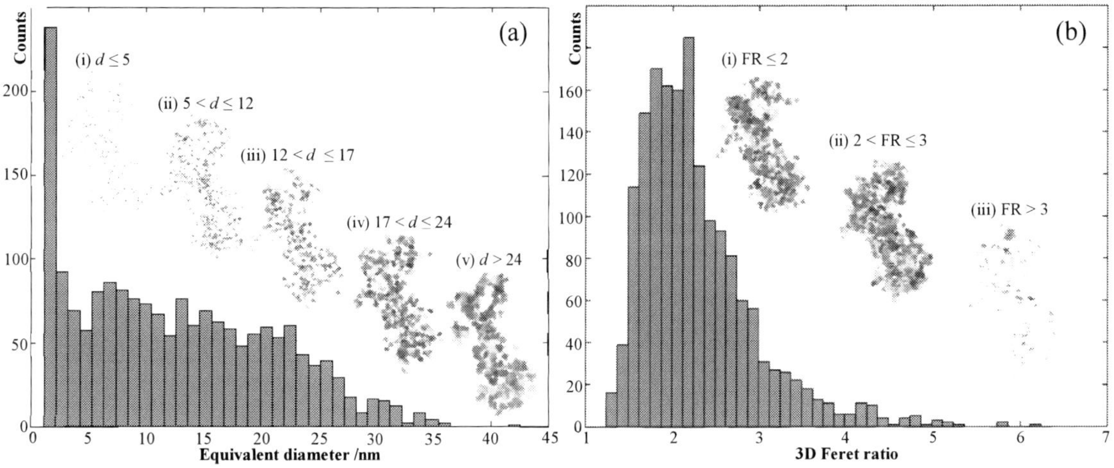

Figure 4. Statistical distributions and size/shape filtered tomograms for (a) equivalent diameter and (b) FR of the particles, based on the segmented data. (Colour online)

Overall, it is clear that the quality of the analysis is limited primarily by artefacts related to the missing wedge. Methods that overcome or negate the effects of limited angular sampling may provide the only routes to truly quantitative electron tomography. Given that the tilt range often cannot be extended for reasons of sample or hardware constraints, advanced reconstruction techniques take on particular importance. In this case, since the material of interest is known to be of single phase, improved reconstruction may be achievable using discrete algorithms [5]. Alternatively, we have recently demonstrated a new approach to reconstruction in electron tomography, known as 'compressed sensing,' which uses the prior knowledge that the object can be represented sparsely in some transform domain [6]. For samples of homogeneous composition we used the total variation transform and enforcement of sparsity in the gradient domain to produce high fidelity reconstructions, even from substantially reduced numbers of projections. Future work will involve the development of these methods, which may be applicable to samples of the type investigated here.

Acknowledgements
This work was funded by the EPSRC and the European Union under the Framework 6 program for an Integrated Infrastructure Initiative, Reference 026019 ESTEEM.

References

[1] Armbrüster M, *et al.* 2010 *J. Am. Chem. Soc.* **132** 14745-47
[2] Okamoto H 2008 *J. Phase Equilib. Diff.* **29** 466-7
[3] Kovnir K, *et al.* 2008 *Z. Kristallogr.* **223** 7-8
[4] Armbrüster M, *et al.* 2011 *J. Am. Chem. Soc.* **133**, 9112-18
[5] Batenburg K J, *et al.* 2009 *Ultramicroscopy* **109** 730-40
[6] Saghi Z, *et al.* 2011 *Submitted.*

TEM studies of stress relaxation in catalytic Au-Pd core-shell nanoparticles

H. Kumarakuru *[1], D. Cherns [1], M. G. Montes de Oca [2] and D.J.Fermin [2].

[1]H.H.Wills Physics Laboratory, School of Physics, Tyndall Avenue, University of Bristol, Bristol. BS8 1TL. United Kingdom.

[2]School of Chemistry, Cantocks Close, University of Bristol, Bristol. BS8 1TS. United Kingdom.

*E-mail: phxkh@bristol.ac.uk

Abstract. Transmission electron microscopy and diffraction are used to measure the lattice strain in catalytic Au-Pd core-shell nanoparticles. In order to simplify analysis, single nanoparticles were oriented to achieve strong two-beam diffracting conditions from the nanoparticle centre, such that diffraction was sensitive to the strain parallel to the Au-Pd interface. Corresponding selected area diffraction patterns were used to measure the in-plane strain as a function of the Pd thickness, which is shown to follow the equilibrium (Matthews) theory. These results suggest that, contrary to expectation, strain is not an important factor accounting for the difference in catalytic activity between nanoparticles and thin films. The significance of these results is discussed.

1. Introduction.

Bimetallic nanoparticles (NPs) have attracted great interest among the scientific and technological community since they lead to many interesting size-dependent electrical, chemical, and optical properties. Specifically they often exhibit enhanced catalytic properties which differ from that of their monometallic counterparts, and from thin films. This paper examines Au-Pd core-shell nanoparticles (CS-NPs). These have been widely studied by several research groups [1-4] in terms of fabrication, stabilization and catalytic behaviour, but the factors controlling catalytic behaviour, in particular the relative contribution of lattice strain and ligand (electronic) effects, remain unclear.

Here, we examine the strain and morphology in Au-Pd CS-NPs as a function of the Pd thickness. Au-Pd CS-NPs were synthesized by colloidal methods for generating Pd shell thickness in a range of 1-10nm over an Au core. Strain in CS-NPs can easily be studied by selected area electron diffraction patterns (SADPs) from particle ensembles, but this method is difficult to interpret as the patterns contain contributions from an in-plane expansion of the Pd (which has the smaller lattice parameter and should be under tensile stress) and an out-of-plane Poisson contraction, This leads to a complex displacement field for the diffracting planes, which are oriented near edge-on to the incoming electron beam (Figure 1). To avoid this problem, we have examined two-beam SADPs from individual CS-NPs, which are oriented by observing the bright field image contrast, to maximise diffraction from the particle centres, and which should, therefore, be sensitive to the strain parallel to the Au-Pd interface (in-plane strain). It is shown that the strain as a function of Pd thickness follows equilibrium theory.

The implications of these results for understanding the catalytic behaviour of CS-NPs are briefly discussed.

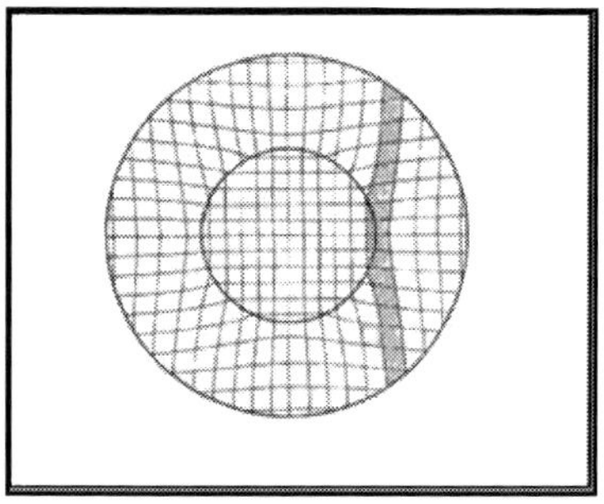

Figure 1: Sketch of the planar displacements for a spherical particle with a coherent interface and a positive misfit for the shell relative to the core [ref 8, figure 3, n.b. Au-Pd CS-NPs have the opposite mismatch].

2. Experimental methods.

The synthesis of Au-Pd core-shell (CS) nanostructures was performed using the following high purity reagents: gold (III) chloride trihydrate ($HAuCl_4 \cdot 3H_2O$, 99.9%), palladium chloride ($PdCl_2$, 99.9999%), sodium citrate dehydrate ($C_2H_5Na_3O_7 \cdot 2H_2O$, 99.5%) and L-ascorbic acid ($C_6H_8O_6$, 99.7%). Milli–Q water (18.2 MΩcm) was used to prepare the solutions. Gold nanoparticles of 19 nm diameter were prepared by the colloidal method [5, 6], reducing $HAuCl_4$ in the presence of sodium citrate. The one step method consisted of adding 10 ml of 1% trisodium citrate to 190 cm^3 of 2.5 x 10^{-4} mol dm^{-3} $HAuCl_4 \cdot 3H_2O$ under reflux and strong stirring. The reaction was allowed to reflux for 1 hour. The preparation method of Au-Pd core-shell nanoparticles is similar to the method previously reported [7], base on a two step process initiated by the nucleation of Au particles. The second step involves the epitaxial deposition of ultrathin Pd onto as-grown Au seeds via reduction of H_2PdCl_4 in the presence of ascorbic acid. The Pd precursor (H_2PdCl_4) was prepared by mixing $PdCl_2$ and HCl. Pd thickness can be effectively tuned by varying the amount of 0.1 mol dm^{-3} H_2PdCl_4, added to 50 ml solutions of Au nanoparticles under vigorous stirring. Then, the solution was cooled to 0°C in an ice bath; this step is followed by adding an excess amount of L-ascorbic acid (0.1 mol dm^{-3}, 6 ml) drop wise, during 1 hour, in order to avoid the formation of isolated Pd clusters. The mixture was stirred for another 30 minutes to ensure a complete reduction of H_2PdCl_4. The volume of Pd precursor (H_2PdCl_4) was adjusted in order to modify Pd thickness from 1 to 10 nm. The solution colour turned from red to brown, indicating that Pd grows on Au cores with different Pd thicknesses.

The synthesized CS-NPs were examined by Philips EM430 TEM operating at 200kV. Samples for TEM images were produced by placing 10 μl drops on a carbon coated copper grid of 3 mm diameter. Excess solution was absorbed by filter paper and the sample was dried in air at room temperature. Images and selected area electron diffraction patterns (SADPs) were recorded using a Philips EM430 TEM operating at 200kV.

3. Results and discussion.

Figure 2a,b illustrate images of CS-NPs with an atomic percentage of 80% Pd. Bright field images taken in weakly diffracting conditions (not shown) displayed a discrete dark core (Au, which is the higher atomic number material) and a lighter shell. Images taken under strongly diffracting conditions, as in Figure 2, showed strong diffraction from both the core and shell regions indicating an epitaxial relationship between the Au and Pd. The bright field image (Figure 2a) shows moiré fringes in the region where the core and shell overlap, suggesting that the Pd has relaxed from the pseudomorphic state expected at very low Pd thicknesses. The dark field image (Figure 2a) shows a complex fringe pattern as expected from the displacement field illustrated in Figure 1. It is also apparent from Figure 2 that irregularities in the particle shape partially reflect those in the inner core, suggesting that the Pd coating is fairly uniform in thickness.

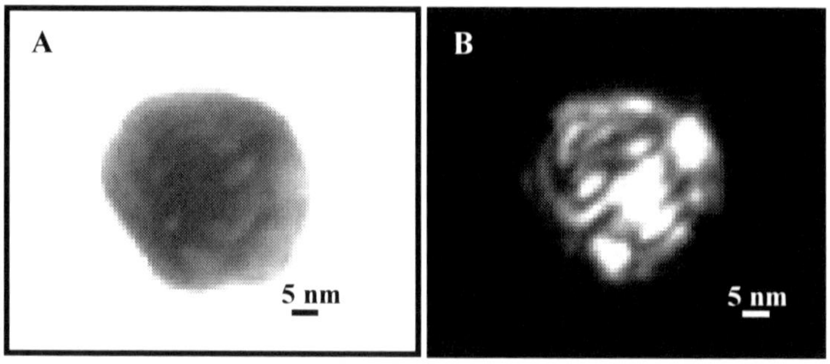

Figure 2: TEM images of the core-shell nanoparticles featuring a 19.3±1.2 nm Au core coated with Pd: (A) Bright field TEM image and (B) Dark field TEM image.

Figure 3 shows a SADP from an ensemble of nanoparticles in random orientation. The SADP shows polycrystalline rings containing individual spots. For a given ring, the spots show some scatter, of about ±3% in the diffraction spacing, i.e. the rings are diffuse. This includes contributions from regions of pure Pd and Au which have different diffraction spacing and some broadening is also due to contributions both from an in-plane expansion of the Pd (which has the smaller lattice parameter and should be under tensile stress) and an out-of-plane Poisson contraction.

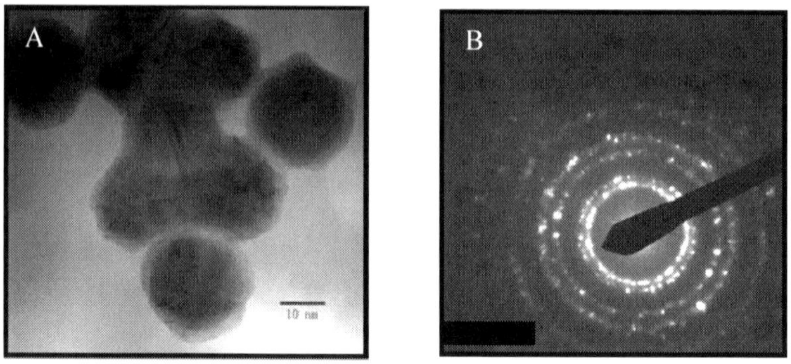

Figure 3: (A) TEM image of CS 5 nanoparticles in random orientations. (B) Its corresponding selected area diffraction pattern (SADP).

To avoid this problem, we have examined two-beam SADPs from individual CS-NPs, which are oriented to maximise diffraction from the particle centres. Under these conditions, we expect diffraction to be relatively sensitive to the in-plane strain, and to be less sensitive to the out-of-plane strain which affects displacements in the diffracting planes towards the edge of the particles; these planes will be partially bent away from the strong diffracting condition as seen in the schematic diagram in Figure 1. A series of two-beam SADPs for Pd atomic percentages, i.e. 20 %(CS1), 40% (CS3), 60% (CS5) and 80% (CS10), are shown in Figure 4. This confirms the epitaxial relationship, with the Pd and Au spots, indicated by lines (Pd is the outer spot) predominantly diverging along the line of the g-vector. As the Pd content increases, the Pd diffraction spots become relatively brighter as expected. The divergence between Pd and Au spots increases with Pd content, indicating a progressive relaxation of strain in the Pd with layer thickness. Tails on the Pd spot in directions obliquely inclined to the g-vector indicate strain relaxation on differently inclined facets. The Pd lattice strain (ε) as a function of the shell thickness is displayed in figure 5. To a first approximation, the experimental results can be rationalized within the framework of the so-called Matthews Theory for the misfit-thickness relationship in thin films, shown by the dotted line [9]. Pd-Au thin films are also known to follow the Matthews theory closely [10].

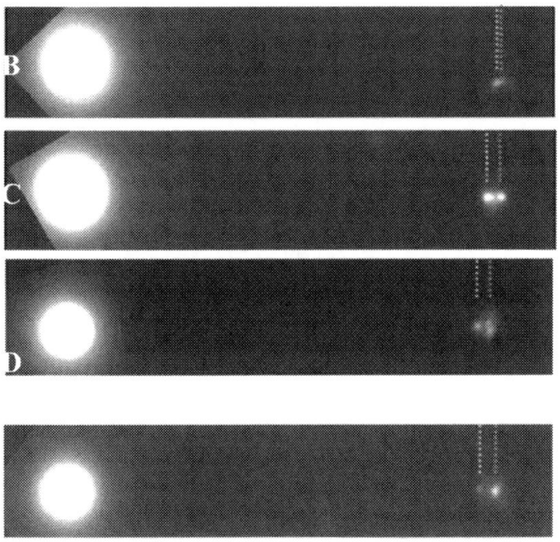

Figure 4: Two beam selected area diffraction pattern from individual CS-NPs (A) CS1, (B) CS3, (C) CS5 and (D) CS10 as shown in figure 2 (A-D) respectively. [from ref 7, figure 1, with permission of ACS]

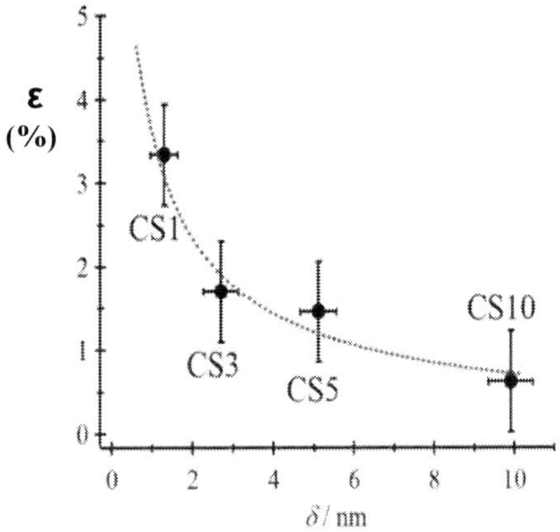

Figure 5: Surface strain (ε) as a function of the average Pd thickness (δ as calculated from SADP. The continuous line describes the strain relaxation for epitaxial layers of Pd on extended Au (111) surfaces. [from ref 7, figure 2, with permission of ACS]

4. Conclusions

The results suggest that the stress relaxation in CS-NPs follows equilibrium theory, as in the case of thin films of Pd on Au, implying that any differences in the catalytic activity between CS-NPs and thin films cannot be attributed to differences in strain.

References

[1] Yudanov I V and Neyman K M 2010 *Phys.Chem.Chem.Phys.* **12** 5094
[2] Hamm G, Becker C and Henry C R 2006 *Nanotechnology.* **17** 1943
[3] Ferrer D, Torres-Castro A, Gao X, Sepúlveda-Guzmán S, Ortiz-Méndez U and José-Yacamán M 2007 *Nano Lett.* **7** 1701
[4] Liu H B, Pal U, Medina A, Maldonado C and Ascencio J A 2005 *Phys. Rev.* B **71** 075403
[5] Turkevich J, Stevenson P C and Hillier J 1951 *Discuss. Faraday Soc.* **11** 55
[6] Frens G 1973 *Nat. Phys. Sci.* **241** 20
[7] Montes de Oca M G, Kumarakuru H, Cherns D and Fermin D J 2011 *J. Phys. Chem.* C. **115** 10489
[8] Vincent R, Preston A R and King M A 1988 *Ultramicroscopy* .**24** 409
[9] Matthews J W, Blakeslee J 1974 *J. Cryst. Growth.* **27** 118
[10] Cherns D, Stowell M J 1975 *Thin Solid Films.* **29** 127

Double-aberration corrected TEM/STEM of solid acid nanocatalysts in the development of pharmaceutical NSAIDS

K. Yoshida[1,2], N. Shiju[3], R. Brown[3], I. Wright[2,4], E.D. Boyes[2,4,5] and P.L. Gai[1,2,4,6]
Departments of [1]Chemistry, [4]Physics and [5]Electronics, [2]The York JEOL Nanocentre, University of York, Heslington, York, YO10 5BR, U.K. ; [3]Department of Chemistry, University of Huddersfield, Huddersfield, HD1 3DH, U.K.

[6]Author to whom any correspondence should be addressed.

*E-mail: pratibha.gai@york.ac.uk

Abstract. We report nanostructural and physico-chemical studies in the development of an efficient low temperature heterogeneous catalytic process for nonsteroidal anti-inflammatory drugs (NSAIDS) such as N-acetyl-p-aminophenol (paracetamol or acetaminophen) on tungstated zirconia nanocatalysts. Using a double-aberration corrected TEM/STEM, modified in-house for in-situ studies at the sub-Angstrom level, we directly observed in real-time, the dynamic precursor transformation to the active catalyst. We quantified the observations with catalytic activity studies for the NSAIDS. The studies have provided the direct evidence for single tungsten promoter atoms and surface WO_x species of ≤ 0.35 nm, with nanoclusters of WO_x (0.6 to 1nm), located at grain boundaries on the surface of the zirconia nanoparticles. The correlation between the nanostructure and catalytic activity indicates that the species create Brønsted acid sites highly active for the low temperature process. The results open up opportunities for developing green heterogeneous methods for pharmaceuticals.

1. Introduction

Catalysts with strong acidic properties are of interest in key catalytic processes including isomerisation of butanes and pentanes and in acylation [1,2]. The rearrangement of ketoximes to amides or lactams in the presence of acid catalysts is known as Beckmann rearrangement. N-acetyl-p-aminophenol (paracetamol or acetamenophen) is one of important non-steroidal anti-inflammatory pharmaceutical drugs (NSAIDS) commonly used by humans because of its analgesic properties. Conventionally it is prepared by acetylation of p-aminophenol with acetic anhydride via homogeneous catalysis. The preparation of N-acetyl-p-aminophenol involves reacting 4-hydroxyacetophenone to obtain the ketoxime (4-hydroxyacetophenone oxime) followed by the Beckmann rearrangement in the presence of an acid catalyst, including fuming sulphuric acid [1,2]. This process however produces unwanted wastes. The use of solid acid catalysts can offer significant advantages over homogeneous acid catalysts for the synthesis of amides/lactams of pharmaceutical importance as in the case of caprolactam synthesis [3]. In this study, we have investigated tungstated zirconia solid acid nanocatalysts to catalyse the Beckmann rearrangement of 4-hydroxyacetophenone oxime to N-acetyl-p-aminophenol at low temperatures. The schematic of the process is shown in Figure 1.

2. Experimental Procedures

We synthesised two sets of samples with different W loadings including 5, 10 and 15 wt% W. The first series of samples (referred to as WZA) were prepared using zirconium oxohydroxide solution and infusing with ammonium metatungstate, followed by drying and calcination. In the second series of samples (WZB) reported in the literature, zirconium oxohydroxide was pre-calcined and then used with ammonium metatungstate, followed by drying and calcinations procedures similar to those used for the WZA series of samples. The two sets of samples exhibited different zirconia grain sizes and locations for active species. The oximes were synthesized from the corresponding ketones by reaction with hydroxylamine hydrochloride in a mixture of ethanol and pyridine at 358 K, by the general method reported in the literature. We used NH_3 adsorption microcalorimetry to estimate the acid strengths of the WZA catalysts as a function W surface density.

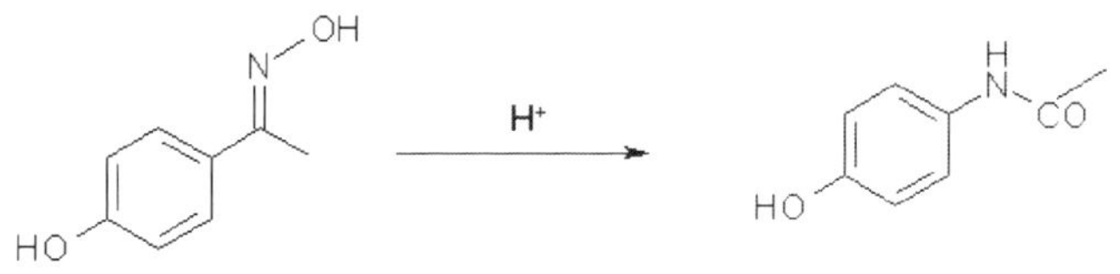

Figure1. Beckmann rearrangement of 4-hydroxyacetophenone oxime to N-acetyl-p-aminophenol.

The development of atomic resolution environmental TEM (atomic resolution-ETEM) [4,5] for the direct visualisation of dynamic gas-solid reactions under controlled reaction conditions of gas pressures (with flowing gas) and temperatures at the atomic level has demonstrated that it is possible to study dynamic catalytic reactions at the atomic scale. In the present study in-situ electron microscopy studies at the sub-Angstrom level were performed using our in-house development of a modified double-spherical aberration corrected JEOL 2200 FS TEM/STEM, operating at 200 kV [6,7]. We modified the instrument with a larger gap (HRP) objective lens polepiece and gas tolerant pumping system [6,7] in the Nanocentre at the University of York, to directly witness live precursor transformation. The in-situ AC TEM studies were complemented by in-situ HAADF STEM. The development enables sample heating using a standard heating holder under controlled conditions and tilting of the sample for alignment into the correct crystal zone axis orientation. The instrument is fitted with an EDX detector for composition analysis at the sub-nm scale. A combination of AC HAADF-STEM and AC-TEM was employed to further analyse the species.

3. Results and Discussion

In WZA sample with 15 wt% tungsten loading (15WZA), W single atoms and surface WO_x species with sizes of $\leq$0.35 nm to slightly larger species of ~0.375 nm were primarily observed using negative defocus imaging, co-existing with nanoclusters and WO_3 nanoparticles (with sizes of up to 0.6 nm to 1 nm), on the surface of nano-zirconia. The single atoms and WO_x species coexisting with nanoclusters are shown in Figure 2 and Figure 3, respectively. The structure and the composition of the nanoparticles were confirmed by measurements of lattice spacings and by sub-nm EDX. Average size of such ZrO_2 crystallites (nanocrystals) in WZA samples was estimated to be 2.6 nm (with sample variance σ^2 =0.334 nm^2). Catalytic activity measurements were performed to correlate the nanostructure of the catalysts containing these active sites, with selectivity and activity to N-acetyl-p-aminophenol at low temperatures [8]. The measurements show very high selectivity and activity of more than 94% and 97%, respectively, for the catalysts at 403K as shown in Figure 4.

Our comparative experiments have elucidated that WZB samples prepared using procedures reported in the literature have larger grains sizes of $\geq$ 30 nm-50 nm. WZB samples and post preparation procedures that deliberately increase the surface density of active species by coimpregnation of ZrO_x and WO_x on zirconia catalysts do not anchor active sites at grain boundaries of zirconia.

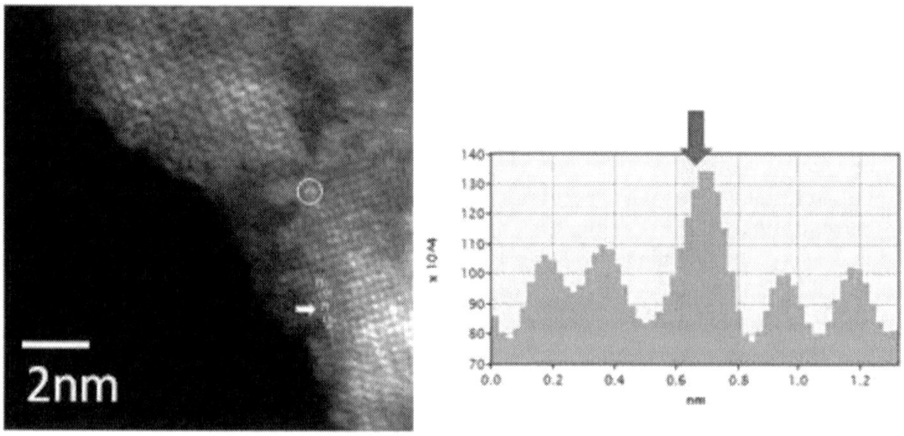

Figure 2. AC-HAADF STEM image indicating the presence of W single atoms from the rectangular area marked by an arrow on the figure. Intensity profile taken from the area is shown on the right. Arrowed peak corresponds to single atom. (Surface WO_x species are indicated by a white circle).

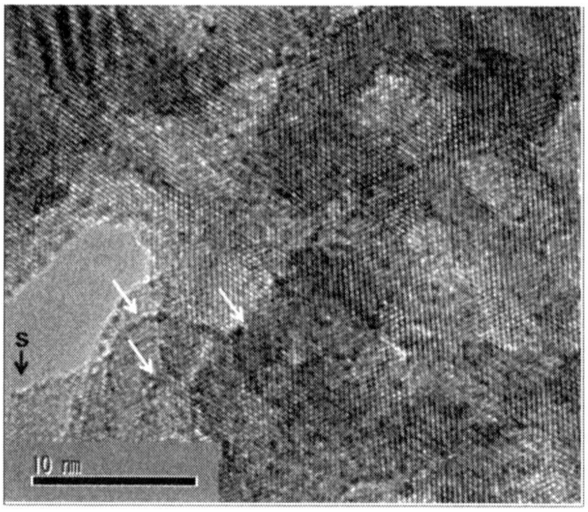

Figure 3. Negative defocus imaging revealing uniformly distributed surface tungsten oxide species (<0.3 nm e.g. at S) and nanocusters anchored at grain boundaries of nano-zirconia (indicated by white arrows).

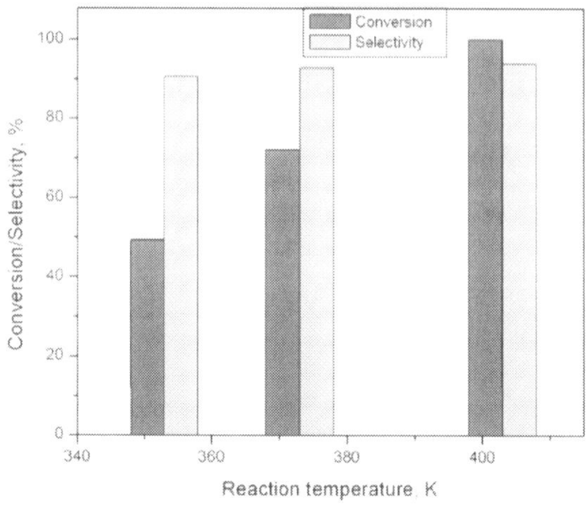

Figure 4. Conversion of 4-hydroxy acetophenone oxime and selectivity of paracetamol as a function of low reaction temperatures. Solvent = Benzonitrile; time= 12 h ; Oxime = 0.9 mmol; Catalyst = 20 mg. Excellent selectivity and activity were observed at 403K.

Figure 5(a) and (b) illustrate the grain size difference between the WZA and WZB preparations, respectively. The direct observation of 15WZA nanocatalysts clearly illustrates the segregation of surface species and clusters to grain boundaries on the nano-zirconia surface as illustrated in Figure 5(a). When 15WZB catalyst was examined in a similar way, a number of differences were apparent in the size and morphology of the catalyst particles compared to 15WZA. The most striking differences were the lack of tungsten oxide species at grain boundaries of zirconia particles and larger grain sizes of zirconia particles in15WZB sample (Fig.5b). Our observations have shown that the lack of anchoring of active species at grain boundaries of catalysts results in the migration and sintering of the species and the subsequent deactivation of the catalysts. The observations are consistent with much lower selectivity and activity observed for the WZB samples.

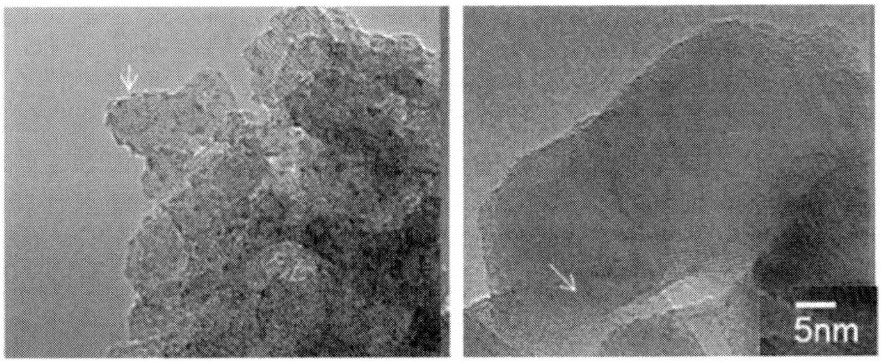

Figure 5. AC-TEM images of: (a) WZA catalyst and (b) WZB catalysts illustrating the difference in zirconia grain sizes. In our preparation (a), zirconia grains are only a few nms compared to about 50 nm in the conventional catalysts. In (b), W-oxide clusters (e.g. arrowed) sintered faster in the conventional catalyst, growing to about 5 nms in about 2.5 hours and many nms after several hours.

4. Conclusions

The results demonstrate the nature and the role of active sites and high activity of tungstated zirconia nanocatalysts in the NSAID reaction. They further provide an environmentally clean alternative to the current homogeneous process.

Acknowledgements

We thank the University of York, JEOL and Yorkshire Forward / European Regional Development fund (ERDF) for sponsoring the York JEOL Nanocentre; Japan Society for Promotion of Science (JSPS) and Japan Fine Ceramic Center (JFCC) for a fellowship (KY) and University of Huddersfield for a fellowship (NS).

References

[1] Davenport K G and Hilton C B 1985 *US Patent* ***4*** 524 217
[2] Hino M and Arata K 1988 *J. Chem.Soc. Chem. Comm.* ***18***, 1259
[3] Bellussi G et al 2000 *CATTECH* ***4***,4
[4] Gai P L et al 1995 *Science* **267** 661
[5] Boyes E D and Gai P L 1997 *Ultramicroscopy* **67** 219
[6] Gai P L and Boyes E D 2009 *Micr.Res.Tech.* **72** 153
[7] Gai P L and Boyes E D 2010 *J. Physics. Conf.Ser.* ***241*** 012055
[8] Shiju N et al. 2011. *Catalysis Science and Tech.* DOI: 10.1039/c0cy00063a.

Journal of Physics: Conference Series **371** (2012) 012027 doi:10.1088/1742-6596/371/1/012027

Quantitative analysis of core-shell catalyst nanoparticles for industrial applications

H E[1], P D Nellist[1], S Lozano-Perez[1] and D Ozkaya[2]

[1] Department of Materials, Oxford University, Parks Road, Oxford OX1 3PH, UK
[2] Johnson Matthey Technology Centre, Blounts Court Road, Sonning Common, Reading RG4 9NH, UK

Email: haibo.e@materials.ox.ac.uk

Abstract. Pd@Pt core-shell designed nanoparticle catalysts have been shown to dramatically increase the activity and selectivity of the oxygen reduction reaction in fuel cells. Aberration corrected electron microscopy offers the spatial resolution and chemical sensitivity to unlock these structures at the atomic scale. Understanding the particle size, shape and the exact nature of the shell coverage (whether it is full, partial or whether the particle is alloyed) is vital to understanding their behaviour. This paves the way for even more effective catalyst designs. We present a semi-statistical investigation into the size, morphology and bimetallic content of various core-shell particle designs, pre- and post- fuel cell cycling, using high resolution HAADF STEM and EDX. In addition, careful quantitative analysis of our datasets will allow us to extract information, not only of the morphology, but also the thickness and coverage of the particle shells. We compare this with chemical findings about activity and selectivity to understand how shell coverage and content affect catalytic activity.

1. Introduction

Though proton exchange membrane fuel cells are one of the promising candidates for meeting the demand of our increasing energy consumption, their effectiveness is limited by the oxygen reduction reaction at the cathode. Pd@Pt core-shell designed nanoparticles electrocatalysts are designed such that the precious Pt metal only comprises a thin outer shell and the core is filled with a different element. The combination alters the electronic structure of the Pt and the catalyst shows an increase in activity and selectivity [1].

Understanding the structural details of these industrial catalysts is vital to understanding their function and aiding in improved designs. However, such small and complicated structured materials are difficult to visualise directly at high resolution, while statistically averaged methods like extended X-ray absorption fine structure (EXAFS) assume homogeneity across a large number of samples and thus lacks detailed information about individual particles. With the common-place usage of aberration correction, it is now readily possible to directly image the atomic configuration of small nano-particles.

While energy dispersive X-ray (EDX) spectroscopy can offer an unambiguous identification of composition, the lack of spatial resolution and poor signal to noise, especially for small beam sensitive samples, makes it insufficient as a technique for detailed analysis of the catalytic surface of the nanoparticles.

High angle annular dark field scanning transmission electron microscopy (HAADF STEM) has often been chosen for such heavy elements on light supports for its strong correlation between atomic number and imaged intensity, known as Z contrast and also the incoherent nature of its detection which greatly simplifies image interpretation [2]. However, though it offers many advantages, meaningful quantitative compositional information is challenging to extract because intensities are dependent on both composition and thickness. Recently, much progress has been made towards quantification of HAADF STEM by calibrating the detector such that the images are scaled to the incident beam [3], enabling full comparisons with simulations. Analysis by cross section can then be applied to identify the composition of individual atomic columns [4]. In this paper we will investigate how quantitative analysis of HAADF STEM images can lead to better understanding of both the size distribution and also structure of various catalysts.

2. Method

Core-shell catalysts were prepared by pre-forming Pd cores of size 5nm on a support of carbon black, followed by Pt deposition from a $Pt(acac)_2$ pre-cursor by proprietary methods [5]. Several thicknesses of Pt shell were considered: ½ monolayer (ML), 2ML and 4ML, where a monolayer is considered to be a complete coverage of the Pd core by one atom thick of Pt. Catalysts were investigated pre-cycling (as made) and post- fuel cell cycling of 1000 hours in H_2SO_4 as described elsewhere [6]. Calibrated HAADF STEM images and EDX data were collected using a JEOL-2200MCO [7] and a JEOL ARM-200F at 200kV and 80kV. Simulations were carried out using an absorptive potential multislice code.

3. Size Distribution

Particle size distribution was estimated from medium magnification HAADF STEM images, each of various catalyst designs pre- and post- cycling (Table 1). Imaging was carried out on a JEOL2200-MCO at 200kV. Both major and minor axes of particles were measured to account for the non-circular shapes. In each sample 200+ particles though those that were out of focus and obfuscated by other overlapping particles were discarded. In total 40+ measurements were made for each sample. The catalysts with thicker shell designs had a larger average diameter but also a larger range of sizes. From the micrographs it is clear that there are clusters of very small (< 2nm diameter) particles in the 2ML and 4ML designed samples. EDX analysis identifies them as pure Pt. Post cycling, the particle size distribution of the 2ML does not change and the small clusters are still present, suggesting that the particles are reasonably stable, as confirmed by stability cycling tests.

Unfortunately, it is difficult to extract compositional information at this low resolution by comparisons of HAADF intensities alone as those particles that oriented on zone axis channel the beam down the columns and thus appear brighter, thus making intensity not only dependent on thickness and composition but also on orientation.

Table 1. Average particle diameter of various catalyst designs pre-cycled and 2ML post-cycled. Low magnification HAADF STEM micrographs were analysed, comprising 200+ particles for each design.

Design / state	Average particle diameter /nm	standard deviation /nm	error /nm
½ ML / pre-cycled	4.9	1.2	0.2
2ML / pre-cycled	5.8	2.3	0.3
4ML / pre-cycled	6.1	1.9	0.3
2 ML / post-cycled	5.5	2.1	0.3

4. Shell Coverage

Catalytic processes are primarily concerned with the surface atoms of nanoparticle catalysts; therefore it is important to understand the precise nature of the shell coverage and thickness.

4.1. HAADF and EDX

EDX of small nanoparticles suffers from low signal to noise and damage due to the long collection times necessary. We carried out EDX maps at 200kV on a JEOL ARM-200F with a 48pA probe and with a fast cumulative scan to minimise damage. Combined with the spatial resolution of HAADF, it is possible to ascertain the distribution of Pt and Pd in the nanoparticle.

A cumulative EDX spectrum (Figure 1b) of the ½ ML catalyst nanoparticle (shown in Figure 1a) showed that both Pt and Pd were present. The higher signal to noise ratio afforded by combining the spectra from each pixel of the whole map allows for a meaningful quantification to be done in order to determine the composition ratios of Pt and Pd in the nanoparticle. A simple Cliff-Lorrimer k-factor quantification suggests an 80/20 atomic% ratio of Pd/Pt, which corresponds to the catalyst design for ½ML coverage. However, there is no information about the uniformity of the coverage because the spectrum was integrated over the whole particle.

The normalised intensity of the HAADF image (Figure 1a) suggests that there is very little Pt in the particle as the intensity is mostly around 5% of the incident beam. There is sufficient intensity at the edges to indicated thin Pt coverage. It is unlikely to be purely Pd as the discontinuous change in intensity seen would mean a large thickness variation over a small lateral distance.

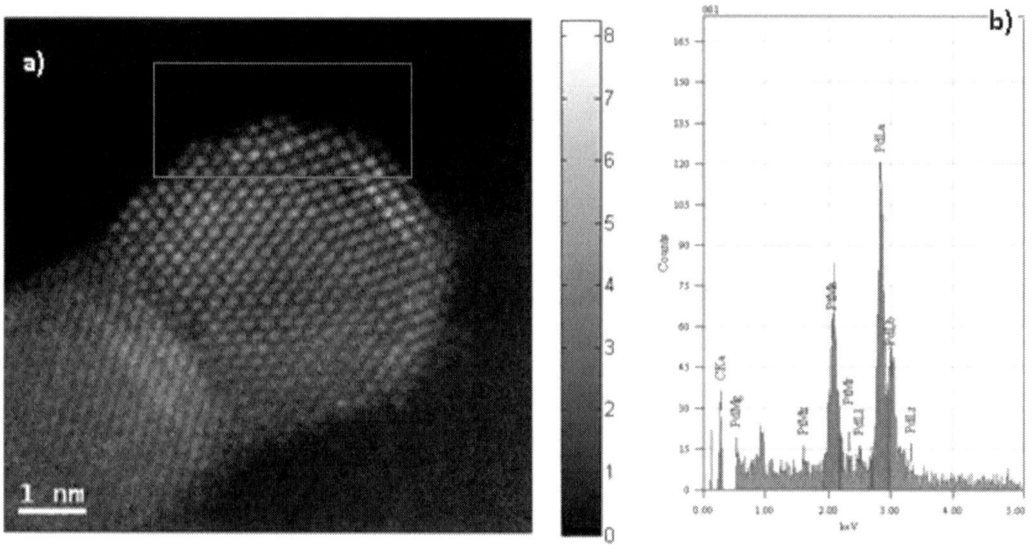

Figure 1a) Calibrated HAADF STEM image of a ½ML designed particle, where the scattering intensity is mapped as a function of incident probe current. The strong scattering of the Pt decoration can be clearly seen on the edges of the particle, the intensity is too strong to be Pd. Step-like edges between the (100) and (111) faces are clearly visible (red box) which suggest non-perfect faceting, commonly seen in nanoparticles. **b)** EDX collected from the whole particle shows the presence of both Pt and Pd. A simple Cliff-Lorrimer k-factor quantification suggests 80/20 atomic% ratio of Pd/Pt which corresponds with the catalyst design for ½ML coverage.

4.2. HAADF and Image Simulations

As all images are 2D projections of 3D objects, even the seemingly clearly core-shell nanoparticle may in fact not be. The 2ML designed particle in Figure 2a shows a large uniform intensity region when viewed along the [110] direction perhaps suggesting large (110) facets. Facets of (110) are thermodynamically less favourable in fcc structures than (111) and (100). However, it is possible that a large (110) facet is terminated by stepped edges onto (100) and (111) faces. Evidence of this is seen in projection (Figure 1a and 2a, red squares)

By imaging as a fraction of the incident beam using a calibrated detector and known probe currents [7], it is possible to compare experiment to simulation absolutely. Since nanoparticles are highly beam-sensitive, calibrating experimental parameters while imaging can lead to damage and makes accurate quantification of the original sample impossible as the sample is changed by the beam. Thus analysis of the individual columns by the absolute cross section of scattering [4] yields a powerful parameter robust method of comparing individual columns to simulations.

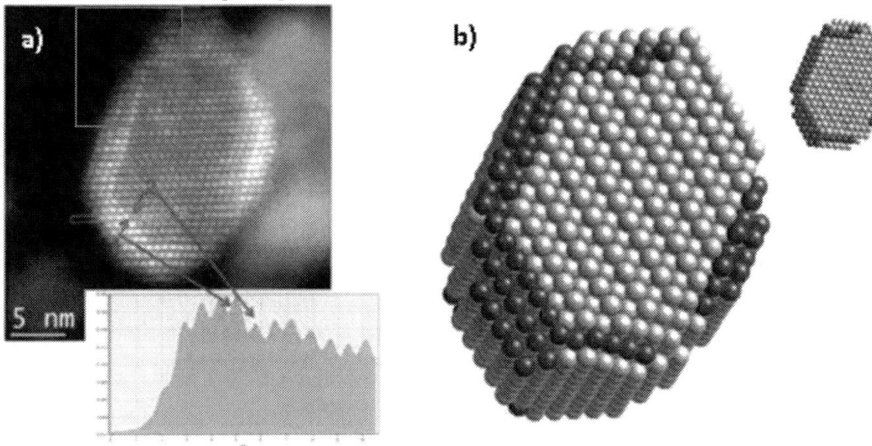

Figure 2a) Calibrated HAADF image of a core-shell nanoparticle with 2ML design pre-cycling showing clear core-shell structure. The two neighbouring columns highlighted by the red arrows suggest a composition change as the difference in intensities (shown in the line profile) rule out a cliff edge due to thickness. Analysis by cross section suggests edge decoration on steps by Pt. Facets terminated by step-like structure can be seen in the red box. **b)** Atom model (scaled down) showing sites of preferential Pt deposition at step-like edges.

Comparison by cross section to absorptive potential multislice simulations [4] reveals possibilities for compositions in neighbouring columns (marked 1 and 2). Column 1 could either be 70 Pd or (40 Pd + 4Pt). Column 2 could be 45Pd or (20Pd + 2Pt). Since it is thermodynamically unlikely that neighbouring columns would differ in height by 25 atoms, it is more reasonable to conclude that the compositions are 45 Pd for column 2 and (40 Pd + 4Pt) for column 1. The intensity comprising the "shell" is likely due to preferential deposition of Pt at step-like edges of the nanoparticle (shown schematically in Figure 2b).

5. Conclusions

Understanding the precise size, morphology and structure of these core-shell nanoparticles can be achieved through a combination of calibrated HAADF imaging, comparison with simulation and low voltage EDX analysis. Medium magnification HAADF STEM yielded particle size distributions for various catalyst designs, with the ½ ML designs being the most uniform in size distribution and the 2ML design the least uniform. The 2ML design did not show significant change in size or distribution post cycling, confirming the design's electrochemical stability. However, we show that though the 2ML Pt shell on Pd core design shows the best electrochemical stability and catalytic activity and selectivity, its functionality is not due to a purely core-shell type structure but a combination of step type edge decoration of Pt and small sub 2nm Pt clusters.

Acknowledgements

This work has been funded by the EPSRC and Johnson Matthey in a CASE-Award studentship.

References

[1] Zhang J et al. 2004 *J. Phys Chem B* **108(30)** 10955
[2] Pennycook SJ, Jesson DE, McGibbon AJ and Nellist PD 1996 *J. Electron Microscopy* **45** 36-43
[3] LeBeau JM and Stemmer S 2008 *Ultramicroscopy* **108** 1653-1658
[4] E H, Nellist PD, D'Alfonso AJ, Allen L *in preparation*
[5] Methods proprietary to Johnson Matthey.
[6] Ball S et al, *ECS Trans.*, 25 (1), 1011-1022 (2009)
[7] E H, Nellist PD, Lozano-Perez S, Ozkaya D 2010 *J. Phys.: Conf. Ser.* **241** 012067

Journal of Physics: Conference Series **371** (2012) 012028 doi:10.1088/1742-6596/371/1/012028

Some recent advances in gold-based catalysis facilitated by aberration corrected analytical electron microscopy

R Tiruvalam[1], Q He[1], A A Herzing[1], J Pritchard[1], N Dimitratos[2], J A Lopez-Sanchez, J K Edwards[2], A F Carley[2], G J Hutchings[2] and C J Kiely[1*]

[1] Department of Materials Science and Engineering, Lehigh University, 5 East Packer Avenue, Bethlehem, PA 18015-3195, USA
[2] Cardiff Catalysis Institute, School of Chemistry, Cardiff University, CF10 3AT, Cardiff, UK

*chk5@lehigh.edu

Abstract. The recent availability of aberration corrected analytical electron microscopes (ACAEM) is revolutionizing our ability to characterize nanostructured catalyst materials. Some recent case studies are presented whereby the application of the high angle annular dark field (HAADF) imaging technique, coupled with STEM-XEDS analysis, has given us a more detailed and realistic view of the catalyst morphology. The examples chosen include supported Au catalysts for low temperature CO oxidation and supported AuPd bimetallic alloy catalysts for the direct production of H_2O_2.

1. Introduction

Supported gold clusters and gold-palladium nanoparticles are intensely studied materials primarily because of their exciting potential applications in catalysis. The recent availability of aberration corrected analytical electron microscopes (ACAEM) is revolutionizing our ability to characterize the morphology, crystallography and chemical composition of such nanoscopic volumes of materials and for the first time are giving us more realistic views of these catalyst systems. In this paper, we present a short review of some of the recent developments in gold catalysis that have been facilitated by ACAEM.

2. Supported gold catalysts for low temperature CO oxidation

The oxidation of CO to CO_2 at low temperature is the most studied reaction in the field of gold catalysis [1]. Bond and Thompson have summarized the most up-to-date understanding of the CO oxidation mechanism over Au in their recent review [2]. This particular reaction is crucial for the removal of carbon monoxide from enclosed spaces such as submarines and space craft, prolonging the life of fuel cells, and extending the usable lifetime of a fire fighters mask. There is a consensus building that a small particle size (which introduces more low coordination number Au atoms), Au oxidation state, the choice of support, particle-support interactions and experimental conditions (*e.g.* moisture level) are all important parameters to control in this reaction. Previously researchers believed that 2-4 nm size Au nanoparticles are the active species for CO oxidation, however experiments in aberration corrected electron analytical electron microscopes are now beginning to alter that view.

2.1 The multiplicity of supported gold species in conventional catalyst preparations

Conventional bright field and phase contrast lattice imaging experiments are effective for characterizing Au particle size distributions, provided that the supported metal particles are greater than ~1nm in size. However HAADF imaging in an aberration corrected STEM indicates that the true nature of the Au distribution on oxide supports is far more complex than previously imagined. For instance, five distinct gold morphologies have been identified by STEM-HAADF to *co-exist* on ZnO supports; namely (i) isolated Au atoms, (ii) sub-nm Au clusters, (iii) ordered monolayer rafts of Au atoms, (iv) assemblies of Au at step edges and (v) epitaxial Au nanoparticles greater than 1nm in diameter, as shown in Fig.1. Interestingly, the relative population density of these various Au species, as well as the overall catalytic activity of the material, can vary considerably with the calcination temperature to which the Au/ZnO material is subjected [3]. The challenge is to devise experiments that will allow us to deduce the relative catalytic activities of these various supported Au species. Once this is known, sample preparation strategies may be developed to maximize the number of the most active entities so that the precious Au metal can be utilized more efficiently. This has been achieved for the Au/FeO_x system where sub-nm bilayer Au clusters have been identified as the key active species for CO oxidation [4].

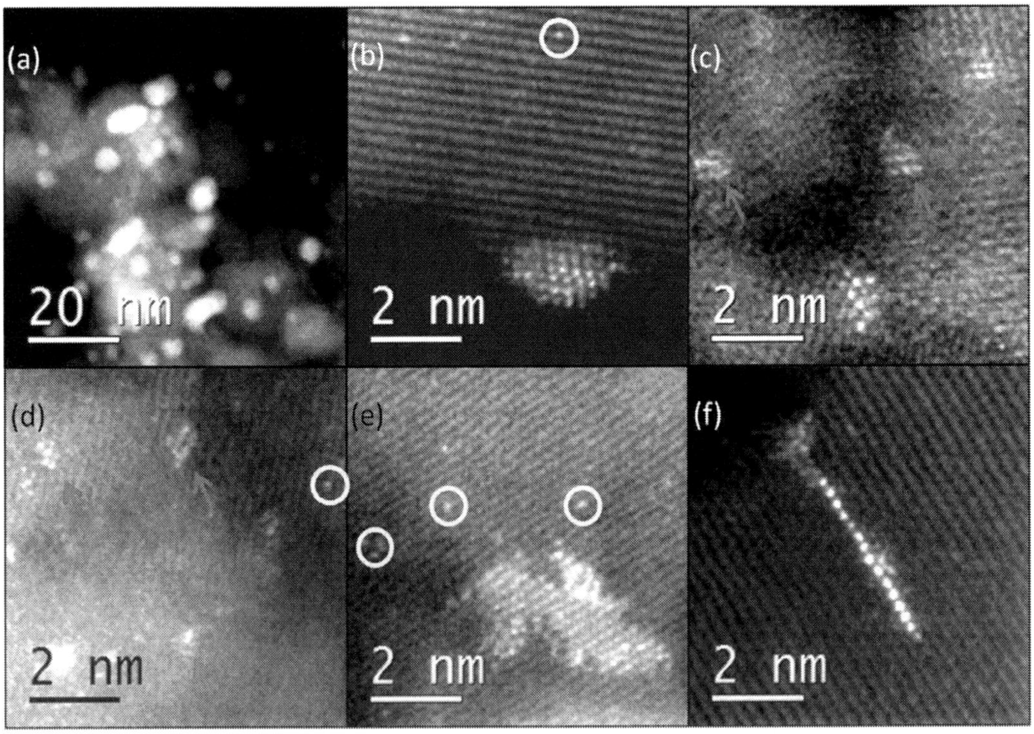

Figure 1. Aberration corrected STEM-HAADF (*z*-contrast) images showing the various Au species co-existing on a ZnO support in a sample made by co-precipitation; (a), (b) 3-dimensional Au nanoparticles; (c) sub-nm Au clusters (arrowed); (d) sub-nm Au clusters and isolated Au atoms (circled); (e) ordered monolayer Au rafts and isolated Au atoms; (f) assemblies of Au at step edges [2].

2.2 Sol-immobilized gold nanoparticles

An alternative technique that allows more precise control over particle size and structure, is to pre-form the Au nanoparticles in a colloidal solution before depositing them onto the support. The only disadvantage to this method arises from the fact that during fabrication the nanoparticles are coated with organic molecules – ligands– which prevent them from clumping together (Fig. 2(a)). Once they are deposited onto a support, these ligands tend to impair the nanoparticle's catalytic performance by blocking the approach of molecules to active sites on the metal surface. Previous methods for stripping away these ligands have involved heat treatments of up to 400°C, which caused sintering and a loss of activity. We have recently discovered a milder alternative for removing the ligands from polyvinyl alcohol (PVA) stabilized gold nanoparticles deposited on a titanium oxide support – a simple hot water wash [5]. HAADF imaging (Fig, 2(b) shows this mild treatment has fewer detrimental effects on the particle morphology or size, and results in a doubling its CO oxidation activity.

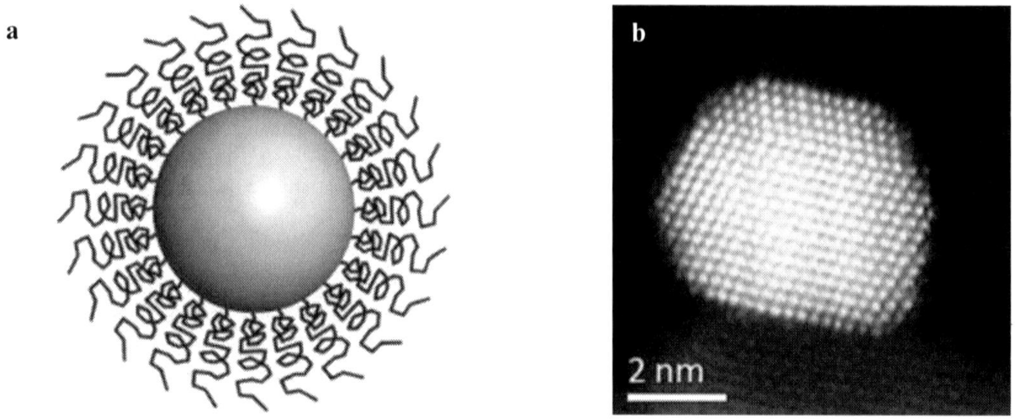

Figure 2. (a) A schematic diagram of a Au particle stabilized with PVA ligands; (b) an HAADF image of a colloidal Au particle after the ligands have been removed by hot water washing. [5]

2.3 Gold re-dispersion by chemical methods

The practical application of gold catalysts has often been limited due to low stability and irreversible deactivation during reaction or thermal pre-treatment. Many studies have investigated the underlying causes behind the loss in activity and have reported surface poisoning, loss of metal support interaction, and nanoparticle sintering as potential contributing factors. Re-activation of the catalysts is problematic, particularly in cases where the gold particles have increased in size, and, therefore, if it were possible to devise an effective method to re-disperse the Au metal, this would be of significant benefit. In a recent study, we have shown that methyl iodide can be used to disperse large (~20 nm)

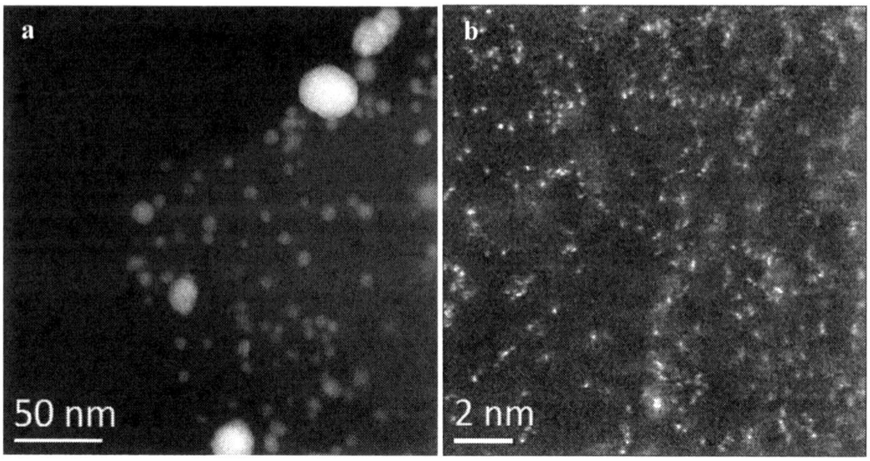

Figure 3. A standard World Gold Council Au/C catalyst displaying large Au nanoparticles (a), which after a mild chemical treatment with CH_3I becomes atomically dispersed (b) on the support [6].

gold nanoparticles supported on carbon into dispersed gold atoms/dimers at low temperature and atmospheric pressure [6]. The process occurs via the progressive removal of gold-halogen entities from the metal nanoparticles and a gradual decrease in the size of the gold on the minute timescale.

3. Bimetallic Au-Pd alloys for the direct production of H_2O_2

There is a growing interest in using supported bimetallic AuPd alloy nanoparticles for the direct production of H_2O_2 from H_2 and O_2. In our first reports on this topic [7, 8] the metal nanoparticles were prepared via a co-precipitation route using $HAuCl_4$ and $PdCl_2$ as precursors. This generated AuPd alloy particles having a broad particle size distribution, with the larger particles being Au-rich and the smaller ones being Pd-rich [9]. Interestingly we found that on oxide supports the AuPd particles tended to adopt a Pd-rich shell/Au-rich core structure, whereas on activated carbon they remained as random homogenous AuPd alloys [10]. The following sub-sections present some specific examples of how STEM-HAADF imaging is increasing our understanding of Au-Pd bimetallic alloy systems.

3.1 Bimetallic Au-Pd colloidal particles with designer morphologies

In a recent study [11], a systematic series of AuPd bimetallic particles have been prepared by colloidal synthesis methods, which gives us better control over the particle size distribution and structure. Particles having random alloy structures, as well as 'designer' particles with Pd-shell/Au-core and Au-shell/Pd-core morphologies, have been prepared and immobilized on both activated carbon and TiO_2 supports. These materials have subsequently been compared as catalysts for the direct production of H_2O_2 in an attempt to elucidate the optimum particle morphology/support combination for the H_2O_2 reaction. The Au-core/Pd-shell nanoparticle variant was found to be the most effective catalyst of the three morphologies for the direct H_2O_2 reaction on both activated C and TiO_2 supports [11]. HAADF imaging can be used to clearly distinguish these three types of colloidal nanoparticle as shown in Figure 4. In fact, with some basic knowledge of the particle structure (*i.e.* cub-octahedral, decahedral or icosahedral) the composition of individual core-shell particles can also be estimated from images such as Figs. 4(b) and (c) by measuring the core diameter and shell thickness [11].

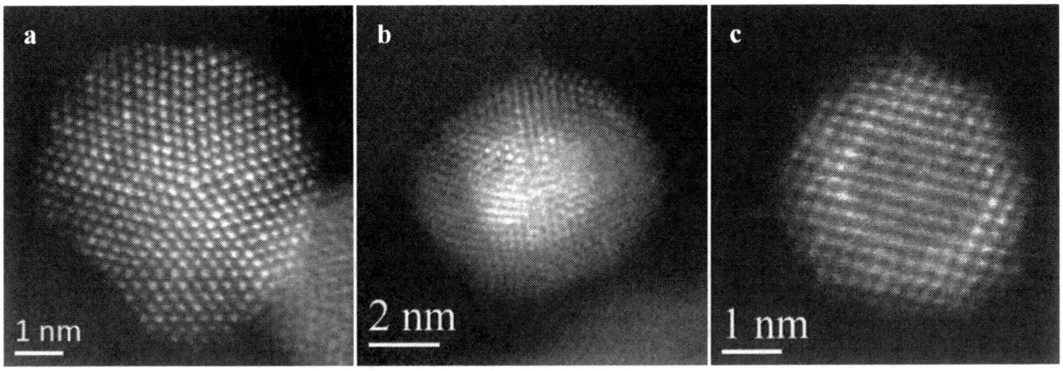

Figure 4. HAADF (*z*-contrast) images of sol-immobilized catalyst particles: (a) a random alloy AuPd nanoparticle; (b) a Au-core/Pd-shell nanoparticle, and (c) a Pd-core/Au-shell nanoparticle [11].

3.2 Interaction of Au-core/Pd-shell colloids with different supports

The effect of a 3h calcination treatment at 200°C on the Au-core/Pd-shell catalyst immobilized on C and TiO_2 supports was also analysed by STEM-HAADF imaging and an interesting support dependant phenomenon was observed [11]. The Au-core/Pd-shell nanoparticle structure was generally well preserved on the activated C support (Fig. 5(a)) as the particles were relatively non-wetting. However, the onset of sintering between neighbouring Au-core/Pd-shell particles was also commonly observed, suggesting that the nanoparticle/activated C interaction was not very strong. By way of contrast, the Au-core/Pd-shell particles on the TiO_2 support showed no strong evidence of sintering, but their Au-core/Pd-shell structure had been compromised (Figs. 5(b & c)). Presumably their better resistance to sintering is associated with the fact that the metal particles had formed extended flat interfaces with the TiO_2 support (Fig. 5(b)), which improved their overall adhesion and thermal stability. It is also fascinating to note that in those particles where a core-shell was still visible, it was not uncommon to find that the Au component had diffused towards, and attached itself preferentially to the TiO_2 support, while the Pd component was displaced outwards forming a hemispherical cap (Fig. 5(c)).

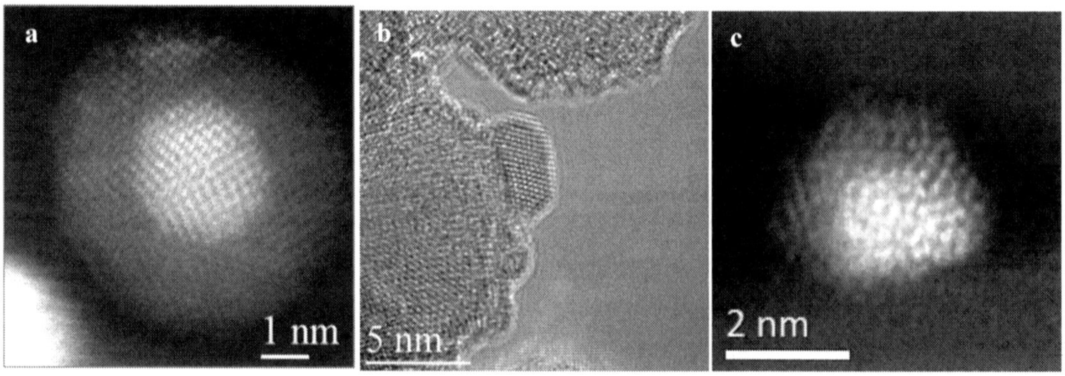

Figure 5. (a) HAADF image of a non-wetting Au-core/Pd-shell particle supported on activated C after 3h calcination at 200°C. An HREM image (b) and HAADF image (c) of a Au-core/Pd-shell particle strongly interacting with a TiO_2 support after 3h calcination at 200°C [11].

3.3 Thermally induced phase separation of Au-core/Pd-shell colloidal particles on TiO_2

The Au-Pd sols immobilized on TiO_2 that had an initial core-shell morphology responded very differently to a 400°C calcination treatment as compared to the random alloy Au-Pd sols on TiO_2. For the latter sample, all the metal particles remained as random AuPd alloys upon calcination. However, no core-shell structures were retained in either of the two core-shell variants on TiO_2 after calcination at 400°C. Instead, three distinct new particle morphologies were found to develop (see Fig. 6(a)), namely; (i) bright particles with extended 1-10 nm faint tails (arrowed blue), (Fig. 6(b)) (ii) isolated faint PdO_x particles (arrowed red) (Fig 6(c)) and (iii) isolated bright Au particles (arrowed yellow) (Fig 6(d)). Localized XEDS measurements on those particles exhibiting extended tails (Fig. 6(b)), showed the faint tail features to be very Pd-rich, whereas the brighter end of the structure was Au-rich [11]. The tailed particles must therefore be a snapshot of ongoing phase separation of the Au and Pd components from the core-shell nanoparticles upon high temperature calcination.

It is indeed curious that Pd and Au components in the core-shell materials readily separate upon calcination at 400°C, whereas such behavior is not observed in the random alloy Au-Pd/TiO_2 variant. This suggests that an initial intimate mixing of the Au and Pd atoms in the nanoparticle (*i.e.* true alloy formation) imparts an additional degree of thermal stability as compared to the situation where one elemental component is simply coated upon the other in the core-shell bimetallic nanoparticles. For the core-shell particles, the Pd component (irrespective of whether it is the shell or the core material) prefers to phase separate to form PdO_x rather than remain associated with the gold. Similarly the Au component prefers to form an extended interface with the TiO_2, rather than retaining it's extensive AuPd interface. The combination of the desire of Pd to oxidize and the tendency of Au to wet the TiO_2 support seem to be the driving force for phase separation in the core-shell morphology particles upon calcination at 400°C.

In summary, the HAADF imaging technique has provided invaluable new (and often unexpected) information on the atomic structure, elemental distribution within particles, and compositional variations between Au-Pd particles. In addition, we have been able to compare their differing thermal stability and sintering behaviors, and to demonstrate that they have quite varying wetting interactions with activated carbon and TiO_2 supports.

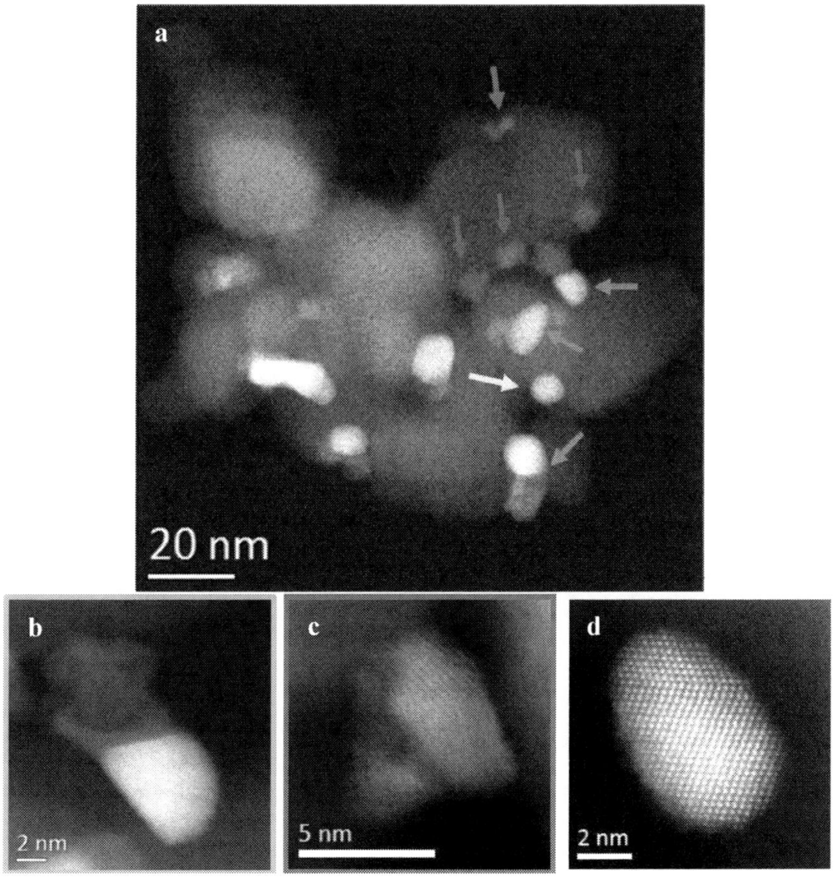

Figure 6. (a) HAADF image of a Au-core/Pd-shell on TiO_2 catalyst after a 3h calcination at 400°C. Three distinct nanoparticle types were found to co-exist after this treatment, namely:- (b) phase separated, but connected, Au-PdO_x particles, (c) PdO_x nanoparticles and (d) Au nanoparticles [11].

References

[1] Haruta M, 2002, *Cattech* **6**, 102.
[2] Bond G and Thompson D, 2009, *Gold Bulletin* **42**, 247.
[3] He Q, Edwards J K, Hutchings G J and Kiely C J, 2010, *Microscopy & Microanalysis,* **16**, 1188.
[4] Herzing A A, Kiely C J, Carley A F, Landon P and Hutchings G J, 2008, *Science,* **321**, 1331.
[5] Lopez-Sanchez J A, Dimitratos N, White S, Brett G, Kosevan L, Miedziak P, Jenkins R L, Tiruvalam R, Carley A F, Kiely C J and Hutchings G J, 2011, *Nature Chemistry*, **3**, 551.
[6] Sá J, Goguet A, Taylor S F R, Tiruvalam R, Kiely C J, Nachtegaal M , Hutchings G J and Hardacre C, 2011, *Agnew.Chem.Int.Ed,* in press, DOI: 10.1002/anie.201102066.
[7] Enache D I, Edwards J, Landon P, Solsana-Espriu B, Carley A F, Herzing AA, Watanabe M, Kiely C J, Knight D W and Hutchings G J, 2006, *Science,* **311**, 362.
[8] Edwards J K, Solsana B, Ntainjua NE, Landon P, Carley A , Herzing A A, Kiely C J and Hutchings G J, 2009 *Science*, **323**, 1037.
[9] Edwards J K, Carley A F, Herzing A A, Kiely C J and Hutchings G J, 2008 *Faraday Discussions*,**138**, 225.
[10] Herzing A A, Watanabe M, Kiely C J, Edwards J K, Conte M, Tang Z R and Hutchings G J, 2008, *Faraday Discussions*, **138**, 337.
[11] Tiruvalam R C, Pritchard J , Dimitratos N, Lopez-Sanchez J A, Edwards J K, Carley A F, Hutchings G J and Kiely C J, 2011, *Faraday Discussions*, DOI: 10.1039/C1FD00020A.

Relationship between Fe^{2+} and Ca^{2+} ions and cyclodextrin in olive trees infected with sooty mold

P. H. A. Aragão; C. G. T. J. Andrade; A. T. Ota and M. F. Costa

Universidade Estadual de Londrina – Laboratório de Microscopia Eletrônica e Microanálise - Campus Universitário - Londrina –Brasil

Abstract. In this work, Energy Dispersive X-ray Fluorescence (EDXRF) was used to observe the peak areas of chemical elements present in healthy and infected samples and a Scanning Electron Microscopy (SEM) to study the damage caused by sooty mold on olive tree leaves from the Mediterranean. Leaves infected with sooty mold presented a high concentration of Fe^{2+} and a low concentration of Ca^{2+}. Our results show that the infected leaves cause a metabolic imbalance in the plants due to an anomalous behavior of macronutrients and micronutrients. Infected leaves start to develop a thin layer of glucose (Cyclodextrin) on their surface. Cyclodextrin (CD) molecules are oligosaccharides consisting of α–D-glucopyranose units linked to glucosides. The most common is β-cyclodextrin (β-CD), which has seven units of α–D-glucopyranose. There are different CDs which are widely used as molecular reactors. In this work, some connections between CD molecules conformations that were obtained in order to observe the relationship of Fe^{2+} and Ca^{2+} in the olive tree infected with sooty mold were studied. The results are discussed in terms of number of ions found inside and outside the cavity formed by the CD molecules.

1. Introduction

The olive tree is one of the plants most often cited in western literature. The olive tree, *Olea europea*, is an evergreen shrub native to the Mediterranean and are among the oldest trees in the world, and were cultivated in Crete around 3000 bC. Olive oil is becoming a very important product thanks to its nutritional value and health benefits compared to other vegetable oils [1]. Therefore, olives and olive oil are very important international commodities; and it is the goal of every grower to get the best out of their olive trees. Sooty mold is a general term applied to several hundred mitosporic and ascosporic species of fungi which inhabit the surface of leaves, giving a dark pigmentation to the cell wall [2]. The sooty mold is a sticky black coating that develops on the surface of plants where the fungi are located. This coating is present in all phases of the mold´s life cycle and it keeps the plant from performing photosynthesis. These minerals are divided into two groups: macronutrients (C, H and O), and micronutrients such as P, N, K, Ca, Mg, S, B, Cl, Cu, Fe, Mn, Mo, Co and Zn [3]. The EDXRF technique was used in this work. The X-ray fluorescence distinguishes the chemical elements present in the sample and can do so very quickly [4].

2. Materials and methods

The study was carried out at the Institute of Mathematics and Physics and the Electron Microscopy Centre in the Sassari University in Sassari, Italy. The leaves used are from the Farm School of Agricultural Sciences Center at Sassari University. The samples were analyzed initially in the EDXRF laboratory and then at the Center for Electron Microscopy [5,6]. The samples were irradiated using a portable X-Ray tube, manufactured by Xraysistem (XRS), XRG/50 model equipped with a tungsten anode. The irradiation time was 100s for each sample and the data stored in a CD and analyzed in real-time using WinQXAS software from International Atomic Energy Agency. The portable X-ray system allows measuring to be done directly on the olive trees leaves, ie *in vivo*. In order to determine the oxidation state of Fe the Mössbauer spectroscopy method was used. Scanning electron microscopy technique was used to obtain images of healthy and infected leaves. In order to use scanning electron microscopy, olive trees leaves were immersed for

60min at 4°C in 2.5% GTA; 0.045M sodium cacodylate buffer (pH 7.2) and post-fixed in 1% osmium tetroxide in the same buffer. After fixation the samples were dehydrated in an ethanol series and critical point-drying technique using CO_2. The leaves were placed onto a sample holder using silver paste, and then sputtered with gold [7]. The material was then analyzed using a scanning electron microscope Phillips/FEI-Quanta 200 SEM operated at high vacuum of 10^{-6}Torr with an accelerating voltage of 30kV, figure 1. In order to observe the behavior of minerals present in olive tree leaves, Cyclodextrin (CDs) molecules were used and their energies were calculated using the program AMBER [8].

FIGURE 1 - Scanning electronic micrograph of olive tree leaves infected with sooty mold.

3. Results

The comparative analysis between peak areas of EDXRF spectrum of healthy and infected olive tree leaves with sooty mold is presented in Table 1. The concentration of the chemical elements present in the samples were obtained using the Student's t-distribution, William Sealy Gosset, for a 95% confidence level (cps). Our results show that the concentrations of the elements Fe, Cu, Ti, Mn and Zn are higher in the infected (inf) leaves when compared to those used as control (ctr). However, the elements Ca and K are significantly reduced in the same samples (Table 1).

Elements (cps)	Control	Infected	Ratio (ctr/inf)
K	6.548	2.534	2.584
Ca	26.555	9.551	2.780
Ti	0.294	0.833	0.352
Mn	0.228	0.783	0.291
Fe	8.081	133.741	0.060
Ni	0.370	0.147	2.517
Cu	0.615	1.230	0.500
Zn	0.145	0.573	0.253

TABLE 1– EDXRF peak areas of minerals in control leaves and infected leaves.

The EDXRF spectrum of olive tree leaves infected with sooty mold shows a high concentration of Ca in olive tree leaves used as control. In the spectrum obtained from olive tree leaves infected with sooty mold, this concentration is reversed. In this case, a high Fe concentration and a low concentration of Ca occur. The oxidation state of Fe was determined to be Fe^{2+} using the Mössbauer spectroscopy method, as shown in Figure 2. This accumulation of Fe^{2+} is observed in plants affected by sooty mold and the action of microorganisms, bacteria, turns glucose naturally present in plants into Cyclodextrin. This molecule grows due to hydrogen bonds, with a hydrophobic internal cavity while the outside is hydrophilic. Our model was calculated using the Monte Carlo method with a structure containing two cyclodextrin molecules placed near each other. The structure which had the minimum energy, meaning that it is more likely be affected by the disease, is the first one depicted. We placed Fe^{2+} and Ca^{2+} ions inside and outside the structure,

Journal of Physics: Conference Series **371** (2012) 012029 doi:10.1088/1742-6596/371/1/012029

respectively. We show that Fe^{2+} is never inside the cyclodextrin molecule while Ca^{2+} agrees to be in this position, as shown in Figure 3, the image in the center. Fe^{2+} was placed in the same position and we calculated the minimum energy in order to show that Fe^{2+} does not stay between the cyclodextrin molecules. Why does this happen? When you have two cyclodextrin molecules they tend to engage, that is, they stay together. Thus, Ca^{2+} does not appear on to the outside of the molecule as it occurs with Fe^{2+}. Therefore, the system could be more complex but is masked by the CD-Ca–CD. Our results are discussed in terms of number of ions found inside and outside the cavity formed by the CDs, Figure 4. The energy difference between the systems is equal to: $\Delta E = (CD + Fe^{2+}) - CD = 2449393.7$ kcal/mol and $\Delta E = (CD + Ca^{2+}) - CD = -130.2$ kcal/mol.

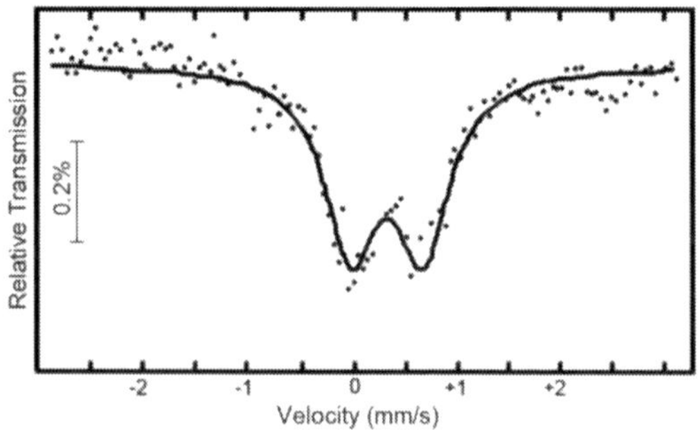

FIGURE 2 – Mössbauer spectrum characteristic of Fe^{2+} on the sample infected leaves

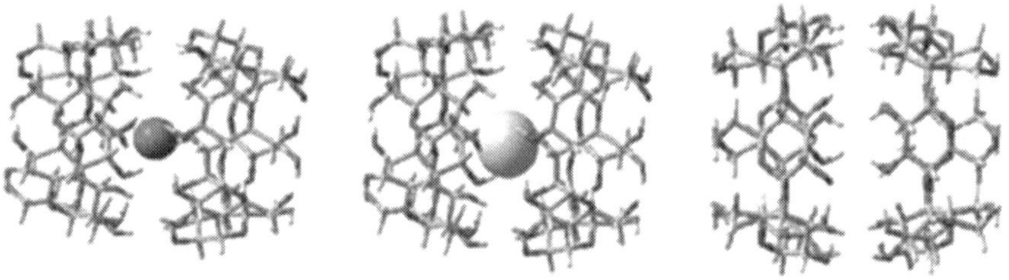

FIGURE 3 - Cyclodextrin with Fe^{2+} (Red), Cyclodextrin with Ca^{2+} (Silver) and Cyclodextrin dimer, respectively.

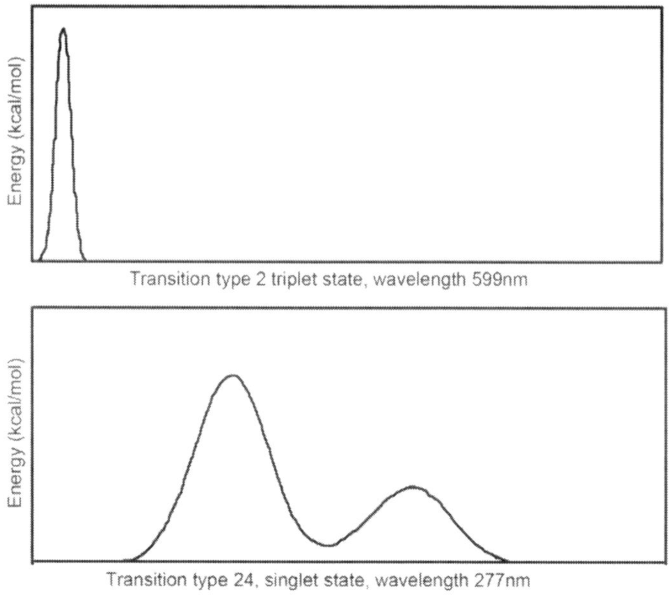

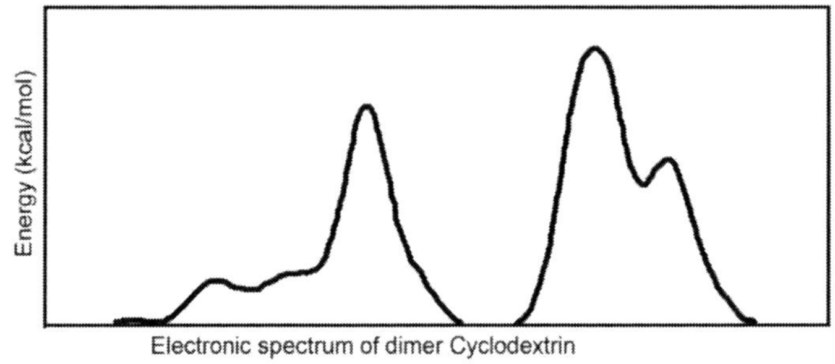

FIGURE 4 - Electronic spectrum using PM3: Fe^{2+} with transition type 2 triplet state, Ca^{2+} with transition type 24 singlet state and Cyclodextrin dimer, respectively.

4. Conclusion

Besides being fast and multielemental, the EDXRF technique was important to identify the minerals present in leaves infected with sooty mold. This technique shows that the nutritional status of plants is altered at the macronutrients (Ca) and micronutrients (Fe) level. Our results show that the Ca^{2+} is located inside the cavity of the Cyclodextrin molecules while Fe^{2+} is located on the outside. Our explanation for this relationship between Fe and Ca which leads to the partial disappearance of the Ca ion peak is due not only to the action of the spores which leads to the development of sooty mold, but also the action of bacteria that cause the formation of cyclodextrin molecules, which end up absorbing part of the calcium that would normally be observed using the EDXRF technique.

Acknowledgement
This research work was supported by CAPES and ProPPG/UEL.

References
[1] P. Rallo *et al*. Development of simple sequence repeats (SSRs) in ole tree (*Olea europaea L.).* Theoretical and Applied Genetics, N° 101, (2000).
[2] D. R. Reynolds. Capnodium citri: The sooty mold fungi comprising the taxon concept. Mycopathologia, N° 148, (1999).
[3] G. C. Vitti and W. Trevisan. POTAFOS, N° 90, (2000).
[4] G.R. Lachance and F. Claisse. Quantitative X-Ray Fluorescence Analysis Theory and Aplication, John Wiley and Sons, (1994).
[5] P.H.A. Aragao *et al.* Iron in olive tree leaves in the Mediterranean area. Journal of Radioanalytical and Nuclear Chemistry, v.249, (2001).
[6] M. S. Blonski *et al.* Elementary Chemical Analysis in Leaves Infected by Fumagina by X-Ray Fluorescence Technique. Brazilian Archives of Biology and Technology, v. 50, N° 5, (2007).
[7] P. Echlin. Handbook of sample preparation for scanning electron microscopy and X-ray microanalysis. Springer Science, (2009).
[8] D. A. Case, *et al.* AMBER 11 and AmberTools 1.5, University of California, San Francisco, (2010).

Influence of Surface Adsorption on Work Function Measurements on Gold-Platinum Interface Using Scanning Kelvin Probe Microscopy

Simon Mugo and Jun Yuan

Department of Physics, University of York, York, YO10 5DD, U.K

E-mail: swm500@york.ac.uk

Abstract. Surface potential difference (SPD) on freshly coated gold and platinum electrodes have been found to be much smaller than bulk work functions consideration and to be dependent on time. We show these discrepancies arise due to formation of surface dipoles caused by adsorbed contaminants in ambient environments. The process is reversible by gentle annealing consistent with contaminant hypothesis. Examination of potential changes on individual electrodes suggest that the Pt surface is more sensitive to ambient conditions than the Au surface in accordance with their relative chemical activities. The result has great implication for interpretation of Kelvin probe measurements obtained on practical devices exposed to ambient environments.

1. Introduction

The nanoscale resolution [1,2] and quantitative measurements achievable with scanning Kelvin probe microscopy (SKPM) makes it attractive in measuring materials' electrical properties and surface reactions [3,4]. The physical principle of the technique is based on contact potentials [5]. A metallic tip is vibrated above the sample and its electrostatic interaction with the sample minimized by application of a bias voltage [3,6]. This voltage corresponds to the contact potential difference (CPD) for clean metals [2]. For metal surfaces with a double layer, the net electric potential is the surface potential difference (SPD) and it gives information about surface charging [2,7]. Gold and platinum have often been used as reference electrodes in the CPD measurements, since they are chemically inert at room temperature [8]. However, various works have reported inconsistency in measured values [9,10], with variations only discussed qualitatively [10,11,12]. These variations present a great challenge for actual Kelvin probe measurements on devices exposed to ambient conditions.

In this work, we not only report discrepancies in measured surface potential values on Au and Pt electrodes, but we also show systematic trends as a function of time and sample treatment. The variations are explained by surface dipoles caused by adsorption of contaminants in ambient environments. The process is reversible by gentle annealing in H_2/Ar environment and is attributed to elimination of volatile adsorbed species. Our result offer simple interpretation of Kelvin probe measurements which is useful in deducing stable electrodes and for examining practical devices exposed to ambient conditions.

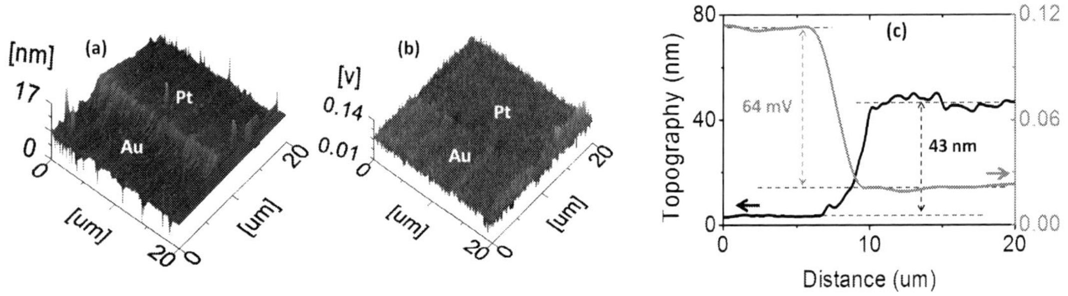

Figure 1. SKPM measurement in vacuum across Au-Pt interface obtained with a Pt-coated tip: (a) Topography showing a Pt step, 43 nm high, deposited on Au surface (b) SPD scan showing contrast along the Au-Pt boundary (c) line profiles across the interface (notice the non-zero potential on Pt electrode)

2. Experimental procedure

A 2 nm chromium film was first deposited on clean 2 x 2 cm^2 n-type silicon. This was followed by deposition of 100 nm gold film. The samples were then half masked and a 50 nm Pt film was subsequently deposited to produce Au-Pt interface. Kelvin probe measurements were immediately obtained in high vacuum and in air with a JSPM 5200 AFM made by JEOL. Scanning was done with Au-coated and Pt-coated AFM tips (resonant frequency = 300 KHz and spring constant = 40 N/M), which had been supplied by Mikromasch. Figure 1 displays results obtained in vacuum with a Pt-coated tip. Topography, figure 1a, shows sharp peaks from contaminating particles from the mask. A clear potential contrast is visible along the Au-Pt boundary, figure 1b. A gradual variation across the interface is observed in line profiles, figure 1c, which is due to a slanting deposition coupled with tip convolution. Measurements are however homogeneous away from the interface. To obtain a representative potential value, we measured several line profiles extending the full scan range. We then repeated the process for two other areas of the same sample and on one other sample prepared in similar conditions. Measurements for samples prepared in similar conditions were reproducible within $\pm$ 8 mV error margin. The final measurement was extracted by averaging SPD profiles across each scan.

3. Results

3.1. Effects of exposing samples to ambient environments on the surface potential

Table 1 displays results for electrodes SPD in air and in vacuum environments. On using both Au-coated and Pt-coated tips to scan across the interface, the residual potential was non-zero and with the Pt electrode recording a higher value. Measurements in high vacuum were slightly higher by about 20 mV compared to measurements in air for all results obtained with both AFM tip materials. However, the measured values are much smaller than theory predicts [13,14].

Table 1. Surface potential measurements at Au-Pt interface in ambient conditions

Reference	In air (mV)	In vacuum (mV)	Theoretical (mV) [13]
Au tip on Au sample	50 $\pm$5	40 $\pm$5	0
Au tip on Pt sample	87 $\pm$5	96 $\pm$5	350-700
SPD	37	56	350-700
Pt tip on Pt sample	75 $\pm$5	56 $\pm$4	0
Pt tip on Au sample	120 $\pm$5	120 $\pm$5	350-700
SPD	45	64	350-700

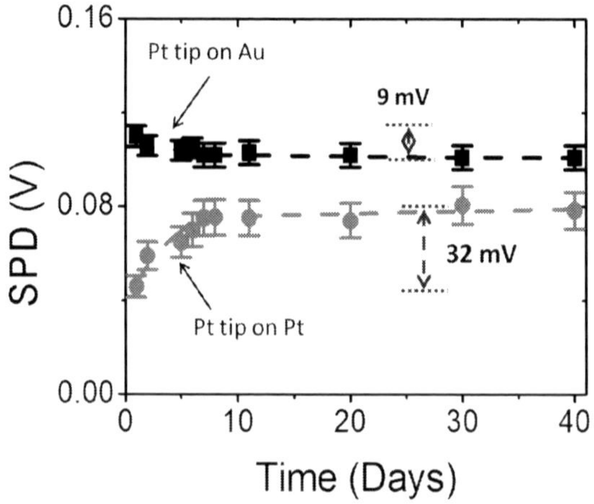

Figure 2. Surface potential trend with time on individual electrodes in vacuum (values converged to a constant value after 14 days with large changes recorded on Pt)

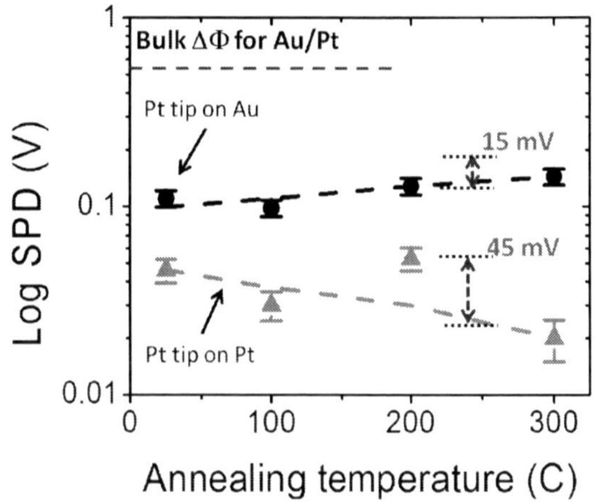

Figure 3. Effect of low temperature annealing on surface potential measurements (reversal of the potential trends is observed on both electrodes)

3.2. Time-dependence SPD on individual electrodes

The time-dependence SPD shows an initial decrease on Au and an increase on Pt, figure 2. The rate of change on Pt is higher compared to Au over the same time and in similar ambient conditions. After 14 days, measurements converged to a constant value which was almost similar for both electrodes. However, potential fluctuations on Pt electrode increased as the sample aged whereas they remained constant on Au electrode, as shown by error bars.

3.3. Effect of temperature on individual electrode SPD

Samples which had been exposed to ambient were then annealed systematically for 30 minutes in H_2/Ar and surface potential measurements repeated. The potential trend was observed to be reversible, figure 3, with an increasing trend on Au electrode and a decrease on Pt electrode. After annealing at 300 ^{0}C, changes of 15 mV and 45 mV were recorded on Au and Pt respectively. Since this result is obtained with a Pt-coated tip, it is clear that the changing trend on Au is towards the Au/Pt metal bulk and on Pt is towards the residual potential. These improvements are however far from bulk work function consideration of 350-700 mV for Au/Pt and zero for the residual potentials [14].

4. Discussion

Work functions of metals are known to be affected by intrinsic surface dipole [14]. However, this change does not depend on time as metal surface remain the same at low temperatures. We believe the observed time dependence is due to adsorbed species on metal electrodes. The AFM cantilevers had been stored in ambient condition for long time and tip adsorption is likely to be saturated. This contamination on tips only affects absolute values and does not contribute to these variations. Proof for adsorption is confirmed by exposing the annealed sample to ambient conditions, where the potential trend became similar to the 'as deposited' sample. Related studies have shown oxygen, water and hydrocarbons to be the most probable contaminants on Au and Pt [15,16]. Such adsorption results in dipole layer [17], given by eqn. (1) [14]: where N_{mol} is density of adsorbed species, P is dipole moment, ϵ is dielectric constant and n is vector normal to surface. Table 2 summarizes molecular adsorption approximated using this relation.

$$V_{dipole} = \frac{N_{mol}(P.n)}{\epsilon} \qquad (1)$$

Table 2. Calculated molecular density

Condition	Electrode	$\Delta V(mV)$	$N_{mol}(m^{-2})$
Aging	Au	9	$2.5x10^{16}$
	Pt	32	1×10^{17}
After annealing	Au	15	$4.3x10^{16}$
	Pt	45	$1.3x10^{17}$

Adsorbed molecular densities on Pt are higher compared to Au, which is consistent with their chemical activities. Initial changes in potential sign (figure 2) is attributed to the fact that Pt oxidizes easily in air, with initial oxygen reorienting adsorbed dipole direction [16]. These values are close to reported hydrocarbons adsorption [18,19], with the slight deviation accounted for by different molecular chains and closest packing imposed by surface [20]. Equalization of SPD with time is not clear from our experiments, but we think it is caused by induced dipoles. This is possible if polarization for physiosorbed or charge transfer for chemically absorbed species occur [21]. The net dipole becomes the sum of individual electric fields due to each of the other dipoles.

5. Conclusion

We have measured large SPD on Au and Pt surfaces far from bulk work function consideration. The time-dependence SPD and which we have shown to be reversible by annealing is explained by surface adsorption forming a dipole layer. Such reproducible experiments in ambient conditions are important in interpreting Kelvin probe measurements on real time devices.

5.1. Acknowledgments

This research has been done using facilities at the York JEOL Nanocentre at the University of York. Funding from Commonwealth scholarships commission is highly appreciated.

References

[1] Jacobs H O, Knapp H F, Muller S and Stemmer A 1997 *Ultramicr.* **69** 39
[2] Zerweck U, Loppacher C, Otto T, Grafstrom S and Eng L 2005 Phys. Rev. B **71** 125424
[3] Nonnenmacher M, O'Boyle M and Wickramasinghe H 1991 *Appl. Phys. Lett.* **58** 2921
[4] Jiang C, Ptak A, Yan B, Moutinho H, Li J and Al-Jassim M 2009 *Ultramicr.* **109** 952
[5] Kelvin L, Fitzgerald G and Lassabatere L 1898 *Philos. Mag.* **46** 80
[6] Nonnenmacher M, O'Boyle M, Wickramasinghe H 1992 *Ultramicr.* **42-44** 268
[7] Weaver J and Abraham D 1991 *J. Vac. Sci. Tech.* B **9** 1559
[8] Sachtler W, Dorgelo G and Holscher A 1966 *Surf. Scie.* **5** 221
[9] Charrier D, Kemerink M, Smalbrugge B, Vries T and Janssen R 2008 *Acsnano* **2** 622
[10] Bhushan B and Goldade A 2000 *Wear* **244** 104
[11] Nony L, Bocquet F, Loppacher C and Glatzel T 2009 *Nanotech* **20** 264014
[12] Palermo V, Palma M and Samori P 2006 *Adv. Mater.* **18** 145
[13] Michaelson H 1978 *Appl. Phys.* **48** 4729
[14] Sze S 1981 *Physics of Semiconductor Devices* (Wiley)
[15] Cherniavskaya O, Chen L, Weng V, Yuditsky L and Brus L 2003 *J. Phys. Chem.* **107** 1525
[16] Leiro T and Laiho J 2006 *Appl. Surf. Scie.* **252** 6304
[17] Hayashi N, Ishii H and Ouchi Y 2002 *Appl. Phys.* **92** 3784
[18] Donnet J, Qin R and Wang M 1992 *Colloid. and Interf. Sci.* **153** 572
[19] Boer B, Hadipour A, Mandoc M, Woudenbergh T and Blom P 2005 *Adv. Mater.* **17** 621
[20] Lu J, Delamarche E, Eng L, Bennewitz R, Meyer E and Guntherrodt H 1999 *Langm.* **15** 8184
[21] Maschhoff B and Cowin J 1994 *J. Chem. Phys.* **101** 8138

Characterisation of Magnetic FeRh Epilayers

M J McLaren[1,2], M A de Vries[2], R M D Brydson[1] and C Marrows[2]

[1]Institute of Materials Research, School of Process, Environmental and Materials Engineering, University of Leeds, LS2 9JT, UK

[2]Department of Physics and Astronomy, University of Leeds, LS2 9JT, UK

E-mail: py06mjm@leeds.ac.uk

Abstract. The lattice structure of roughly equiatomic iron-rhodium epilayers was studied using transmission electron microscopy. Thin films were grown on magnesium oxide substrates using molecular beam epitaxy with thicknesses of 50 nm and capped with a 2-3 nm thick MgO or Al cap. The samples were prepared into cross-sections for characterisation by TEM. Observation of the interfaces FeRh makes with the cap and the substrate are of interest, due to the potential for strain and interfacial diffusion, which will affect the magnetic properties. TEM imaging combined with X-ray photoelectron spectroscopy depth profiling data give an insight into how strain and diffusion at the interface can affect the magnetic transition.

1. Introduction

The magnetic properties of iron-rhodium (FeRh) alloys are extremely unusual, making them an interesting area of research. When in a roughly equiatomic state (i.e. Fe_{48}-Rh_{52} to Fe_{56}-Rh_{44}), the alloy is in an ordered α' state and arranged in a CsCl structure[1]. At room temperature, FeRh exhibits antiferromagnetic (AFM) behaviour. It undergoes a first order phase transition to ferromagnetism (FM) at ~350 K[2] and has a Curie temperature of 675 K[3].

The transition also involves a uniform 1% increase[4] in unit lattice volume when entering the ferromagnetic phase, which is a lattice parameter jump from 2.987 Å to 2.997 Å[1] (a change of 0.03%). Other characteristics of the phase change include an increase in entropy and resistivity when changing to the FM phase as well as a ~10 K temperature hysteresis[5]. The properties of the phase change can be altered and thus controlled by changing the structure of the FeRh through several techniques, including ion beam irradiation[6] and doping with various elements[7]. One other such method of altering the magnetic transition discussed here is growing FeRh epitaxially onto a substrate with a small (<5%) lattice mismatch. The epitaxial growth causes a compressive or tensile strain on the FeRh lattice (depending on whether the substrate is lattice parameter is larger or smaller than that of FeRh), thereby causing an increase in the transition temperature with a compressed lattice and vice versa.

Due to the magnetic phase change and its ability to be tuned to specific temperatures, FeRh has potential uses for spin valves in magnetic data storage or microelectromechanical systems. The ability to constrain the lattice via a substrate could yield nanoscale switchable magnets if grown on a piezoelectric substrate.

There is an interest in using electron microscopy to fully characterise the FeRh system in a thin film state. Much work has been done on the magnetic properties of the transition but little has been done to relate these properties to the nanostructure of the thin films. In particular, there is much interest in the regions surrounding the interfaces because of the changes in structure and magnetism.

Experiments with polarised neutron reflectometry[8] revealed, for 50 nm FeRh epilayers grown on MgO and capped with 2 nm of MgO, after cooling through the transition to the AFM phase there were still latent FM moments in the film, as demonstrated in figure 1 by the increase in magnetic scattering length density (SLD). The measurement of SLD arises from measuring the reflectivity as a function of the spin eigenstate of a neutron being parallel or antiparallel to a direction defined by an applied magnetic field. For the magnetic SLD profiles, the value is zero for AFM domains and non-zero for FM domains. The structural SLD changes depending on the density, meaning different elements and isotopes will yield different values. The results show that the difference in magnetic phase is constrained to 6-8 nm from both cap and substrate interfaces. It is theorised (by Fan et al)[8] that diffusion at the interfaces and imbalances in the Fe:Rh ratio could influence this metastable phase. Structural scattering length-density information shows there could be a change in composition or density close to the interface, as shown in figure 1.

Transmission electron microscopy is an ideal method to characterise this as it can provide direct structural information as well as combine it with compositional data from energy dispersive X-ray spectroscopy (EDX) and electron energy loss spectroscopy (EELS).

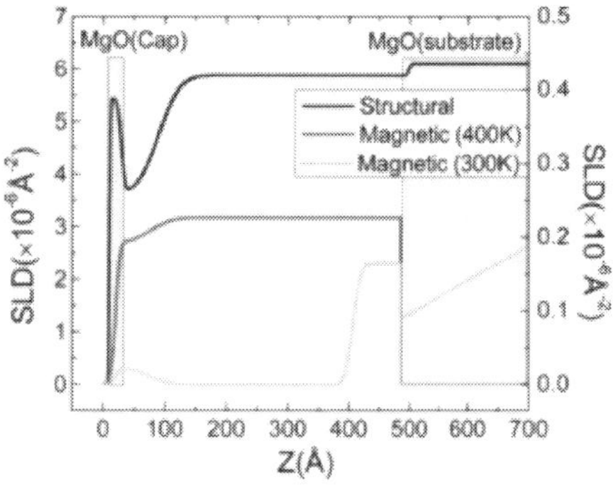

Figure 1. Structural and magnetic scattering length-density (SLD) depth profiles for FeRh above and below the transition temperature obtained from the CRISP reflectometer at ISIS. The increase in the magnetic SLD at 300 K shows the presence of ferromagnetic domains close to both interfaces. The change in the structural SLD suggests a Fe-deficient (Rh-rich) area towards the cap which could explain the FM region[8].

2. Experimental

The FeRh films were grown via molecular beam epitaxy onto an MgO substrate, codeposited from separate Fe and Rh sources at a rate of 0.3 Å/s. The pressure during growth was 5×10^{-10} Torr and the substrate temperature was kept at 300 K. The samples were then annealed at 800°C for an hour then capped in-situ with a 20-30 Å layer of MgO or Al.

For observation in the TEM, the samples were cut into cross-sections. The thin films were cleaved in half and fixed together with epoxy, then encapsulated in a molybdenum wire surrounded by brass tubing. 800 μm discs were cut using a low-rpm saw (Buehler IsoMet) and hand-ground to a thickness of 80 μm. Using a dimple grinder (Gatan 656), the middle of the samples were thinned to 20 μm. The final step utilised a precision ion polishing system (Gatan 691) with an Ar ion source. The sample was thinned with a 5 keV beam until a hole formed in the middle and then smoothed at 2 keV. The typical thickness around this hole should be ~20 nm. The samples were characterised in a FEI Tecnai TF20 at 200 kV.

Cross-sectional samples were also prepared for TEM observation by focused ion beam with a FEI Nova 200 NanoLab FEGSEM with a Ga ion source. A layer of carbon ~30 nm thick was deposited on the film to prevent the beam damaging the crystalline structure. A layer of platinum was deposited over the area of interest to protect the area during milling. A 30 kV beam at 20 nA excavated on both sides of the area of interest to a depth of 5 μm. The sample was then thinned further using a beam current of 100 pA. The sample was lifted out in-situ by welding to a micromanipulator and moved to a TEM sample holder.

In order to provide elemental analysis of the cap-film interface, X-ray photoelectron spectroscopy measurements were performed on a VG Escalab 250 with an Ar-source ion beam attachment to allow for depth profiling in plan view through the sample. Measurements were taken at varying intervals in the sample from the surface up to 12 nm.

The results of two samples are discussed in this paper. The first, used for TEM images, is a 50 nm FeRh epilayer grown on an MgO substrate with a 2 nm MgO cap. The second is another 50 nm FeRh epilayer on MgO, doped with a small amount of Pd and a 3 nm Al cap, used for XPS interfacial analysis.

3. Results and Discussion

Initially, it is interesting to compare the focused ion beam and conventional sample preparation methods. There was difficulty preparing samples using FIB as the material would redeposit in structures in front of the sample, blocking the beam until it was ablated. A potential reason for this would be that ion beam irradiation is causing areas to become ferromagnetic. A method by Aikoh et al[6] has shown how Ga ions in a FIB could lower the effective transition temperature of the material, forcing it into a ferromagnetic phase. Once the sample was prepared with FIB, the major difference compared to conventional sample preparation was the thickness. The samples prepared with the FIB were ~60-90 nm in thickness whereas ones prepared via the conventional method were ~20 nm thick. The images obtained from conventional TEM cross-sections demonstrate the high level of coherence (figure 2) at the FeRhMgO substrate interface, which is to be expected considering the epitaxial method of growth. The expected epitaxial relationship is FeRh[100](001)||MgO[110](001). For FeRh, the (110) direction is 4.224 Å, while the MgO lattice parameter is 4.21 Å(100), a 0.24% difference. It is evident from the images shown in figure 2 that the lattice matches over the interface although there is little evidence of any dislocations in the structure. Comparing to the interface at the MgO cap in figure 3, there is still a degree of coherence although it is less pronounced. Also, there is a ~2-3 nm thick contrasted area along the interface in the FeRh film that could suggest that the lattice is being extensively strained or there is a difference in the material composition in this area.

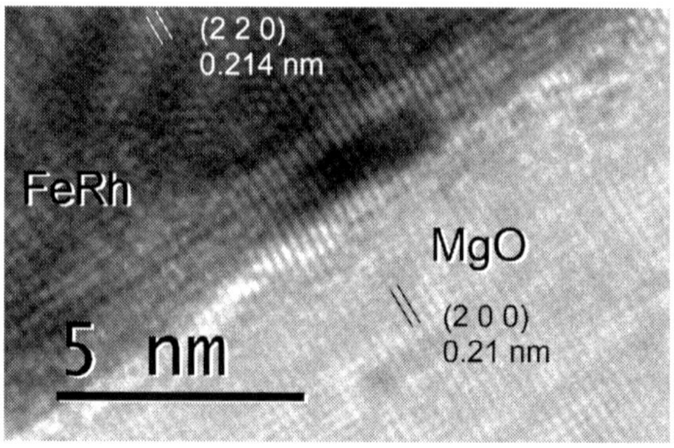

Figure 2. TEM images of the FeRh/MgO substrate interface. The image demonstrates the coherence of this interface, due to the epitaxial nature of the thin film growth. The FeRh film is the darker area towards the top while the MgO area is the lighter area at the bottom.

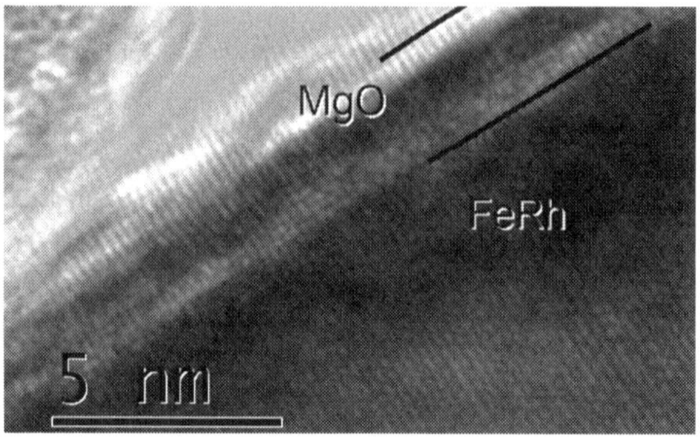

Figure 3. TEM images of the FeRh/MgO cap interface. The coherence at this interface is far less pronounced than at the substrate interface shown in figure 2. There is a contrast ~2-3 nm thick that runs along the FeRh side of the interface, suggesting potentially a lattice strain or a change in composition along this region.

Depth profile results from the XPS analysis of a sample with an Al cap, although only focused on the FeRh:Al cap interface, give a good insight into the latent ferromagnetic region as seen by Fan et al. Even past the interface there is still a presence of Al in the FeRh film, as shown in figure 4. The depth of this area is ~2 nm, which is the same as that of the contrasted area found in the TEM sample. This could also, however, be due to redeposition of the Al that was sputtered from the sample during depth profiling. There is also a change in the Fe:Rh ratio within this area, as well as evidence for diffusion of the FeRh through into the Al cap.

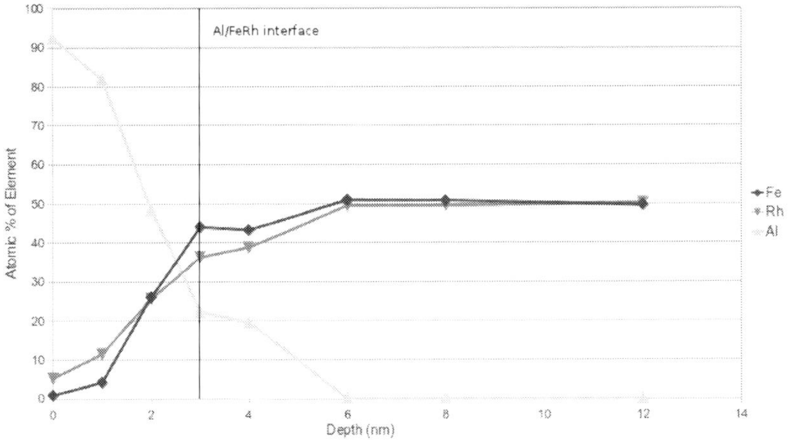

Figure 4. Depth profile data taken using XPS of an FeRh epilayer with an Al cap. The FeRh/Al interface is shown at 3 nm. It is clear that quantities of Al have diffused across the interface as significant quantities of Al are detected beyond the interface.

4. Conclusions

The results of the XPS give backing to the theory that interfacial diffusion, potentially driven by strain on the lattice, is affecting the magnetic transition at the interface. The TEM images of this interface show a high contrast that could be explained by a large amount of strain being applied to the FeRh lattice at the interface or by a change in composition. Further research with TEM cross-sections could give information into how the interfacial diffusion is affecting the lattice structure of the FeRh and, ultimately, how it is driving changes to the phase transition.

Further research into the FeRh system can focus on the lattice structure during the transition. By using a heating system in-situ, the strain on the lattice can be observed in real time to see how the transition nucleates and grows in a constrained thin film sample.

Acknowledgments

This grant has been supported by EPSRC (EP/G065640/1).

References

[1] Swartzendruber L J 1984 Bulletin of Alloy Phase Diagrams 5 45662
[2] Kouvel J S and Hartelius C C 1962 *J. Appl. Phys.* **33** 1343
[3] McKinnon J, Melville D and Lee E 1970 *J. Phys. C: Solid* **46** S46
[4] Khan M, J. 1979 *Phys. F: Metal Phys.* **9** 457
[5] Ibarra M and Algarabel P 1994 Phys. Rev. B 50 4196
[6] Aikoh K, Kosugi S, Matsui T and Iwase A, 2011 *J. App. Phys.* **109** 07E311
[7] Schinkel C, Hartog R, and Hochstenbach F 1974 *J. Phys. F: Metal Phys.* **4** 1412
[8] Fan R et al 2010 *Phys. Rev. B* **82** 2-6

Characterization of InMnSb epitaxial films for spintronics

L Lari[1,2], S Lea[1], C Feeser[3], B W Wessels[4,5], and V K Lazarov[1,2]

[1]Department of Physics, The University of York, YO10 5DD, York, UK
[2]The York JEOL Nanocentre, York Science Park, Heslington,YO10 5BR, York ,UK
[3]Department of Chemical Engineering, Northwestern University, Evanston, IL 60208, USA
[4]Dept. of Material Science & Engineering, Northwestern University, Evanston, IL 60208, USA
[5]Material Research Centre, Northwestern University, Evanston, IL 60208, USA

E-mail: leonardo.lari@york.ac.uk

Abstract. Dilute (III, Mn)V ferromagnetic magnetic semiconductors have potential applications in spintronic devices as magnetic field sensors, spin transistors and reconfigurable logic devices. For such applications ferromagnetism at room temperature is a practical requirement. Previous work has shown that $In_{1-x}Mn_xSb$ films grown on GaAs (100) substrates by atmospheric pressure metalorganic vapour phase epitaxy were single phase and had high temperature ferromagnetism for Mn concentrations up to 2%. Here we present a study of InMnSb thin films at higher Mn concentration (10 at%) showing ferromagnetic properties at room temperature. Aberration corrected and analytical (scanning) transmission electron microscopy techniques were used to study the structure and elemental distribution of the $In_{1-x}Mn_xSb$/GaAs system. The crystalline quality of the film and the presence of phase separation is evaluated in the context of a cluster-mediated ferromagnetism model.

1. Introduction

Dilute magnetic semiconductors have attracted a lot of attention in recent years for their properties that potentially enables spin electronic applications including, spin transistors, logic devices and magnetic field sensors [1]. In fact, the replacement of transition metals atoms within a semiconductive matrix results in adding magnetic properties to the electronic properties of the matrix. In particular, the substitution of Mn atoms to group III in a III-V semiconductor introduces at the same time free holes for conduction as well as local magnetic moments. The $In_{1-x}Mn_xSb$ alloys are of special interest due to their light holes, small energy gap, and considerably higher carrier mobility than other III-Mn-V ferromagnetic semiconductors. These properties make $In_{1-x}Mn_xSb$ alloys ideal candidates for infrared spin photonics [2]. A fundamental requirement for the practical realization of spintronic devices is the Curie temperature (*Tc*) of the magnetic semiconductor to be above room temperature. Recent theory predicts that (III-V)Mn semiconductors, where Mn forms a shallow level, can lead to high *Tc* [3]. The natural candidate for a ferromagnetic alloy grown on GaAs would be the (Ga, Mn)As alloy. This system avoids the strain effects at the film/substrate interface, however its *Tc* is below 180K [4]. (In, Mn)Sb alloys, instead, have been reported to have higher *Tc* than (Ga, Mn)As. In particular single phase InMnSb films have been demonstrated to have high temperature ferromagnetism for Mn concentrations up to 2% [5]. The low solubility of Mn has been already shown in (Ga, Mn)As dilute semiconductors where at high Mn concentration several secondary phase precipitates are formed [6]. Similarly, the (In, Mn)Sb system is expected to phase separate at high Mn concentrations. While in general secondary phase formation is detrimental to the functional properties of spintronic devices, for the case of InMnSb some of the secondary phases, such as MnSb and Mn_2Sb, are still ferromagnetic with *Tc* well above room temperature. In such semiconductors, following the quasi-chemical model for cluster-mediated ferromagnetism described in [3], it is predicted that the overall *Tc* of the alloy should tend towards the highest *Tc* of its constituent phases. In this work we study the crystalline structure of highly doped (10 at% of Mn) InMnSb films grown on GaAs and showing a *Tc* above 500K [7].

2. Experimental

InMnSb thin film were deposited on semi-insulating GaAs (100) substrates in an atmospheric pressure metalorganic vapour phase epitaxy (MOVPE) chamber modified as reported in [8]. Reactants were optimized to obtain a nominal 10at% Mn concentration at the growth temperature of 420°C. TEM cross-section samples were obtained by mechanical polishing and room temperature Ar+ Ion milling using Gatan PIPS up to electron transparency. Microstructural and chemical analyses were undertaken using a conventional 200kV JEOL 2011 Transmission electron microscope and a CEOS double aberration corrected JEOL 2200FS (scanning) transmission electron microscope [9] operating at 200keV, equipped with an in-column Omega type electron energy filter and a Thermo energy dispersive X-Ray (EDX) Si(Li) detector.

3. Results and Discussion

The Bright Field (BF) TEM image in figure 1(a) shows an overview along the [110] zone axis of the InMnSb film. The film has a cubic structure, with the fallowing epitaxial relationship with respect to the GaAs substrate: $[001]_{film} \parallel [001]_{GaAs}$ and $[110]_{film} \parallel [110]_{GaAs}$, as shown by the selected area diffraction (SAD) pattern in figure 1(b) taken from an area that includes both the film and the substrate.

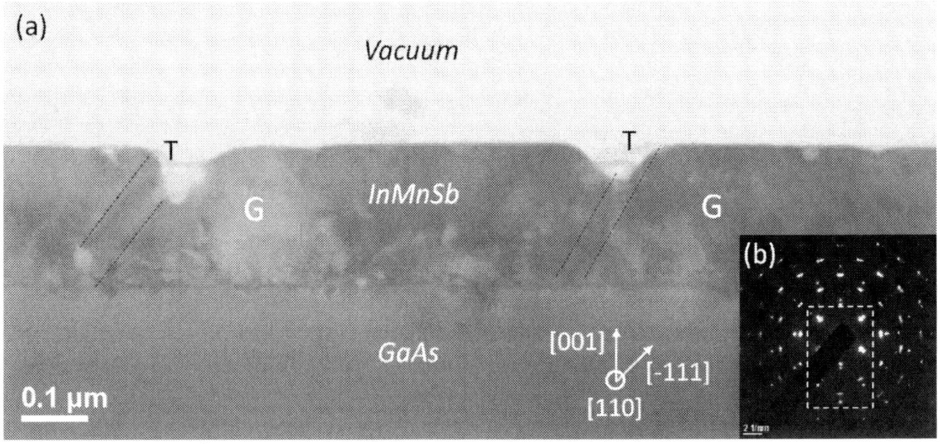

Figure 1. (a) Overview BF image taken along the [110] zone axis. Labels G and T indicates film grains and twin crystal sections. **(b)** SAED pattern from an area including both the film and the substrate. The white dashed box refers to figure 2(a).

The film consists of micrometer size grains separated by twin crystal regions as indicated in figure 1(a). Each one of the grains has a slightly different tilt respect to the substrate. This results in a small angular shift of the Bragg spots in the diffraction pattern. Figure 2(a) clearly shows, close to each other, the InMnSb multiple Bragg spots that correspond to the same Bragg reflection. Figure 2(b) shows an intensity profile across the (00-4) spots of InMnSb along a direction perpendicular to the [001] direction of GaAs. Each one of these peaks, labeled A, B, C, corresponds to different grains with +1.2 (A), -0.5 (B), and 2.5 (C) degrees tilt respect to the substrate.

The high resolution transmission electron microscopy (HREM) image in figure 2(c) shows the edges of two grains labelled as grain 1 (G1) and grain 2 (G2). The grains are separated by a twin boundary region with two mirror planes 10 nm apart and are rotated by 3 degrees with respect to each other. The twin boundary structure is highlighted in the diffractogram inset in figure 3(d). The tilt of the grains, the twin crystal sections together with a high density of misfit dislocations at the film/substrate interface contribute to reduce the strain caused by the 14% lattice mismatch between the substrate and the film. As expected, due the high level of Mn doping, the InMnSb film was found to have undergone phase separation. In particular, two main types of precipitates were found, interfacial structures extending into the substrate and structures buried within the film. Typical secondary phases are shown in the BF field TEM image in figure 3(a). V-shaped structures at the interface are indicated by white arrows. In figure 3(b) is reported an energy-filtered TEM (EFTEM) map of Mn. The map was obtained using the three window method [10] applied to the Mn-$M_{2,3}$ electron energy loss image electron energy loss edge at 49eV. Two pre-edge images (at 35 eV and 43) and one post-edge image (at 59 eV) were acquired. The width of the omega filter energy slit was set to 8 eV. Each 512 by 512 pixel image was recorded with an acquisition time of 45s. The images were cross-correlated to correct for spatial drift of the sample during the acquisition. The background component was calculated from the pre-edge images and removed from the post-edge image using the EFTEM analysis routine of Gatan Digital Micrograph™. Both interfacial and buried secondary phases are Mn-rich. This result is confirmed by annular dark field (ADF) STEM imaging and EDX analyses as shown in figure 3(c). The atomic number (Z) contrast dependence of the intensity in ADF images highlights the presence of secondary phases. Given the difference in atomic number of the InMnSb alloy components (Z_{In}= 49, Z_{Mn}=25 and Z_{Sb}=51), Mn-rich areas appear darker (box1 in figure3(c)) than InSb, and Sb-rich areas (box3 in figure3(c)) brighter that the GaAs substrate (Z_{Ga}=31 and Z_{As}=33). EDX analyses of the film were quantified using the thin film approximation with calculated Cliff-Lorimer factors. Elemental concentrations were obtained from the intensity of respectively Mn-K, Ga-K, As-K, In-L, and Sb-L peaks. Quantified results are reported in the table of figure 3(c). Mn was mainly found to be contained in the secondary phases while the surrounding matrix consists mainly of InSb. Interfacial secondary phases extending into the substrate were found to contain both Mn and Sb indicating Ga substitution with Mn and Sb, and formation of non-stoichiometric MnAsSb.

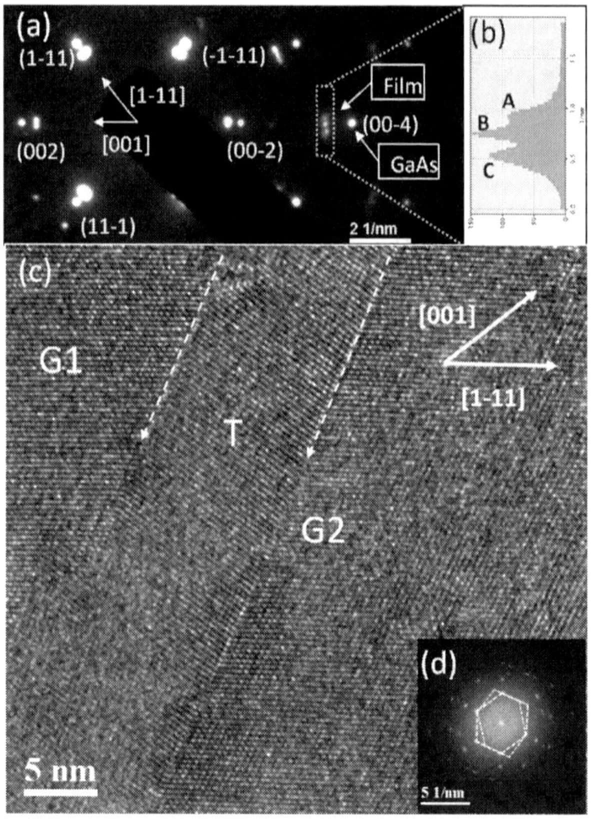

Figure 2. **(a)** SAD pattern analysis from the area indicated in figure 1(b). **(b)** Intensity profile across the InMnSb (00-4) Bragg reflection. **(c)** HREM image of an area including the edges of two Grains G1 and G2 separated by a twin boundary section labelled T. **(d)** Diffractogram with highlighted the mirrored structure characteristic of the twin boundary.

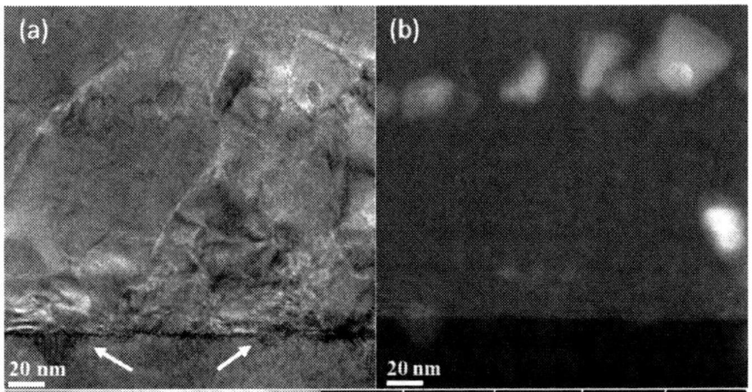

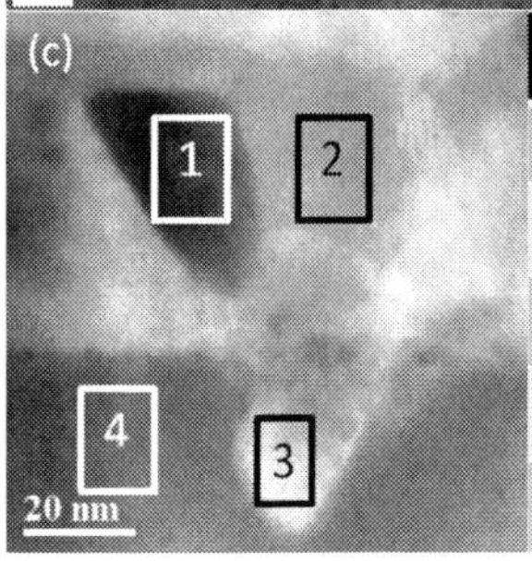

Wt%	1	2	3	4
Mn K	27±1	-	18±1	-
Ga K	4±1	3±1	24±1	49±2
As K	3±1	2±1	38±3	50±3
In L	32±4	39±2	2±1	-
Sb L	33 ±4	55±5	18±3	-

Figure 3. (a) BF field TEM image showing secondary phases buried within the film and at the substrate interface (indicated by white arrows). **(b)** Mn EFTEM map showing secondary phase.**(c)** ADF STEM image showing the two types of secondary phases. Boxes refer to the EDX analyses reported in the table on the right

4. Conclusions

Analytical electron microscopy techniques were applied to study the structure and the elemental distribution in $In_{0.8}Mn_{0.2}Sb/GaAs$. The lattice mismatch between the deposited film and the substrate was found to be compensated by organizing the film in micrometer size grains interconnected by thin twin boundaries and through misfit dislocations at the substrate interface. The high Mn concentration was found to lead to Mn-rich secondary phases buried within an InSb matrix and at the interface with the substrate. Here the secondary phases are formed by Ga substitution with Mn and Sb resulting in non-stoichiometric MnAsSb. These findings indicate that Mn-rich precipitates contribute to the high Curie temperature observed in the two-phase semiconductor films.

Acknowlegments

We acknowledge the University of York, JEOL, Yorkshire Forward (ERDF) fund for sponsoring the York JEOL Nanocentre and the Royal Academy Of Engineering/EPSRC for partial funds.

References

[1] Zutic I *et al.*, Reviews of Modern Physics, 76 (2004) 323-410.
[2] Panguluri R P *et al.*, Applied Physics Letters, 84(2004) 4947-4949.
[3] Wessels B W, New J. of Phys., 10 (2007) 055008.
[4] Wang K *et al.*, AIP Conf. Proc. 772 (2005) 333.
[5] Parashar N D *et al.*, Physical Review B. 81(2010) 115321.
[6] Boeck J D *et al.*, Appl. Phys. Lett.,. 68(1996) 2744-2746.
[7] Feeser, C. *et al.*, to be submitted.
[8] Qian L Q *et al.*, Applied Physics Letters, 63 (1993) 628-630.
[9] Hutchison *et al.*, Ultramicroscopy 103(2005) 7-15.
[10] R.F. Egerton, Electron Energy-Loss Spectroscopy in the Electron Microscope, Plenum Press, New York, 1996]

Journal of Physics: Conference Series **371** (2012) 012033 doi:10.1088/1742-6596/371/1/012033

Structural Variations of $BiMnO_{3+x}$ Revealed by Electron Diffraction

A S Eggeman[1], A Sundaresan[2], C N R Rao[2] and P A Midgley[1]

[1] Department of Materials Science and Metallurgy, University of Cambridge, CB2 3QZ, UK.

[2] Jawarhalal Nehru Centre for Advanced Materials Research, Bangalore, 560064, India.

E-mail: ase25@cam.ac.uk, pam33@cam.ac.uk

Abstract. Electron diffraction studies have been performed on a number of samples within the series $BiMnO_{3+x}$ (for the range $-0.11 < x < 0.16$). Samples with a large value of x (either positive or negative) crystallize as a monoclinic perovskite supercell structure not dissimilar from the stoichiometric ideal. However, the variation of oxygen composition manifests itself as disorder in either the cation or anion sub-lattices, leading to a reduction in symmetry (to $C2$, $P2_1$ or $P2$). In addition, three new structures have been identified at compositions closer to the perovskite ideal. One structure is similar to a mixed cation phase reported by Hughes *et al.*, while another is an n=2 Ruddlesden-Popper phase; importantly both of these phases are acentric and so could support multiferroic properties. The third is as yet not fully characterised but seems to crystallise into a different monoclinic structure than the stoichiometric ideal phase.

1. Introduction

Bismuth manganite is a perovskite-based material that has been proposed as a possible alternative to bismuth ferrite for multiferroic applications [1, 2]. While the ferromagnetic properties of iron and manganese oxides are well known it is the presence of bismuth that makes these materials candidates also for ferroelectricity. A lone pair of electrons allows the Bi^{3+} ion to reside away from the centre of the perovskite unit, thus allowing possible ferroelectric polarisation. While this is true for bismuth ferrite, studies on stoichiometric bismuth manganite seem to suggest that the structure forms a stable monoclinic structure with cell parameters $a = 9.532$Å, $b = 5.606$Å, $c = 9.854$Å, $\beta = 110.67°$, and space-group $C2/c$. This space-group is centrosymmetric and so cannot support ferroelectricity, however there have been conflicting reports of bismuth manganite samples exhibiting ferroelectric polarisation [3].

One possible explanation from Belik *et al.* [4] is that the structure has a space-group of $C2/c$ but that there are local domains within this structure where the symmetry is lowered to $C2$ or $P2_1$, shown by previously forbidden reflections existing at high temperatures and after the incorporation of additional elements into the structure. These mechanisms suggest that by incorporating vacancies into the manganese or oxygen sub-lattices of the structure, it is possible to reduce the overall symmetry of the structure, potentially allowing a non-centrosymmetric structure to be stable for the material.

2. Experimental Details

Electron diffraction studies were performed on a number of bismuth managanite $BiMnO_{3+x}$ samples with oxygen deviation (x) varying from -0.11 to 0.16. The majority of experiments,

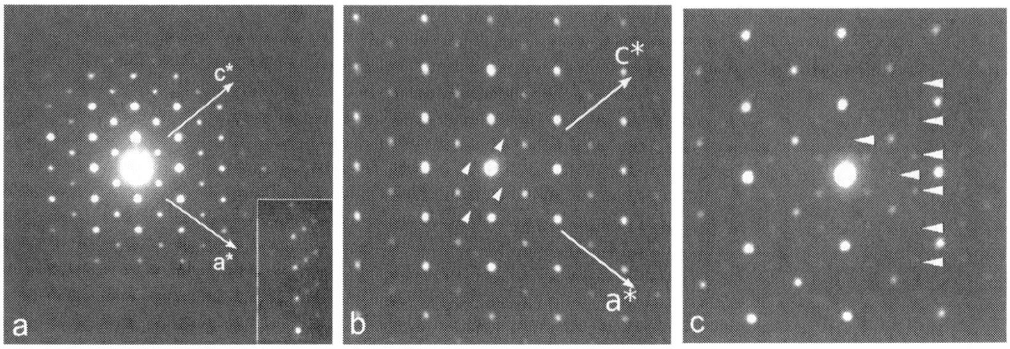

Figure 1. SAED patterns recorded parallel to the [010] axis of bismuth manganite samples. a) $BiMnO_{3.00}$. HOLZ reflections are shown as inset and identify the c-glide present. b) $BiMnO_{2.89}$. The indicated reflections show the $h0l, l = 2n$ condition for allowed reflections. c) $BiMn_{3.16}$. Weak reflections show that there are no forbidden reflection conditions in this projection.

including all precession electron diffraction (PED), were undertaken on a Philips CM30 TEM with a Nanomegas Spinningstar Precession device. Some additional work performed on a Philips CM300 FEGTEM. There have been reports of bismuth manganite undergoing structural transformations under high energy electron irradiation [5] so all TEM operation was performed at 150kV. All diffraction patterns were recorded on Ditabis Imaging Plates

3. Results and Discussion

3.1. Monoclinic Bismuth Manganite Samples

3.1.1. $BiMnO_{3.00}$ As a comparison with the expected structure, a sample of stoichiometrically ideal $BiMnO_3$ was examined. Figure 1a shows a SAED pattern recorded parallel to the monoclinic [010] axis of the structure. The basis vectors are indicated and from this pattern it is possible to determine the cell parameters of the material to be $a = 8.63$Å, $b = 5.85$Å, $c = 8.97$Å, $\beta = 110.1°$, showing good agreement with the accepted values. The reflection condition for the ZOLZ layer is $h0l, h, l = 2n$. Given the offset of the HOLZ reflections, the allowed reflections in the l-direction in the HOLZ and a diad rotation axis (indicated by the whole pattern symmetry) the only allowed space-group is $C2/c$.

3.1.2. $BiMnO_{2.89}$ To determine the symmetry of the sample with composition $BiMnO_{2.89}$ the [010] zone axis was investigated. The SAED pattern for this is shown in Figure 1b, unlike the stoichiometric ideal phase, this bismuth manganite structure has been distorted (by vacancy ordering on the oxygen sub-lattice) and the reflection condition for $h0l$ has changed to $l = 2n$ only. Without a full structural analysis it is not possible to identify the space-group although it is restricted to $C2$ or $P2_1$, and so is likely to be non-centrosymmetric despite the similarity to the parent structure. As with the reported study [4], the weakness of these reflections suggest that these additional reflections are not present throughout the sample but might be caused by some sort of local domain structure.

3.1.3. $BiMnO_{3.16}$ As with the previous samples the zone axis pattern parallel to [010] was used to identify the major features of the space-group symmetry of the sample. In the case where the oxygen stoichiometry is greater than the ideal value of 3.0 the diffraction pattern (shown in Figure 1c) shows a further reduction in the overall symmetry, in this case there are no reflections forbidden in the [010] zone axis diffraction pattern allowing only $P2$ or $P2_1$ as possible

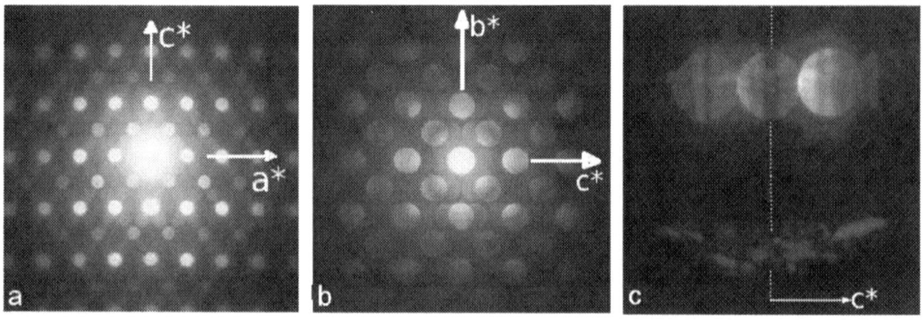

Figure 2. CBED patterns from the $BiMnO_{2.99}$ sample recorded parallel to a) [010], b) [100] and c) approximately 21° along [010] from [100]

space-groups, suggesting a possible acentric structure. The 'extra' reflections (compared to the stoichiometric ideal pattern Figure 1a) in this pattern are much more strongly defined which suggests that any ordering of vacancies on the manganese sub-lattice is not confined to small domains, as the $BiMnO_{2.89}$ sample, but is a more general feature of the samples structure.

3.2. Orthorhombic and Other Structure Variants of Bismuth Manganite

3.2.1. $BiMnO_{2.99}$ Despite its compositional similarity to the ideal bismuth manganite structure, a slightly deoxygenated sample, $BiMnO_{2.99}$, crystallized into a large orthorhombic cell with approximate cell parameters, a = 15.6Å, b = 5.52Å, c = 11.1Å, (determined from CBED patterns shown in Figures 2a and b). These cell parameters suggest that the structure is similar to a mixed cation perovskite $Bi(Mn/Ni)O_3$ [6]. There are compositional similarities that could explain why this structure is common to both systems: the removal of oxygen from the perovskite structure will lead to a reduction in the total oxidation state of the manganese, which is comparable to the incorporation of a divalent ion such as nickel in the place of a trivalent manganese ion. While the reported structure has equal amounts of manganese and nickel on the sublattice it seems that a small oxygen deficiency is also capable of stabilising this structure [7].

Figure 2c shows a CBED pattern recorded after tilting by approximately 21° along [010] from the [100] zone axis of the $BiMnO_{2.99}$ sample. The long extinction lengths of HOLZ reflections make them more sensitive to subtle symmetry breaking and this, combined with the reflection conditions from Figures 2a and b, allows the space-group of the structure to be identified as $Pnm2_1$, a non-centrosymmetric space-group. While this structure does seem to be quite different to the monoclinic perovskite superlattice seen for the first three samples, it is interesting to note that the [010] zone-axis diffraction pattern from this sample is similar to the [010] zone axis pattern from the $BiMnO_{3.16}$ sample. A small change in the symmetry of ions occupying the manganese sublattice can account for this change and so there seems to be a strong structural link between these two phases.

3.2.2. $BiMnO_{3.10}$ The superoxygenated sample ($BiMnO_{3.10}$) is a mixed phase sample. One component phase is a non-centrosymmetric n=2 Ruddlesden-Popper structure (seen in other mixed phase bismuth manganite samples [7]) identified from Figures 3a, a PED pattern recorded parallel to the [001] zone-axis of this structure. The structure of the secondary phase in this structure has not yet been fully determined, the evidence in Figure 3b suggests that this is a monoclinic supercell with cell parameters a = 9.60Å, b = 5.32Å, c = 9.81Å, β = 109.7°. The CBED pattern in Figure 3c has clear Gjonnes-Moodie lines indicating $p2gg$ planegroup symmetry for this projection, which only allows $P2_1/c$ as the space-group. To remain consistent

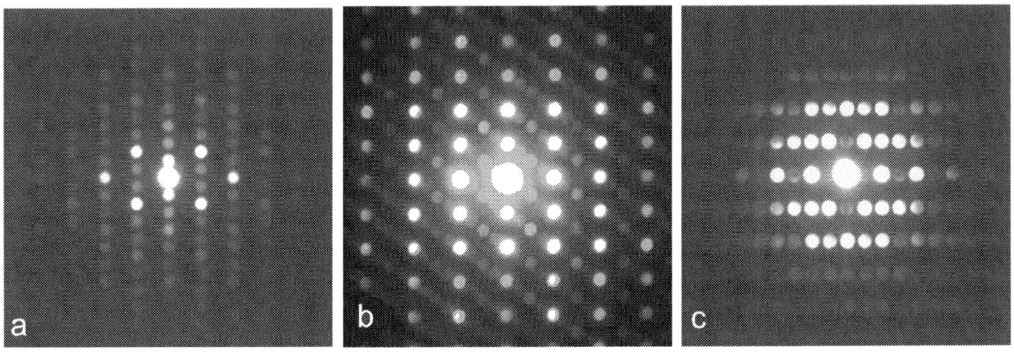

Figure 3. Electron diffraction patterns from the $BiMnO_{3.10}$ sample. a) is a PED pattern from an n=2 Ruddlesden-Popper phase recorded parallel to [001], b) PED and c) CBED are from a new monoclinic phase; patterns were recorded parallel to [010] and [100] respectively

.

with the [100] zone-axis pattern the reflection conditions in the [010] zone axis pattern have to be $h0l, l = 2n$.

4. Conclusions

For multiferroic properties to exist in a sample it is vital that a non-centrosymmetric crystal structure can be supported. While it is clear that stoichiometric bismuth manganite cannot (under ambient conditions) support ferroelectricity because of the centrosymmetric space-group, it is clear that small variations in composition can lead to significant changes in the structure of the material. It is clear that full understanding of the structural transitions between the linked $C2/c$, $P2$ (or $P2_1$) and $Pnm2_1$ phases is important and could enable the oxygen composition to be tuned to create phases with a non-centrosymmetric structure, and possible multiferroic behaviour. The Ruddlesden-Popper phase seen in this series is also of great interest as it is non-centrosymmetric and to date seen only in two-phase samples. Understanding how the two phases arise from a single precursor will be important to enable specific structures to be synthesized from within the bismuth manganite system. Further work is needed to refine the structures of the monoclinic phases present in the $BiMnO_{2.89}$ and $BiMnO_{3.10}$ samples.

References

[1] Atou T, Chiba H, Ohoyama K, Yamaguchi Y and Syono Y 1997 *J. Solid State Chem.* **145** 639-42
[2] Moreira dos Santos A, Cheetham A K, Atou T, Syono Y, Yamaguchi Y, Ohoyama K, Chiba H and Rao C N R 2002, *Phys. Rev.* B **66** 064425
[3] Sunderesan A, Mangalam R V K, Iyo A, Tanaka Y and Rao C N R 2008 *J. Mater. Chem.* **18** 2191-95
[4] Belik A A, Iikubo S, Yokosawa T, Kodama K, Igawa N, Shamoto S, Azuma M, Takano M, Kimoto K, Matsui Y and Takayama-Muromachi E 2007 *J. Am. Chem. Soc.* **129** 971-77
[5] Yang H, Chi Z H, Li F Y, Jin C Q and Yu R C 2006 *Phys. Rev.* B **73** 024114
[6] Hughes H, Allix M M B, Bridges C A, Claridge J B, Kuang X, Niu H, Taylor S, Song W and Rosseinsky M J 2005 *J. Am. Chem. Soc.* **127** 13790-91
[7] Eggeman A S, Sundaresan A, Rao C N R and Midgley P A 2011 *J. Mater. Chem.* doi: 10.1039/C1JM12840J

Atomic-resolution STEM imaging and EELS-SI of defects in $BiFeO_3$ ceramics co-doped with Nd and Ti

L Q Wang[1], B Schaffer[2], I MacLaren[1], S Miao[3,4], A J Craven[1] and I M Reaney[3]
[1]SUPA, School of Physics and Astronomy, University of Glasgow, Glasgow G12 8QQ, UK
[2]SuperSTEM, Daresbury Laboratories, Keckwick Lane, Warrington WA4 4AD, UK
[3]Department of Materials Science and Engineering, University of Sheffield, Sheffield S1 3JD, UK
[4]Now at Zhongshan Road 457, Dalian Institute of Chemical Physics, Chinese Academy of Sciences, Dalian, China (116023)

E-mail: ian.maclaren@glasgow.ac.uk

Abstract. High resolution scanning transmission electron microscope (HRSTEM) imaging and electron energy loss spectroscopy spectrum imaging (EELS SI) methods have been successfully applied to the analysis of $(Bi_{0.85}Nd_{0.15})(Fe_{0.9}Ti_{0.1})O_3$ (BNFT) ceramics. The atomic scale structural and chemistry of defects in these ceramics have been determined, leading to the discovery of a new class of Nd-rich nanoprecipitates in these ceramics.

1. Introduction

In recent years, $BeFeO_3$ (BFO), which is the only multiferroic material possessing magnetoelectric effects at room temperature, has gained considerable attention. At room temperature, BFO has rhombohedrally distorted ABO_3 perovskite-like structure with R3c symmetry. Below the Curie temperature ($T_c \approx 830^{\circ}C$), BFO is ferroelectric [1]. Neutron diffraction measurements [2] show that BFO is antiferromagnetic (AFM) below Neel temperature ($T_N \approx 370^{\circ}C$) with a G-type ordering along $[001]_h$ (where h=hexagonal) with each Fe^{3+} magnetic ion surrounded by six Fe^{3+} nearest neighbours with anti-parallel magnetic moments. High resolution neutron diffraction studies [3] and line-shape analysis of nuclear magnetic resonance spectra [4] have confirmed that this G-type AFM is actually modified by a long-range cycloidal spiral modulation in the $[110]_h$ direction with a period, $\lambda = 620Å$. Due to this space modulated spin structure (SMSS), it is difficult to observe the linear magnetoelectric effect (ME) in BFO, and recent studies focused on using rare earth A-site doping to suppress the SMSS by stabilising a collinear AFM structure [5-9]. In addition to the suppression of the SMSS, part substitutions of Bi^{3+} can also minimise the occurrence of oxygen or bismuth vacancies, together with secondary phases [7-9]. However, the leakage current caused by charged vacancies is still too high for the A-site doped BFO to be used as a dielectric or ferroelectric, and recent work has used A and B site co-doping to reduce the conductivity. [10-13]. Kalantari *et al.* [10] reported that Ti^{4+} doping reduced the conductivity of $PbZrO_3$-structured $Bi_{0.85}Nd_{0.15}FeO_3$ (BNFO) significantly. However, at 10% Ti doping unusual nanoscale features are observed. This study reports the first combined structural and chemical characterisation of these features and they are shown to be Nd-rich nanorod precipitates.

2. Experimental details

BNFT ceramics were prepared at the University of Sheffield using conventional mixed oxide synthesis [10]. Samples for transmission electron microscopy (TEM) were prepared by conventional methods culminating in ion milling with a Gatan precision ion-polishing system. TEM samples were also coated with a few nm of carbon on both sides to prevent charging from electron beam. Areas for study by HRSTEM were identified using FEI Tecnai T20 operated at 200 kV with images and diffraction patterns recorded using an Olympus SIS Megaview III camera. HRSTEM was performed at the SuperSTEM laboratory using a Nion UltraSTEM equipped with a 3rd generation quadrupole-octupole C_s-corrector and a Gatan Enfina spectrometer operating at an accelerating voltage of 100 kV. Both high angle annular dark field (HAADF) and EEL-SI data were taken from the same areas of the sample. The beam convergence semiangle and spectrometer collection semiangle were set at 30mrad and 31mrad, respectively. The SI data was post-processed by principal component analysis (PCA) [14] using the MSA (multivariate statistical analysis) plug-in for Digital Micrograph prior to the generation of the elemental maps to remove random noise from the spectra. A low pass filter consisting of the following matrix was used on the final elemental maps to remove a little more noise:

$$\begin{pmatrix} 1 & 1 & 1 \\ 1 & 4 & 1 \\ 1 & 1 & 1 \end{pmatrix}$$

3. Results, Analysis and Discussion

A dark field TEM image and its related diffraction pattern along the $[001]_p$ direction of one area of interest are shown in Figures 1(a) and 1b), respectively. The diffraction pattern of Figure 1(b) shows that the structure of BNFT is still the $PbZrO_3$-like structure as reported in Ref. [15], even if the quadrupled reflections between the strong perovskite sublattice reflections are somewhat streaky and diffuse. The dark field image of Figure 1a) also shows 3 diagonal boundaries across the field of the image, and in all regions to the top left of the lowest boundary there is a highly unusual contrast in the material of apparently randomly scattered white dots. Further investigation using HRSTEM with HAADF imaging [Figure 1(c)] reveals that those bright spots are actually defects with a clearly recognisable pattern of 8 atoms (where there would be 9 in a perfect perovskite) with two columns of atoms in the centre surrounded by three other columns atoms on each side of it.

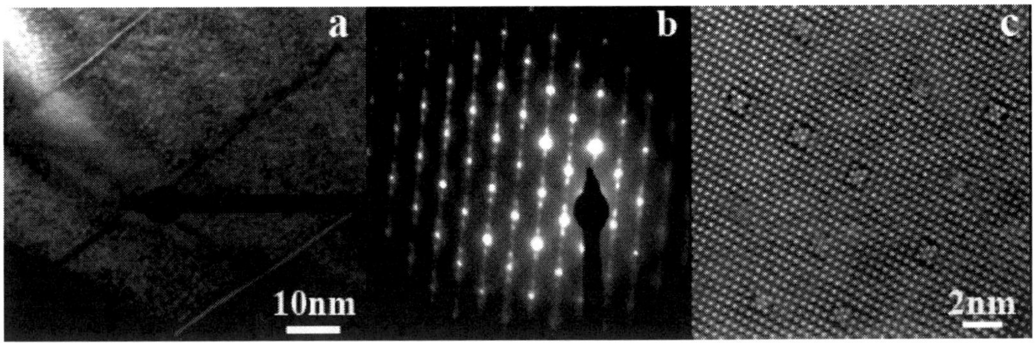

Figure 1. (a) Dark-field image shows defects as bright spots at lower magnification; (b) $<001>_p$ diffraction shows $\frac{1}{2}\{011\}_p$ reflections due to the rotation of FeO_6 octahedra and $\frac{1}{4}\{011\}_p$ reflections relating to the ordered cation displacement; (c) HAADF image reveals defects with an 8-atom pattern.

In order to determine the structure of the defects quantitatively and thereby to build accurate defect models, the atom positions were determined by the following procedure. Firstly, series of HAADF images were taken with very short dwell times of 5μs for each pixel, to reduce the effects of drift on individual images; figure 2(a) is one slice of such an image series. In order to enhance the image contrast and suppress the noise, spatial drift correction was performed on the image series using a specially designed plug-in for Digital Micrograph [16]; figure 2(b) is a sum of the images after spatial drift correction. The positions of the intensity peaks for the A-site columns are determined using a specific image-processing tool entitled iMtools [17]. The peak position list is then exported to a spreadsheet to do the data processing. In order to determine the A-site atoms positions as accurately as possible, horizontal and shear corrections were carried out on the atomic position data, since the original image is slightly distorted due to imperfections and miscalibrations in the scanning system. This can be seen in the image as a distortion of the simple square perovskite cell. Images of six different 8-atom defects (in both possible orientations) have been processed and all have very similar atom positions with standard deviations on the average column positions averaging about 8.5 pm for A-site atoms. The atom positions together with their error bars are shown on Figure 3.

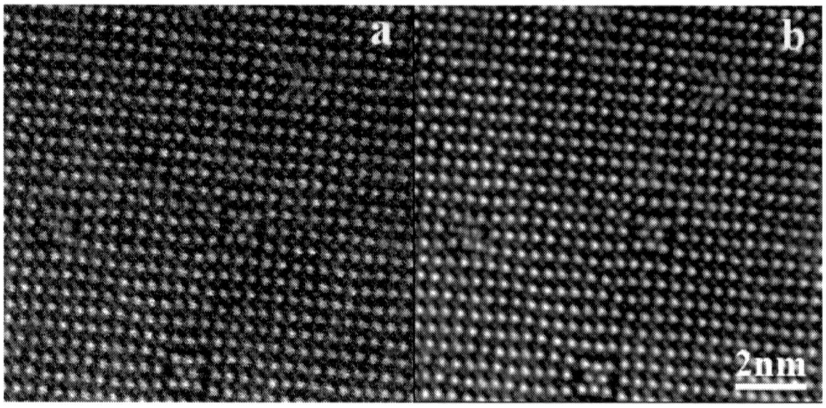

Figure 2. (a) one slice from a series of images taken using a dwell time of 5μs/pixel; (b) a sum of the image series after spatial drift correction.

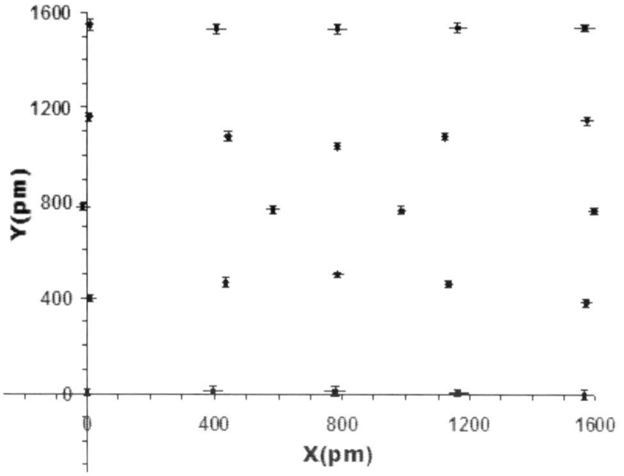

Figure 3. A-site column positions for the defect and its surrounding atoms, including errors on the column positions, expressed as 95% confidence intervals (1.96σ) to make the error bars more visible.

Atomic resolution EELS SI and elemental maps have been carried out with one of these 8-atom defects. Figure 4 shows the elemental maps after PCA has been applied to the original spectrum image to remove random noise and after application of a weak low pass filter to remove remaining random noise. These maps show that the two central atoms are very rich in Nd while the B-site ions like Fe are strongly depleted in the defect core, Ti is also enriched on some B-site columns at the edges of the defect. Although there is some contrast in the oxygen map, this mainly corresponds to signal depletion due to heavy scattering by the very high Z A-site columns, where there is no oxygen, thus there is little information in the map about oxygen distribution within the defect. The form of defects is likely to be nanorod precipitates since the columns are clearly defined and totally different from the surrounding perovskite lattice in images such as in Figure 4. Early evidence from side view images suggests that these defects have lengths of about 5 nm.

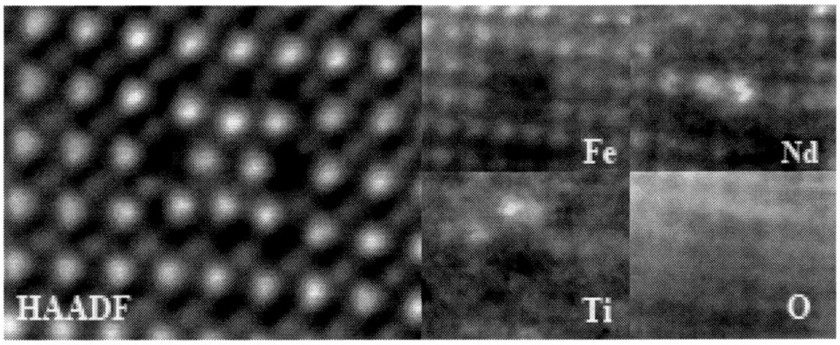

Figure 4. Atomic-resolution EELS SI mapping the elements in the 8 atom defects.

4. Conclusion

These fascinating, and previously unknown, structures of nanorod precipitates show us that they are clearly results of the co-doping of Nd and Ti. The nanorod precipitates act as sinks for excess Nd beyond the solid solubility limit, but also attract some Ti. As stated in the introduction, the purpose of doping with Nd and Ti is to compensate oxygen vacancies due to the volatility of Bi or the partial reduction of Fe^{3+} to Fe^{2+}. In this study, however, it is clearly shown that over-doping with Ti and Nd together has destabilized the structure of the material, resulting in the formation of a high density of Nd-rich nanorod precipitates, as shown in Figure 1(c). The precipitation of Nd together with the segregation of Ti reduces the fraction of these elements in solid solution in the perovskite phase, and thus the influence of doping on properties is likely to deviate from linearity, which may partly explain why T_C decreases non-linearly with increasing Ti doping as was observed by Kalantari *et al.* [10]

Acknowledgement

The authors gratefully acknowledge the funding for this work by the EPSRC including the funding of the SuperSTEM facility (EP/D040205/1, EP/D040396/1, EP/D040566/1), access to SuperSTEM for this work (EP/I000879/1), funding for the multiferroic ceramics work at Sheffield (EP/G069069/1and EP/G005001/1), and the funding of a DTA studentship for LQ Wang.

References

[1] Teague J R, Gerson R and James W J 1970 *Solid State Commun.* **8** 1973
[2] Kiselev S V, Ozerov R P and Zhdanov G S 1963 *Sov. Phys.-Dokl.* **7** 742
[3] Sosnowska I, Neumaier T P and Steichele E 1982 *J. Phys. C: Solid State Phys.* **15** 4835
[4] Zalesskii AV, Zvezdin A, Frolov A and Bush A 2000 *JETP Lett.* **71** 465
[5] Zalesskii AV, Frolov A A, Khimich T A and Bush A A 2003 *Phys. Solid State* **45** 141
[6] Lee Y H, Wu J M, and Lai C H 2006 *Appl. Phys. Lett.* **88** 042903
[7] Yuan G L, Wing Or S, Liu J M, Liu Z G 2006 *Appl. Phys. Lett.* **89** 052905
[8] Yuan G L and Wing Or S 2006 *J. Appl. Phys.* **100** 024109
[9] Das S R, Choudhary R N P, Bhattacharya P, Katiyar R S, Dutta P, Manivannan A and Seehra M S 2007 *J. Appl. Phys.* **101** 034104
[10] Kalantari K, Sterianou I, Karimi S, Ferrarelli M C, Miao S, Sinclair D C and Reaney I M 2011 *Advanced Functional Materials* doi: 10.1002/adfm.201100191
[11] Cui Y F, Zhao Y G, Luo L B, Yang J J, Chang H, Zhu M H, Xie D and Ren T L 2010 *Appl. Phys. Lett.* **97** 222904
[12] Yu B F, Li M, Wang J, Pei L, Guo D and Zhao X 2008 *J. Phys. D: Appl. Phys.* **41** 185401
[13] Pradhan D K, Choudhary R N P, Rinaldi C, Katiyar R S 2009 *J. Appl. Phys.* **106** 024102
[14] Bosman M, Watanabe M, Alexander D T L and Keast V J 2006 *Ultramicroscopy* **106** 1024
[15] Karimi S, Reaney I M, Levin I and Sterianou I 2009 *Appl. Phys. Lett.* **94** 112903
[16] Schaffer B, Grogger W and Kothleitner G, 2004 *Ultramicroscopy* **102** 27
[17] Houben L, iMtools imaging processing software, http://www.er-c.org/methods/software.htm

The effect of Ta_2O_5- and ZnO-doping on the Curie temperature of $BaTiO_3$

Yaseen Iqbal and Asad Jamal

Materials Research Laboratory, Institute of Physics & Electronics, University of Peshawar, Pakistan

dryaseeniqbal@yahoo.co.uk

Abstract The phase, microstructure and Curie temperature of $Ba_{1-2x}Ta_{2x}Ti_{1-x}Zn_xO_3$ ($0.005 \le x \le 0.1$) have been investigated in an attempt to widen its temperature stable region. The compositions formed dense, single-phase, fine-grained $BaTiO_3$ ceramics at $x \le 0.01$ and multiphase ceramics at $x \ge 0.015$. The dielectric constant decreased with an increase in the dopant concentration due to second phase formation. The Curie temperature increased from ~123°C at x = 0.005 to ~148°C at x = 0.1 and the optimum sintering temperature decreased from ~1430°C at x = 0.005 to 1276°C at x = 0.1.

1. Introduction

A number of ABO_3-type compounds have excellent electrical properties making them suitable for industrial applications e.g. multilayer ceramic capacitors (MLCCs), transducers, sensors and resonators. To expand the temperature range of their applications, the effect of various dopants on the A and/or B site of the perovskite structure have been tried and investigated [1-3]. Barium meta-titanate ($BaTiO_3$), with ABO_3 type perovskite structure, is one of the most widely used dielectric materials, e.g. as a dielectric in MLCCs due to its high permittivity and low losses, and in electro-optic devices and thermistors because of its good dielectric characteristics [1,2]. $BaTiO_3$ has five isomorphs namely, hexagonal >1430°C, cubic >120°C, tetragonal below 120°C to 5°C, orthorhombic below 5°C to -90°C and rhombohedral below -90°C. Its cubic form is paraelectric while in the other forms its behavior is ferroelectric [4,5]. The low Curie temperature 'T_c' (~120°C) of $BaTiO_3$ limits its temperature stable region for applications like MLCCs. Various dopants have been tried on the A and/or B sites of $BaTiO_3$ in an attempt to overcome this disadvantage. Doping of Ce^{3+} or Y^{3+} on the Ba^{2+} site and Ce^{4+} on Ti^{4+} sites are known to lower T_c [6,7]. Co-doping of $BaTiO_3$ with rare earth ions of relatively smaller radii has been reported to shift T_c towards higher values with increased tetragonality [8,9]. Of all the single element dopants, only Pb doping on the A-site of $BaTiO_3$ could increase T_c to ~490°C but Pb is avoided due to its toxic nature. Recently, Kumada *et al* [10] reported a rise in T_c of $BaTiO_3$ up to 140°C as a result of co-doping with Bi and Cu. Here we report our findings regarding the effect of Ta and Zn co-doping on the phase, microstructure and Curie temperature of $BaTiO_3$.

Journal of Physics: Conference Series **371** (2012) 012035 doi:10.1088/1742-6596/371/1/012035

2. Experimental

$BaTiO_3$ samples were prepared by solid state mix-oxide route using laboratory reagent grade $BaCO_3$, TiO_2, Ta_2O_5, and ZnO. Batches of $Ba_{1-2x}Ta_{2x}Ti_{1-x}Zn_xO_3$ ($0.005 \leq x \leq 0.1$) were mix-milled for 24h in a horizontal ball mill with Y-toughend ZrO_2 balls as milling media and 2-propanol as lubricant. The resulting slurries were dried overnight and sieved. Powder samples were calcined at 1210°C for 15h, with a cooling/heating rate of 10°C/min. The calcined powders were re-milled for 30 min to dissociate agglomerates, dried, sieved and pressed into 2-3 mm high 13 mm diameter pellets at ~100 MPa. The pellets were sintered for 2h at temperatures ranging from 1276°C to 1430°C for different compositions at 5°C/min. Density of sintered pellets was measured using a MD-300S Electronic densitometer. A Philips 1700 series X-ray diffractometer (Cukα radiations with λ ~1.54Å) operating at 40 kV and 30 mA was used for phase analysis. Microstructural characterization of thermally etched gold-coated samples was performed using a JEOL JSM-5910 SEM operating at 15-20 kV. Electrical properties were measured using a HP 4287A LCR meter from RT-200°C at 1 kHz to 1 MHz.

3. Results and Discussion

XRD revealed the presence of tetragonal $BaTiO_3$ (PDF# 5-626) only in the samples calcined at 1210°C and sintered at 1430°C and 1400°C at low ($x \leq 0.01$) doping concentrations (figure 1); however, with an increase in the amount of dopant >0.01, additional XRD peaks due to secondary hexagonal $Ba_3Ta_{3.2}Ti_5O_{21}$ (PDF# 83-2318) phase emerged (figure 1). A few more low intensity peaks (labeled as '?') were also present on the XRD patterns from $x \geq 0.015$ samples which could not be identified.

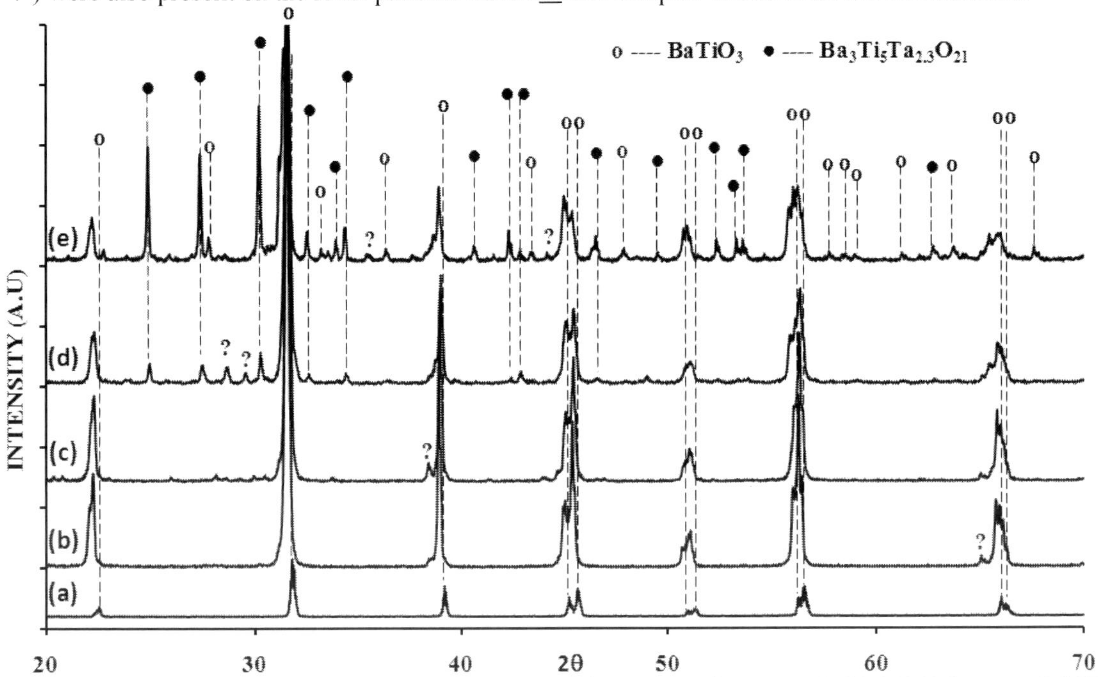

Figure 1. XRD patterns of $Ba_{1-2x}Ta_{2x}Ti_{1-x}Zn_xO_3$ sintered at a) 1250°C at x = 0, b) 1430°C at x = 0.005, c) 1400°C at x = 0.01, d) 1320°C at x = 0.015 and e) 1276°C at x = 0.1 showing almost single phase $BaTiO_3$ at $x \leq 0.01$ and multi-phase ceramics at $x \geq 0.015$.

SEM revealed the presence of somewhat hexagonal-shaped $BaTiO_3$ grains of size ~0.3x0.3µm² to <1x1µm² in the sintered microstructure of $Ba_{1-2x}Ta_{2x}Ti_{1-x}Zn_xO_3$ (x=0.005) (figure 2a) and the grain size decreased up to ~0.2x0.2µm² to 0.5x0.5µm² with an increase in x to 0.01(figure 2b). At x = 0.015, the general morphology of the grains appeared unchanged; however, a few relatively larger grains labeled as 'S' containing ~67wt% $BaCO_3$, 25wt% TiO_2 and 8wt% Ta_2O_5 indicative of second phase formation could be seen (figure 2c). The microstructure of the samples with x = 0.1 (figure 2d) demonstrated a multiphase microstructure with elongated grains labeled as 'R' containing ~51wt% $BaCO_3$, 41wt% TiO_2, 6.5wt% Ta_2O_5 and 1.5 wt% ZnO and deformed irregular-shaped massive grains labeled as 'M' containing ~34wt% $BaCO_3$, 20wt% TiO_2 and 46wt% Ta_2O_5 in a fine grained matrix of composition close to $BaTiO_3$. BF-TEM images and corresponding SADPs from sintered $Ba_{1-2x}Ta_{2x}Ti_{1-x}Zn_xO_3$ samples showed single-phase compound iso-structural with tetragonal $BaTiO_3$ at x=0.005 (figure 3a) and an additional $Ba_3Ta_{3.2}Ti_5O_{21}$ phase at x=0.1 (figure 3b) consistent with XRD results.

Figure 2. SEIs from $Ba_{1-2x}Ta_{2x}Ti_{1-x}Zn_xO_3$ samples showing a-b) visibly fine-grained, single phase microstructure at x≤0.01 and multiphase microstructure at x ≥0.015.

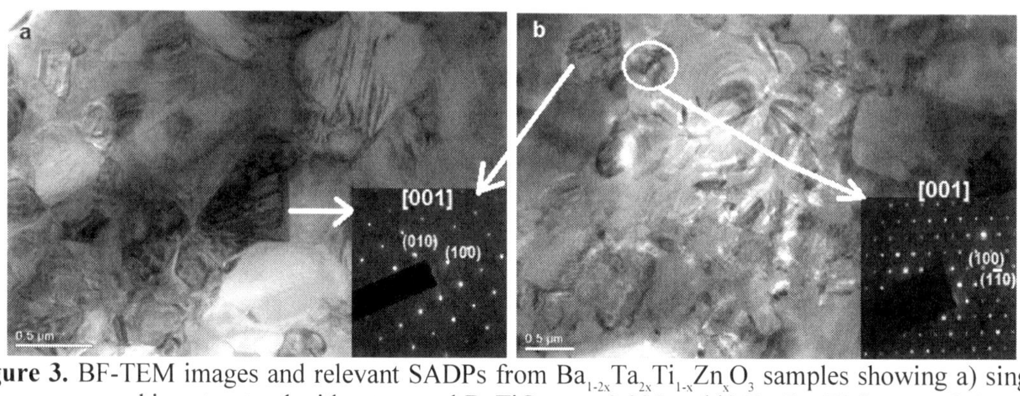

Figure 3. BF-TEM images and relevant SADPs from $Ba_{1-2x}Ta_{2x}Ti_{1-x}Zn_xO_3$ samples showing a) single-phase compound iso-structural with tetragonal $BaTiO_3$ at x=0.005 and b) $Ba_3Ta_{3.2}Ti_5O_{21}$ at x=0.1.

The maximum ε_r~3000 at 10 kHz to 1 MHz (figure 3a) measured in the present study corresponded to the ~95% dense sample with minimum dopants concentration (x = 0.005). The relative density slightly increased from 95% to 97% with an increase in dopant concentration from x=0.005 to 0.1; however, ε_r decreased (figure 4a-b) probably due to second phase formation. Keeping in view the very small variation in density with increase in x from 0.005 to 0.1, the observed increase in Curie temperature from ~123°C to 148°C may be due to the increase in dopant concentration (figure 4a-b). This indicates that Ta and Zn are more effective dopants at low concentrations than Bi and Cu [10] and can be used to widen the temperature stable region of $BaTiO_3$.

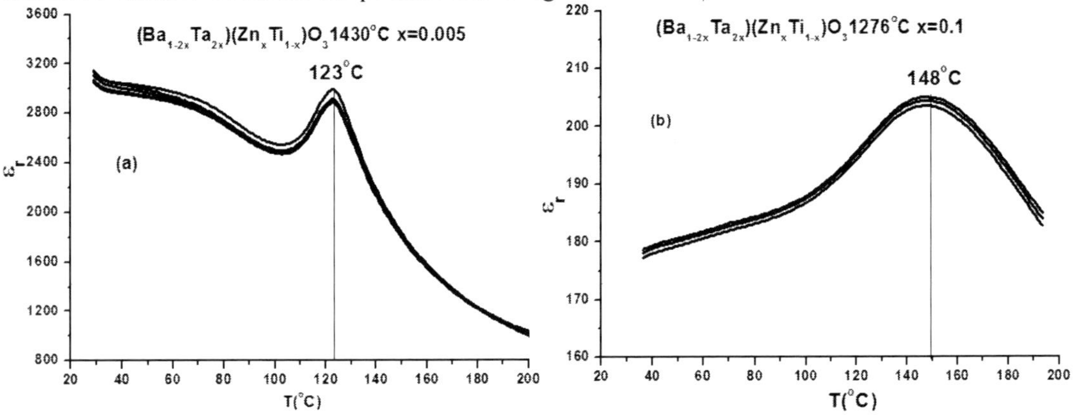

Figure 4. The plots of ε_r at 10 kHz to 1 MHz versus temperature for co-doped $BaTiO_3$ showing an increase in T_c from ~123°C for x = 0.005 to ~148°C for x = 0.1.

4. Conclusions

$Ba_{1-2x}Ta_{2x}Ti_{1-x}Zn_xO_3$ sintered to single phase dense ceramics with ε_r~3000 at $x \leq 0.01$. The highest T_c (~148°C) was achieved for the composition with x=0.1 but with decrease in ε_r due to second phase formation at $x \geq 0.015$. The optimum sintering temperature decreased from ~1430 to 1276°C with increase in x from 0.005 to 0.1.

Acknowledgement

The authors acknowledge the financial support of the Higher Education Commission Pakistan and US Academy of Sciences for the Development of Materials Connection (Project ID 131).

References

[1] Buscagliaa V, Viviania M, Buscagliaa M T, Nannib P, Mitoseriub L, Testinob A, Stytsenkoc E, Daglishc M, Zhaod Z and Nygrend M 2004 *Powder Technology* **148** 24-27.
[2] Lee B I, Badheka P, Yoon D H, Magadala V and Wang M 2004 *J. Ceram. Proces. Res.* **5** 127-32.
[3] Shukla A, Choudhary R N P, Thakur A K, Pradhan D K 2010 *Physica B* **405** 99-106.
[4] Miller C A 1967 *Br. J. Appl. Phys.***18**(12) 1689- 97.
[5] Langhammera H T, Muller T, Bottcher R, Mueller V and Abicht H P 2004 *J. Eur. Ceram. Soc.* **24** 1489-92.
[6] Kim J H, Yoon S H and Han Y H 2007 *J. Eur. Ceram. Soc.* **27** 1113-16.
[7] Hwang J H and Han Y H 2001 *J. Am. Ceram. Soc.* **84(**8) 1750-54.
[8] Jo S K, Park J S and Han Y H 2010 *J. Alloys Compds* 501 259-64.
[9] Jung Y S, Na E S, Paik U, Lee J and Kim J 2002 *Mat. Res. Bull.* **37** 1633-40.
[10] Kumada N, Ogiso H, Shiroki K, Wada S, Yonesaki Y, Takei T and Kinomura N 2010 *Mat. Lett.* **64** 383-85.

Atomic scale structure and chemistry of anti-phase boundaries in $(Bi_{0.85}Nd_{0.15})(Fe_{0.9}Ti_{0.1})O_3$ ceramics

L Q Wang[1], B Schaffer[1,2], I MacLaren[1], S Miao[3,4], A J Craven[1] and I M Reaney[3]
[1]SUPA, School of Physics and Astronomy, University of Glasgow, Glasgow G12 8QQ, UK
[2]SuperSTEM, Daresbury Laboratories, Keckwick Lane, Warrington WA4 4AD, UK
[3]Department of Materials Science and Engineering, University of Sheffield, Sheffield S1 3JD, UK
[4]Now at Zhongshan Road 457, Dalian Institute of Chemical Physics, Chinese Academy of Sciences, Dalian, China (116023)

E-mail: ian.maclaren@glasgow.ac.uk

Abstract. High resolution scanning transmission electron microscope (HRSTEM) and electron energy loss spectroscopy spectrum imaging (EELS SI) methods have been successfully applied to the analysis of planar antiphase boundaries in $(Bi_{0.85}Nd_{0.15})(Fe_{0.9}Ti_{0.1})O_3$ (BNFT) ceramics. Two distinct structures for these boundaries have been identified, one of which has never previously been reported for perovskites. This latter form has a characteristic ladder structure in the core, and comes in stepped and unstepped forms. Atomic resolution spectrum imaging reveals that excess Ti is key to the formation of these ladder-structured antiphase boundaries.

1. Introduction

There has been great interest in bismuth ferrite, $BiFeO_3$ (BFO), in recent years because of its purported multiferroic properties at room temperature. Nevertheless, there are significant issues to be addressed before applications for bulk ceramics may be countenanced; the magnetic ordering is antiferromagnetic (AFM) but canted over long range, and the electrical conductivity is too high to access the dielectric or ferroelectric properties. Recent work has concentrated on using donor doping on the B-site of the ABO_3 perovskite sublattice to prevent oxygen vacancy (V_O) formation and thereby reduce conductivity [1-3], together with rare earth (RE) A-site site to modify the magnetic ordering [1-5]. Combining such A- and B-site doping results in the formation of a low conductivity ceramic with stable AFM ordering [6-9]. Some of the best results to-date [6] have been achieved by a combination of Nd and Ti doping in the solid solution $Bi_{1-y}Nd_yFe_{1-x}Ti_xO_3$ but it was noted that although T_N varies linearly with Ti concentration from 361°C at 1% Ti to 300°C at 10% Ti, T_C does not (reaching a value of 250°C at 10% Ti). The present work uses a combination of high angle annular dark field (HAADF) imaging and electron energy loss spectroscopy spectrum imaging (EELS SI) in an aberration corrected scanning transmission electron microscope to study nanoscale features of the microstructure of these ceramics. It is found that planar or stepped antiphase boundaries (APBs) are frequently observed with a unique structure that has never previously been observed in perovskites. The details of the structure and chemistry of such boundaries are elucidated.

2. Experimental details

BNFT ceramics were prepared using conventional mixed oxide synthesis as detailed elsewhere [10]. Samples for transmission electron microscopy (TEM) were prepared by conventional mechanical methods, followed by ion milling and carbon coating. Preliminary TEM imaging and diffraction investigations were preformed using a FEI Tecnai T20 operated at 200 kV. Crystal directions referred to with a subscript $_p$ (e.g. $[100]_p$) refer to the primitive perovskite cell and not the larger orthorhombic antiferroelectric cell. HRSTEM was performed using a Nion UltraSTEM equipped with a 3rd generation quadrupole-octupole C_s-corrector and a Gatan ENFINA spectrometer operating at an accelerating voltage of 100 kV at the SuperSTEM laboratory. The beam convergence and spectrometer collection semiangles were set at 30mrad and 31mrad, respectively. To overcome image distortions due to drift, series of HAADF images were taken at very short dwell times of 5μs for each pixel, the images in the series aligned, and then summed using the SDSD plugin for Digital Micrograph [10] to produce drift free images with high signal to noise ratios. Prior to the generation of the elemental maps the SI data was post-processed by principal component analysis (PCA) [11] using the multivariate statistical analysis plug-in for Digital Micrograph to remove random noise from the spectra. A light low pass filter (a convolution with the following matrix) was used on the final maps to reduce the residual noise:

$$\begin{pmatrix} 1 & 1 & 1 \\ 1 & 4 & 1 \\ 1 & 1 & 1 \end{pmatrix}$$

3. Results, Analysis and Discussion

Planar or near-planar antiphase boundaries are frequently observed in these ceramics and one example is shown in Figure 1 of a stepped APB. So far, at least three different types of APBs have been observed in this material. Figure 2(a) is a higher magnification image of the boundary of Figure 1, which is a stepped APB with a ladder-like structure with pairs of bright, high-Z A-site atoms interspersed with pairs of less bright medium-Z B site atoms on each terrace. The length of each terrace varies and terraces are always shifted by one perovskite subcell with respect their neighbouring terraces. There is a horizontal and vertical shift of half a perovskite cell across this boundary, as required for an antiphase boundary in a perovskite; presumably there is a half cell shift in the viewing direction too, but this is, of course, undetectable. The width of this antiphase boundary structure is about 2 perovskite unit cells. As is clearly visible in Figures 1(a) and (b), there are unusual 8-atom defects to the left of the boundary and not to the right of it. Such defects are Nd-rich nanorod precipitates and are discussed in detail in a separate paper in this Proceedings, but they are characteristically observed along the $[001]_p$ direction of the crystal structure, which has been confirmed been confirmed by selected area diffraction. To the right of the boundary, no such defects were observed, and a diffraction pattern corresponding to a $[100]_p$ or $[010]_p$ direction was observed, without the additional ¼$\{011\}_p$ reflections present in the $[001]_p$ pattern. This means that the **c**-axis of the crystal is rotated by 90° across the boundary. Such an effect

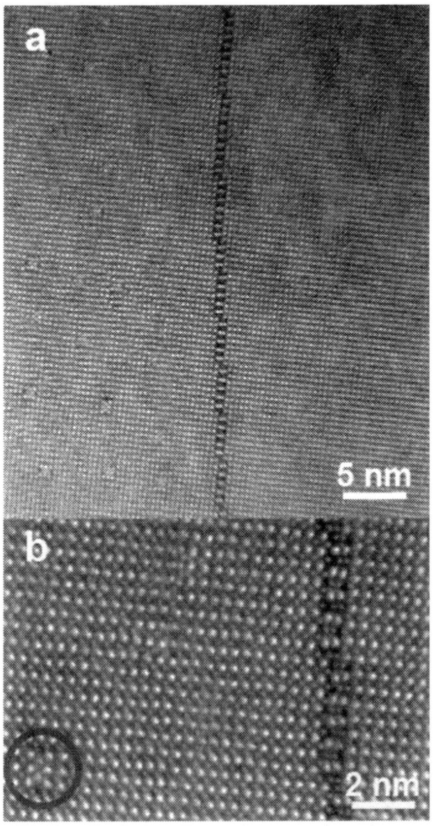

Figure 1. (a) HAADF image showing a large area of the stepped APB; (b) higher magnification HAADF image of with 8-atom nanorod precipitates to the left as circled.

typically occurs for the 60° domain boundary [12,13] in orthorhombic antiferroelectrics. In this case the crystal orientation is rotated by 120° about a $[111]_p$ direction resulting in a cyclic permutation of the perovskite cell vectors (i.e. **c** switches to **a**, and so on), and a rotation of the polarisation direction of 60°. Thus, this boundary is both an APB and a 60° domain boundary.

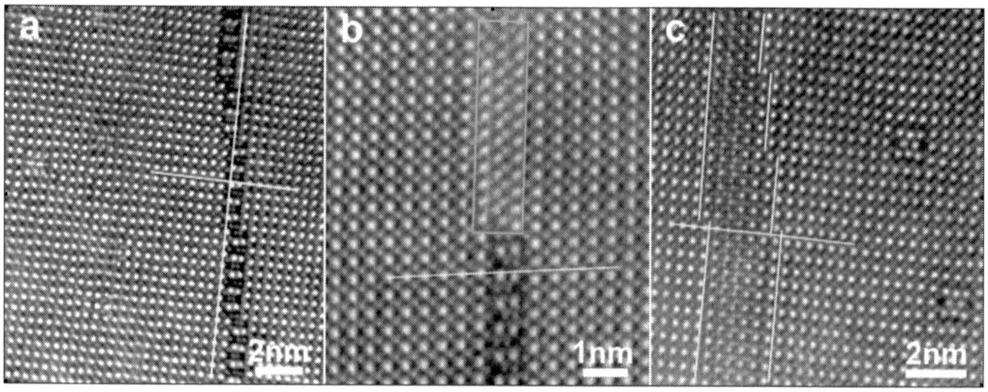

Figure 2. Three different types of APBs revealed by HRSTEM; the half-cell shifts are indicated by yellow lines: (a) stepped APB with ladder structure; (b) straight APB with ladder structure transforming to a double layer rock salt structure (in the pink box); (c) a different structured antiphase boundary (on the left side of the image), but the detailed structure is undetermined (boundary indicated within the cyan lines). (Colour online, in black and white the coloured lines appear as light grey).

Figure 2(b) shows another APB, with the lower part having a similar internal structure to Figure 2(a), but without steps, and without functioning as a domain boundary. The top part of this boundary (indicated with a pink box) transforms to the APB structure more frequently observed in perovskites with two layers of A-site ions and oxygen in a rock salt structure [14], which in this case would be BiO. The area above this (not shown in Figure 2(b)) is perfect crystal without any antiphase boundary, which means that the APB must end at the top of Figure 2(b) in a partial dislocation. A Burgers circuit can be made around the end of the APB and this produces a Burgers vector of type ½<1?0>; this is probably ½<110> but it is impossible to determine the vertical component of **b** from a single projection. Thus, this antiphase boundary transforms from the newly observed ladder structure (without steps) to a classic rock-salt structure before ending at a partial dislocation.

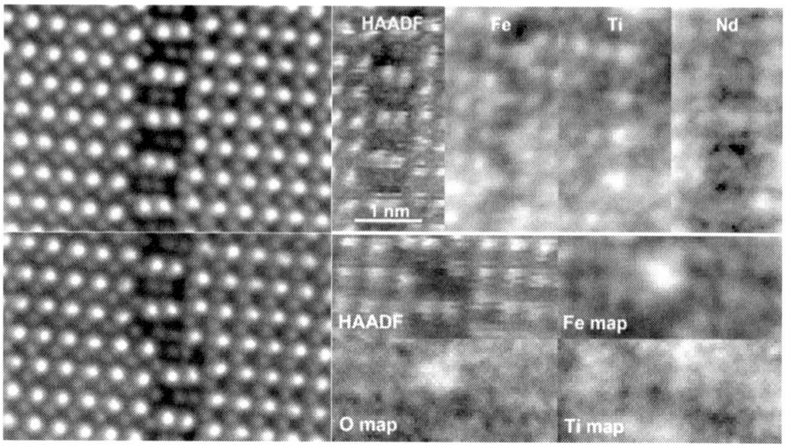

Figure 3. Detailed HAADF images (left) and elemental maps (right), together with a simultaneously recorded HAADF image for a terrace of the boundary of Figure 2(a) (upper half) and a step on the boundary of Figure 2(a) (lower half).

Figure 2(c) shows a third kind of APB. Its detailed structure is unclear, except that its image spread across several cells in the horizontal direction suggesting that the boundary is inclined to the z-direction. It may be that this is similar to the boundary in Fig 2(a), rotated by 90° about the boundary normal: this would lead to an appearance of being spread across 3-4 cells, due to the stepping through the material thickness. Nevertheless, no final conclusions can be reached on this point without further evidence.

Figure 3 shows the elemental maps calculated from spectrum images. These reveal that within the terraces the central B-sites are strongly enriched in Ti, indicating that Ti is critical to the formation of these ladder structures. On each side of the boundary, some columns are enriched in Nd. At the steps between terraces in the lower part of Figure 3, a bright signal for both of Fe and O indicates that an iron oxide structure just a couple of atoms wide forms these steps. Around the edges of the steps, Ti is also enriched. The oxygen map for the terrace and the Nd map for the step, which are not shown here, are relatively featureless and reveal little of interest.

4. Conclusion

A unique structure for an antiphase boundary has been observed in Ti and Nd co-doped $BiFeO_3$ consisting of a ladder-like arrangement of pairs of A-site Bi/Nd atoms interspersed with pairs of B-site Fe/Ti atoms. Two forms of this boundary have been seen, one having steps which form around a group of four Fe-O rich columns, and serving as a 60° domain boundary, whereas the other is a pure antiphase boundary, with no steps, which can finish at a partial dislocation within the crystal. The fact that these unusual planar features are enriched in Ti, suggests that their formation is promoted by a large excess concentration of Ti in the ceramic and that, since they will remove Ti from solid solution, they will account for the observed non-linear dependence of T_C as a function of on Ti concentration [6].

Acknowledgement

The authors gratefully acknowledge the funding for this work by the EPSRC including the funding of the SuperSTEM facility (EP/D040205/1, EP/D040396/1, EP/D040566/1), access to SuperSTEM for this work (EP/I000879/1), funding for the multiferroic ceramics work at Sheffield (EP/G069069/1 and EP/G005001/1), and the funding of a DTA studentship for LQ Wang.

References

[1] Yuan G L, Wing Or S, Liu J M, Liu Z G 2006 *Appl. Phys. Lett.* **89** 052905
[2] Yuan G L and Wing Or S 2006 *J. Appl. Phys.* **100** 024109
[3] Das S R, Choudhary R N P, Bhattacharya P, Katiyar R S, Dutta P, Manivannan A and Seehra M S 2007 *J. Appl. Phys.* **101** 034104
[4] Kiselev S V, Ozerov R P and Zhdanov G S 1963 *Sov. Phys.-Dokl.* **7** 742
[5] Sosnowska I, Neumaier T P and Steichele E 1982 *J. Phys. C: Solid State Phys.* **15** 4835
[6] Kalantari K, Sterianou I, Karimi S, Ferrarelli M C, Miao S, Sinclair D C and Reaney I M 2011 *Advanced Functional Materials* doi: 10.1002/adfm.201100191
[7] Cui Y F, Zhao Y G, Luo L B, Yang J J, Chang H, Zhu M H, Xie D and Ren T L 2010 *Appl. Phys. Lett.* **97** 222904
[8] Yu B F, Li M, Wang J, Pei L, Guo D and Zhao X 2008 *J. Phys. D: Appl. Phys.* **41** 185401
[9] Pradhan D K, Choudhary R N P, Rinaldi C, Katiyar R S 2009 *J. Appl. Phys.* **106** 024102
[10] Schaffer B, Grogger W and Kothleitner G, 2004 *Ultramicroscopy* **102** 27
[11] Bosman M, Watanabe M, Alexander D T L and Keast V J 2006 *Ultramicroscopy* **106** 1024
[12] Tanaka M, Saito R and Tsuzuki K 1982 *Jpn. J. Appl. Phys. Pt. 1* **21** 291
[13] MacLaren I, Villaurrutia R and Pelaiz-Barranco A 2010 *J. Appl. Phys.* **108** 034109
[14] Lebedev OI, Van Tendeloo G, Abakumov AM, Amelinckx S, Leibold B, Habermeier HU 1999 *Philos. Mag. A* **79** 1461

(S)TEM analysis of the interdiffusion and barrier layer formation in Mn/Cu heterostructures on SiO_2 for interconnect technologies

J. G. Lozano[1,2], S. Lozano-Perez[1], J. Bogan[2], Y. C. Wang[3], B. Brennan[2], P. D. Nellist[1] and G. Hughes[3]

[1]Department of Materials, University of Oxford, Parks Road, OX13PH, Oxford, UK
[2]School of Physical Science, Dublin City University, Glasnevin, Dublin 2, Dublin, Ireland
[3]FEI Company, Hilsboro, Oregon, USA

Email: juan.lozano@materials.ox.ac.uk

Abstract. Mn/Cu heterostructures thermally evaporated onto SiO_2 and subsequently annealed were investigated by transmission electron microscopy (TEM) related techniques in order to study the diffusion interactions which lead to barrier layer formation. Electron energy loss spectroscopy provide evidence for the interdiffusion between the Mn and Cu layers following a 450 °C anneal, where the Mn diffuses towards the surface of the structure, while Cu diffuses towards the Mn/SiO_2, surrounding metallic Mn clusters but not propagating into the dielectric. The chemical composition of the 2-3 nm interfacial layer is primarily a mixture of +2 and +3 Mn valencies, in good agreement with previously reported results.

1. Introduction

Copper has progressively become the material of choice for transistor interconnects in the current processing technology for semiconductor devices; mainly due to its lower resistivity and improved electromigration damage compared with aluminium. However, the high diffusion rates between Cu and the silicon oxides or polymers used as insulating layers requires the introduction of physical barriers to prevent interdiffusion across the interface. These barriers are commonly made of a stack of Ta/TaN layers, but they present a poor electrical conductivity and increase the total resistance of the interconnect. Therefore, there is a need to reduce the thickness of the barriers while still preserving their efficiency, high quality interfaces and adhesion to the adjacent layers. Moreover, as the semiconductor industry moves towards the 32 nm technological node, these barriers are no longer a viable option and other approaches have to be developed.

Cu(Mn) self-forming diffusion barriers have become a promising option to overcome these problems. Their formation process involves the deposition of a Cu(Mn) alloy directly onto the SiO_2 layer and after annealing, the alloying element segregates towards the interface chemically reacting with the insulator. The result is the formation of a $Mn_x(Si)O_y$ thin barrier layer (2-3 nm[1]) that can avoid the diffusion of Cu into the SiO_2. These layers have already be proven to be effective, but unfortunately the information about their exact nature and composition still remains limited and in some cases contradictory [2-4]. All this discrepancy may arise from the fact that all these studies have been performed so far on Cu(Mn) alloys, but the basic elemental interactions, rather than the alloy interactions, between pure Mn and the substrate layer have not been considered yet. Therefore, the chemical interaction between pure metallic Mn and SiO_2, and the basic interaction between metallic Cu and Mn as a function of annealing could reveal important information.

2. Experimental

A series of samples consisting of 30 nm of metallic Mn capped with 20 nm of metallic Cu were grown by thermal evaporation in a Leybold Univex Chamber at a vacuum pressure of 2×10^{-6} mbar. The substrates employed were 600 nm of amorphous SiO_2 on Si(001) and the samples were studied after annealing at 450 °C for 2 hours at a pressure of 10^{-8} mbar. Conventional transmission electron microscopy (TEM), high resolution TEM (HRTEM), high angle annular dark field (HAADF) and EELS were carried out in a JEOL 3000 and a Tecnai Osiris electron microscopes operating at 300 kV and 200 kV respectively. Samples in cross section geometry were prepared by the conventional method of grinding and polishing followed by Ar+ ion milling in a Gatan PIPS until electron transparent.

3. Results

In Figure 1, a low magnification conventional TEM micrograph of the heterostructure is displayed. In all the areas of the sample, continuous layers with smooth interfaces and good adhesion to the substrate are observed. The Mn layers have a more uniform thickness while the Cu layer displays a higher degree of roughness which is attributed to its bigger average grain size. HAADF micrographs of this sample show an unexpected bright contrast in a region approximately 10 nm thick above the interface between the Mn layer and the SiO_2. Since the difference in contrast in the images obtained with this technique is directly related to the difference in Z number of the elements that are present, these brighter regions must contain an element or compound with an average Z number higher than Mn.

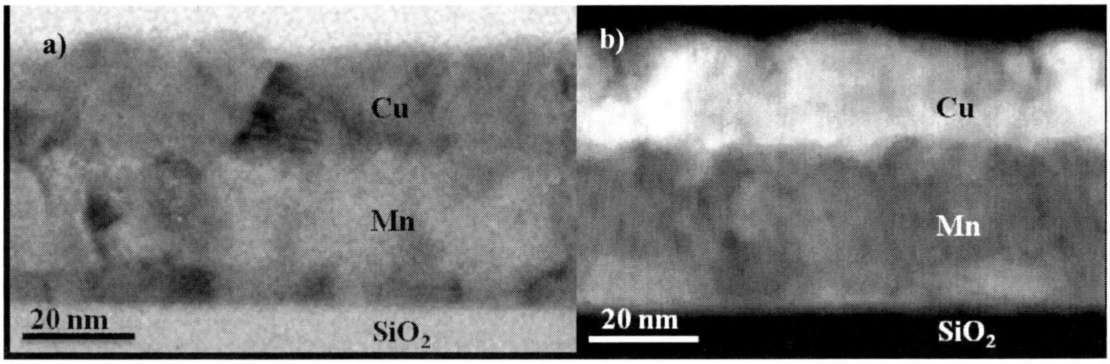

Figure 1: a) TEM image of the Cu/Mn heterostructure on SiO_2; b) HAADF micrograph of the same sample showing the unexpected brighter contrast at the interface Mn/SiO_2

To confirm this, EELS elemental maps were recorded in STEM mode of the different layers. The nominal size of the probe used is 1 nm and the maps contain 100x100 pixels. The extracted signal of the different elements is obtained after background subtraction and noise reduction by a multivariate statistical analysis [5]. A Hartree-Slater step function was considered to take into account the effects of sample thickness and the remaining signal under the peaks was integrated. The results of the application of this procedure for Cu *L*, Mn *L* and O *K* peaks are shown in Figure 2 a)-c). A relatively large amount of Mn can be found in the Cu layer as expected; but also segregation of Cu into the Mn layer has occurred. This segregated Cu tends to encapsulate areas of the Mn layer close to the interface, where a higher Mn signal and lower O is found. Since the oxygen arises from the exposure of the TEM sample to environmental conditions, and is more accentuated in the Mn layer with the exception of these regions, this means that the segregated Cu is preventing these encapsulated areas from being as heavily oxidized as the rest, and as a consequence the metallic Mn signal is stronger.

An important fact to highlight is that no trace of Cu is found in the SiO_2 layer, which can be attributed to the formation of an effective diffusion barrier layer.

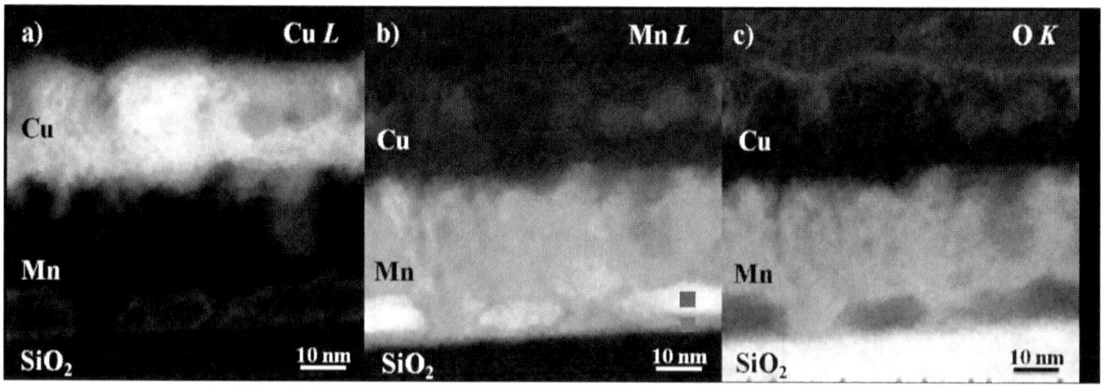

Figure 2: EELS extracted signal elemental maps for a) Cu L b) Mn L and c) O K edges. Squares in b) indicate the position where the spectra shown in Figure 3 were taken.

Finally, the nature of the chemical reactions at the Mn/SiO_2 interface to form a diffusion barrier are now explored by EELS. Here we have estimated the Mn peaks $I(L_3)/I(L_2)$ intensity ratio after background subtraction to investigate possible variations of the Mn valence state along the layer. Firstly, due to the lack of available data in the literature for typical EELS spectra of metallic manganese, reference spectra were taken from a sample consisting of deposited Mn kept under vacuum conditions during the whole process to avoid oxidation. An example is displayed in Figure 4, where an average intensity ratio of 1.23±0.2 was calculated. For the valence state of Mn in different oxides, we rely on the data from Schmidt et al [6].

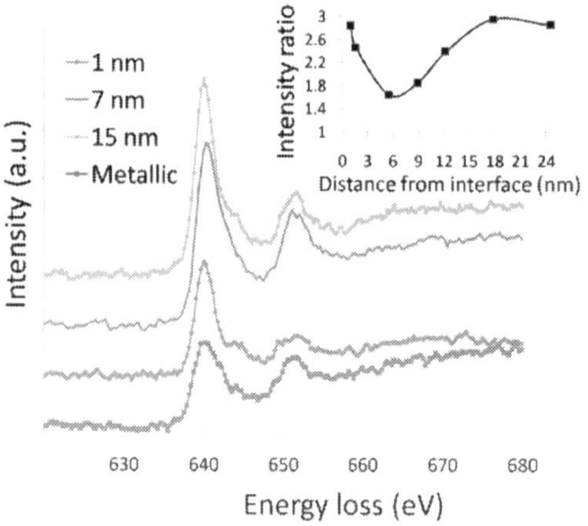

Figure 3. EEL spectra recorded in the Mn layer at different positions from the interface, and metallic Mn taken as reference. In the inset, variation of the intensity ratio I_2/I_3 in the interfacial region indicating different valence states for Mn.

Figure 3 displays several point spectra which were recorded at different positions through the lower half of the Mn layer in the sample starting from the Mn/SiO_2 interface. From the calculated intensity ratios displayed in the inset in Figure 3, clear variations in the Mn oxidation state can be deduced. We can distinguish three different regions in the layer: the first one, with a thickness of 2-3 nm situated just above the interface, where the intensity ratio suggests a mixture of Mn valence of +2 and +3, consistent with the results reported by Otsuka et al. [4]. A second region (I_3/I_2~2.9) corresponds to the bulk Mn layer. Taking into account the results from Schmidt et al. [6], the relative intensities of the two peaks in this area should contain Mn_2O_3 (I_3/I_2=2.8) as the main constituent. However, as the surface of the original metallic Mn layer was oxidized due to the exposure of the sample to environmental conditions, we believe that the EEL spectra in that region do not correspond to those of a layer of Mn_2O_3 but to the overlapping of the spectra corresponding to metallic Mn (I_3/I_2=1.2) sandwiched between the layers of Mn oxides (I3/I2=3.9) along the beam path in XTEM. Finally, the third region is located in the encapsulated Mn area, where the measured intensity ratio (I_3/I_2~1.7) suggests a higher content of metallic manganese in agreement with the EELS results.

4. Discussion

The diffusion of Mn in Cu in Cu(Mn) alloys have been widely reported in the literature, but to the best of our knowledge, no published data exists about diffusion of Cu in Mn in thick layered structures. We therefore propose that similar to copper diffusion in other metals, the polycrystalline nature of the Mn layer is the reason for the presence of Cu in this layer. The small average Mn grain size creates numerous paths for the migration of Cu atoms into the layer. Similarly, Mn would migrate to the Cu layer following the same mechanism. The 450°C anneal of the Cu/Mn layered structure results in the formation of metallic Mn clusters surrounded by a thin layer of Cu. For the upper layer, the migration of Mn atoms towards the surface of a Cu matrix is mainly enhanced by the presence of residual oxygen contamination [7] introduced in the deposition process and subsequent handling, to form MnO via the displacement reaction $Cu_2O+Mn \rightarrow 2MnO$ during the vacuum annealing. Equivalently, for the solute Cu in the Mn matrix (lower layer), it is expected that Cu atoms will migrate to the interface Mn/SiO_2 to form a more stable oxide. However, these Cu atoms do not reach the dielectric, since the reaction between the Mn atoms and the SiO_2 at the interface leads to the formation of a $MnSiO_3$ self-formed barrier layer that prevents interdiffusion.

Acknowledgements

J. G. Lozano would like to acknowledge the support of the European Commission under the Marie Curie Programme 2009. Financial support from the SFI PI Program Science under the Grant No. 08/IN.1/I2052 is also acknowledged.

5. Conclusions

In summary, Mn/Cu heterostructures thermally evaporated onto SiO_2 and subsequently annealing were investigated by TEM related techniques. EELS results indicate the occurrence of interdiffusion between the Mn and Cu layers, where the Mn tends to diffuse towards the surface of the structure, and the Cu atoms diffuse towards the Mn/SiO_2 interface. EELS analysis would also indicate that the chemical composition of the 2-3nm interfacial layer is primarily Mn in +2 and +3 oxidation states, which is effective at preventing Cu diffusion into the dielectric layer.

References

[1] Koike J and Wada M 2005 *Appl. Phys. Lett.* **87**, 041911
[2] Koike J, Haneda M, Iijima J, Otsuka Y, Sako H, Neishi K 2007 *J. Appl. Phys.* **102**, 043527
[3] Ablett J M, Woicik J C, Tokei Zs, List S, Dimasi E 2009 *Appl. Phys. Lett.* **94**, 042112
[4] Otsuka Y, Koike J, Sako H., Ishibashi K, Kawasaki N, Chung S M, Tinaka I 2010 *Appl. Phys. Lett.* **96**, 012101
[5] Lozano-Perez S, de Castro V, Nichols R J 2009 *Ultramicroscopy* **109**, 1217
[6] Schmid H K and Mader W 2006 *Micron* **37**, 426
[7] Katayama I, Oura K, Shoji F and Hanawa T 1998 *Jap. J. Appl. Phys.* **27**, L1822

Low resistivity Pt interconnects developed by electron beam assisted deposition using novel gas injector system.

R. J. Dias[1], C. O'Regan[2], P. Thrompenaars[3], A. Romano-Rodriguez[1], J. D Holmes[2], J. J. L. (Hans) Mulder[3] and N. Petkov[2*]

[1] Departament d'Electronica Universitat de Barcelona c/ Marti i Franques 1, 08028 Barcelona, Spian.

[2] Electron Microscopy and Analysis Facility (EMAF), Materials Chemistry and Analysis Group, Tyndall National Institute, Lee Maltings, Prospect Row, Cork, Ireland.

[3] FEI Electron Optics, Eindhoven, The Netherlands.

* Corresponding author: nikolay.petkov@tyndall.ie

Abstract. Electron beam-induced deposition (EBID) is a direct write process where an electron beam locally decomposes a precursor gas leaving behind non-volatile deposits. It is a fast and relatively in-expensive method designed to develop conductive (metal) or isolating (oxide) nanostructures. Unfortunately the EBID process results in deposition of metal nanostructures with relatively high resistivity because the gas precursors employed are hydrocarbon based. We have developed deposition protocols using novel gas-injector system (GIS) with a carbon free Pt precursor. Interconnect type structures were deposited on preformed metal architectures. The obtained structures were analysed by cross-sectional TEM and their electrical properties were analysed ex-situ using four point probe electrical tests. The results suggest that both the structural and electrical characteristics differ significantly from those of Pt interconnects deposited by conventional hydrocarbon based precursors, and show great promise for the development of low resistivity electrical contacts.

1. Introduction

The fabrication of nano-scale structures at desired locations, with ultimate precision and resolution is normally achieved by "top-down" nano-lithography processes such as electron beam lithography, nano-imprint lithography (NIL), scanning probe lithography and electron or ion beam induced depositions (IBID or EBID) [1]. Among them EBID is a mask-less lithography process in which an electron beam is focused on a substrate surface and it is used to decompose gas precursor molecules leaving behind non-volatile deposits [2]. The decomposition process is a result of the interaction of surface emitted secondary electrons (SEs) in the irradiated area with locally absorbed gas molecules delivered by a gas injector system (GIS).

Major advantage of the EBID process over EBL and NIL techniques is that it is a direct write process that offers a fast and inexpensive route to direct prototyping and contacting of device architectures as well as the ability to fabricate 3D nano-structures [3]. Therefore FIB/SEM systems equipped with GIS have now numerous important applications such as carrying out device edits on prototype devices to fix design errors or to probe circuits for failure analysis.

Unfortunately, a major disadvantage of the EBID process with regards to metal deposition is that the deposits are highly resistive [2]. For example Pt structures developed by EBID show resistivities up to 5 orders of magnitude higher than that of bulk Pt which is at 10.6 μΩ cm. This is normally accompanied by very high variation in the reported values, which can only be explained by large variation in the deposition conditions including contamination within the vacuum chamber. Nevertheless, it is well accepted that the main reason for the high resistivity of the Pt interconnects developed by EBID originates from the type of precursor used [4]. The standard precursor used in the GIS is an organometallic Pt compound, methylcyclopentadienyl platinum trimethyl (MeCpPtMe3), consisting of a large hydrocarbon backbone. As a result, the Pt deposits normally suffer from a very large (up to 80%) amorphous carbon content, and show inferior structural integrity. In an effort to reduce the resistivity of such deposits, post-deposition treatments such as high temperature annealing or additional irradiation under high electron beam current have been explored [2]. Unfortunately such steps introduce additional complexity to the process and can be counteractive to fast and inexpensive prototyping *via* EBID. It is postulated that, reducing the hydrocarbon amount within the precursor molecule or using fully carbon-free Pt precursors can lead to depositions with reduced carbon content and improved structural integrity, and therefore improved resistivity, approaching that of bulk Pt. It has been suggested that one such precursor is $Pt(PF_3)_4$ [5].

In this study we examine the applicability of a newly developed GIS for EBID, fitted onto a Helios NanoLab Dual Beam system that operates with $Pt(PF_3)_4$ precursor molecules. A detailed study of the deposition parameters and GIS operation guide lines are presented. Structural and electrical characterization of the obtained interconnects are compared for varying deposition conditions.

2. Experimental

2.1. Pt EBID using $Pt(PF_3)_4$ precursor

Pt interconnects were developed by EBID on FEI's Nova 600i and Helios Dual Beam systems using a GIS specifically designed to incorporate $Pt(PF_3)_4$ precursor. The $Pt(PF_3)_4$ precursor was purchased from Strem Chemicals and introduced into the GIS using a glove box. The depositions were performed after overnight evacuation of the microscope chamber (normally reaching pressures of 10^{-6} mbar). Before deposition the corresponding beam currents were measured using a Faraday cup at various apertures. Changes in the chamber pressure upon dosing $Pt(PF_3)_4$ were strictly monitored and normally resulted in pressure increase of one order of magnitude after opening the GIS valve. The pressure equilibrium is normally reached within less than 5 s after opening the GIS valve. Successive depositions were performed after evacuating the chamber back to its base pressure of about 10^{-6} mbar. Before developing Pt interconnect lines, the performance of the GIS and corresponding deposition rates were monitored by developing 1 μm^2 square patterns at 5 kV and varying beam currents at constant deposition times. Notably, the performance of the GIS system started to deteriorate (e.g. the dosing of the precursor was considerably hindered) after 3 months of usage reflecting the corrosive nature of the precursor that resulted in blockage of the GIS aperture. A new improved designed of the GIS aperture was developed and resulted in reliable performance of the GIS. The improvements required choice of a new material for the manufacture of the isolating O-rings and apertures.

Pt interconnects were deposited at 5 kV on pre-fabricated Au contacts developed by optical lithography on 200 nm thermally grown SiO_2 on Si substrates. Various deposition conditions were explored and the obtained Pt interconnect structures and their electrical responses are discussed below. For comparison, similar depositions were obtained using standard Pt precursor, $MeCpPtMe_3$.

2.2. Structural and electrical characterisation

The morphology of the obtained Pt structures was investigated directly after disposition using the SEM on the Dual Beam systems. Additionally, cross-sections for TEM of the obtained interconnects were prepared and examined by JEOL 2100 TEM. The electrical response of the as-deposited interconnects was examined by four point electrical testing with a parameter analyzer on a Cascade Probe-station 1200 series.

3. Results and Discussion

3.1. Testing the GIS performance using $Pt(PF_3)_4$ precursor.

The performance of the GIS fitted to FEI's Nova NanoLab systems was monitored for 3 months by writing standard 1 μm^2 square patterns. One such example is shown in Figure 1 (a), where depositions were performed at increasing beam currents. Such depositions were used to calculate the deposition rate after accurately determining the thickness of the deposits. Tilt-view SEM imaging was used as a fast and simple method to determine the thickness of the as-deposited structures. Unfortunately, after comparing the thickness of the structures determined by tilt-view SEM to the thickness determined by cross-sectional SEM, a discrepancy of about 10 % was found. This is due to difficulties in accurately identifying the edges of the deposited structures under tilt-view SEM.

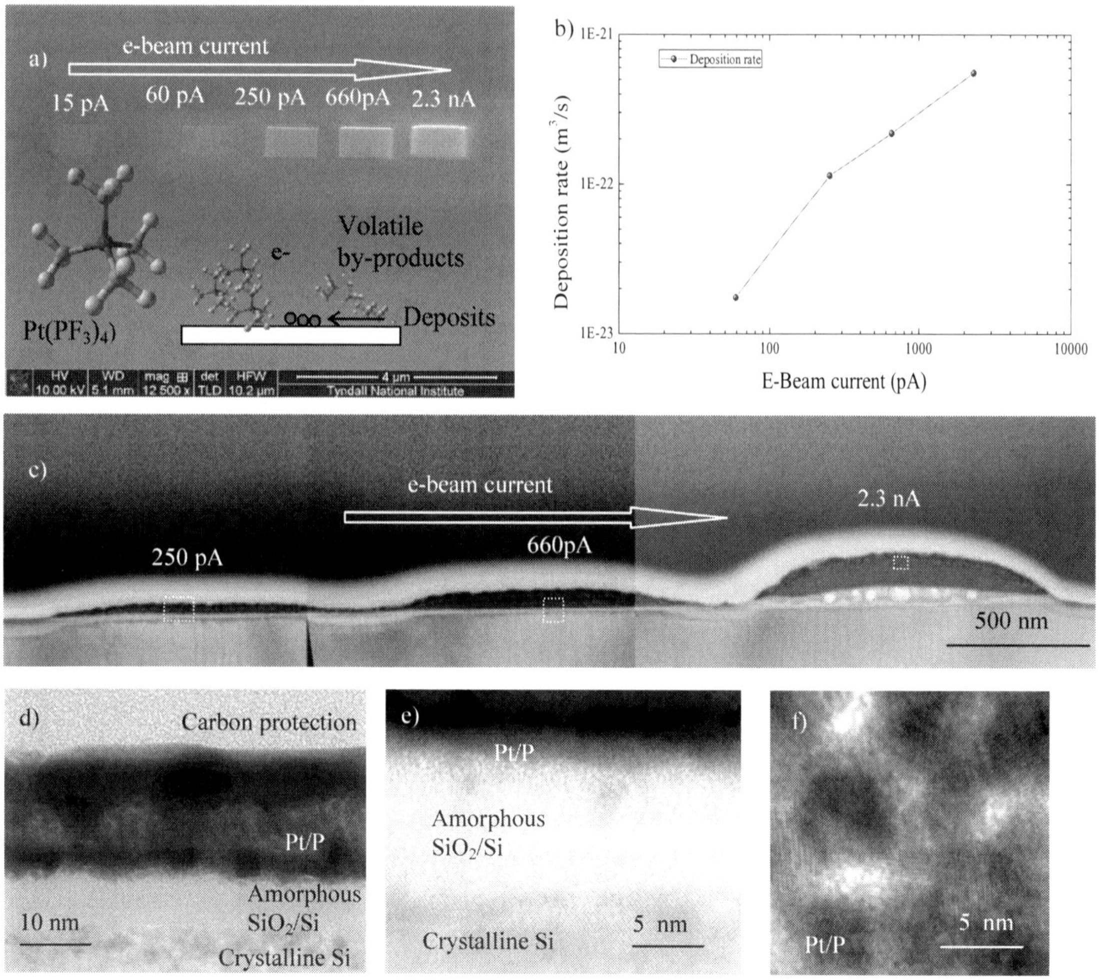

Figure 1. a) Tilt-view SEM image of a set of depositions performed at increasing beam current (in-set: schematics of the deposition process using $Pt(PF_3)_4$ molecules), b) calculated deposition rate at increasing beam current, c) TEM cross-sections of deposits at increasing beam current, d), e) and f) corresponding HRTEM images taken at the mark areas. The elemental content was determined by EDX.

More accurate determination of the thickness of the deposited structures and therefore the depositions rates was achieved be performing cross-sectional TEM analysis (Figure 1(c)). Furthermore, cross-sectional imaging supplied important information regarding the shape of the Pt deposits. The images clearly show that the structures have no sharp edges and that the overall cross-section is not rectangular. Using the thickness values, measured at the center of the cross-sections the deposition rate was calculated (Figure 1(b)). The deposition rates show linear increase with the beam current and reach 5.6 10^{-4} $\mu m^3/s$ at 2.6 nA. It should be mentioned that this rate is overestimated since the accurate dimensions of the deposited structures deviate from perfectly rectangular boxes. The deposits are composed of Pt nanoparticles within amorphous phosphorous-containing matrix.

3.2. Electrical properties of Pt interconnects developed by EBID.
Figure 2 shows electrical characterisation of the as-deposited interconnects over pre-fabricated Au contact fingers. The depositions were performed at varying thickness and cross-sections of the interconnects at a set beam current of 880 pA. The length of the interconnects was determined by the spacing between the Au contact fingers (one such device is shown in Figure 2 (b)). The obtained mean resistivity value is at 1160 ±550 (μohms cm) which is about two orders of magnitude higher than that of bulk Pt but at the lower range of resistivity values reported in the literature for as-deposited Pt interconnects by EBID. The *IV*-curves show ohmic dependence (Figure 2 (c)), for voltages between -10 mV and +10 mV. This is different from the obtained non-ohmic behaviour of Pt interconnects developed by using standard Pt EBID precursors.

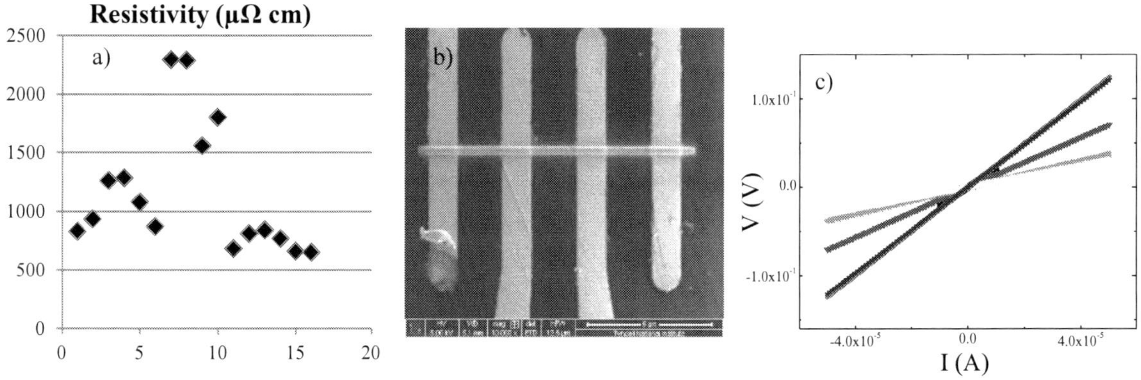

Figure 2. a) Calculated resistivity values for 16 different devices, b) example of one such device, c) experimental *IV*-curves for the corresponding devices.

4. Conclusions
Relatively low resistivity, ohmic Pt interconnects were developed by EBID using novel GIS and characterized structurally and electrically. Post-deposition treatments including gas annealing or in-situ carrier gas deposition might further improve the characteristics of the obtained structures.

Acknowledgements
This work was supported through the SFI 09/SIRG/I1621 funding.

References
[1] McCord M A M, Rooks J *Handbook of Microlithography, Micromachining and Microfabrication,* 2000.
[2] Botman A, Mulders J J L, Hagen CW, 2009 *Nanotechnology* **20** 372001.
[3] Furuya K, 2008 *Sci. Technol. Adv. Mater.* **9** 014110.
[4] Botman A, Mulders J J L, Weemaes R, Mentink S, 2006 *Nanotechnology* **17** 3779–3785.
[5] Takeguchi M,·Shimojo M,·Furuya K, 2008 *Appl Phys A* **93** 439–442

Phase, microstructural characterization and microwave dielectric properties of $SrLa_4Ti_{5-x}Sn_xO_{17}$ (x=0-3) ceramics

Yaseen Iqbal and Abdul Manan

Materials Research Laboratory, Institute of Physics and Electronics, University of Peshawar Pakistan

dryaseeniqbal@yahoo.co.uk

Abstract. The phase, microstructure and microwave dielectric properties of $SrLa_4Ti_{5-x}Sn_xO_{17}$ ($0 \leq x \leq 3$) were investigated in an attempt to tune their temperature co-efficient of resonant frequency (τ_f) to zero. The compositions sintered to single phase $SrLa_4Ti_5O_{17}$ ceramics at x=0, and $SrLa_4Ti_{4-x}Sn_xO_{17}$ along with a small amount of $La_2Ti_2O_7$ at x=1. The major phase observed at x=2 was $La_2Ti_2O_7$ but along with $SrLa_4Ti_4SnO_{17}$ and $SrLa_4Ti_4O_{15}$. At x=3, the major phases observed were $La_2Sn_2O_7$ and $Sr_2La_3Ti_3SnO_{15}$. The observed variation in the chemical composition of the grains combined with the morphological change from rod-shaped grains to cubical and irregular-shaped grains with increase in Sn content appeared in agreement with the phase constitution of samples identified by X-ray diffraction. τ_f decreased from 117 to 9.5 ppm/°C with increase in x from 0 to 3 but at the cost of ε_r and Qf_o which decreased from ~61 to 23 and 9969 to 2084GHz respectively due to the formation of secondary phases.

1. Introduction

The rapid growth in the wireless communication became possible due to the development of dielectric resonators (DRs). Many ceramics have been developed during the last 30 years in which very few have got practical applications as DRs in cell phone handsets and base station antenna [1]. There is a continuous demand for materials with ε_r >50 for size reduction and high Q_u-factor (~1/Tanδ) for the stability of the electronic components utilizing DRs. Microwave dielectric properties of several $A_nB_nO_{3n+2}$ (n=4.5 and 5) type ceramics such as $CaLa_8Ti_9O_{31}$ and $CaLa_4Ti_5O_{17}$, $Sr_{5-x}Ca_xNb_4TiO_{17}$, $La_9Mg_{0.5}Ti_{8.5}O_{31}$ and $CaLa_{3.995}Nd_{0.005}Ti_5O_{17}$ have been investigated [2-5]. $SrLa_4Ti_5O_7$ has been reported to have ε_r ~39, quality factor Qf_o~14200 GHz and τ_f ~58 ppm/°C [6]. Iqbal *et al* [7] reported ε_r~60.8, Qf_o~9969 GHz and τ_f~117 ppm/°C for $SrLa_4Ti_5O_{17}$ which crystallize into an orthorhombic crystal structure with lattice parameters *a*=5.549(5) Å, b=31.299(9) Å, c=3.9104(13) Å, z=2 and V=679.1(8) $Å^3$; however, its high positive τ_f precluded its use as a DR. More recently [8], the substitution of Sm^{3+} for La^{3+} enabled tuning the τ_f to zero with a reasonable ε_r (~53) but was accompanied by a decrease in Qf_o to ~3000 GHz. The substitution of Sn^{4+} for Ti^{4+} has been reported to lower τ_f of some compounds, for example, $Ba_4LaMNb_3O_{15}$ (M=Ti, Sn) [9] without any significant effect on the Qf_o; therefore, in the present study, the effect of Sn^{4+}-substitution for Ti^{4+} on the phase, microstructure and microwave dielectric properties of $SrLa_4Ti_{5-x}Sn_xO_{17}$ ($0 \leq x \leq 3$) was investigated in an attempt to lower τ_f of the resulting compounds.

2. Experimental

$SrLa_4Ti_{5-x}Sn_xO_{17}$ ($0 \leq x \leq 3$) batch compositions were prepared by mixing the required amounts of $SrCO_3$ (Aldrich, 99+%), La_2O_3 (Aldrich, 99.95%), TiO_2 (Aldrich, Anatase, 99+%) and SnO_2 (Aldrich, 99.95%). The batches were ball milled for 24h in polyethylene disposable mill-jars using Y-toughened ZrO_2 balls as grinding media and iso-propanol as lubricant. The resulting slurry was dried overnight and calcined for 6h at 1250°C for $SrLa_4Ti_5O_{17}$ and 1300°C for the compositions with x = 1, 2 or 3 at a heating / cooling rate of 5°C/min. The calcined powder was finely ground in pestle and mortar for ~45 min. and pressed into 4-5 mm high 10 mm diameter pellets at ~80 MPa and sintered at 1450-1625°C for 4h at a heating / cooling rate of 5°C/min. Densities of the sintered pellets were measured using Archimedes method. Phase analysis of the samples was carried using a Philips X-ray diffractometer (XRD) with CuK_α radiations (λ~1.5406Å). Microstructural characterization was performed using a JEOL 6400 scanning electron microscope (SEM) operating at 20 kV. The microwave dielectric properties were measured using a R3767CH Agilent network analyzer by placing the cylindrical pellets on a low loss quartz single crystal in the centre of a gold-coated brass cavity away from the cavity walls. τ_f was measured by noting the temperature variation of TE_{011} resonance in the temperature range 20-80°C.

3. Results and Discussion

XRD patterns (not shown) of the sintered compositions with x=0 and 1 were identical and matched PDF# 57-0094 for $SrLa_4Ti_5O_{17}$ but with an appropriate shift in the peak positions towards lower angles for the sample with x=1. The observed shifting may be due to the presence of relatively larger Sn ions at the B-site of the perovskite unit cell. This indicated the formation of a $SrLa_4Ti_4SnO_{17}$-like phase iso-structural with $SrLa_4Ti_5O_{17}$. The presence of a few low intensity XRD peaks due to $La_2Ti_2O_7$(PDF# 70-1690) indicated the beginning of second phase formation at x=1. At x=2, XRD peak intensities due to $La_2Ti_2O_7$ suggested that it was the major phase and the presence of peaks due to $SrLa_4Ti_4SnO_{17}$ and $SrLa_4Ti_4O_{15}$ (PDF# 49-0254) indicated the formation of secondary phases. At x=3, $La_2Sn_2O_7$ (PDF# 73-1688) was observed as the major phase but along with another phase iso-structural with $Sr_2La_3Ti_3NbO_{15}$ (PDF# 57-737). The observed increase in the intensity of XRD peaks due to $La_2Ti_2O_7$ and $La_2Sn_2O_7$ indicated an increase in their amount with increase in x.

Figure 1 shows secondary electron SEM images (SEIs) from thermally etched, gold-coated sample of $SrLa_4Ti_{5-x}Sn_xO_{17}$ with x=0 sintered for 4h at 1500°C and x=1, 2 and 3 sintered for 4h at 1625°C. The microstructure of the composition with x=0 comprised elongated grains of size ranging from ~1x1 μm^2 to 5x7 μm^2 (figure 1a). The molar elemental composition calculated from SEM EDS data (table 1) collected from grains e.g. labeled as "A" indicated that its composition was close to $SrLa_4Ti_5O_{17}$. The sintered microstructure of the composition with x=1 also comprised elongated rod- and plat-shaped grains e.g. labeled as "B" (figure 1b) with molar elemental composition (table 1) close to $SrLa_4Ti_4SnO_{17}$. The microstructure of the composition with x=2 comprised of micro-regions comprising cubical and spheroidal grains labeled as "C" (figure 1c). The composition of these grains was close to $La_2Ti_2O_7$ phase as EDS detected little Sr or Sn in these grains. Similarly, the microstructure of the composition with x=3 comprised cubical grains along with a few elongated grains (figure 1d). The approximate molar elemental composition (table 1) recorded for the grains labeled as "D" and "E" indicated that these grains could be from the $La_2Sn_2O_7$ phase as EDS detected no or little Sr or Ti in these grains. A few irregular-shaped deformed grains in the microstructure of the composition with x=3 contained ~2, 3, 3 and 1 mole of Sr, La, Ti and Sn respectively. This suggested the presence of grains due to $Sr_2La_3Ti_3SnO_{15}$ phase as observed by XRD of the same sample. The detection of all the constituent elements in the grain labeled as "F" suggested that it was probably from some $Sr_xLa_{4-x}Ti_{4-x}Sn_xO_{15}$–type phase isostructural with $Sr_2La_3Ti_3NbO_{15}$ consistent with XRD findings. The observed variation in the composition of the grains combined with the morphological change from rod-shaped grains to cubical grains with increase in Sn content indicated the formation of secondary phases.

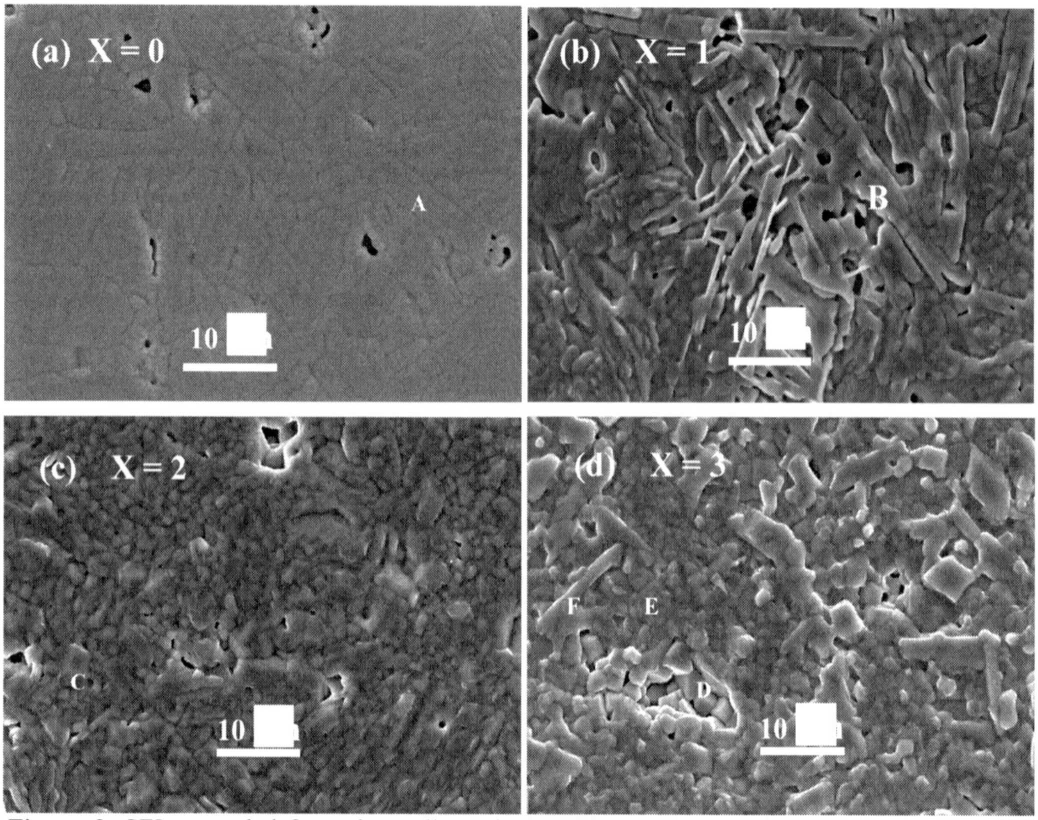

Figure 2. SEIs recorded from thermally etched, gold-coated sintered $SrLa_4Ti_{5-x}Sn_xO_{17}$ compositions showing a) elongated grains at x=0, b) elongated grains at x=1, c) cuboidal / spheroidal grains at x=2 and d) a mixture of cuboidal, elongated and irregular-shaped grains at x=3.

Table1. Approximate elemental composition (in moles) of the grains labelled in figure 2.

Grain	Sr	La	Ti	Sn
A	1±0.1	4.1±0.1	4.3±0.1	0
B	2.9±0.1	10.2±0.1	9.4±0.1	1±0.1
C	1.3±0.1	13.9±0.1	15.1±0.1	1±0.1
D	1±0.1	22.6±0.1	2.2±0.1	21±0.1
E	0	12.6±0.1	1±0.1	12.5±0.1
F	1.4±0.1	11.3±0.1	5.2±0.1	1±0.1

The microwave (MW) dielectric properties of $SrLa_4Ti_{5-x}Sn_xO_{17}$ ceramics sintered at their optimum sintering temperatures are compared in table 2. ε_r was observed to decrease from ~60.8 to 23.5, τ_f from ~117 ppm/°C to 9.6 ppm/°C and Q_uf_o from ~9969 GHz to 2084 GHz with an increase in Sn^{4+} content from 0 to 3. $CaLa_4Ti_5O_{17}$ and $SrLa_4Ti_5O_{17}$ have been reported to exhibit ε_r in the range of 39-51 with a Q_uf_o ranging from 13140-14200 GHz; however, $SrLa_4Ti_5O_{17}$ has a positive τ_f (+58ppm/°C) in contrast to its Ca analog with a negative τ_f of -29 ppm/°C [6] which explains the observed decrease in τ_f. The relative permittivity of $La_2Ti_2O_7$ is ~22 [5] and it has been reported that the substitution of Sn^{4+}

for Ti^{4+} causes a decrease in ε_r and hence τ_f in other compounds [9]; therefore, $La_2Sn_2O_7$ can be expected to exhibit ε_r < 22 and the observed decrease in ε_r of $SrLa_4Ti_{5-x}Sn_xO_{17}$ may be due to the presence of $La_2Ti_2O_7$ at x=1-2 and $La_2Sn_2O_7$ at x =3. The observed decrease in Qf_o upon increasing Sn content may be due to the formation of secondary phases, where the interface causes additional dielectric loss and also due to the relatively higher ionic conductivity of $La_2Sn_2O_7$ [10]. Thus the optimum MW dielectric properties (τ_f~39 ppm/°C, ε_r ~46 and Qf_o ~7900 GHz), corresponded to the $SrLa_4Ti_{5-x}Sn_xO_{17}$ composition with x = 1 with minimum Sn content and hence secondary phases.

Table 2. Preparation conditions, apparent density and MW dielectric properties of $SrLa_4Ti_{5-x}Sn_xO_{17}$ ($0 \le x \le 3$)

X	C.T(°C)	S.T(°C)	ρ_{ap}(g/cm^3)	ε_r	Q_uf_o (GHz)	τ_f (ppm/°C)
0	1250/6h	1500/4h	5.36	60.8	9969	+117
1	1300/6h	1625/4h	5.69	45.5	7919	+38.6
2	1300/6h	1625/4h	5.55	31.6	4191	+23.0
3	1300/6h	1625/4h	5.44	23.5	2084	+9.6

C.T = Calcination temperature, S.T = Sintering temperature, ρ_{ap} = Apparent density

4. Conclusions

$SrLa_4Ti_{5-x}Sn_xO_{17}$ compositions crystallize into single phase ceramics at x = 0 and although in small amounts but second phase formation begins at x=1. The substitution of Sn for Ti causes substantial decrease in τ_f but at the cost of ε_r and Qf_o due to second phase formation which makes higher concentrations of Sn unfavourable. The optimum MW dielectric properties i.e. τ_f~ 39 ppm/°C, ε_r ~ 46 and Qf_o ~ 7900 GHz, corresponds to the $SrLa_4Ti_{5-x}Sn_xO_{17}$ composition with x = 1.

Acknowledgement

The authors acknowledge the financial support of the Higher Education Commission of Pakistan (NRPU 20-569 and IRSIP) and Development of Materials Connection Centre under the Pak-US Joint S&T Cooperation Program (Project ID 131).

References

[1] Reaney I M, Idles D 2006 *J. Am. Ceram. Soc.* **89**(7) 2068-2072.
[2] Jawahar I N, Santha N, Sebastian M T and Mohanan P 2002 *J. Mater. Res.* **17** 3084–3089.
[3] Manan A, Iqbal Y and Qazi I 2011 *J. Mater. Sci*, **46**(10) 3415-3423.
[4] Vanderah T A, Miller V L, Levin I, Bell S M and Negas T 2004 *J. Solid. State Chem.* **177** 2023-2038.
[5] Chen Y C Chen K C Wang S C 2010 *Ferroelectrics Lett.* **37**(5) 83-89.
[6] Demsar K Skapin S D Meden A and Suvorov D 2008 *Acta Chim. Slov.* **55** 966-972.
[7] Iqbal Y Manan A Reaney I M 2011 *Mater. Res. Bull.* **46**(7) 1092-096.
[8] Manan A and Iqbal Y 2011 *J. Mater. Sci: Mater. Elect.* DOI 10.1007/s10854-011-0372-3.
[9] Fang L, Zhang H, Huang T H, Yuan R Z and Dronskowski R 2004 *Mater. Res. Bull.* **39** 1649-654.
[10] Yu T H 1996 *Electrical Properties and Structural Disorder in Stannet Pyrochlores*, PhD thesis, MIT pp-50.

Structural study of Ge/GaAs thin films

V K Lazarov[1,2], L Lari[1,2], P M Lytvyn[3], V V Kholevchuk[3] and V F Mitin[3]
[1] Physics Department, University of York, Heslilngton, YO10 5DD, UK
[2] York-JEOL Nanoncentre, University of York, Heslington, YO10 5DD, UK
[3] V. Lashkaryov Institute of Semiconductor Physics, National Academy of Sciences of Ukraine, 03028 Kyiv, Ukraine

E-mail: vlado.lazarov@york.ac.uk

Abstract. Ge/GaAs heterostructure research is largely motivated by the application of this material in solar cells, metal–oxide–semiconductor field-effect transistors, mm-wave mixer diodes, temperature sensors and photodetectors. Therefore, understanding of how the properties of Ge/GaAs heterostructure depend on its preparation (growth) is of importance for various high-efficiency devices. In this work, by using thermal Ge evaporation on GaAs(100), we studied structural properties of these films as a function of the deposition rate. Film grains size and morphology show strong dependence of the deposition rate. Low deposition rates results in films with large crystal grains and rough surface. At high deposition rates films become flatter and their crystal grains size decreases, while at very high deposition rates films become amorphous. Cross-sectional TEM of the films show that the Ge films are granular single crystal epitaxially grown on GaAs. The Ge/GaAs interface is atomically abrupt and free from misfit dislocations. Stacking faults along the [111] directions that originate at the interface were also observed. Finally by using the Kelvin probe microscopy we show that work function changes are related to the grain structure of the film.

1. Introduction

The study of Ge/GaAs heterostructure is largely motivated by application of this material in various solar cells, metal–oxide–semiconductor field-effect transistors, mm-wave mixer diodes, temperature sensors and photodetectors. Therefore, information on how the properties of Ge/GaAs heterostructure depend on its preparation conditions is of importance for the application of this system in various high-efficiency devices. The studies of growth mechanisms of thin Ge films on GaAs [1-5] show that Ge films grow on GaAs (100) in the layer-by-layer (Franck-van der Merwe) mode at temperatures below about 400 °C, whereas at higher temperatures growth occurs by the island (Stranski-Krastanov) growth mechanism. Numerous studies have been reported on the influence of the substrate temperature and the GaAs surface composition on the electrical properties of Ge films and heterojunctions [1, 6–9]. However, the effect of the deposition rate on the properties of thin Ge films on GaAs has not been investigated thoroughly; in most of the reports rather high film deposition rates have been chosen as a rule. Recently we have demonstrated that by varying the deposition rate of Ge/GaAs one can control the transport properties effectively [10]. The high deposition rates produce films that are n-type doped with low resistivity (10^{-2} Ω·cm) and high carrier concentration (4×10^{18} cm^{3}). The temperature dependence of conductivity in these films is very weak and of non-activation type, indicating metallic like transport with degenerate charge carriers. In contrast, low deposition rates produce single crystal films with larger grains that are p-type doped with high resistivity (10^2 Ω·cm) and low carrier concentration (10^{15} cm^3) [10]. In this work we present a structural study of thermally deposited Ge/GaAs films by transmission electron microscopy (TEM) and atomic force microscopy (AFM).

2. Experiment

The films were deposited on GaAs(100) substrates using thermal evaporation of Ge in the vacuum. The temperature of GaAs substrate remained constant (490 ± 10 °C) during the Ge film deposition. The deposition rate (from 0.025 to 0.37 nm/s) was controlled by varying the temperature of Ge crucible. Films surface morphology has been investigated by AFM. Bright field (BF) TEM imaging and selected area diffraction (SAD) has been performed by JEOL 2011. The high resolution TEM (HRTEM) and high angle annular dark filed imaging has been done by the double aberration corrected JEOL 2200 FS microscope. Scan-line energy dispersive X-ray analysis was performed by using the Thermo-EDX system. TEM cross-section specimens were prepared by gluing the films' sides with a Gatan G1 epoxy resin. Mechanical thinning followed by polishing was carried out until ~20μm specimen thickness. The final step included Ar ion milling by using Gatan PIPS ion miller in order to achieve electron transparency. Initial ion-milling parameters were set as the following: ion beam energy 3keV with 5^0 angles with respect to the bottom and top of specimen surface; at the final stage of ion milling the energy of the beams was decreased to 2.5 keV and angles to 4^0.

3. Results and Discussion

Both GaAs and Ge have cubic (zinc-blende) structure with lattice constant of 5.6533 Å and 5.6461 Å, respectively. The same crystal structure and the very small difference in the lattice constants make these two materials almost an ideal epitaxial match, with mismatch of only -0.12 %. Indeed the deposited Ge films are single crystal, as confirmed by the SAD from the Ge film (Figure 1). The epitaxial relation between the film and the substrate is a simple cube on cube relation; Ge(001)||GaAs(001) and Ge(100)||GaAs(100). The AFM surface morphology and the BF-TEM clearly show that the growth is in Stransky-Krastanov (3D) mode with the grain size of ~75 nm (Figure 1a and b).

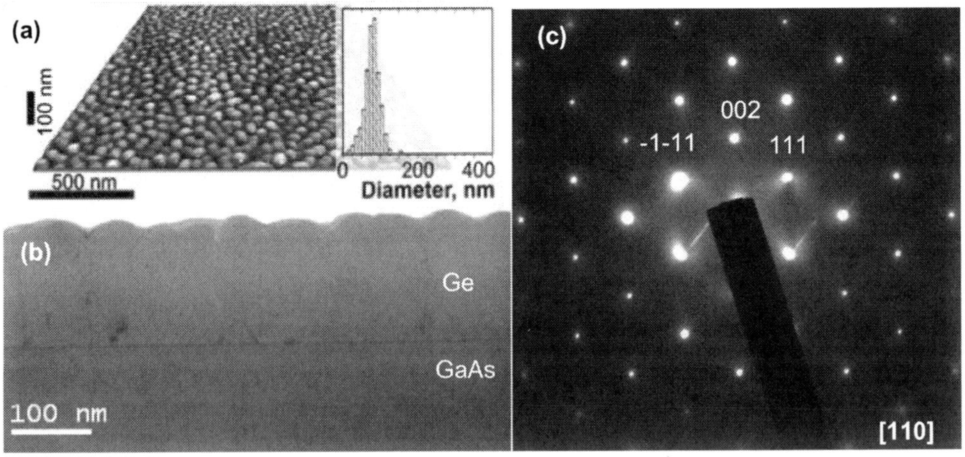

Figure 1. a) AFM showing surface topology of Ge/GaAs films deposited at 0.37 nm/s. b) BF cross-section TEM of Ge/GaAs, and (c) SAD from the Ge film.

The interface between Ge and GaAs appears to be well defined; interface roughness (up to 1 nm) can be seen from the HTREM images (Figure 2a). The high resolution (HR) HAADF imaging supports this observation (Figure 2b); though a small amount of waviness is present at the interface. This indicates possible atomic mixing at the interface region. The scan line EDX across the interface also suggest that such mixing could happen, though due to the beam spread and the size of interface region this should be examined by additional methods in order to fully verify the presence of atomic mixing at the Ge/GaAs interface. The BF-TEM (Figure 1b) as well as HAADF (Figure 3a) show presence of linearly shaped defects that starts at the interface and extend in the Ge film for several tens of nm. The HAADF (Figure 3c) and the HRTEM (not shown here) clearly show that these defects are twin stacking faults (SF). The SF occurs on the {111} type of planes, and they vary in their width from several to tens of atomic planes. All the SF faults originate at the interface and at small number of SF we observe presence of dislocation(s), either at the Ge/GaAs interface or along the twin mirror plane of the SF.

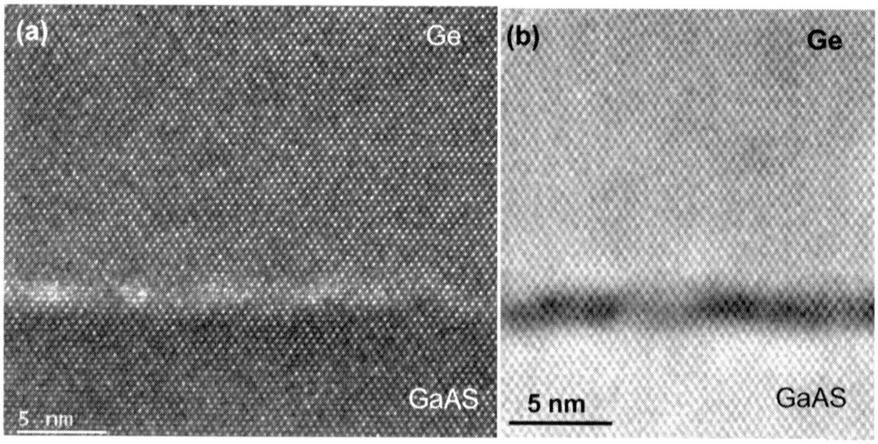

Figure 2. HRTEM (a) and HAADF (b) from the Ge/GaAs(100) interface.

In addition to the SFs, Ge grain boundaries are clearly seen in BF-TEM and HAADF (Figure 1b and Figure 3a, respectively). Regions with white (dark) contrast are observed along these boundaries in BF (HAADF) images. The HR-HAADF (Figure 3c) shows that these dark contrast regions are fully crystalline; the darker contrast most likely originates from the presence of voids along the grain boundaries, thus these regions appear darker in the Z-contrast imaging. The void formation is believed to be a result of the 3D columnar growth of the Ge film.

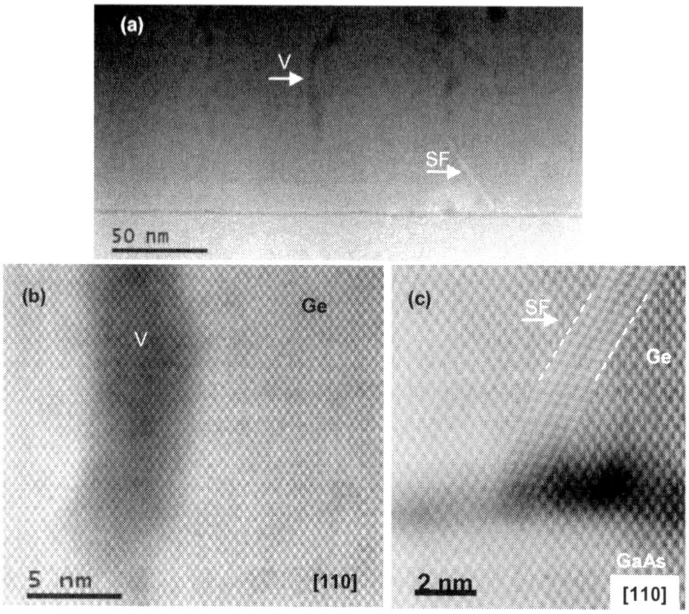

Figure 3. a) HAADF from the the Ge/GaAs film. b) HR- HAADF from the grain boundary regions (b) where voids (V) are present. c) HR-HAADF from twin stacking fault that originates at the Ge/GaAs interface.

The importance of the Ge grain boundaries and their possible role in the outdiffusion of Ge/As in the film (and ultimately the Ge films conductivity) is illustrated in Figure 4. The surface topography of low rate deposition Ge film (Figure 4a) and the corresponding surface potential mapping by Kelvin probe AFM (Figure 4b) show good correlation. The changes of the surface potential coincide with the Ge grain boundaries indicating that the outdiffusion of Ga(As) can be significant along the grain boundaries.

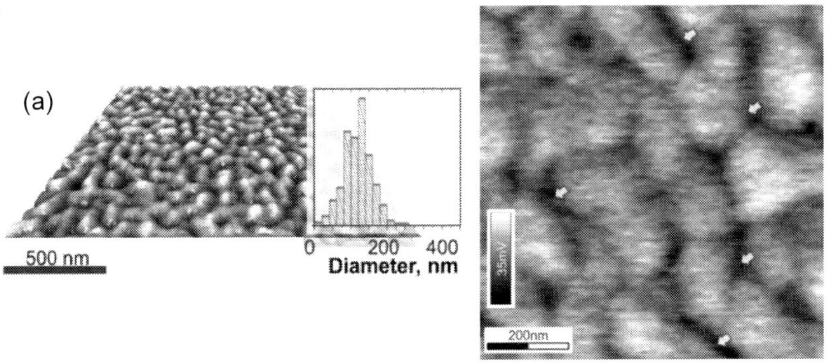

Figure 4. a) Surface topology of low deposition rate (0.04 nm/s) Ge film. b) Surface potential map, the regions with lower work function are marked with arrows.

4. Conclusions

We present a TEM/AFM study of the Ge/GaAs films. The 3D growth of Ge film results in single crystal epitaxial films on GaAs(100). The film-substrate interface is sharp and presences of twin stacking faults have been found in the interface region. No defects have been observed at Ge grain boundaries, the darker HAADF contrast at these boundaries is most likely due to the voids. These boundaries can have significant role in the Ga(As) outdiffusion from the substrate into the Ge film.

Acknowledgments

This work has been partially supported by the RAEng/EPSRC fellowship and by Ukrainian State Program "Nanotechnologies and nanomaterials", Projects No. 2.2.6.15 and 3.6.1.10.

References

[1] Luo GL, You Han Z, Chien C-H, Ko C-H, Wann CH, Lin H-Y, Shen Y-L, Chung C-T, Huang SC, Cheng CC, and Chang CY, J. Electrochem. Soc.**157**, H27 (2010).
[2] Wang XS, Self KW, Weinberg WH, J. Vac. Sci. Technol. A **12**, 1920 (1994).
[3] Wang XS, Self KW, V. Bressler-Hill, Maboudian R, and Weinberg WH, Phys. Rev. B **49**, 4775 (1994).
[4] Neave JH, Larsen PK, Joyce BA, Gowers JP, van der Veen JF, J. Vac. Sci. Technol. B **1**, 668 (1983).
[5] Emiliani V, Shkrebtii AI, Goletti C, Frisch AM, Fimland BO, Esser N, Richter W, Phys. Rev. B **59**, 10657 (1999).
[6] Kawanaka M, Sone J, J. Cryst. Growth 95, 421 (1989).
[7] Shiota I, Motoya K, Ohmi T, Miyamoto N, and. Nishizawa J, J. Electrochem. Soc. **124**, 155 (1977).
[8] Kawanaka M, Sone J, J. Electron. Materials **19**, 575 (1990).
[9] Salazar-Hernández B, Vidal MA, Navarro-Contreras H, and Vázquez-López C, Thin Solid Films **352**, 269 (1999).
[10] Mitin VF, Lazarov VK, Lytvyn PM, Hasnip PJ, Kholevchuk VV, Kotenko IE, Mitin VV, Venger EF, Phys. Rev. B **84**, 125316 (2011).

Journal of Physics: Conference Series **371** (2012) 012041 doi:10.1088/1742-6596/371/1/012041

Microstructural and analytical analysis of plasma dissociated zircon

E.G. Minnaar[1], J.H. Neethling[1], M.E. Lee[1] and E.Snyders[2]

[1]Department of Physics, Nelson Mandela Metropolitan University, Port Elizabeth
[2]Materials Programme, South African Nuclear Energy Corporation (Necsa), South Africa

Email: ettienne.minnaar@live.nmmu.ac.za

Abstract. The investigation of the microstructure and distribution of impurities was carried out on plasma dissociated zircon (PDZ). The morphology of the PDZ and crystalline nature of the zirconia was determined by scanning electron microscopy (SEM) and transmission electron microscopy (TEM) while energy dispersive X-ray spectrometry (EDS) provided compositional information. The SEM and TEM results indicate that the morphology and crystalline nature of the zirconia varies as a result of the existence of a thermal gradient. The EDS results show that the majority of the impurities segregates to the silica phase of the PDZ.

1. Introduction

Zirconium is a major component in many alloys due to its thermal stability and resistance to corrosion [1]. These alloys, known as zircaloys are used in the nuclear power industry for structural parts in the core; such as pressure tubes, channels, guide tubes or fuel cladding [2] due to its low thermal neutron cross section [3]. The mineral zircon ($ZrSiO_4$) is the primary source of zirconium metal. Zircon is mined as one of several products from heavy-mineral sand deposits associated with ancient shorelines [4].

The chemical inertness of zircon makes the separation of the zirconium and silicon components by chemical techniques difficult. The application of thermal plasma technology has shown to be an effective technique for the separation of the zirconia and silica phases [5]. In this process the zircon can be made chemically more tractable by dissociation in a plasma flame above the melting point (1687°C) into zirconia (ZrO_2) and silica (SiO_2).This product is known as plasma dissociated zircon (PDZ). The structure of PDZ material can be described as finely dispersed zirconia crystals (m.p. 2700°C) embedded in a glassy silica matrix (m.p. 1650°C). The amorphous silica can be dissolved by acids or bases to obtain pure zirconia.

The zirconia in PDZ exists in three crystalline structures: monoclinic which is stable below 1180°C, tetragonal which is stable in the range of 1180-2370°C and cubic which is stable above 2370°C [6]. Characterization of the size and distribution of zirconia within the silica produces important information regarding the efficiency of the dissociation process. Microanalysis of the zirconia phase is thus important since the presence of impurities is detrimental to the physical properties of zircaloys.

The present work investigates the phase morphology of zirconia crystals as well as the redistribution of impurities within the PDZ material.

2. Materials and methods

South African zircon mined at the Namakwa Sands project was used in this study. The PDZ material as-supplied by the South African Nuclear Energy Corporation (NECSA) was produced using a non-transfer arc plasma system. The as-received samples were resin embedded and polished to produce cross sections which were analysed in a scanning electron microscope (SEM) using the backscattered electron (BSE) imaging mode. Analysis of the impurity distribution within each phase was determined by energy dispersive X-ray spectrometry (EDS).

Material for analysis in the transmission electron microscope (TEM) was prepared from a suspension of finely powdered PDZ material in ethanol and placed in an ultrasonic water bath to ensure a uniform dispersion of particles. The fine powder was then collected on a carbon coated copper grid.

3. Results and discussion

3.1. Microstructural analysis

An SEM BSE micrograph showing the two distinct zones (A and B) within a partially formed PDZ grain is shown in figure 1. Zone A consists of undissociated zircon, while zone B contains two distinct phases namely zirconia (bright) and silica (dark). Figure 2 shows a PDZ particle consisting entirely of zirconia spherulites in a silica matrix. Spherulites are non-crystallographic branching arms that extend in all directions [7].

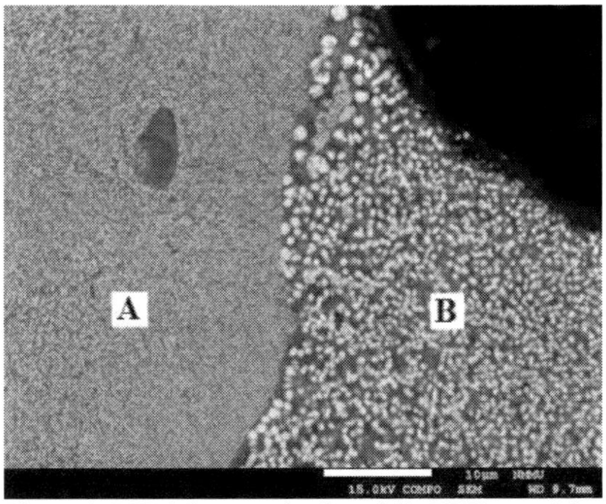

Figure 1. BSE Micrograph of partially PDZ material.

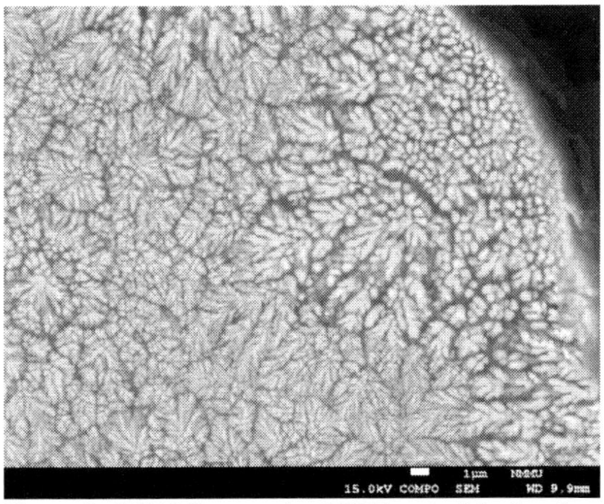

Figure 2. BSE Micrograph of PDZ material, showing the spherulite growth morphology.

The observed morphologies has been previously described, [7, 8], and will be discussed. According to the phase diagram of the ZrO_2-SiO_2 system based on the work of Butterman and Foster [9], the zirconia which has dissociated out of the zircon will be stable in the tetragonal or cubic crystal structure. The quenching of these particles forces its crystalline structure to change to the monoclinc structure resulting in a volume increase. The zirconia grows in a spherulitic fashion, thus they are called spherulites. Spherulites forms from large zirconia crystals due to the low surface heat transfer rate of these drops, as a result the temperature remains sufficiently high for crystal growth to take place.

The zirconia crystals that dissociated out of the zircon possessing a small surface area retain the tetragonal and cubic crystal structure. Due to the high heat transfer rate that these small zirconia particles possess, the formation of the spherulites is prevented. These particles crystallises into oval and spherical shaped zirconia particles as shown in figure 3.

The ZrO_2-SiO_2 phase diagram from McPherson and Shafer [7] shows that molten zircon could enter the miscibility gap if cooled rapidly to inhibit the formation of zirconia crystals. In this gap the zircon separates into a zirconia rich and silica rich liquid. This is known as spinodal decomposition as shown in figure 4. The zirconia rich liquid would rapidly crystallise during cooling producing a dispersion of very fine spherical particles of tetragonal and cubic zirconia in glass. The silica rich liquid would form a glassy matrix which also undergoes further separation to produce zirconia rich droplets in the glass.

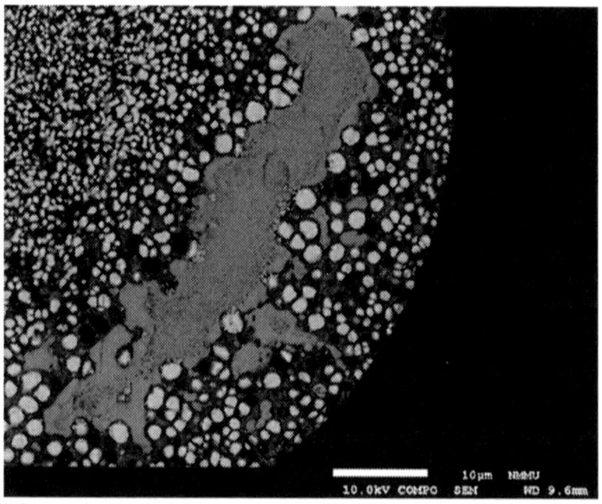

Figure 3. BSE Micrograph of PDZ material, showing spherical zirconia crystals.

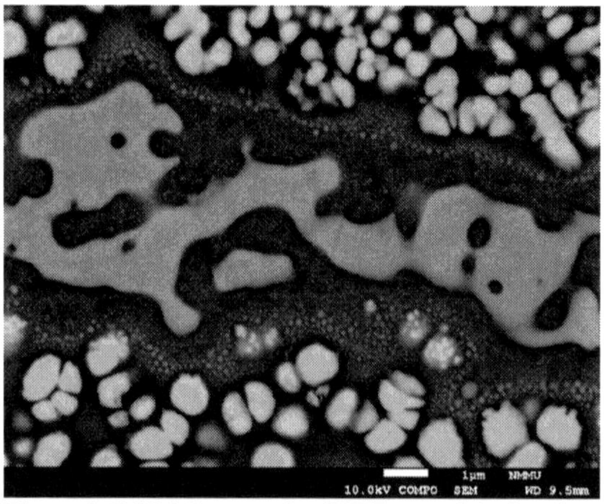

Figure 4. BSE Micrograph of PDZ material, showing spinodal decomposition.

The microstructure of the PDZ as analysed in the TEM revealed amorphous (light) and nanocrystalline (dark) regions as shown in figure 5. Measurements of the d-spacings of the fast Fourier transform (FFT) in figure 6 show the zirconia to have the monoclinic and tetragonal crystal structures. Reflections A and B corresponds to that of monoclinic zirconia with Miller indices 100 and $\bar{1}10$ respectively. Reflections C and D corresponds to that of tetragonal zirconia with Miller indices 101 and 211 respectively. The cubic crystalline structure could not unambiguously be identified as X-ray diffraction (XRD) analysis found the ratio of the of monoclinic to tetragonal to cubic zirconia present in PDZ to be 100:10:1 [9].

Figure 5. TEM micrograph of PDZ material

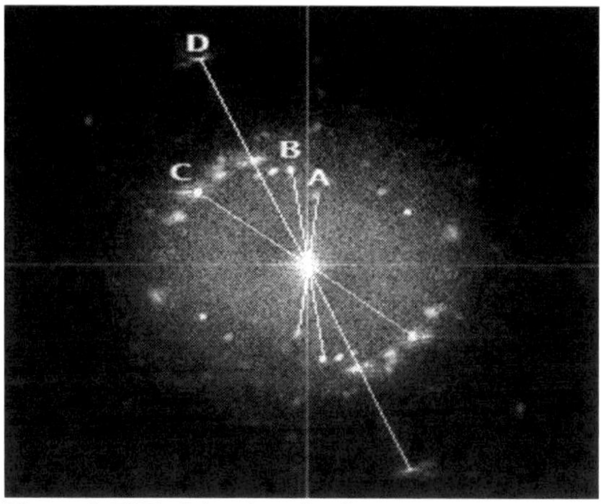

Figure 6. FFT of the zirconia phase.

3.2. *Quantitative analysis*

The results of the impurity analysis of the PDZ as determined by SEM-EDS are summarized in the table.

Table. Impurity analysis of PDZ (weight percentage)

	Al	Fe	Ti	Ca	Hf
Undissociated zircon	2.0	0.5	0.3	0.2	0.1
PDZ zirconia phase	0.2	-	0.1	-	0.4
PDZ silica phase	1.3	0.7	-	0.3	-

The analysis revealed that majority of the impurities present in zircon segregates to the silica phase during dissociation. However the hafnium and titanium segregated to the zirconia phase. The hafnium concentration is higher than the standard specification for nuclear grade zirconium metal (100ppm) [10], but is lowered to the required amount by sublimation and reduction reactions with magnesium.

4. Conclusion

The nucleation of the zirconia crystals from zircon arises from the solidification that occurs during the cooling process of the dissociated zircon. There exists a range of possible thermal histories and the morphologies of the zirconia observed depend upon the cooling rate. Large zirconia crystals grow in a spherulitic manner and have the monoclinic crystal structure. Some smaller crystals formed via a rapid cooling process appear to retain the tetragonal or cubic crystal structures rather than relaxing to the monoclinic structure, and this may be because such nanocrystals are stabilised by small surface area or constraints from surrounding materials. Spinodal decomposition occurs at lower temperatures which results in the formation of a zirconia and silica rich liquid.

Impurity distribution analysis of the PDZ showed that majority of the impurities present in zircon segregates to the silica phase during the dissociation process.

References

[1] Biswas R K, Habib A K, Karmakar A K and Islam M R 2010 Hydrometallurgy **103** 124-129
[2] Dupin N, Ansara In Servant C, Toffolon C, Lemaignan C and Brachet J C 1999 J. Nucl. Mater. **275** 287-295
[3] Lee S J, Kwon H S, Kim W, Choi B H 1999 Mater. Sci. Eng., A **263** 23-31
[4] Pirkle F L and Podmeyer D A 1993 Transactions **292** Society for Mining, Metallurgy and Exploration, inc
[5] Yugeswaran S, Ananthepadmanabhan P V, Thiyagarajan T K, Selvarajan V and Janardhanan Nair 2010 J. Phys. Conference Series **208** 012122
[6] Christensen A, Carter E A 1998 Phys. Rev. A **58** 12
[7] McPherson R and Shafer B V 1984 J. Mater. Sci. **19** 2696-2704
[8] Ai Ming Wong and McPherson R 1967 Am. Mineral.**52** 880-885
[9] Unpublished work
[10] Hui Li, Nersisyan H H, Park K, Park S, Kim J, Lee J 2011 J. Nucl. Mater. **413** 107-113

Journal of Physics: Conference Series **371** (2012) 012042 doi:10.1088/1742-6596/371/1/012042

Using TEM and XRD to probe crystal orientation in organic thin films grown with OMBD

J B Gilchrist, S Heutz and D W McComb

Department of Materials, Imperial College London, Exhibition Road, London, SW7 2AZ, UK

j.gilchrist10@imperial.ac.uk

Abstract. Correlating X-ray diffraction data with images of lattice fringes obtained in the transmission electron microscope allows diffraction from planes in crystalline materials to be studied from different perspectives. The focus of this work is on the characterisation of thin films of copper phthalocyanine prepared using organic molecular beam deposition. Films grown on silicon have been studied using X-ray diffraction, which identifies that the (100) plane lies parallel to the substrate surface. Films grown on amorphous carbon have been studied using transmission electron microscopy which confirms that the (001) plane is close to perpendicular to the substrate surface. The promotion of a different crystal orientation can be achieved by deposition of copper phthalocyanine on to a film of perylene tetracarboxylic dianhydride. The X-ray diffraction data of these films shows that the (11-2) plane is now parallel to the substrate surface and this is confirmed by the observation of lattice fringes corresponding to the (100) plane in the transmission electron microscope. The information obtained can be used to correctly predict the angles needed to tilt the sample in the transmission electron microscope to observe lattice fringes from both the (100) and (001) planes in the double layer films.

1. Introduction

Organic semiconducting materials can be used for electronic and optoelectronic applications such as transistors and solar cells and, because they are relatively cheap and significantly lighter [1], they could provide an alternative to conventional silicon based semiconductors. These possible applications have driven considerable research in this area. Organic photovoltaic devices are a combination of electron donor and acceptor materials, the interface between which results in the dissociation of excitons (bound electron-hole pairs, formed by the absorption of a photon in the organic layer) [2]. The small, macrocyclic, aromatic molecule copper phthalocyanine (CuPc), has shown to be a good donor layer for photovoltaic devices when combined with fullerene (C_{60}) as an acceptor material [3]. One method used to prepare these organic photovoltaic devices is organic molecular beam deposition (OMBD), which involves the sublimation under vacuum of the source material from a Knudsen cell [4]. The sublimed species is collimated by an aperture into a beam and deposited onto the substrate located above the source. On weakly-interacting substrates (non-chemisorbing), CuPc grows as crystalline grains in the α-phase [5]. The crystal structure of these grains can be probed with X-ray diffraction (XRD) to estimate both crystal phase and texture, *i.e.* which planes will be lying parallel to the substrate surface. Transmission electron microscopy (TEM) can be used to identify which planes are close to perpendicular to the substrate surface in top views of thinner but otherwise equivalent films grown on carbon coated copper TEM grids, by measuring the interplanar spacing of lattice fringes found in the crystalline grains. This information can be correlated with XRD results. The orientation of the crystallographic planes will then indicate molecular orientation. Since charge transport in CuPc is anisotropic with respect to molecular orientation, this information is important when growing films for electronic applications. On weakly-interacting substrates at room temperature, α-CuPc will grow with the molecular plane close to perpendicular to the substrate surface, *i.e.* the undesirable orientation for electronic applications. The promotion of a more desirable molecular orientation can be achieved by growing the CuPc film on a previously deposited film of perylene tetracarboxylic dianhydride (PTCDA), another small, aromatic molecule. Molecules of PTCDA are known to lie close to parallel to the substrate surface. This will cause the CuPc molecules to be oriented so that the molecular plane is also close to parallel to the substrate surface. This promotion is attributed to the strong π-π interaction that would occur at the interface between these two materials [6].

2. Method

Films of CuPc (395 nm) and CuPc (436 nm) on PTCDA (247 nm) were grown on a Si (100) substrate in a Kurt J. Lesker high vacuum chamber (base pressure about 10^{-8} mbar, with a growth rate set at 1 $Ås^{-1}$. Films of CuPc and CuPc on PTCDA were also grown directly onto carbon coated copper TEM grids (Agar) with the same conditions as for the Si substrates but thicknesses set to 50 nm for both CuPc and PTCDA. SEM imaging was performed in a Gemini Leo 1525 microscope with a field emission gun (FEG), operated at an accelerating voltage of 10 kV and a 30 μm aperture. CuPc and CuPc/PTCDA films grown on the silicon substrate were probed with XRD using a Philips PW1700 series diffractometer operated in theta-2-theta using Cu Kα radiation at 40 kV and 40 mA. Each scan was performed from $2\theta = 5$ - 30° with a step size of 0.033° and a time per step of 99.7 s. All XRD data has been interpreted using the model proposed by Hoshino *et al.* [5], although this is different from work done by others e.g. Sullivan *et al.* [6]. The films of CuPc and CuPc/PTCDA that were grown on carbon coated copper TEM grids were examined using a JEOL 2010F with a LaB_6 crystal source and operated at 120 kV. Measurements were obtained using the line profile tool in Gatan's digital micrograph program.

3. Results

3.1. CuPc Single Layer

The growth of grain structures of CuPc on Si (100) can be observed (figure 1a). XRD data of this film confirms that the film is crystalline, adopting the α-phase model proposed by Hoshino *et al.* [5] and that diffraction from the (100) plane (at $2\theta = 6.9°$) is the predominant diffracting plane (figure 2a). Diffraction from this plane means that the orientation of the CuPc molecules will be almost perpendicular to the substrate surface (figure 2b). The presence of peaks at $2\theta = 13.7°$ and 27.7° which arise from diffraction from the (200) and (400) planes also confirms this. Diffraction from other planes (the (001) plane at $2\theta = 7.3°$, (01-1) at 24.0° and (111) at 26.7°) suggest that the preferred orientation of the film is not perfect, this is attributed to the grains not all having the same growth direction at high thicknesses. The films grown on carbon coated copper TEM grids are thin enough to be transparent to the electron beam. The presence of lattice fringes confirm that the films are crystalline and the distance between these fringes suggests diffraction from the (001) plane, 1.2 nm (figures 2c and d). Diffraction from these planes occurs when they are close to parallel to the electron beam, meaning that the molecules are almost perpendicular to the substrate surface, confirming the XRD data.

3.2. CuPc/PTCDA Double Layer

The grain growth seen in the CuPc single layer cross section appears to be preserved when grown on a film of PTCDA (figure 1b). However, the XRD data suggests a large deviation from the single layer crystallinity. The peak resulting from diffraction from the (100) plane is not present, and a large peak at 27.5° suggests diffraction from the (11-2) plane (figure 3a). When the (11-2) plane lies parallel to the substrate surface the molecular plane will be almost parallel to the substrate surface (figure 3b).

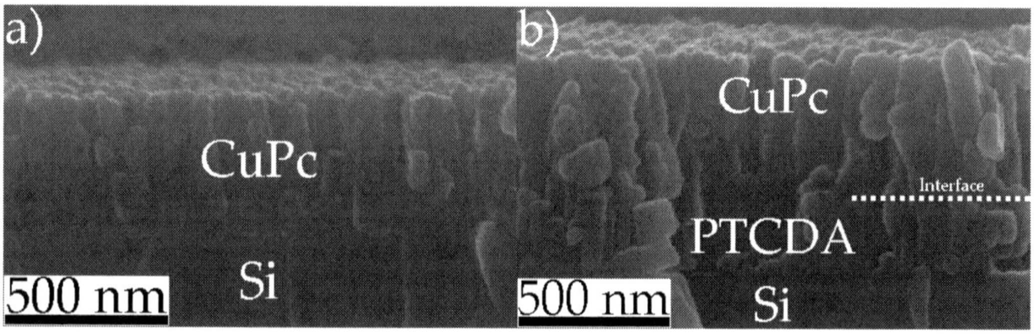

Figure 1 a) An SEM image of a 395 nm CuPc film grown on silicon which has been cleaved to reveal a cross section. b) A 436 nm CuPc film grown on a 247 nm PTCDA film on silicon.

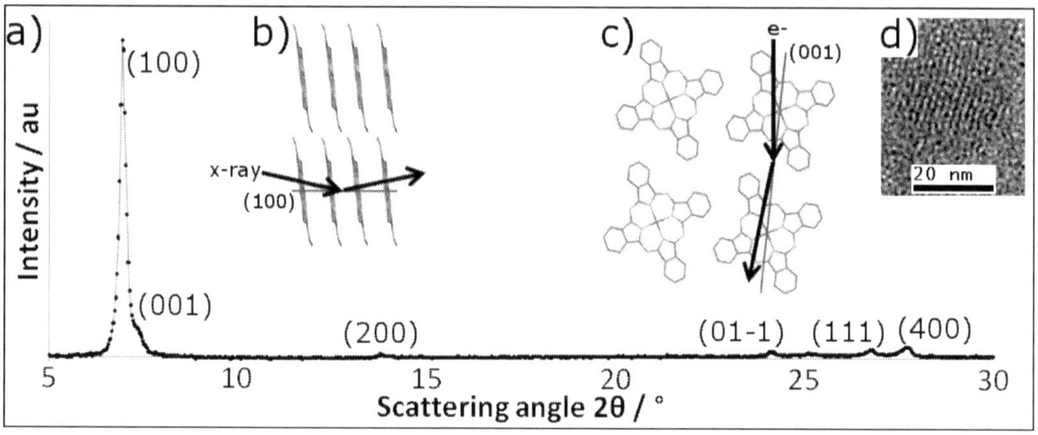

Figure 2 a) XRD from a 395 nm CuPc film grown on Si (100). b) Schematic of X-Ray diffraction occurring on planes parallel to the substrate surface. c) Schematic of electron diffraction occurring on planes more perpendicular to the substrate surface (plane tilt exaggerated for clarity). d) TEM image showing lattice fringes corresponding to diffraction from the (001) plane (1.2 nm) in thinner films grown on carbon coated TEM grids.

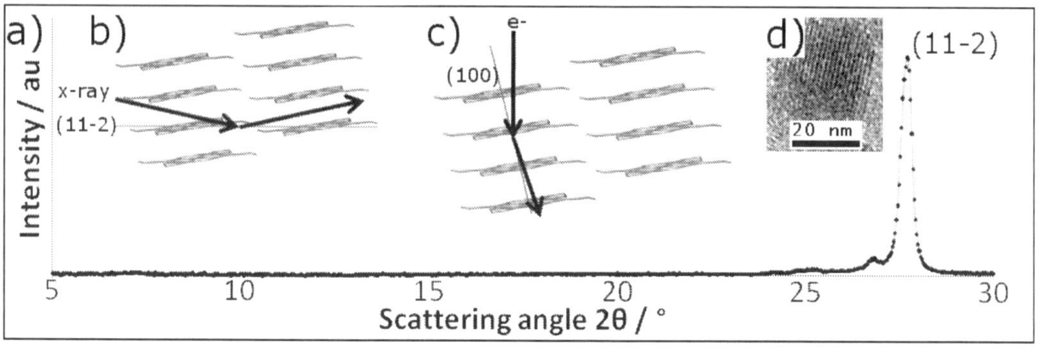

Figure 3 a) XRD from a 436 nm CuPc film grown on 247 nm PTCDA film. The main peak is now due to diffraction from the (11-2) plane which shows that the molecular plane is now nearly parallel to the substrate surface. b) X-ray diffraction schematic. c) Electron beam diffraction schematic when the (11-2) plane is parallel to the substrate surface. d) Fringes corresponding to diffraction from the (100) plane (1.3 nm) visible when the (11-2) plane lies parallel to the substrate surface.

The change in crystal orientation will therefore cause the (100) plane to be close to perpendicular to the substrate surface, and therefore close to parallel to the electron beam when imaging the films grown on carbon coated copper TEM grids. The presence of lattice fringes with a spacing of 1.3 nm confirms that the (100) plane does lie almost parallel to the electron beam, therefore confirming data from the XRD (figures 3c and d). Diffraction patterns of the single and double layer films cannot currently be obtained, this being attributed to the sensitivity of the material to the electron beam and the short time that the fringes exist before being degraded.

On examining the crystal structure of α-CuPc, it can be seen that when the (11-2) plane is parallel to the substrate surface the (100) plane will diffract the electron beam. It can also be seen that it would be possible to bring the both the (100) and the (001) planes to a position where diffraction will occur by tilting the sample 28° and 13° in perpendicular directions, giving rise to lattice fringes from both the (100) and (001) planes that are perpendicular to each other (figure 4).

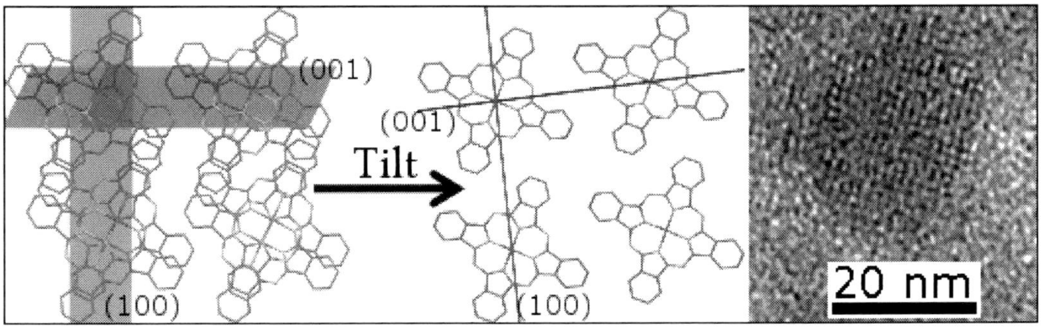

Figure 4 A CuPc unit cell viewed perpendicular to the (11-2) plane, showing how by tilting the sample crystalline regions can be brought into a position where both the (100) and (001) planes show up as lattice fringes perpendicular to each other in the images.

4. Conclusions

The results show that films of CuPc grown by OMBD form grain structures which are crystalline. For CuPc films grown on silicon substrates at room temperature the α-CuPc crystal structure arises. XRD data shows that the preferred crystal orientation is one in which the (100) plane is parallel to the substrate surface. This is confirmed by observing lattice fringes in CuPc films grown directly onto TEM grids which have a spacing corresponding to diffraction from the (001) planes. When CuPc is grown on a film of PTCDA, a different crystal orientation arises in which the (11-2) plane is now parallel to the substrate surface, shown by XRD data as a peak occurring at 27.5°. This is again confirmed by observing lattice fringes within grains that correspond to diffraction from the (100) plane, which would occur when the (11-2) plane is parallel to the substrate surface. The (001) as well as the (100) planes can both be imaged by tilting the sample, which brings both these planes into a position where diffraction will occur. This results in lattice fringes that are perpendicular to each other being observed. The results show that XRD data can be confirmed with data obtained in the TEM.

References

[1] Forrest S R 2004 *Nature* 911-918
[2] Gregg B A 2003 *J. Phys. Chem.* B **107** 4688-4698
[3] Peumans P, Forrest S R 2001 *Appl. Phys. Lett.* **79** 126
[4] Forrest S R 1997 *Chem. Rev.* **97** 1793-1896
[5] Hoshino A, Takenaka Y and Miyaji H 2003 *Acta Crystallogr.* B 393-403
[6] Sullivan P, Jones T S, Ferguson A J and Heutz S 2007 *Appl. Phys. Lett.* **91** 233114

In-situ gas study and 3D quantitation of titania photocatalysts by advanced electron microscopy

N. Tanaka[1,2], K. Yoshida[1,2] and S. Arai[1]

[1] Department of Crystalline Materials Science and Ecotopia Science Institute, Nagoya University, Chikusa-ku, Nagoya, 464-8603, Japan

[2] Nanostructures Research Laboratory, Japan Fine Ceramics Center, Atsuta-ku, Nagoya, 456-8587, Japan
E-mail: a41263a@nucc.cc.nagoya-u.ac.jp

Abstract. Titanium dioxide (TiO_2) with metal nanoparticles exhibits a high photocatalytic activity by a charge separation. Recently the size of the nanoparticles has been reduced to less than 1 nm. In contrast, the size of TiO_2 support particles has been kept at sub-micron sizes to ensure stability for practical use at high temperatures. For visualizing and analyzing the metal nanoparticles which are supported on different positions of those complex substrates, a high voltage transmission electron microscope(HVEM) shows a strong performance because of its high transmission capability to view whole areas of the catalysts. Electron tomography is also important to quantify the three-dimensional morphology such as size, density, surface area and nearest particle distance. Here we report the data obtained by using He-cooled 3D TEM/STEM, a newly developed 1MV high voltage environmental TEM/STEM and a 200kV aberration corrected TEM/STEM. Nanostructures of catalytic samples are characterized using these EMs as well as measurement of their chemical activity.

1. Introduction

Oxides supporting precious metals act as oxidation catalysts. Platinum supported on titania (Pt/TiO_2) is an especially active oxidation photocatalyst. Examples of photocatalysis such as aqueous photo-degradation, antibacterial activity and environmental cleanup are already seen in many products familiar to the public. Since the early 19th century, many researchers have tried to develop methods to achieve high photocatalytic activity by controlling nano-structures of platinum on titania, e.g., producing fine particles or porous configurations with large surface areas and controlled sizes of Pt clusters on the TiO_2 surface as the effective reaction sites [1]. Although Pt dots on TiO_2 particles of several hundred nm's diameter can be analyzed by spectroscopic methods including X-ray diffraction, it is rather difficult to obtain useful data on the dots of less than a few nm.

Transmission electron microscopy (TEM) is one of the most effective methods to analyze such supported nanodot structures, although information obtained from the images is limited to two-dimensions (2D). In 1992, "electron tomography" was proposed in order to obtain three-dimensional (3D) structure information from 2D TEM images [2]. In electron tomography, 3D structures of samples are reconstructed from many TEM images taken in various orientations, in the same manner as X-ray computed tomography (CT). In recent years, electron tomography using high angle annular dark field scanning TEM (HAADF-STEM) images has been expected to be a more powerful method to observe 3D structures of crystalline materials than using TEM images [3,4]. In the present study, cryogenic high-angle annular dark field scanning transmission electron microscopy (Cryo-HAADF STEM) tomography has been used to obtain 3D structural information about metallic nanodots on TiO_2 particle surfaces as well as high-resolution SEM.

2. Experimental procedure

The platinum (Pt) nanoparticles were prepared on titanium dioxide by photo-electrocrystallization in water and 5 wt% tetrachloroauric acid. Titanium oxide powder (Ishihara Sangyou Co.) and tetrachloroauric acid were aged after adding in water. Then UV light (10W $\times$ 2, λ = 310 ~ 400nm) was used to irradiate the mixture stirring at 300K for 5 minutes. The mixture was filtered off and the recovered solid was thoroughly washed with ethanol and acetone and oven-dried with UV irradiation for 12 hours.

The morphology was evaluated by high-resolution scanning electron microscope (Hitachi, SU8000) at 800V. Then electron tomography was performed for analysis of Pt nanoparticles. Scanning transmission electron microscopy (STEM) images were recorded with tilt angles ranging from -76 to +74 degrees in 1 degree increments using 300kV He-cooled TEM/STEM (FEI, Tecnai Polara G2). A high-angle annular darak field(HAADF) detector (100 mrad inner collection angle) was used for obtaining Z-contrast images. Due to Howie's detector[5], one can use incoherent thermal diffuse scattering (TDS) electrons and obtain Z-contrast STEM tilt images. The IMOD suite of programs [6,7] was used for 3D re-construction of the tilt series. After adjusting experimental data such as image sizes, orientation and image drift, 3D reconstruction was performed. By calculating the three-dimensional coordinates and intensities of the images, the pore size, volume fraction of pores and specific surface areas of the porous structures could be analyzed using IMOD and in-house programs. The atomic structure of Pt clusters on the TiO_2 surface and the interface structure were investigated by using double aberration corrected(AC) TEM/STEM (JEOL, JEM-2100CS). In-situ observations in gas atmospheres were also performed with a 1MV high voltage environmental TEM/STEM (JEM-1000K RS) in Nagoya University.

3. Results and discussion

Figure 1 is a SEM image of the present Pt/TiO_2 catalyst. The arrowed particles are typical octahedra consisting of anatase TiO_2{101} planes. Average particle size of the octahedron

Figure 1. SEM image of the Pt/TiO_2 catalyst on the edge of a micrograid.

nanoparticles was estimated as 174±18nm. On such three-dimensional substrates, the spatial resolution and image contrast of deposited Pt nanoparticles are strongly affected by the location and orientation of the TiO_2. Pt particles were not visible using the present SEM with a high-resolution thermal field emission gun system.

Figure 2 is a spherical aberration corrected STEM image of the Pt/TiO_2. The crystal structure of the Pt nanoparticles and atomic interfaces with the TiO_2 substrate were characterized by aberration corrected high-resolution STEM. The HAADF-STEM observations have clarified the single crystalline nature of Pt nanoparticles (atomic number; Z = 78), which show stronger contrast than the anatase substrate (atomic number in "gray atom approximation"; Z = 12.7). Furthermore, localized growth was also visualized by the HAADF-STEM analysis. On the corner site of the octahedron particle (white-arrowed in Fig. 2) Pt nanoparticles were hardly detected.

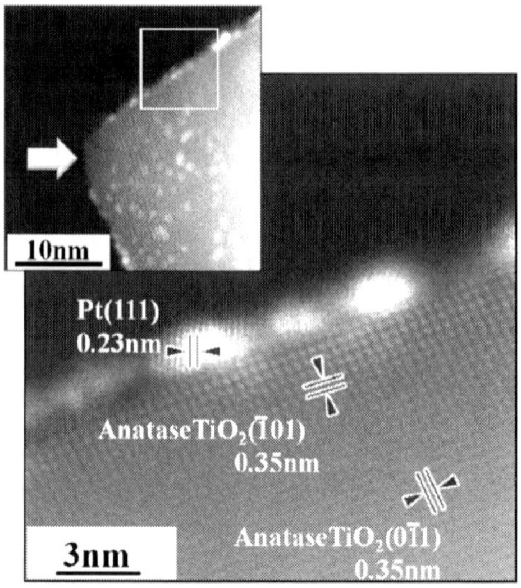

Figure 2. Aberration-corrected(AC) HAADF-STEM image of the Pt/TiO_2 interface.

The growth site of the Pt nanoparticles was quantitatively studied by electron tomography. Figure 3 shows a tilt series and reconstructed tomogram of a Pt/TiO_2 prepared in water solution. Fig. 3(a)-(g) are HAADF-STEM images with tilt angles of -60, -40, -20, 0, 20, 40 and 60 degrees. Fig. 3(h) shows the reconstructed 3D structure of the Pt/TiO_2 particle. We can clearly observe fine Pt nanoparticles about 1-3 nm diameter located on the TiO_2 surface without the artifacts which result from a missing wedge or digital quantum noise. After 3D reconstruction based on the back-projection method, we have analyzed quantitatively the size, volume and number of Pt nanoparticles for each unit area on the titania surface by using the IMOD and in-house measurement programs [8].

In Table 1, the morphology of the Pt/TiO_2 is summarised. The present study is the first application of electron tomography to analysis of the 3D shape and distribution of Pt nanoparticles (2.86 nm in diameter) on the TiO_2 particle surface. Pt volume, surface area and nearest neighbour particle distance were accurately measured in real space.

Figure 4(a) is a histogram of the intensity distribution of the reconstructed tomogram. This histogram shows that there are three types of area. We set two threshold values of 864 and 892 between vacuum, TiO_2 and Pt. All voxels corresponding to Pt intensity were categorized by distance from edge of the octahedron (d_n) as shown in Figure 4(b). Localized growth of Pt nanoparticles is described by Pt volume fraction and density at each distance from the edge of the particle as shown in Figs. 5 and 6.

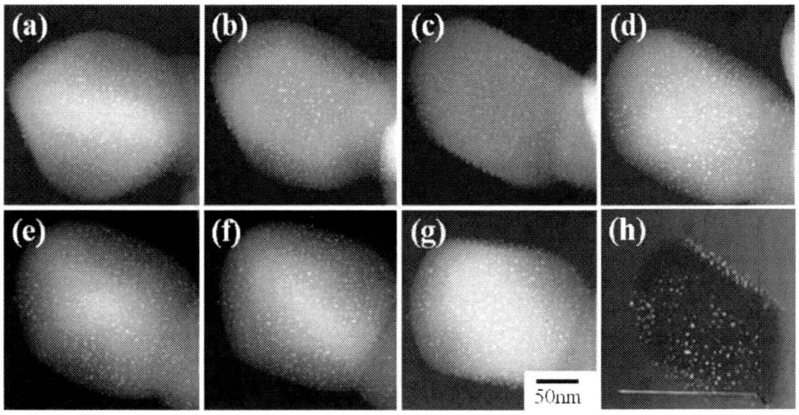

Figure 3. Tilt-series(a-g) and reconstructed tomogram(h) of the Pt/TiO_2.

Number of particle	Average volume [nm^3]	Average surface area [nm^2]	Nearest neighbour distance [nm]
664	9.8	26	5.2

Table 1. Quantitative morphology of the Pt nanoparticles on a TiO_2 support.

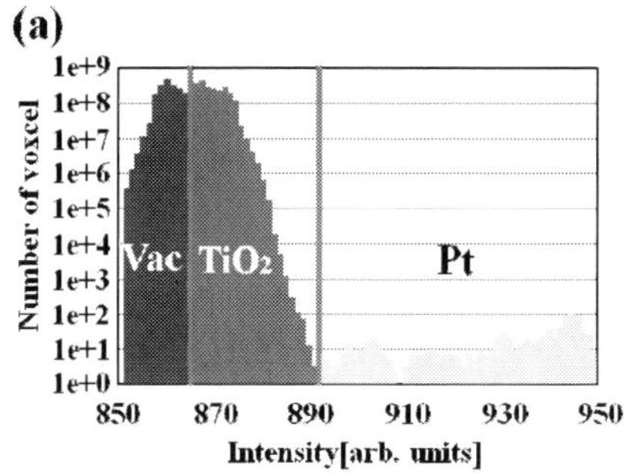

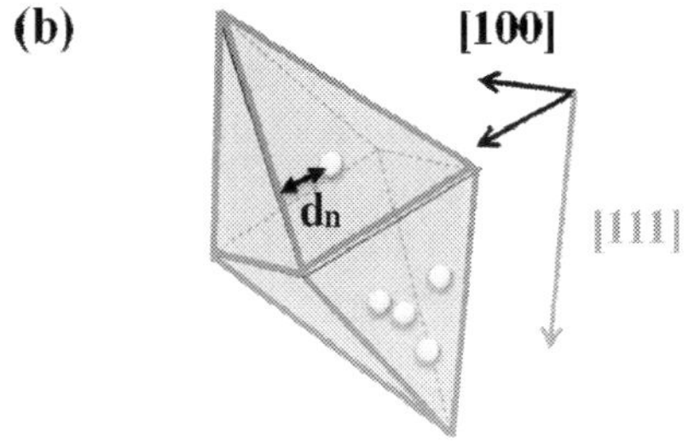

Figure 4. (a) Intensity histogram of the tomogram in Fig 3(h), and (b) diagram of a TiO_2 particle with Pt nanoparticles (small dots), where d_n is the distance from the edge.

These results indicate that in the present photo-electrocrystallization process larger Pt particles are present with higher density on the TiO_2 surface away from the corners or edges of octahedron. Although we need further experiments to discuss the influence of the TiO_2 support morphology on the growth process, especially in complex liquid phase reaction, electron tomography is undoubtedly the best method for clarifying the process.

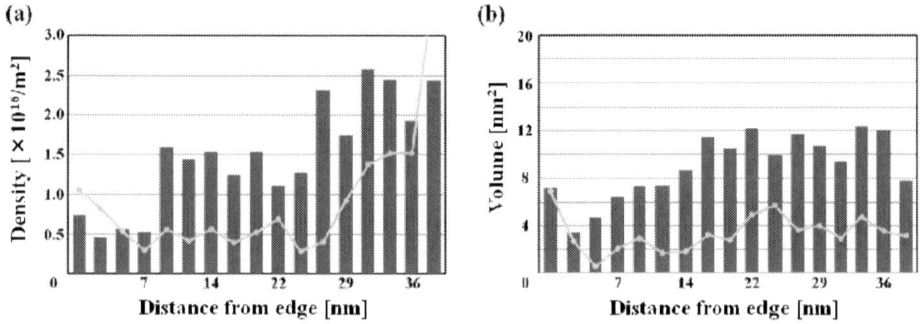

Figure 5. Local density (a) and volume (b) of Pt nanoparticles in a Pt/TiO_2 catalyst. The polygonal graphs show variation of the standard deviation.

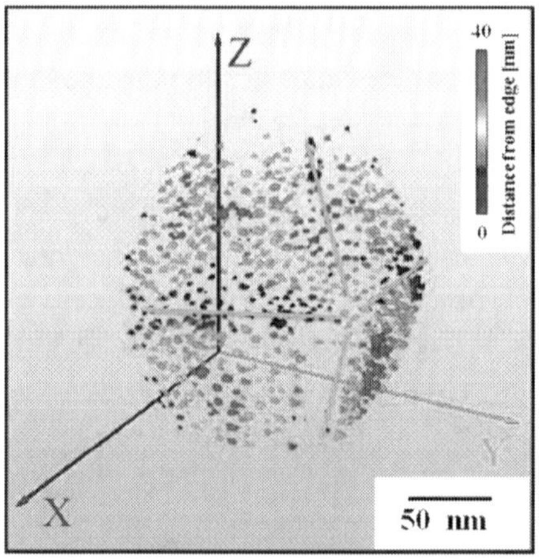

Figure 6. Spatial distribution of Pt nanoparticles on TiO_2 surfaces measured from the tomography.

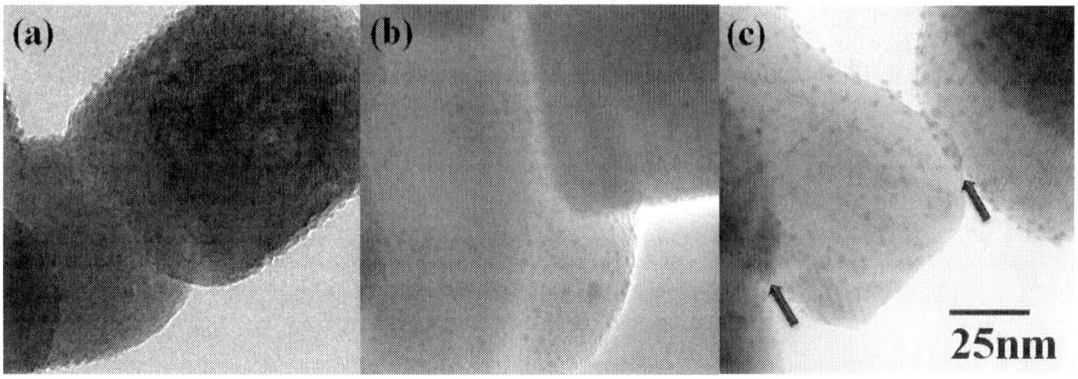

Figure 7. TEM images of the Pt/TiO_2 with non-aberration corrected(n-AC) 200kV(a), AC 200kV(b) and n-AC 1MV TEMs(c). The arrowed areas in (c) are overlapped areas, even where small Pt clusters are observed clearly due to the use of HVEM.

Finally, we will discuss future prospects for transmission electron microscopy (TEM) in heterogeneous nanocatalysis. As you can see in Figure 7(a), Pt nanoparticles (2-5nm in diameter) are very difficult to visualise with conventional TEM when supported on sintered oxide substrates such as TiO_2, CeO_2, and Al_2O_3. Recently spherical aberration correction has enabled improvement of spatial resolution and image contrast around zero focus. Figure 7(b) is an AC-TEM image obtained with 200kV accelerating voltage. In suitably thin areas, this image contains enough information to measure the Pt distribution. In the present study we also used ultra high voltage TEM for visualizing complex nanostructures of a beam sensitive photocatalyst. Figure 7(c) is a TEM image obtained with 1MeV electrons[9]. The result clarified that improved transmittance makes it much easier to observe Pt nanoparticles, which aggregated on grain boundaries of sintered TiO_2 (arrowed in Fig. 7(c)). Careful consideration about irradiation damage is important when performing ultra high voltage TEM (and for STEM even at 200kV), and the relatively low exposure time is required (less than 0.5 seconds) particularly in gas atmosphere of O_2 and N_2[9]. Aberration correction in such ultra high voltage transmission electron microscopes would contribute drastically to in-situ study of nanostructured and heterogeneous catalysts, as clearly seen by our in-situ observation also in 300kV[10].

4. Conclusion

In the present study, we have performed three-dimensional electron tomography of catalytic nanoparticles by using the HAADF-STEM and 1MV TEM newly installed in Nagoya University. One of the recent key topics in catalysis is metal particles on TiO_2 and CeO_2 and their interfaces, which drastically enhance the catalytic activity. For visualizing and analyzing the metal nanoparticles which are supported on different positions of those complex substrates, high voltage transmission electron microscopy is well performed because of its ability to image thick samples. Electron tomography was successfully used to quantify the three-dimensional morphology of Pt nanoparticles on TiO_2, including the size, density, surface area and nearest particle distance. We are also performing in-situ gas observation by using a newly developed 1MV high voltage environmental TEM/STEM and 300kV environmental TEM. Nanostructures of samples are characterized in gas atmospheres, and the catalytic activity for a CO oxidation reaction and a methylene blue (MB) degradation reaction are evaluated for understanding of the relationship between their 3D structures and catalytic performance.

Acknowledgements

The present authors thank Drs. K. Saitoh and J. Yamasaki of Nagoya University and Drs. Y. Sasaki A. Bright of JFCC for valuable support through the present study. The present study is partly supported by Grant-in-Aid for Scientific Research of MEXT, Japan (#17201022)

References

[1] M. Haruta 1997 *Catalysis Today* **36** 15.
[2] J. Frank, "*Electron Tomography:Three Dimensional Imaging with the Transmission Electron Microscope*", Plenum, New York, 1992.
[3] P. A. Midgley et al. 2003 *Ultramicroscopy* **96** 413.
[4] K. Yoshida, M. Makihara, N. Tanaka, S. Aoyagi, E. Nishibori, M. Sakata, E. D. Boyes and P. L. Gai 2011 *Microsc. Microanal.* **17** 264.
[5] A. Howie 1979 *J. Microsc.* **117** 11.
[6] J. R. Kremer, D.N. Mastronarde and J.R. McIntosh 1996 *J. Struct. Biol.* **116** 71.
[7] D. N. Mastronarde 1997 *J. Struct. Biol.* **120** 343.
[8] S. Sueda, K. Yoshida, N. Tanaka 2010 *Ultramicrosc.* **110** 1120.
[9] N. Tanaka, J. Usukura, M. Kusunoki., K. Kuroda, Y. Saito, T. Tanji, S. Muto and S. Arai (2010) Proceedings of IMC17 (Rio de Janeiro) I9.2.
[10] K. Yoshida, A. Bright and N. Tanaka, 2012, J. Electron Microsc. **62**, in print.

In situ TEM study of platinum electrode catalysts with 80kV-1MV accelerating voltage

K. Yoshida[1,2,3], A. Carlsson[4], J. R. Jinschek[4], D. J. Stokes[4], P. L. Gai[2], S. Arai[3], K. Saitoh[3] and N. Tanaka[1,3]

[1] Nanostructures Research Laboratory, Japan Fine Ceramics Center, Atuta-ku, Nagoya, 456-8587, Japan
[2] Department of Chemistry and The York JEOL Nanocentre, University of York, Heslington, York, YO10 5DD, UK
[3] Ecotopia Science Institute, Nagoya University, Chikusa-ku, Nagoya, 464-8603, Japan
[4] FEI COMPANY, PO Box 80066, 5600 KA, Eindhoven, The Netherlands

E-mail: k_yoshida@jfcc.or.jp

Abstract. We perform transmission electron microscopy (TEM) at different accelerating voltages for analysis of complex nano-structures of platinum (Pt) nanoparticle catalysts synthesized on titanium dioxide. TEM images obtained at 80, 200, 300, and 1000kV are carefully compared. The particle size, density and volume of the Pt nanoparticles deposited under different conditions are evaluated using electron tomography. The atomic structures of Pt nanoparticles on the TiO_2 and amorphous carbon a surfaces are also investigated using spherical aberration corrected TEM in order to study the diffusion of these Pt nanoparticles and atoms.

1. Introduction

Platinum supported on titanium dioxide (Pt/TiO_2) exhibits remarkably high activity towards the decomposition reaction of organic molecules such as acetaldehyde and carboxylic acid. Recently, Pt/TiO_2–Carbon catalyst has been synthesized for proton exchange membrane fuel cells [1]. In such catalytic applications, the importance of reducing size and increasing density of Pt nanoclusters on anatase titanium dioxide has attracted strong interest. In contrast, the size of the TiO_2 support particles has been kept at sub-micron sizes to ensure stability at high temperature. The TiO_2 surface has also been modified to form a heterogeneous nanostructure. Transmission electron microscopy (TEM) is one of the most effective methods to analyze such complex nanostructures. Electron tomography allows reconstruction and quantification of the three-dimensional (3D) morphology [2-4]. Spherical aberration corrected high-angle annular dark-field scanning transmission electron microscopy (AC-HAADF-STEM) imaging has been widely used for measuring size distributions of the catalyst nanoparticles, clusters and single atoms active in various reaction processes [5]. Although it is clear that these complex nanocatalysts can be visualized at typical voltages such as 200kV, switching between several accelerating voltages (progressive electron microscopy) provides much better data about the sample structure.

Here we report a progressive electron microscopy study of such heterogeneous catalysts using accelerating voltages of 80 kV to 1 MV. Switching voltages is also essential for estimating the beam effect when performing environmental transmission electron microscopy (ETEM). We present aberration-corrected ETEM images of Pt/TiO_2 photocatalysts taken over this range of accelerating voltages.

2. Experimental Procedure

The platinum (Pt) nanoparticles were prepared on titanium dioxide by the photo-electrocrystallization in the two-phase liquid solution of water and hydrophobic ionic liquid. Titanium oxide powder (Ishihara Sanngyou Co.), hydrophobic ionic liquid N,N,N-Trimethyl-N-propylammomium bis(trifluoromethanesulfonyl)imide (TMPA TFSI, Kanto Reagent) and tetrachloroauric acid were aged after adding to water. UV light (10W × 2, λ = 310 ~ 400nm) was used to irradiate the mixture under stirring at 300K for 5 minutes. The mixture was filtered off and the recovered solid was thoroughly washed with ethanol and acetone and oven dried with UV irradiation for 12 hours.

The samples were characterised using a double spherical aberration corrected environmental TEM/STEM (JEM-2200FS, The University of York) [6], a spherical aberration corrected environmental TEM (Titan ETEM, FEI COMPANY) [7] and Ultra-High Voltage environmental TEM (JEM-1000K RS, Nagoya University) [8].

3. Results and discussion

Figure 1 is a schematic diagram showing the heterogeneous catalyst with wide range in size. Electron tomogram visualizes the complicated structure of the present Pt/TiO_2 samples. Figure 2 shows reconstructed 3D structures of Pt/TiO_2 prepared in water solution (PTA) and water/ionic liquid solution (PTB), respectively. We can observe clearly fine Pt nanodots about 1-3 nm diameter located on the TiO_2 surface without showing artefacts, which occur from a missing wedge or digital quantum noise. After 3D reconstruction based on the back-projection method, we have analyzed quantitatively Pt nanoparticle structures in terms of size, volume and number of nanodots for each unit area on the titania surface by using the IMOD [9] and in-house measurement programs [3]. In Table 1, the morphology of the Pt/TiO_2 samples are summarised. In the present study, electron tomography was firstly applied to the 3D analysis of the shapes and 3D distribution of Pt nanoparticles (PTA; 2.89 nm and PTB; 2.20nm in diameter) on the TiO_2 surface. Density and volume were accurately measured in real space.

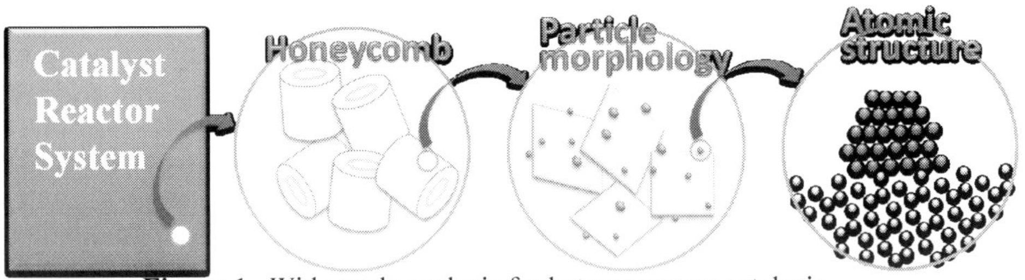

Figure 1. Wide-scale analysis for heterogeneous catalysis.

	Particle size	Density	Volume
PTA	2.89nm	$2.75 \times 10^{16}/m^2$	$17.7 \times 10^{-3} g/m^2$
PTB	2.20nm	$4.50 \times 10^{16}/m^2$	$2.68 \times 10^{-3} g/m^2$

Table 1. Morphology of Pt nanoparticles on TiO_2 support.

Figure 3 are spherical aberration corrected TEM (AC-TEM) images of PTB. The surface structure of the Pt particle and an organic network (adsorbed IL molecules) are successfully observed by 80kV

AC-TEM and agreed well with image simulation. On the other hand, a 300 kV accelerating voltage provided clearer lattice imaging of the TiO_2 supporting particles. For *in situ* study of reduction/deactivation process of the Pt/TiO_2, transmittance of Ultra-High Voltage EM showed big advantage for visualizing nanoparticles, which aggregated on grain boundaries of sintered TiO_2.as shown in Figure 4 We also present an *in situ* AC-TEM study of Pt nanoparticle on anatase TiO_2, amorphous carbon and amorphous silicon nitride surfaces (Figure 5). The nano-step structure working as trapping site of Pt nanoparticle and reactionsite had been observed with deposited nanoparticle at same time. The present results indicated that adsorbed ionic molecules prevent sintering of nanoparticle and improve robustness.

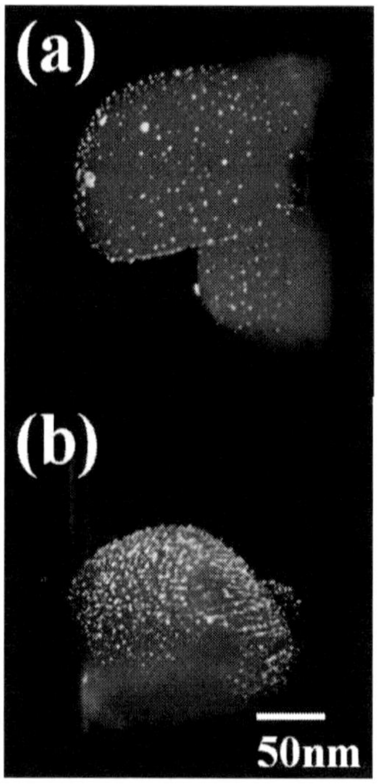

Figure 2. Reconstructed 3D structure of the Pt/TiO_2 of (a) PTA and (b) PTB. The scale bar is an estimate only.

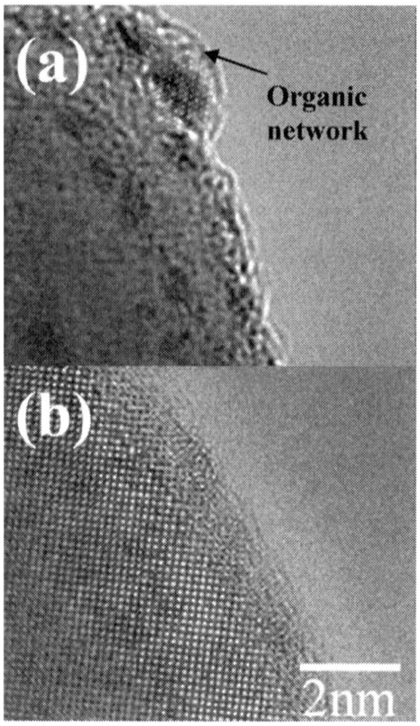

Figure 3. AC-TEM image of Pt/TiO_2 obtained with (a) 80 keV and (b) 300 keV by Titan ETEM.

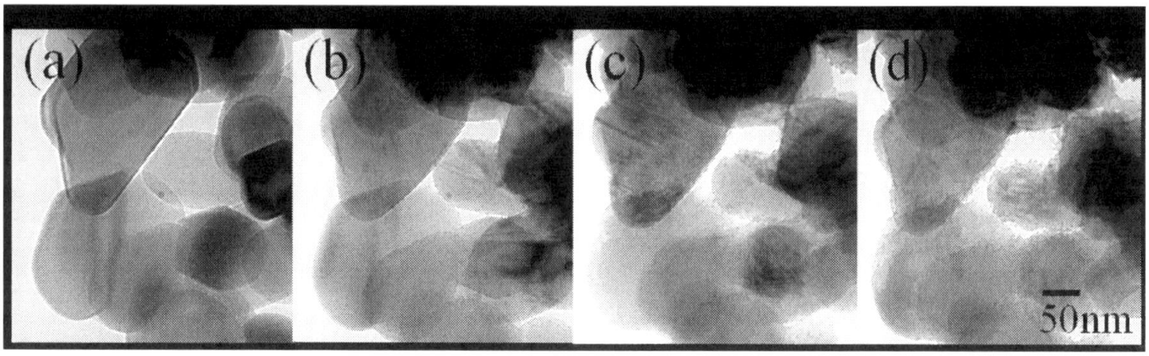

Figure 4. *In situ* observation of reduction process of Pt/TiO_2 by Ultra-High Voltage EM.

(a) **(b)**

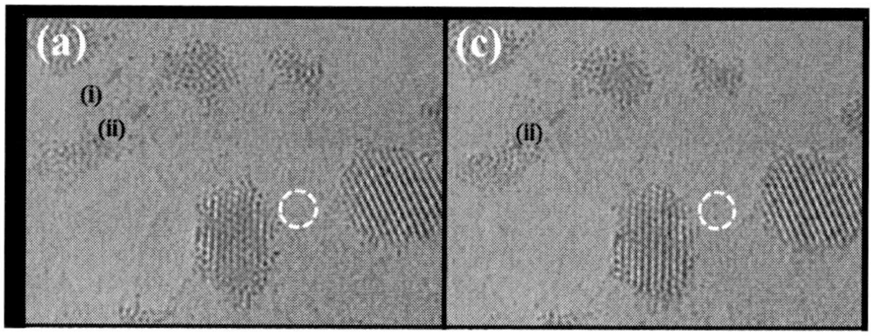

Figure 5. *In situ* AC-TEM images of single Pt atoms (i) & (ii) on a-Carbon obtained after 3 and 6s.

Figure 5 are a Selected Area Captured (SAC) images from the original video file. Points (i) and (ii) in Fig. 5(a) show intensities of a single Pt atom. Those dots with sharp contrast moving between particles were successfully visualized. By such a transfer of the surface atoms one particle shrinks while the other grows. This result indicates that in-situ AC-TEM observation is useful for visualizing and identifying sintering mechanisms such as coalescence and Ostwald ripening.

4. Conclusion

In order to analyze such complex nanostructures of Pt/TiO_2–C catalysts, Cs corrected TEM with various accelerating voltage and electron tomography was applied.

The Small and uniform Pt crystallites 2-3nm in diameter were measured in real space using electron tomography. A numbers of nanoclusters and nanoparticles were indicated on PTB. In the 80 kV TEM image, selective lattice imaging of the Pt nanoparticle was achieved. Greater contrast was obtained in comparison with a conventional 200 kV image. On the other hand, a 300 kV accelerating voltage provided clearer lattice imaging of the TiO_2 supporting particles. In addition, we achieved single atom imaging of platinum on amorphous carbon using 300kV AC-TEM and studied the behavior of Pt atomic species dynamically. Single atoms, clusters (<1nm) and nanoparticles (<3nm) were successfully visualized in the same region at the same time. The time resolution (0.2s) is similar to that reported in a previous in-situ AC-STEM study [10].

5. Acknowledgments

This study was partly supported by a Grant of JSPS Research Fellowship for Young Scientists (K. Y.). We thank Mr. S. Sueda and Mr. H. Sugiura for assistance with the image processing. We thank Yorkshire Forward, the European Union through the European Regional Development Fund and JEOL (UK) Ltd and FEI COMPANY for the EN facilities.

6. References

[1] Z. Z. Jiang, Z. B. Wang, Y. Y. Chu, D. M. Gu and G. P. Yin 2011 *Energy Environ. Sci.* **4** 728.
[2] P. A. Midglay et al. 2003 *Ultramicroscopy* **96** 413.
[3] S. Sueda, K. Yoshida and N. Tanaka 2010 *Ultramicrosc.* **110** 1120.
[4] K. Yoshida, M. Makihara, N. Tanaka, S. Aoyagi, E. Nishibori, M. Sakata, E. D. Boyes and P. L. Gai 2011 *Microsc. Microanal.* **17** 264.
[5] A. A. Herzing, C. J. Kiely, A. F. Carley, P. Landon and G. J. Hutchings 2008 *Science* **321** 1331
[6] P.L. Gai, E.D. Boyes 2009 *Microsc. Res. Tech.* **72** 153
[7] T. W. Hansen, J. B. Wagner and R. E. Dunin-Borkowski., 2010 *Mater. Sci. Technol..* **26** 1338
[8] N. Tanaka et al. Proceeding IMC17, I9-2
[9] J. R. Kremer, D.N. Mastronarde and J.R. McIntosh 1996 *J. Struct. Biol.* **116** 71
[10] P. E. Batson, *Microsc. Microanal.* **14** (2008) 89.

Journal of Physics: Conference Series **371** (2012) 012045 doi:10.1088/1742-6596/371/1/012045

An *in-situ* TEM study of the effects of 6 keV He ion irradiation on Si and SiO_2

S E Donnelly[1], J A Hinks[1], C J Pawley[1,2], K J Abrams[2] and J A van den Berg[2]

[1]Electron Microscopy and Materials Analysis Group,
School of Computing and Engineering, University of Huddersfield,
Huddersfield HD1 3DH, UK.

[2]Centre for Materials and Physics,
School of Computing, Science and Engineering, University of Salford,
Salford M5 4WT, UK.

Email: s.e.donnelly@hud.ac.uk

Abstract. Using the new MIAMI facility (**M**icroscope and **I**on **A**ccelerators for **M**aterials **I**nvestigations) at the University of Huddersfield, an in-situ TEM study has been carried out on the effects of helium implantation on a tri-layer system consisting of monocrystalline silicon (c-Si), silicon dioxide (SiO_2) and polycrystalline silicon (poly-Si). Co-irradiation of these three components enabled direct comparisons to be made of differences between the response of c-Si and poly-Si to the He irradiation; in particular, differences in the development of interstitial clusters and bubbles and eventual amorphisation. For implantations to high fluences ($>10^{17}$ ions/cm^2), very significant levels of porosity were observed to build up in the Si, leading to changes in the width of the Si layers of up to 29% Although no helium bubbles were formed in the SiO_2 layer, a very significant dimensional change (but in this case an observed shrinkage) also occurred in this material. Finally, room temperature amorphisation of the Si was observed at high fluences, beginning at somewhat lower fluences in the poly-Si than in the c-Si. A brief discussion of the origins of these effects is presented.

1. Introduction

In order to compare the development of porosity in c-Si and poly-Si as a result of implantation with helium ions, we have prepared specimens from a tri-layer structure consisting of poly-Si (114 nm) on a SiO_2 layer (57 nm) on bulk crystalline silicon using standard cross-sectional TEM specimen preparation techniques. (Interest in this system stemmed from a desire to modify the mechanical properties of Si for use in MEMS RF resonator structures). These thin tri-layers were then irradiated at 6 keV with He^+ at room temperature *in situ* in a TEM at the MIAMI facility. This approach enabled us to subject the two materials to identical treatments (flux, fluence, temperature) and thus make accurate comparative measurements. We report on the development of radiation damage in the form of interstitial clusters and He bubbles, on the swelling due to the radiation damage and the differences in the amorphisation of the poly-Si and c-Si at high fluences. We also present and discuss changes in the SiO_2 layer (shrinkage) due to the ion irradiation which indicate that viscoelastic processes previously observed for irradiation at higher energies (MeV) also occur at low energies. Note that the viewing direction in these experimenst is approximately parallel to the interfaces between the three layers and the ion beam is incident at an angle of 25° to the electron beam.

2. Experimental

The tri-layer specimens were prepared at the Scottish Microelectronic Centre at the University of Edinburgh by depositing a 114nm layer of poly-Si onto a 57 nm thermally grown amorphous SiO_2 layer on a {100} Cz crystalline Si wafer [1]. Cross-sectional specimens were then prepared by means of mechanical (tripod) polishing and ion-beam thinning using a Gatan Model 691 PIPS system with 4 keV Ar^+ ions incident at 4° to the specimen surface. Note that no defects attributable to the Ar ion beam were observed in the Si following this preparation. Details of the MIAMI in-situ TEM/ion-accelerator facility can be found elsewhere [2]. The experiments presented here were performed at room temperature, with 6 keV He^+ ions at an ion flux of $\approx 10^{13}$ ions/cm^2/s. Note that the specimens were irradiated in this cross-sectional, tri-layer form so that each of the layers was exposed to an identical flux of He ions. (The angle between ion and electron beams is 25°). Images were recorded using a 1.4 megapixel Gatan ES500W digital camera mounted above the viewing chamber. The real space images were post-processed into time-lapse video sequences of the experiments. Diffraction patterns were recorded using a bottom-mounted 4 megapixel Gatan Orius SC200 digital camera.

3. Results and Discussion

Figure 1a) is a bright-field micrograph of the specimen prior to irradiation in which the 114 nm wide poly-Si layer can be seen at the top of the image and is separated from the c-Si layer by a 57 nm layer of SiO_2. The micrographs in figure 1b) to 1e) show the changes that resulted from irradiation with 6 keV He^+ ions.

3.1. Interstitial clusters

The implantation of He into c-Si at room temperature leads to the formation of interstitial clusters. These are visible in bright-field TEM in all orientations and, due to the fact that these defects are made visible by the interaction of the electron beam with their strain field whose extent is much larger than the cluster itself, they are generally visible by TEM at fluences lower than those at which bubbles are visible. The clusters can be seen in the c-Si layer in figure 1b) as small regions of dark contrast. However in the poly-Si layer, clusters are only seen to form in the largest grains and in much of this layer are never observed. Grains larger than approximately 40 nm behave similarly to monocrystalline silicon and interstitial clusters form in them; in grains below this size interstitial clusters do not form. The implication of this observation is that, in small grains, interstitials diffuse to the nearby grain boundary before encountering other interstitials and nucleating interstitial clusters.

3.2. He bubbles

Nanometre-sized He bubbles, identified by contrast reversal on passing through focus, are first discernible at fluences of $\approx 7\times10^{16}$ ions/cm^2 in both Si layers and grow in size with increasing fluence; however, the diffraction contrast from interstitial clusters renders it more difficult to observe the bubbles in the c-Si layer. Figure 1c shows the bubble morphology following irradiation to a fluence of 1.3 x10^{17} ions/cm^2. It is not possible to measure the bubble diameters with any degree of accuracy at this stage as the size observed in the TEM varies significantly with degree of defocus and the bubbles are not visible at focus — but they are approximately 1– 2 nm in diameter.

There is no measurable difference in bubble size in the two layers and no grain boundary decoration (formation of larger bubbles) is observed. Also, unlike the case reported for bubbles in some metals [3,4], a denuded region close to grain boundaries does not occur. Figures 1d and 1e show the specimen following implantation to fluences of 2.3 and 3.1×10^{17} ions/cm^2 respectively and in these images also, no significant differences can be seen in the morphology of the poly-Si and c-Si layers.

3.3. Swelling of the Si layers

In addition to the contrast due to the interstitial clusters and helium bubbles, larger scale changes are also visible in figure 1. Specifically, changes to the width of the poly-Si and SiO_2 layers occur as the fluence is increased – the poly-Si layer increases and the SiO_2 layer decreases in width.

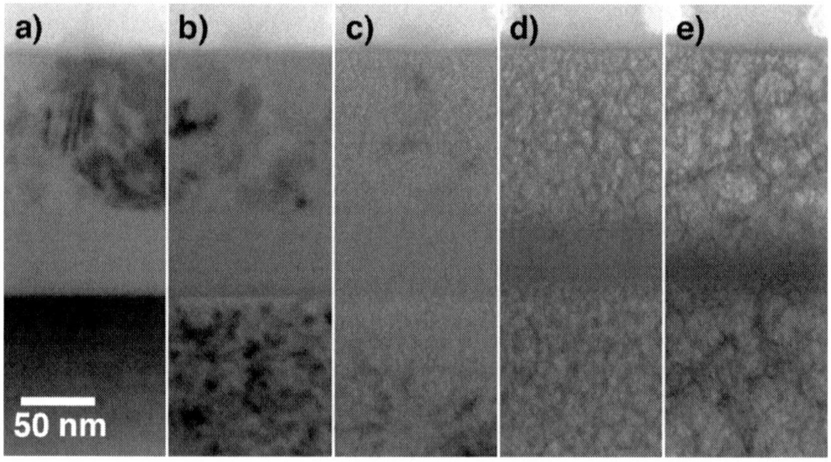

Figure 1. TEM micrographs of the c-Si/SiO_2/poly-Si tri-layer (the c-Si layer is at the bottom of the images) following irradiation to nominal fluences of 6 keV He ions of: a) 0; b) 5.0×10^{16}; c) 1.3×10^{17}; d) 2.3×10^{17} and e) 3.1×10^{17} ions/cm^2. (Bright-field images).

The swelling is clearly measurable in the poly-Si layer and in figure 1e) the poly-Si layer is 29% wider than in the unirradiated material. Assuming an isotropic swelling, this corresponds to a volume swelling of 115%. If the swelling results entirely from the displaced material from the He bubbles, with no additional contribution from elastic distortion around interstitial defects, then this corresponds to a porosity of 53%. This may be a reasonable assumption as, at this stage, the silicon has been amorphised (see below). Although the swelling cannot be measured in the c-Si layer, the similar bubble morphology leads us to the conclusion that a similar level of swelling occurs in this layer also.

3.4. Shrinkage of the SiO_2 layer

Helium bubbles do not form in the SiO_2 due to the high mobility of He in SiO_2 which enables the He to diffuse out of the layer [5]; however, in figure 1 the SiO_2 layer can be clearly seen to shrink with increasing He fluence. The shrinkage, at the highest fluence of 3.1×10^{17} ions/cm^2 is 68%.

There is a body of literature on the effects of ion irradiation on SiO_2 with ions of much higher energies (MeV) [6,7]. It is well established that SiO_2, under mechanical stress and under ion irradiation, undergoes viscoelastic flow so as to reduce the applied stress to zero. The dimensional changes to the SiO_2 layer in the current experiments can be explained in the following manner: the swelling in the silicon layers on either side of the oxide layer then applies stress to the oxide layer at both interfaces. The SiO_2 layer will, therefore, expand in both orthogonal directions parallel to the interface and conservation of mass requires that the width of the layer decrease in the unconstrained orthogonal direction, giving rise to the observed shrinkage. We, therefore, believe that our experiments provide evidence for the occurrence of ion-beam induced viscoelastic flow processes of stressed SiO_2 under irradiation with 6 keV helium ions. Assuming isotropic swelling in both of the Si layers, the observed shrinkage of the SiO_2 layer width would be expected to be twice the observed linear swelling in the Si layers. The actual values (29% and 68%) may indicate a degree of anisotropy in the Si swelling (greater in the direction of ion incidence) but further studies will be required to test this.

3.5. Si Amorphisation

Experiments in which selected area diffraction patterns were recorded as a function of fluence indicate a difference in the amorphisation behaviour of c-Si and the poly-Si with the later amorphising at lower fluences. This can be seen in figure 2.

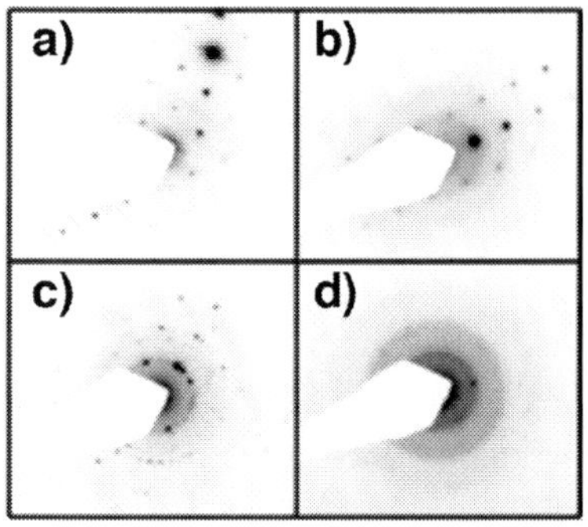

Figure 2. Selected area electron diffraction patterns before and after irradiation with 6 keV He to a fluence of 2.5×10^{17} ions/cm^2. a) c-Si and c) poly-Si before irradiation; b) c-Si and d) poly-Si after irradiation. A much greater degree of amorphisation is seen in the poly-Si than in the c-Si as evidenced by the intensity of the amorphous rings in d) compared with those in b) and by the greater intensity of the crystalline spots in b) compared with those in d).

Although there is ambiguity in the scientific literature regarding the possibility of amorphising Si using light ions such as He, both Reutov and Sokhatskii [8] and Siegele and Weatherly [9] have reported such amorphisation with higher energy ions (17 and 20 keV respectively). The present results constitute the first evidence for amorphisation of silicon at an energy as low as 6 keV.

4. Conclusions

Irradiation of poly-Si and c-Si with 6 keV He^+ ions at room temperature results in the formation of identical helium bubble distributions in both materials and, at high fluences, leads to significant porosity and swelling (53% and 115% respectively). The formation of interstitial clusters is, however, suppressed in poly-Si and this is assumed to be due to absorption of Si interstitials by the grain boundaries. A surprising conclusion is thus that the presence of interstitial clusters appears to have no effect on the nucleation and growth of helium bubbles in Si.

The large degree of swelling in the two Si layers exerts a tensile stress in both orthogonal directions at the SiO_2 interfaces and this effect, coupled with ion-beam-induced viscoelastic flow of the oxide so as to mimimise the stress leads to a shrinkage of the width of the SiO_2 layer in the direction directly observed in our experiments. These experiments constitute the first indication that ion-beam-induced viscoelastic processes in SiO_2 can occur during irradiation using ions with energies below the MeV range.

Both c-Si and poly-Si have been observed to amorphise as a result of room temperature irradiation with 6 keV He ions with the poly-Si amorphising at lower fluences. This is assumed to be due to preferential nucleation of the amorphous material at the grain boundaries.

Acknowledgements

The authors acknowledge the UK Engineering and Physical Sciences Research Council (grant number EP/E017266/1) for provision of the funding that enabled the development of the MIAMI facility used for this work. We thank L I Haworth and J G Terry at the Scottish Microelectronic Centre for provision of the Si/SiO_2 trilayers.

References

[1] Abrams, K J, 2011, PhD Thesis, University of Salford.
[2] Hinks, J A, Donnelly S E, van den Berg J A 2011 *J. Vac. Sci. Technol.* **A29**(2) 021003-1–6
[3] Rose M, Bough A G, Hahn H, 1997 *Nucl. Instrum. & Meth. in Phys. Res.* **B127/128** 119–22
[4] Kesternich W, Schwahn D, and Ullmaier H 1984 *Scripta Metallurgica* **18** 1011–1016
[5] Szakács G, Szilágyi E, Pászti F, and Kótai E 2008 *Nucl. Instrum. & Meth. in Phys. Res* **B266**, 1382-1385
[6] Brongersma M L, Snoeks E, and Polman A 1997 *Appl. Phys Lett.* **71** 1628–31
[7] Snoeks E and Polman A, Volkert C A 1994 *Appl. Phys Lett.* **65**, 2487–90
[8] Reutov V F and Sokhatskii A S 2002 Tech. Phys. Lett. **28**, 615–17
[9] Siegele R and Weatherly G C 1995 *Appl. Phys Lett.* **66**, 1319–21

Transmission Electron Microscopy Study of Graphite under *in situ* Ion Irradiation

J.A. Hinks*[1], A.N. Jones[2], A. Theodosiou[3], J.A. van den Berg[4] and S.E. Donnelly[1]

[1] Electron Microscopy and Materials Analysis Group, School of Computing and Engineering, University of Huddersfield, Huddersfield, HD1 3DH, UK

[2] School of Mechanical, Aerospace and Civil Engineering, University of Manchester, Manchester, M60 1QD, UK

[3] Chemistry Department, Cardiff University, Cardiff, CF10 3AT, UK

[4] Centre for Materials and Physics, School of Computing, Science and Engineering, University of Salford, Salford, M5 4WT, UK

* Email: j.a.hinks@hud.ac.uk

Abstract. Graphite is employed as a moderator and structural component in 18 of the United Kingdom's fleet of Magnox and Advanced Gas-cooled Reactors (AGRs). During the operational lifetime of a reactor, graphite undergoes complex physical and mechanical property changes including dimensional modification, owing to the effects of temperature, oxidation and irradiation-induced atomic displacements. In order to safely extend the lifetime of the current fleet of AGRs, and also to develop materials for GenIV concepts such as the Very-High-Temperature Reactor (VHTR), it is important to gain a better understanding of the fundamental atomic processes which underpin the behaviour of graphite under current and future operational conditions.

This study has focused on the effects of temperature and displacing radiation on the evolution of Mrozowski cracks in highly-orientated pyrolytic graphite (HOPG) using the new Microscope and Ion Accelerator for Materials Investigations (MIAMI) facility. This instrument allows transmission electron microscopy to be performed *in situ* whilst simultaneously ion irradiating to radiation damage levels typically reached in a reactor. By using this technique, it is possible to explore the development of radiation damage under a range of different conditions continuously from start-to-finish rather than just observing the end-states accessible in ex situ studies.

1. Introduction

Graphite is used as both a moderator and structural component in Magnox and Advanced Gas-cooled Reactors (AGRs). The combined effects of the neutron flux, temperature and oxidation which the graphite experiences cause it to undergo significant physical modification including dimensional change [1–3]. These effects are not fully understood in terms of the underlying fundamental atomistic processes or the implications for continued operation of these reactors. In order to safely extend the operating lifetime of the existing fleet of AGRs and Magnox reactors – as well as develop materials for GenIV concepts such as the Very-High-Temperature Reactor (VHTR) – it is important to gain a better understanding into the behaviour of graphite under these conditions.

One of the features of nuclear graphite which is known to change as a consequence of both displacing radiation and temperature are Mrozowski cracks – the opening and closing of these cracks are two of the phenomena behind dimensional change as well as the modification of other physical properties [4]. In this study, the behaviour of these cracks has been monitored under irradiation with 30 kV He ions and subsequent annealing.

Transmission electron microscopy (TEM) with *in situ* ion irradiation allows the dynamic effects of radiation damage on the structure of materials to be explored at the nanoscale whilst controlling experimental parameters such as temperature. These effects are the result of a complex competition between multiple processes and so it is important to observe them as they occur rather than simply examining the end-states accessible in *ex situ* studies.

2. The MIAMI Facility

The Microscope and Ion Accelerator for Materials Investigations (MIAMI) facility was originally constructed at the University of Salford and has recently moved to the University of Huddersfield [5]. It consists of a JEOL JEM-2000FX 200 kV TEM interfaced to a 100 kV ion accelerator. A summary of the key specifications of the MIAMI facility is given in table 1.

Table 1. Specifications of the MIAMI facility.

Ion accelerating voltage	0.5–100 kV	Resolution (lattice)	0.14 nm
Ion species	H to Xe	Resolution (point)	0.31 nm
Typical flux (10 kV He)	3×10^{13} ions.cm^{-2}.s^{-1}	Sample tilt	±20°
Ion/electron beam angle[a]	30°	Digital cameras	Gatan ES500W 1.4 MP
Electron beam voltage	80–200 kV		Gatan ORIUS 4 MP
Magnification range	50–800 k	Sample temperature	100–373 K or RT to 1273 K

[a] Previously 25°.

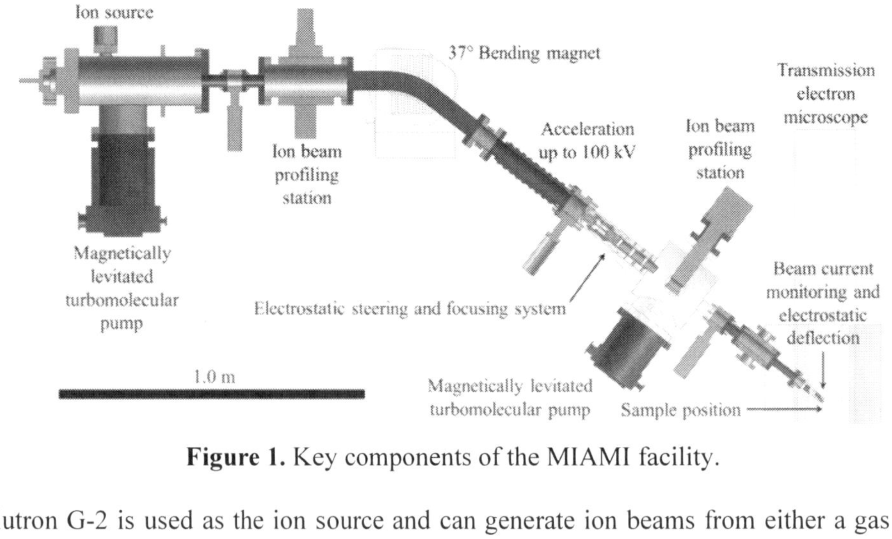

Figure 1. Key components of the MIAMI facility.

A Colutron G-2 is used as the ion source and can generate ion beams from either a gas supply or solid charge with an accelerating voltage of up to 10 kV. Whilst this is adequate for many experiments including the implantation of light atoms such as H or He into most samples, the source can also be floated up to 100 kV with post acceleration performed after the bending magnet – see figure 1. The ion beam then enters the TEM via a high take-off angle x-ray port at an angle of 53° to the vertical. Electrostatic deflection is then performed inside the microscope column in order to make the ion beam incident upon the sample at angle of 30° to the vertical.

Journal of Physics: Conference Series **371** (2012) 012046 doi:10.1088/1742-6596/371/1/012046

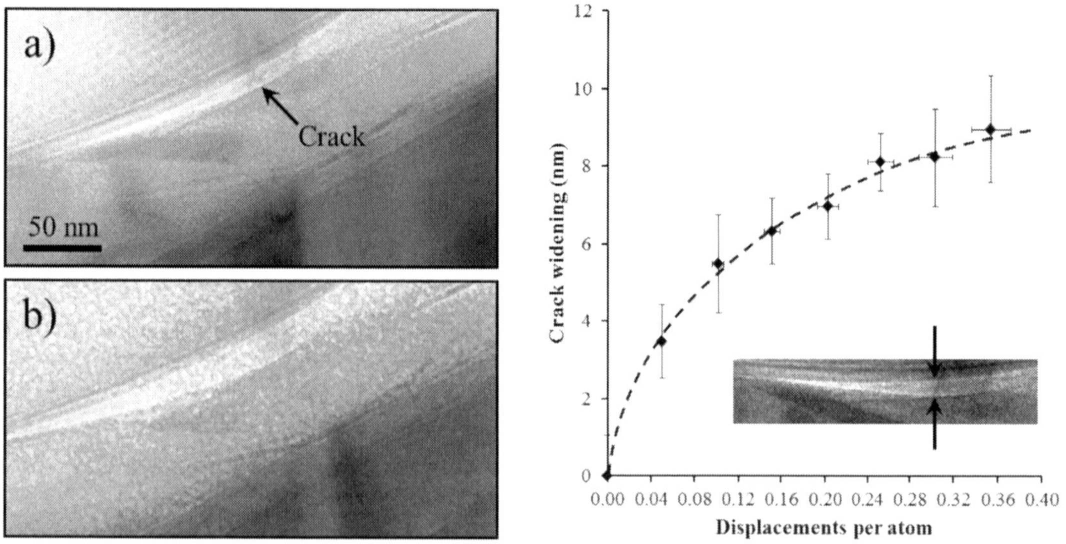

Figure 2. Bright-field transmission electron micrographs (underfocus ≈ 1 μm) showing the widening of a Mrozowski crack in HOPG under 30 keV He irradiation at room temperature: a) virgin; and b) after a fluence of 4.5×10^{16} ions.cm^{-2} (0.35 dpa). Scale marker applies to both micrographs. Trendline provided as guide to the eye and is not fitted to data.

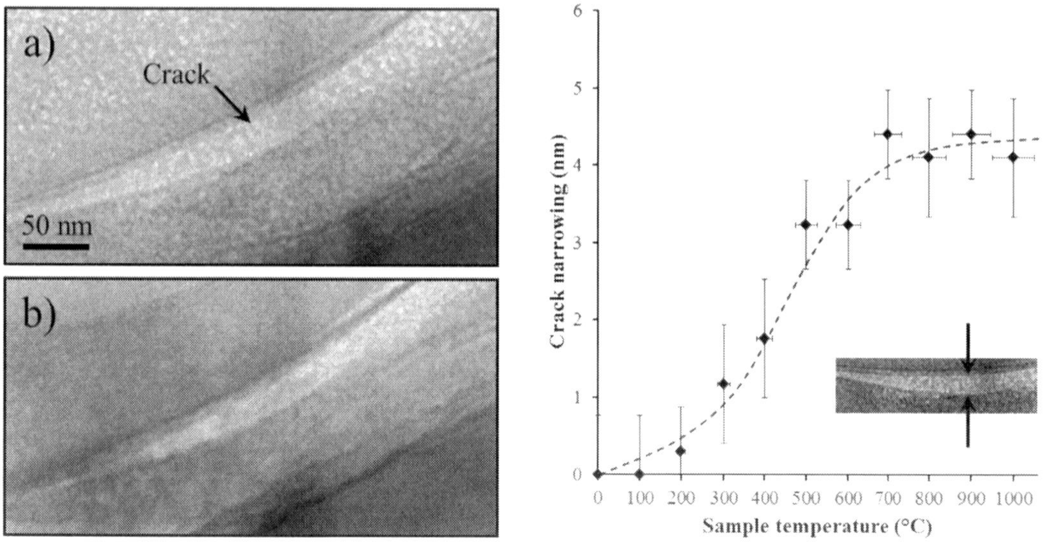

Figure 3. Bright-field transmission electron micrographs (underfocus ≈ 1 μm) showing the narrowing of an irradiation-widened Mrozowski crack in HOPG during temperature-ramp to 1000°C: a) room temperature; and b) after ramp to 1000°C. Scale marker applies to both micrographs. Trendline provided as guide to the eye and is not fitted to data.

3. Experimental Method

Electron-transparent samples of graphite for TEM were produced by exfoliation of bulk highly-orientated pyrolytic graphite (HOPG) using adhesive tape. The adhesive was dissolved using acetone and the thin layers of graphite were captured in copper mesh oyster-grids.

TEM of *in situ* ion irradiation and annealing was performed at room temperature using the MIAMI facility [5] using an electron beam accelerating voltage of 200 kV. Irradiation was performed using 30 keV He ions with a flux of 2×10^{13} ions.cm^{-2}.s^{-1}. Using this ion species and energy, 1 displacement per atom (dpa) can be achieved in graphite at a fluence of approximately 5×10^{16} ions.cm^{-2} whilst introducing only 0.1% additional atoms to a 100 nm thick sample according to calculations performed using the Stopping and Range of Ions in Matter (SRIM) 2008 computer program [6]. The electron beam was kept on during both irradiation and annealing to allow video to be captured. Still images and video were both recorded using a Gatan ORIUS SC200 CCD digital camera and sample annealing was performed using a Gatan Model 652 double-tilt heating holder with a ramp rate of 1°C.s^{-1}.

4. Results

Mrozowski cracks were observed to widen under irradiation by 30 keV He at room temperature (see figure 2) and to narrow during annealing to 1000°C (see figure 3). The crack shown in figures 2 and 3 was observed to widen significantly to 0.2 dpa and to widen more slowly at higher doses; during annealing it was observed to narrow during annealing to 800°C. No significant effects were observed as a result of electron beam irradiation (without ion beam) and the sample did not undergo any detectable tilting which could otherwise account for the observed changes to crack width

5. Discussion and Conclusions

The widening of Mrozowski cracks under displacing irradiation could be caused by either the rearrangement of the material inside the cracks (possibly under strain applied by the adjoining crystal) or by the transport of displaced carbon atoms from the surrounding bulk. Assuming the structure factor of the material inside the crack was relatively constant during the irradiation, the mass-thickness contrast within the crack in figure 2 appears to remain constant suggesting that the latter is the more likely mechanism.

As the narrowing of the crack was observed to proceed rapidly above 400°C and to be complete by 800°C, it would appear that the mechanism(s) responsible were activated in this temperature range. Previous work by Wen *et al* concluded that crack closure in HOPG during annealing to 800°C was caused by dimensional changes in the surrounding crystals due to thermal expansion [7].

The crack followed in this study was parallel to the edge of the sample and far from the support grid which was in good contact with, but not physically attached to, the HOPG. Therefore, it is not expected that differing coefficients of thermal expansion between the grid and sample could have contributed to the narrowing of the crack.

It should be noted that the crack widened by approximately 9 nm under irradiation but narrowed by only a little more than 4 nm under annealing to 1000°C. It is possible that further narrowing may occur at temperatures higher than those available in the MIAMI facility. Further exploration of both phenomena is required to ascertain the mechanisms and activation energies for crack growth and shrinkage and will be the subject of further study.

Acknowledgments

The authors wish to thank the Engineering and Physics Sciences Research Council for funding under grant numbers EP/C549465/1, EP/E017266/1, EP/I003169/1 and EP/I003223/1.

References

[1] B.T. Kelly 1981 *Physics of Graphite* (Barking, England: Applied Science Publishers)
[2] R.E. Nightingale 1962 *Nuclear Graphite* (London, England: Academic Press)
[3] J.H.W. Simmons 1965 *Radiation Damage in Graphite* (London, England: Pergamon Press)
[4] B.J. Marsden, G.N. Hall, O. Wouters, J.A. Vreeling and J. van der Laan 2008 *J. Nucl. Mater.* **381** 62
[5] J.A. Hinks, J.A. van den Berg, and S.E. Donnelly 2011 *J. Vac. Sci. Technol. A* **29(2)** 021003
[6] J.F. Ziegler 2004 *Nucl. Instrum. Methods Phys. Res. B* **219–220** 1027
[7] K. Wen, J. Marrow and B. Marsden 2008 *J. Nucl. Mater.* **381** 199

In-situ cobalt nanocrystal synthesis by intense electron beams in TEM

T Gnanavel and G Möbus
Department of Materials Science and Engineering, University of Sheffield, Mappin Street, Sheffield, S1 3JD, UK

E-mail: t.gnanavel@sheffield.ac.uk

Abstract. We report the synthesis of cobalt nanoparticles of a range of sizes from raw cobalt fluoride powder using high intensity focused electron beams and their characterisation *in-situ* in a TEM. A variety of beam intensities with different focussing level in a JEOL JEM 3010 TEM at an acceleration voltage of 300 kV is used. Depending on the beam intensity either a distribution of a large number of nanoparticles of 1-150 nm size or singular/a small number of fine metal particle(s) of ~150 nm up to micron size can be produced. Mechanisms of the nanoparticle formation are discussed. This simple, organic precursor free, non-hazardous and oxide-free fabrication method offers great potential for basic research, as it allows the fabrication of fresh particles "on demand" in a location with a distribution as required. As the size range is tuneable, the full magnetic properties from superparamagnetic, over single domain ferromagnetic, to multi-domain are covered.

1. Introduction
Halide materials have long been identified as one of the most sensitive compounds under electron beam irradiation with the resulting end product of transformation being the reduced metal as the rate of depletion of anions is faster compared to cations [1, 2]. Previous research with a motivation on patterning by resist exposure or fabrication of nanoparticles mostly used thin films as starting material [1-3], and sometimes precursor/powder processing and/or long irradiation durations with both extended and convergence beams were applied [4-10]. It was also suggested that this method may be particularly suitable for nanoparticle production of comparatively low melting point metals [7, 9, 10] or semiconducting lead chalcogenides (IV-VI) [7] while production of high melting point metal nanoparticles is challenging. In this work, we report a cheap and fast one-step synthesis method for the production of either a distribution of multiply facetted magnetic nanocrystals of versatile shape or singular/a small number of fine metal particle(s) of relatively high melting point with a range of 1-1000 nm size using electrons beams placed at various focusing levels onto the micropowders in a conventional TEM.

2. Experimental method
Both $CoF_2.4H_2O$ and anhydrous CoF_2 crystalline powder (Alfa Aesar, 99.99%) were dispersed on a TEM grid covered with carbon film (Agar Scientific, UK) after ultrasonication in isopropyl alcohol

(IPA). TEM observation of ambient dried specimens were carried out on some well separated 1-5 μm sized 3D fragments, well different from the continuous thin films used in other work.

The fabrication of particles was carried out in a LaB_6 gun JEOL JEM 3010 TEM at an acceleration voltage of 300 kV with variable electron beam focusing levels. The maximum intensity on the screen for a fully focused beam at the largest condenser aperture was ~180 $pAcm^{-2}$. We distinguish two irradiation schedules:

(i) To produce a small number of metal nanoparticles or a singular large particle of >150 nm in size, the electron beam is initially spread to below ~130 $pAcm^{-2}$ followed by the maximum intensity with fully focused beam at the largest condenser aperture. The process is observed in real time on the screen, and the threshold focusing level can be adjusted to whatever gun brightness is active in a particular situation. The resulting metal particles are of roundish "molten-like" appearance which corresponds to our earlier studies of NiF_2 irradiation [11], and we call them *nano-beads* (3.1. below) to distinguish them from crystalline facetted nanoparticles.

(ii) If the highly focused electron beam (~130-180 $pAcm^{-2}$) with the largest condenser aperture inserted or without any condenser aperture (>180 pA/cm^{-2}) is directed at the fluorides ($CoF_2.4H_2O$ or CoF_2) without medium-level pre-irradiation, then a distribution of well dispersed nanocrystals in huge numbers and in particular histograms of sizes can be fabricated (3.2. below). The focused electron beam was directed, usually along one edge/corner of a 3D micron sized fragment for a range of sub-second to 5s durations. This results in a distribution of cobalt nanocrystals in the vicinity of the fragment, isotropic in all radial directions. All TEM images were taken after spreading the beam to such reduced intensity as needed to avoid any further dynamic changes of the produced structures. The imaging of all resulting nanostructures was done in the same JEOL JEM 3010 LaB_6 gun, while for analytical TEM, samples were transferred to a JEM 2010F FEG-TEM with EDX and GIF-EELS, with irradiated regions re-centred with the help of mesh-counting and low-magnification mapping.

3. Results and analysis

3.1 Nanobead formation

The initial micron sized CoF_2 fragment (Fig. 1a) was irradiated with a slightly spread electron beam (~130 $pAcm^{-2}$), resulting in an intermediate state of nanocomposite material, which is a mixture of metal cobalt nanoparticles distributed within the amorphous porous CoF_2 (Fig. 1b). Followed by a highly focused electron beam (>180 $pAcm^{-2}$), this mixture turned mostly into a singular roundish or spherical Co metal bead by sweeping out the nearby composite material at the expense of fluorine (Fig. 1c).

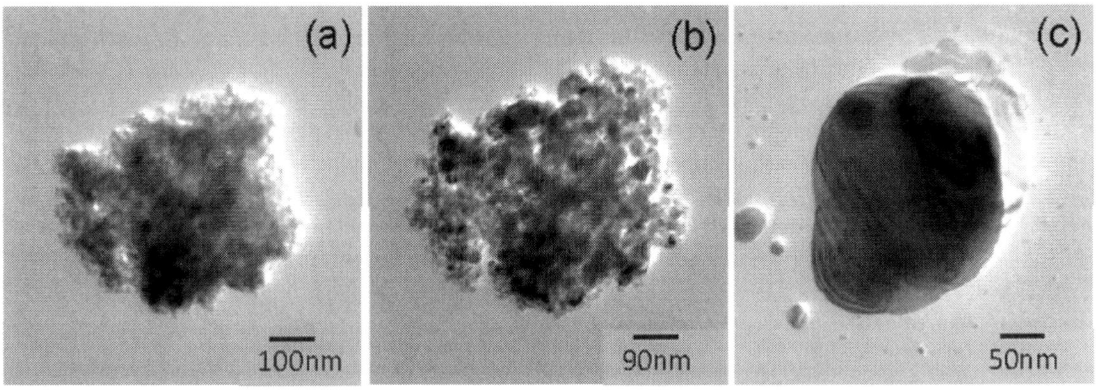

Figure 1. (a) Initial CoF_2 fragment, (b) partially fluorine depleted nanocomposite, (c) end product metallic Co nanobead with stacking fault fringes.

3.2 Distribution of nanocrystal formation and diversity of shapes

Various distributions of *in-situ* electron beam synthesised Co nanocrystals are shown in figure 2(a-d). In projection, they are mostly hexagonally (Fig. 2(c,d)), octagonally (Fig. 2b) or less likely heptagonally (Fig. 2a) faceted nanocrystals, with regular or irregular shape. These structure differences are believed to be dependent on the beam parameters. Examples for various projected shapes of individual nanocrystals, selected from distributions of the kind of figure 2(a-d), are listed in figure 2(e-l). From 2D TEM images, only a footprint (support or shadow) of particle shape can be extracted, while tilting experiments using a high-tilt tomographic holder can reveal the 3D shape. However, sometimes (e.g. Fig. 2(g,h)), the 3D shapes can be estimated without a tilt series because edges appear as dark lines and top/side facets differ in grey values due to scattering contrast.

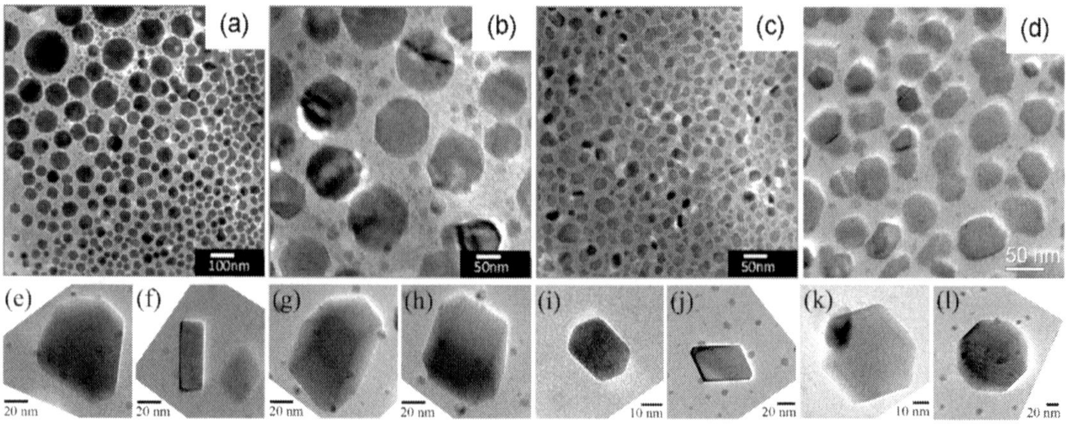

Figure 2. (a-d) Various distributions of the synthesised Co nanocrystals. (e-l) Variously shaped Co nanocrystals produced by electron beam (JEM 3010 at 300 kV).

3.3 Particle size distribution analysis

Two dispersions of nanocrystals, along with their size distribution histogram analysis chart in insets with fitted Gaussian in red, are shown in figure 3(a,b). These distributions correspond to mean sizes of 12.0 nm and 6.4 nm particles, respectively, with (a) being closer and (b) further away from the fluoride edge and electron beam centre. The two standard deviations indicating the spread of sizes were $\sigma = 6.3$ nm and $\sigma = 4.4$ nm.

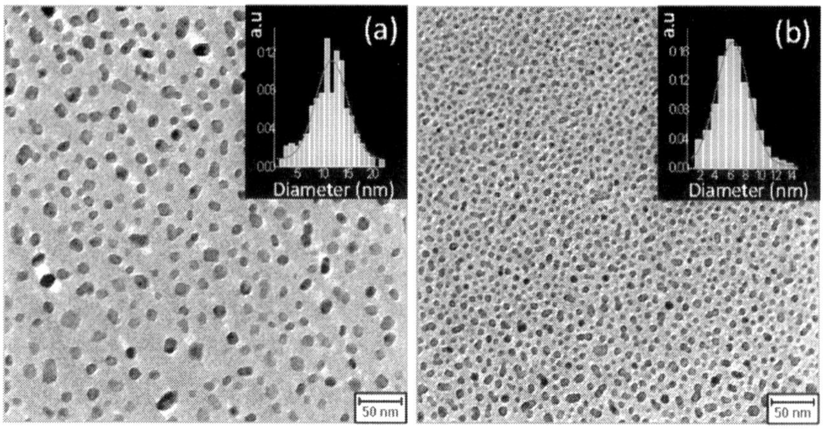

Figure 3. (a,b) Size distribution analysis of the synthesised Co nanocrystals.

For this, first, a numerical histogram was built using the data obtained from ImageJ [12] analysis by counting the number of well defined particles with diameters falling into given size ranges. Then, the numerical histogram was calculated into the normalized discrete function $F(D_i)$ of the size distribution and a bar chart representing the size distribution was plotted (insets, Fig. 3(a,b)). Finally, this discrete function was approximated to the continuous normalized Gaussian profile (with D_o the mean size and σ the standard deviation): $G(D)=\frac{1}{\sigma\sqrt{\frac{\pi}{2}}}\exp\left[-2\left(\frac{D-D_o}{\sigma}\right)^2\right]$.

3.4 Co nanocrystals decorated by nanoclusters on their surface

Selected images of a tilt series of a ~100 nm sized cluster-decorated Co nanocrystal, acquired between +30° and -45° at 5° steps, are shown in figure 4. This nanocrystal shows ~2-5 nm sized nanoclusters on the surface with a quite regular distance indicating self-organisation. Two-stage irradiation of differently centred and focused beams is likely to cause the two hierarchy levels of Co nucleation.

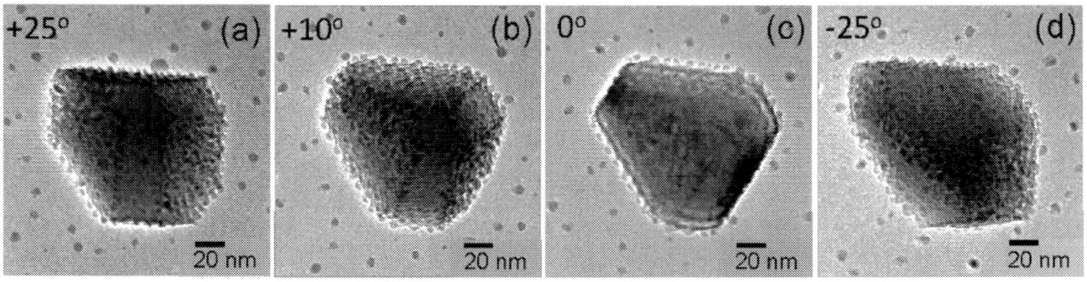

Figure 4. (a-d) Tilt series of a ~100 nm Co crystal itself decorated by ~2-5 nm Co nanoclusters, selected from a distribution of nanocrystals.

4. Discussions and conclusions

In-situ synthesis of porous composite and metallic nanoparticles/crystals by electron beam irradiation is demonstrated. The intensity of the beam and its focusing level are the main factors contributing to the formation of singular/a small number of nanoparticle(s), a distribution of nanocrystals or a hierarchical two-stage assembly of clusters on particles. It is believed that a highly focused electron beam induces sputtering with evaporation/sublimation and subsequent far-reaching diffusion followed by nucleation/growth, while a moderate intensity beam induces "quasi-melting" and local conversion of salt to metal via intermediate composite materials stages. Prospective applications of the phenomena described can be envisaged in the fields of magnetic data storage, nanoparticle sensors, and for surface-plasmon enhanced light coupling or wave-guiding in the context of photonic devices.

Acknowledgments

This work was supported in part by EPSRC, UK, # EP/G036748/1.

References

[1] Muray A, Isaacson M and Adesida I 1984 *Appl. Phys. Lett.* **45** 589-91
[2] Streblechenko D and Scheinfein M R 1998 *J. Vac. Sci. Technol., A* **16** 1374-79
[3] Malac M, Schoefield M, Zhu Y M and Egerton R 2002 *J. Appl. Phys.* **92** 1112-21
[4] Sepulveda-Guzman S, Elizondo-Villarreal N, Torres-Castro D F A, Gao X, Zhou J P and Jose-Yacaman M 2007 *Nanotechnology* **18** 335604
[5] Kim J U, Cha S H, Shin K, Jho J Y and Lee J C 2005 *J. Am. Chem. Soc.* **127** 9962-63
[6] Mahapatra S K, Bogle K A, Dhole S D and Bhoraskar V N 2007 *Nanotechnology* **18** 135602
[7] Hu J Q, Sun Y G and Chen Z G 2009 *J. Phys. Chem. C* **113** 5201-05
[8] Kim S H, Choi Y S, Kang K and Yang S I 2007 *J. Alloys Compd.* **427** 330-32
[9] Pan Z W, Dai Z R and Wang Z L 2002 *Appl. Phys. Lett.* **80** 309-11
[10] Yan Y F, Liu P, Wen J G, To B and Al-Jassim M M 2003 *J. Phys. Chem. B* **107** 9701-04
[11] Gnanavel T, Kumar S and Möbus G 2011 *J. Nanosci. Nanotechnol.* **11** 1019-24
[12] ImageJ. (accessed on 26th Oct. 2010 from http://rsb.info.nih.gov/ij/)

Journal of Physics: Conference Series **371** (2012) 012048 doi:10.1088/1742-6596/371/1/012048

On the effect of atomic structure on the deactivation of catalytic gold nanoparticles

M J Walsh[1, 4], P L Gai[1, 2] and E D Boyes[1, 3]

Departments of Physics[1], Chemistry[2] and Electronics[3], York JEOL Nanocentre, University of York, York, YO10 5DD, UK

[4]Author to whom any correspondence should be addressed.

Email: mw527@york.ac.uk

Abstract. Here we present atomic scale studies into the nature of both the internal structure and external surfaces of catalytic Au nanoparticles using aberration corrected *in-situ* electron microscopy. The activity of catalytic nanoparticles is thought to be highly sensitive to the particles' structure, meaning typical local atomic rearrangements are likely to significantly affect the overall performance of the catalyst. As-deposited Au nanoparticles are found to exhibit a variety of morphologies, with many being internally strained or highly stepped at the surface. Upon heating, surface atoms are observed to minimise the particles' surface energy by restructuring towards planar (111) facets, resulting in the removal of low co-ordinated sites thought to be crucial in catalysis by Au nanoparticles. These results suggest the process of surface energy minimisation made possible by heating may lead to a loss of active sites and consequently contribute to the deactivation of the catalyst.

1. Introduction

Traditionally believed to be inert, it was not until the mid 1980s and the pioneering work of Haruta *et al* [1] and Hutchings *et al* [2] that gold was discovered to be catalytically active. Haruta *et al* went on to show that when reduced to less than 5nm in diameter, gold is in fact a more active catalyst than other noble metals for many applications, and that even at low temperatures, small Au nanoparticles exhibit extremely high activity for CO oxidation [3]. As the activity of gold has been shown to be strongly dependent on particle size, properties that vary as a function of size have been proposed as the fundamental cause behind gold's activity. In particular quantum size effects [4], perimeter sites [5], particle-support interfaces [6] and low co-ordinated atoms [7-9] have all been suggested to be the reason behind the catalytic properties of small gold nanoparticles.

The chemical activity of gold has been described as being strongly dependent on the co-ordination number of the gold surface atoms, and it has been shown that the strength of the Au-CO and Au-O bond varies with Au co-ordination number [10]. This is explained by the weak bonding between Au(111) and O relative to the stronger interaction between O and Au atoms of lower co-ordination number found at steps and corners of small particles. The *d* states of atoms in the Au(111) surface are too low in energy to interact with the O 2*p* valence states, and are therefore unable to disassociate O_2. In contrast, the lower co-ordination atoms have *d* states closer to the Fermi level, resulting in a stronger interaction. This effect can be further explained by the *d* band model [11-13], in which both

Journal of Physics: Conference Series **371** (2012) 012048 doi:10.1088/1742-6596/371/1/012048

tensile strain and reducing co-ordination numbers can modify the *d* band, causing a narrowing and subsequent increase in the band's population.

The inherent instability of small nanoparticles is one of the major problems with the widespread application of nano-catalysts, with sintering via particle coalescence or Ostwald Ripening processes being well documented [14-16]. Whilst sophisticated supporting materials can be used to minimise such processes; in time a loss in overall surface area can cause significant deactivation. However, it seems clear that there is a strong structural sensitivity of the catalytic properties of gold nanoparticles, and that any induced change in an individual nanoparticle's internal structure or external surfaces may also result in a loss of activity.

2. Experimental Procedure

In-situ aberration corrected electron microscopy is a powerful tool for studying catalytic nanoparticles at the atomic scale under simulated reaction conditions. The removal of spherical aberration allows for a consistent phase contrast throughout the image, whilst operating at close to zero defocus helps to minimise delocalisation effects; allowing direct interpretation of inter-atomic distances, internal structure and external surfaces from a single image. This is crucial for *in-situ* studies, as the dynamical nature of the system means it is rarely possible to obtain a through-focal series. Furthermore, the exposure times necessary for imaging and hence the effects of the beam on the sample are minimised. In the present study a double aberration corrected JEOL 2200FS microscope has been used, with the wider gap pole piece made possible by correction of the spherical aberration allowing for an extended tilt range and the use of a regular hot stage [17].

Colloidal gold nanoparticles in the size range of 1–10 nm have been deposited on to 3.5nm thin amorphous carbon supports. The continuous amorphous carbon films provide a well defined substrate whilst the 3.5nm thickness allows for reduced noise and improved resolution of weak phase objects such as small nanoparticles.

3. Results and Discussion

The as-deposited particles exhibit a variety of internal structures, often multiply twinned, and it is clear that the small difference in energy terms between such structures [18] means that small variations in the number of atoms or the local environment can result in vastly different morphologies. An example of a multiply twinned particle can be seen in figure 1. This crystallographically forbidden Marks decahedral structure [19] consists of five tetrahedra adjoined by {111} twin planes, meaning significant internal strain has to be incorporated into the non-space filling structure. In the Marks decahedral, {111} re-entrant facets act to further minimise the particles surface energy, even at the expense of an increase in surface area.

Figure 2 is again as-deposited and shows a single crystal of gold with a (twin) dislocation in the AC-HRTEM image in [011] orientation. Such dislocations may arise due to internal stresses during the crystal growth or during the initial crystallisation. Preliminary displacement vector (Burgers vector) of the dislocation can be inferred based on the atomic scale planar image. Due to the small size of the nanoparticle, tilting the sample to different zone axes is difficult and requires prolonged exposure to the electron beam. Such exposures are avoided due to concerns over beam induced surface atom movement and possible modifications of the internal structure, meaning determination of the component of the displacement vector in the third crystal direction (z) has not been possible. The measurements suggest that the Burgers vector of the dislocation is of the type 1/6 [211]. Such dislocations lie on adjacent twinning planes of {111} type leading to stacking faults.

Also clear from the image is the highly stepped nature of the surface. The internal dislocation terminates in a partial terrace at the surface, whilst the rest of the surface is regulary stepped throughout the perimeter of the particle in the image. Such a disordered surface structure generates an abundance of low co-ordinated surface sites thought to be crucial for the catalytic activity of gold nanoparticles. However, whilst surface atoms with reduced co-ordination are likely to provide active

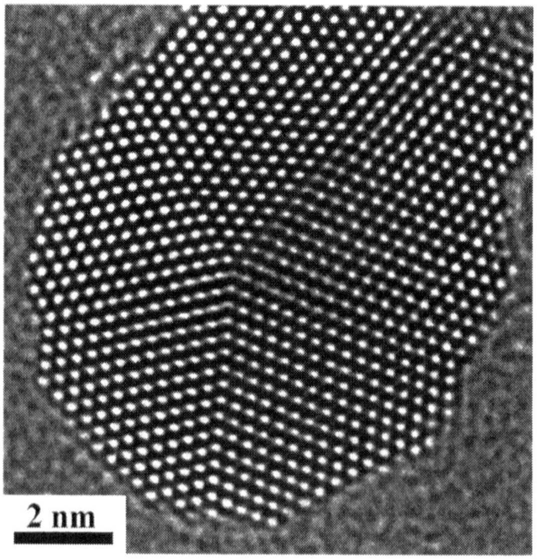

Figure 1. As-deposited Marks Decahedral Au nanoparticle with {111} re-entrant facets.

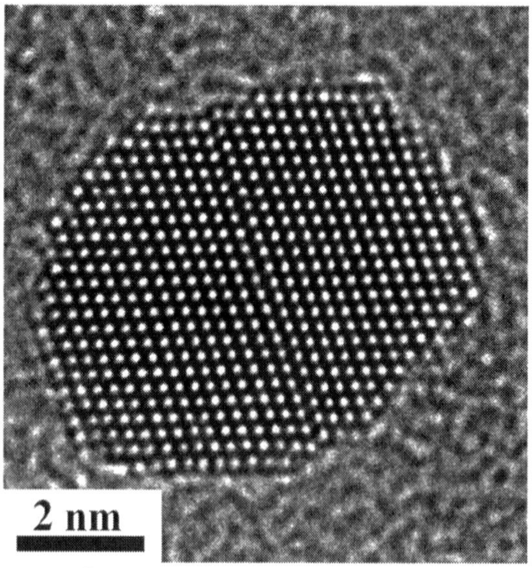

Figure 2. As-deposited single crystal Au nano-particle with internal dislocation and surface steps.

sites for catalysis, they are also likely to be inherently unstable as they result in an increase in total surface energy. This can be represented by the expression for surface energy [20]:

$$\sigma_{hkl} = \frac{1}{2}\frac{N_{hkl}}{Z_B}E_{coh}$$

where Z_B is the bulk co-ordination number, E_{coh} is the cohesive energy between two neighbouring atoms and N_{hkl} is the number of broken bonds. Clearly low co-ordinated ad-atoms, steps and terraces result in a greater number of broken bonds and hence increased surface energy.

The effect of heating on these low co-ordinated sites is shown in figure 3, where exposure to the electron beam over the course of a minute causes movement of the surface atoms and minimistation of surface energy terms via restructuring towards planar (111) surfaces. These low energy surfaces exhibit much higher co-ordination, with density functional theory calculations predicting very weak adsorption of CO and consequently much reduced activity [13].

In-situ heating to 600°C provides further evidence of such effects, as is demonstrated in figure 4. Here we can see a single crystal cuboctahedral particle bound predominantly by {111} surface facets. Such large single crystal structures with almost atomically clean facets were only produced as a result of heating to elevated temperatures. Nearly all surface step sites have been removed; with heating having an annealing effect on both the surfaces and the internal structure, as any twin boundaries or dislocations have been removed. Added effects due to gas adsorption may further modify the nature

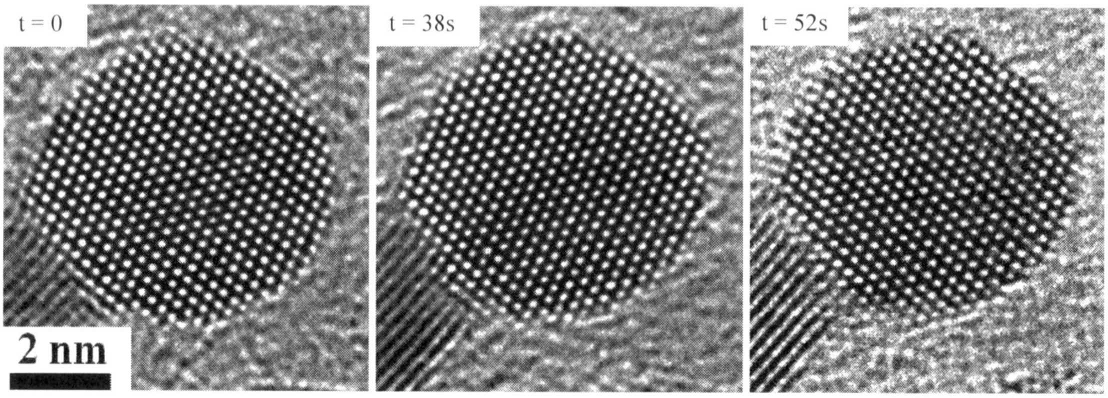

Figure 3. Re-arrangement of Au surface atoms towards planar {111} surface facets and increased co-ordination as a result of heating by 200kV electron beam.

of both the surface and internal structure [21].

Whilst the origin of gold's remarkable catalytic activity is still being debated in the literature, a consensus seems to have been reached that under co-ordinated surface atoms provide highly active binding sites for incoming adsorbates, and clearly any reduction in the number of such sites will have a detrimental effect on the nanoparticles activity. Here we have demonstrated the effect of heating on these sites, showing a tendency for the nanoparticle to minimise its surface energy by restructuring towards planar {111} surface facets. These studies suggest that as well as maintaining the catalysts surface to volume ratio by preventing conventional sintering processes, it is necessary to identify the actual active sites and consider their evolution under typical operating conditions.

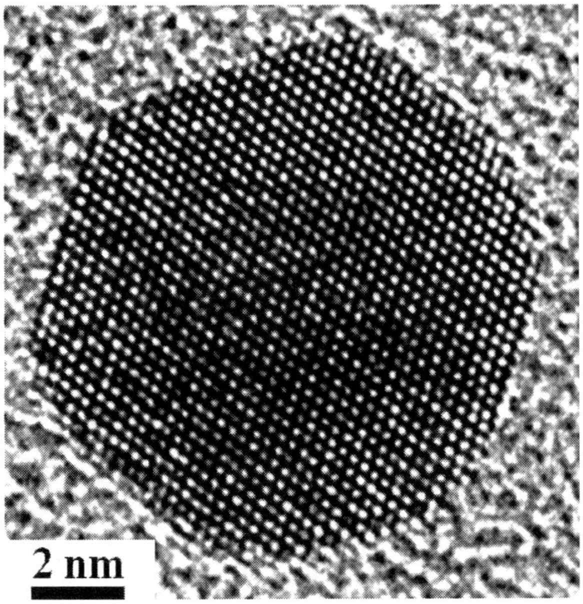

Figure 4. A single crystal cuboctahedral Au nanoparticle produced by heating to 600°C.

Acknowledgements
The authors would like to thank the University of York, Yorkshire Forward, European Regional Development Fund and JEOL (UK) Ltd for the facilities at the Nanocentre at the University of York. MJW would like to thank the EPSRC and Johnson Matthey Plc for funding.

References

[1] Haruta M, Koybayasi T, Sano H and Yamada M 1987 *Chemistry Letters*, 405- 08
[2] Hutchings G J *J. Catal* 1985, **96**, 292-95.
[3] Haruta M, Tsubota S, Koybayashi T, Kageyama H, Genet M J and Delmon B 1993 *J. Catal.* **44**.
[4] Valden M, Lai X and Goodman D W 1998 *Science*, **281**, 1647
[5] Haruta M 2004, *J. New. Mater. Electrochem. Syst.* **163,** 7
[6] Henry C R 2000 *Appl. Surf. Sci.* **164**, 252-259.
[7] Remediakis I N, Lopez N and Norskov J K. 2005 *Angew. Chem.* **44**, 1824
[8] Lopez N, Norskov J K, Janssens T V W, Carlsson A, Puig-Molina A, Claussen B S and Grunwaldt J D 2004 *J. Catal*, **86**, 225
[9] Rashkeev S N, Lupini A R, Overbury S H, Pennycook S J and Pantelides S T 2007 *Phys. Rev. B*, **76**, 035438
[10] Lopez N, Janssens T V W, Clausen B S, Xu Y, Mavrikakis M, Bligaard T, Norskov J K 2004 *J. Catal.* **223**, 232-235
[11] Hammer B and Norskov J K. 1995 *Surf. Sci.* **343**. 211
[12] Ruban A, Hammer B, Stoltze P, Skriver H L and Nørskov J K 1997 *J. Mol. Catal. A: Chem.* **115**, 421
[13] Schnur S and Groß A 2010 *Phys. Rev. B.* **81**, 033402
[14] Campbell C T 1997 *Surf. Sci. Rep.* **1**, 227.
[15] Gjostein N A and Wynblatt P 1976 *Acta Metall.* **24.** 1165
[16] Walsh M J, Yoshida K, Gai P L and Boyes E D. *J. Phys.: Conf. Ser.* **241** 012058
[17] Gai P L and Boyes E D 2009 *Microsc. Res. Tech.***72**, 153
[18] Marks L D 1984 *Phil.Mag. A* **49**, 81 - 93
[19] Howie A and Marks L D 1984 *Phil. Mag. A* **49**, 95-109
[20] Galanakis I, Papanikaolaou N and Dederichs P H 2002 *Surf. Sci.* **511,** 1-12
[21] McKenna K P 2009 *Physical Chemistry Chemical Physics* **11**, 4145-4151

Journal of Physics: Conference Series **371** (2012) 012049 doi:10.1088/1742-6596/371/1/012049

Environmental TEM investigation of the reduction of α-Fe_2O_3 nanorods under H_2 atmosphere

Trevor P. Almeida[1], Michael W. Fay[2], Yanqiu Zhu[3], Thomas W. Hansen[4], Rafal Dunin-Borkowski[4,5] and Paul D. Brown[1*]

[1]Division of Materials, Mechanics and Structures, Department of Mechanical, Materials and Manufacturing Engineering, Faculty of Engineering; &
[2]Nottingham Nanotechnology and Nanoscience Centre;
University of Nottingham, University Park, Nottingham, NG7 2RD, UK.
[3]College of Engineering, Mathematics and Physical Sciences, University of Exeter, Streatham Campus, Exeter, EX4 4QF, UK.
[4]Center for Electron Nanoscopy, Technical University of Denmark, DK-2800 Kgs. Lyngby, Denmark.
[5]Ernst Ruska-Centre for Microscopy and Spectroscopy with Electrons, Institute for Microstructure Research, Research Centre Jülich, D-52425 Jülich, Germany.

Abstract. The thermal reduction of hydrothermally synthesised α-Fe_2O_3 nanorods (NRs) to Fe_3O_4 NRs under hydrogen is investigated. Complete reduction of α-Fe_2O_3 NRs to Fe_3O_4 NRs was achieved during *in situ* XRD under 1 bar H_2 atmosphere at 360°C. Complementary environmental transmission electron microscope investigation at high resolution, during *in situ* heating under an H_2 pressure of 5 mbar at 500°C, provided evidence for the very first stages of transformation, supporting a model for the migration of oxygen along favoured α-Fe_2O_3 lattice planes during the templated thermal reduction of α-Fe_2O_3 NRs to Fe_3O_4 NRs.

1. Introduction

One-dimensional (1D) nanostructures have attracted considerable attention due to their novel magnetic properties which are greatly dependent on nanostructure size and shape [1]. Ferrimagnetic magnetite (Fe_3O_4) is of particular interest as a multifunctional nanostructured material with potential application in ferrofluids, magnetic recording and magnetic resonance imaging [2]. Acicular hematite (α-Fe_2O_3) nanorods (NRs) may be used as a precursor for the synthesis of Fe_3O_4 NRs, through a process of partial reduction [3]. Hydrothermal synthesis (HS) constitutes an ideal method for the production of α-Fe_2O_3 NRs from dilute iron (III) chloride ($FeCl_3$) precursor solution, via the formation of an intermediate β-FeOOH (akaganeite) phase. The addition of phosphate anions has been shown to play a decisive role in the anisotropic growth of α-Fe_2O_3 NRs, defining their acicular shape by mediating the process of oriented attachment of primary α-Fe_2O_3 nanoparticles (NPs) [4]. X-ray diffractometry (XRD) investigation of α-Fe_2O_3 NRs heated *in situ* under a reducing atmosphere may be used to gain insight into the process of bulk phase transformation to Fe_3O_4. However, in order to elucidate the intricate details of this phase transformation process, it is necessary to visualise the reduction process at high resolution. Spherical aberration (C_S) correction combined with environmental TEM (ETEM) enables the detailed investigation of localised chemical reactions under gas atmospheres with interpretable spatial resolution at the sub-angstrom level [5, 6]. In this context, a high resolution ETEM investigation of the partial reduction of α-Fe_2O_3 NRs to Fe_3O_4 under conditions of high temperature and reducing atmosphere is presented.

*Author to whom correspondence should be addressed: paul.brown@nottingham.ac.uk

Figure 1 (a-c) BF diffraction contrast TEM images of: (a) α-Fe_2O_3 NRs (Sample H) formed from the HS of 0.2 ml $FeCl_3$ at 200°C for 2 hours in the presence of 3 mg $NH_4H_2PO_4$, with indexed SAED (inset); (b,c) Fe_3O_4 NRs (Sample M) formed following the annealing of Sample H at 360°C under ~ 1 bar H_2 atmosphere for 30 min, with indexed SAED (inset). (d) Associated XRD patterns of Samples H and M, indexed to α-Fe_2O_3 (JCPDS 72-469) and Fe_3O_4 (JCPDS 88-315), respectively.

2. Experimental

Acicular α-Fe_2O_3 NRs (Sample H) were prepared through the hydrothermal synthesis of 0.2 ml $FeCl_3$ aqueous solution (45% pure $FeCl_3$; Riedel-de Haen, Germany) in the presence of ammonium dihydrogen-phosphate (99.999% $NH_4H_2PO_4$; Sigma-Aldrich, UK) surfactant, further diluted in 40 ml distilled water and mechanically stirred in a 125 ml Teflon-lined steel autoclave. The autoclave was sealed and inserted into a temperature controlled furnace at a reaction temperature of 200°C for 2 hours. The autoclave, once removed from the furnace, was allowed to cool down to room temperature naturally. For the purpose of *in situ* XRD investigation, the HS product suspensions were centrifuged for 6 minutes at 6000 rpm, cleaned with acetone, deposited onto single crystal silicon substrates and analysed using a Bruker D8 Advance X-ray diffractometer (Cu Kα radiation; θ/2θ diffraction geometry). Sample H was examined at room temperature and then heated to 360°C in a sealed XRD sample chamber under an H_2 atmosphere of 1 bar for 30 min (Sample M). For the purpose of TEM investigation, the centrifuged HS and XRD reaction product samples were cleaned with acetone and dispersed using an ultrasonic bath before deposition onto carbon film / gold mesh support grids (Agar Scientific Ltd, UK). Conventional diffraction contrast bright field (BF) imaging of the reaction products was performed using a JEOL 2000FX electron microscope. Further, ETEM investigation of the reduction of Sample H as a function of temperature (room temperature to 500°C at 20°C / min under 5 mbar H_2 atmosphere) was performed using a Gatan double tilt heating holder within a FEI Titan E-Cell TEM with a C_S corrector on the objective lens, operated at 300 kV (Centre for Electron Nanoscopy (CEN), Technical University of Denmark). High angle annular dark field (HAADF) imaging was performed using a FEI Titan (S)TEM with a C_S corrector on the condenser system, operated at 300 kV (CEN).

3. Results

The BF diffraction contrast TEM images of Figures 1a-c illustrate the morphology of the initial HS reaction products (Sample H) and the products formed following annealing at 360°C under ~ 1 bar H_2 atmosphere for 30 min during *in situ* XRD investigation (Sample M). The associated XRD patterns of Samples H and M (Figure 1d) are consistent with the presence of crystalline α-Fe_2O_3 (JCPDS 72-469) and Fe_3O_4 (JCPDS 88-315), respectively. TEM investigation of Sample H (Figure 1a) confirmed the initial HS suspension to comprise acicular α-Fe_2O_3 NRs (~ 420 nm long, ~ 60 wide nm), whilst Sample M (Figures 1b,c) consisted of similar sized acicular Fe_3O_4 NRs.

The heating of Sample H under a reducing atmosphere in the ETEM provided insight into the progressive reduction of α-Fe_2O_3 on the localised scale. No change occurred after 30 min at 360°C, as seen during *in situ* XRD, and hence the temperature was increased to 500°C, with a view to accelerate the reduction process. Figure 2 presents BF diffraction and phase contrast TEM images of Sample H acquired at 500°C under an H_2 atmosphere of 5 mbar. The BF TEM image of Figure 2a shows large α-Fe_2O_3 NRs located near the edge of the gold TEM grid. The phase contrast image of Figure 2b illustrates the fine details of an individual α-Fe_2O_3 NR surface (Figure 2a, boxed region), as identified by (d_{104} = 2.70Å) (inset). Selected images from a time-lapse series acquired at 500°C provided evidence for the localised development of an Fe_3O_4 NP on the α-Fe_2O_3 NR surface (Figures 2c–i), as evidenced by the {220} (d_{220} = 2.96Å) and {111} (d_{111} = 4.83Å) lattice fringes (Figures 2e & f). Figure 2j shows the newly

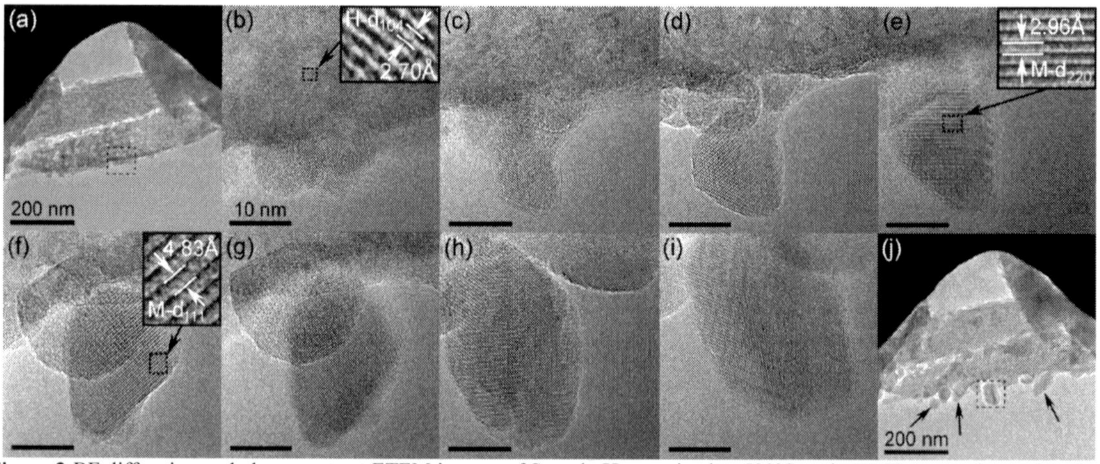

Figure 2 BF diffraction and phase contrast ETEM images of Sample H, examined at 500°C under an H_2 atmosphere of 5 mbar. (a) Large α-Fe_2O_3 NRs located at the edge of the gold TEM grid. (b) High resolution image of the surface of an α-Fe_2O_3 NR (boxed region of Figure 2a). (c-i) Time-lapse phase contrast images showing the development of an Fe_3O_4 NP, as identified by lattice fringes (e,f inset). Scale bars correspond to 10 nm. (j) α-Fe_2O_3 NRs decorated with the developed Fe_3O_4 NPs (arrowed).

developed Fe_3O_4 NP (outlined) and reveals the presence of additional Fe_3O_4 NPs (arrowed) developed on the α-Fe_2O_3 NR surface during heat treatment.

Figure 3 similarly presents phase contrast ETEM images from Sample H acquired at 500°C under an H_2 atmosphere of 5 mbar. Figure 3a presents a close examination of the tip of an α-Fe_2O_3 NR, as identified by {104} lattice fringes (d_{104} = 2.70Å) (inset). The defects in the α-Fe_2O_3 NR are an effect of the coalescence of α-Fe_2O_3 particles during growth [4]. Figures 3b,c show the localised development of an Fe_3O_4 NP (arrowed), as identified by {220} lattice fringes (d_{220} = 2.96Å, Figure 3c, inset).

Figure 4 presents BF diffraction contrast and HAADF images of Sample H acquired after 5 hours of heating under a reducing atmosphere within the ETEM. The BF TEM of Figure 4a shows a single crystalline α-Fe_2O_3 NR (~ 420 nm long, ~ 70 nm wide), as identified by SAED (inset). Figures 4b & 4c present HAADF images of α-Fe_2O_3 NRs before and after heating in the ETEM, respectively, with a porous structure of the annealed α-Fe_2O_3 NR revealed in Figure 4c.

4. Discussion

In situ heating of α-Fe_2O_3 NRs in the XRD at 360°C under an H_2 atmosphere of 1 bar for 30 minutes promoted their complete transition to Fe_3O_4. This is consistent with the reduction of precursor α-Fe_2O_3, being a recognised method for the production of Fe_3O_4, as summarised by the equation: $3Fe_2O_3 + H_2 \rightarrow 2Fe_3O_4 + H_2O$. The H_2 atmosphere interacts with the oxygen of the α-Fe_2O_3 NR surfaces to form water, whilst heating to 360°C provides sufficient thermal energy to promote oxygen diffusion from inside the α-Fe_2O_3 NRs to restore the equilibrium oxygen concentration at the surface, resulting in partial reduction to Fe_3O_4. SAED combined with TEM imaging of individual NRs showed that the major axes of the α-Fe_2O_3 (Figure 4a) and Fe_3O_4 NRs (Figure 1c) were $\langle 006 \rangle_H$ and $\langle 111 \rangle_M$, respectively. This orientation relationship between α-Fe_2O_3 and Fe_3O_4 is consistent with previous reduction studies, with oxygen considered to migrate along {006} planes [7]. Here, the restacking to Fe_3O_4 is not so severe as to affect the NR acicular morphology (Figure 1c), hence, the α-Fe_2O_3 NRs act as templates for the topotactic transformation to Fe_3O_4 NRs.

BF TEM imaging and SAED (Figure 4a) during ETEM investigation demonstrated that α-Fe_2O_3 NRs were only reduced to Fe_3O_4 at the NR surfaces, under an H_2 atmosphere of 5 mbar at 500°C for 5 hours, with the caveat that the achievable H_2 pressure of 5 mbar in ETEM was significantly lower than the 1 bar used during *in situ* XRD investigation. This low H_2 pressure was found to be insufficient to promote the thermal reduction of complete α-Fe_2O_3 NRs during *in situ* heating in the ETEM. Thermal effects, however, did lead to the development of porosity in the α-Fe_2O_3 NRs through the release of trapped water and phosphate [8], as revealed by HAADF imaging (Figure 4c). Nevertheless, it is recognised that the time-lapse imaging at 500°C during ETEM provided for direct observation of the very first stages of

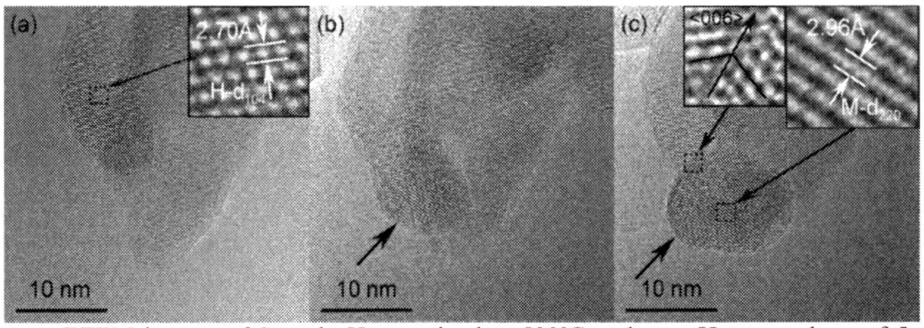

Figure 3 Phase contrast ETEM images of Sample H, examined at 500°C under an H_2 atmosphere of 5 mbar. (a) Tip of an individual α-Fe_2O_3 NR, identified by lattice fringes (inset). (b,c) Time-lapse images showing the development of an Fe_3O_4 NP, identified by lattice fringes in (c) (inset).

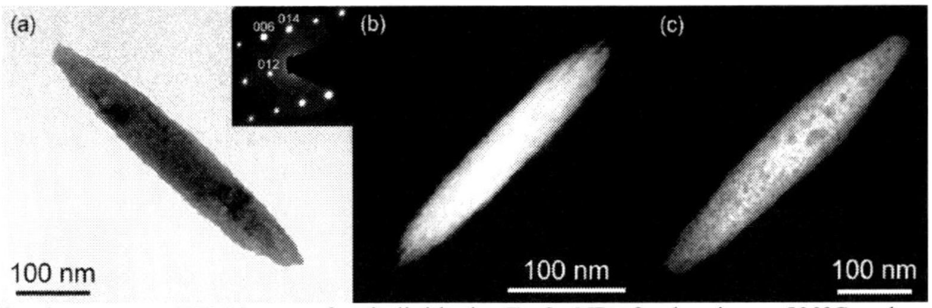

Figure 4 (a) BF diffraction contrast TEM image of an individual α-Fe_2O_3 NR after heating at 500°C under an H_2 atmosphere of 5 mbar for 5 hours, as identified by SAED (inset, indexed). (b,c) HAADF images of α-Fe_2O_3 NRs examined (b) before and (c) after heating at 500°C under an H_2 atmosphere of 5 mbar for 5 hours.

the localised development of Fe_3O_4 NPs on the side and tip of α-Fe_2O_3 NRs (Figures 2 and 3, respectively). It is likely that focusing the 300keV electron beam, during imaging (Figures 2 & 3) contributed to the development of the Fe_3O_4 NPs in this low pressure H_2 environment. In Figures 2b-i, the electron beam was converged on an area slightly larger than the field of view, which included the location of the newly formed NPs (Figure 2j, arrowed). Likewise, formation of the Fe_3O_4 NP on the tip of the α-Fe_2O_3 NR (Figure 3c) occurred during high resolution imaging. Further, the surface of the α-Fe_2O_3 NR shown in Figure 2j became noticeably eroded. Hence, the high current density of the electron beam acted to destabilise the α-Fe_2O_3 NR surface, allowing oxygen to react with H_2 and form water, leaving an oxygen deficiency in the α-Fe_2O_3 NR surface, whilst remaining unstable iron and oxygen supplied the growth of Fe_3O_4 NPs.

Considering the relationship between the orientation of the α-Fe_2O_3 NRs and the growth of the initial Fe_3O_4 NPs, Figures 2 and 3 illustrate the formation of Fe_3O_4 NPs on the side and tip of α-Fe_2O_3 NRs, respectively. Closer examination reveals the growth interface of the tip Fe_3O_4 NP to be parallel to the major axis of the α-Fe_2O_3 NR (Figure 3c, inset), and hence, in both cases, the Fe_3O_4 NPs grow on surfaces parallel to the α-Fe_2O_3 <006> direction. Elemental supply to developing Fe_3O_4 NPs through an interface parallel to the α-Fe_2O_3 c-axis reinforces the notion that α-Fe_2O_3 {006} planes are favourable for oxygen diffusion during the thermal reduction of α-Fe_2O_3 NRs to Fe_3O_4 NRs. The α-Fe_2O_3 {006} or Fe_3O_4 {111} planes may become unstable and collapse during oxygen diffusion, resulting in defects.

In summary, fine details of the reduction of α-Fe_2O_3 NRs to Fe_3O_4 NRs during *in situ* XRD were examined. ETEM experiment provided evidence to support a model for oxygen migration along favoured α-Fe_2O_3 lattice planes during templated thermal reduction of α-Fe_2O_3 NRs to Fe_3O_4 NRs.

References

[1] Bate G, 1991 *J. Magn. Magn. Mater.* **100**, 413.
[2] Mornet S, Vasseur S, Grasset F, Duguet E, 2004 *J. Mater. Chem.* **14**, 2161-2175.
[3] Itoh H and Sugimoto T, 2003 *J. Colloid Interface Sci.* **265,** 283-295.
[4] Almeida T P, Fay M W, Zhu Y Q and Brown P D, 2010 *Nanoscale* **2**, 2390–2399.
[5] Hansen T W, Wagner J B and Dunin-Borkowski R E, 2010, *Materials Science and Technology* **26,** 1338-1344.
[6] Boyes, E D, Gai, P L, 1997 *Ultramicroscopy* **67,** 219-232.
[7] Watanabey Y, Takemuraz S, Kashiwayaz Y and Ishii, K, 1996, *J. Phys. D: Appl. Phys.* **29**, 8–13.
[8] Suber L, Fiorani D, Imperatori P, Foglia S, Montone A and Zysler, R, 1999, *NanoStruct. Mater.* **11**, 797-803.

Metals on BN Studied by High Resolution Transmission Electron Microscopy

U Bangert[1], R Zan[1,2], Q Ramasse[3], Rashid Jalil[2], Ibstam Riaz[2] and K S Novoselov[2]

[1] School of Materials, The University of Manchester, Manchester, UK
[2] School of Physics and Astronomy, The University of Manchester, Manchester, UK
[3] SuperSTEM, STFC Daresbury Laboratory, Warrington, UK

ursel.bangert@manchester.ac.uk

Abstract. Metal impurities, gold and nickel, have been deliberately introduced into boron-nitride (BN) sheets. The structural and topographic properties of doped BN have been studied by aberration corrected scanning transmission electron microscopy (STEM). Analysis revealed that metal atoms cluster preferentially in/on contaminated areas. The metal coverage on BN is almost the same for the same evaporated amount of 1 Å.

1. Introduction

Since the isolation of graphene, 2D nanomaterials, like BN, have attracted considerable attention due to their unique physical, chemical, and structural properties as well as their great potential for applications [1]. BN is known to be a wide-gap semiconductor in contrast to gapless graphene. Owing to the unique properties of BN, much effort has been put into exploring its potential use in electronic and optoelectronic devices such as field emitters and ultraviolet or radiation detectors [2, 3] It has also been recently found that BN is a much better substrate (inert) for graphene studies than the SiO_2, as it improves the graphene electronic properties [4]. To date, only little research has been carried out on the BN-metal system [5-7], compared to the suspended graphene-metal system [8]. In general such studies have up to now been mostly theoretical works [9,10]. Our previous investigation of the graphene-metal interaction via scanning transmission electron microscopy (STEM) has inspired us to carry out microstructural studies of the metal-BN system, another technologically important 2D material.

Here we present a STEM study of the BN-graphene interaction, which have revealed that no Au or Ni atoms were seen to adhere to clean patches of BN, similar to the case of graphene.

2. Experimental

In order to obtain thin sheets of BN, the micromechanical cleavage technique has been used, which is well known for obtaining graphene [1,11]. h-BN powders with an average crystal size of ~20 μm were 'peeled' several times using adhesive tape and then placed onto a silicon wafer with a 90 nm silicon oxide layer. Thin flakes were identified under the optical microscope, coated with PMMA and transferred to a TEM grid (quantifoil, 400mesh gold grid), after having etched-off the Si/SiO_2. The PMMA was dissolved by dipping the grid with the film into acetone for ~10 minutes. Finally, the flakes were dried in critical point dryer. Metal impurities, Au and Ni, have been deposited at room temperature to 1 Å nominal thickness in both cases onto the BN sheets in a thermal evaporator.

The structural and topographic properties of doped-BN have then been studied by using aberration corrected STEM (Daresbury SuperSTEM). Atomic sites of singular metal atoms and clusters on BN have been directly revealed in high angle annular dark field (HAADF) images (by comparing the latter with bright field (BF) images). The samples were studied at 60 and 100 kV acceleration voltages for the Ni and Au case, respectively.

3. Results and Discussion

Figure 1 shows results of 1 Å gold evaporated on few-layer BN, at different magnifications. Individual gold atoms or clusters were not observed on pristine parts of the BN sheets. However, ad-atoms and nano-gold-crystals can be seen in the hydrocarbon contamination, which resembles the behavior of Au on graphene surfaces [8]. This indicates the extreme mobility of gold on the clean BN surface, and can be seen in fig.1a), a low magnification HAADF overview image. Single gold atoms have been observed in contaminated areas, as a result of dissociation from bigger gold cluster by the scanning beam in the STEM. These can be seen in fig.1c) around the middle cluster, which is the magnified area (indicated by the white square) of fig.1b). Another observation is the 'peeling' of BN layers at 100kV. This has previously been shown [12]: the holes that occur are triangular in shape as indicated by the dashed white triangle in the HAADF image in fig1c) and in the corresponding BF image in fig.1d). The BN lattice has much higher clarity in the BF- than in the HAADF image.

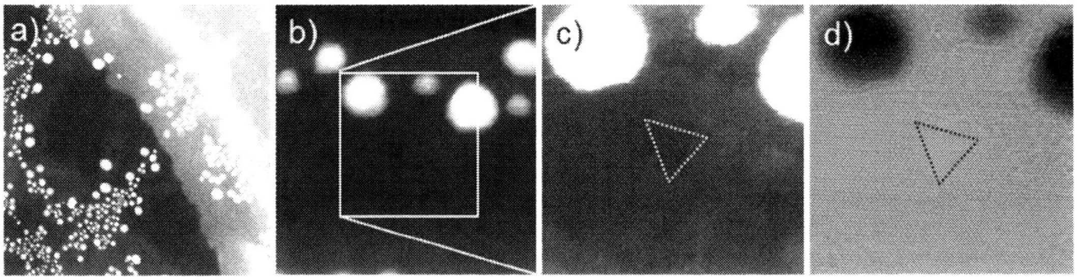

Figure 1. a) HAADF image of 1Å Au doped BN at low magnification, obtained at 100kV. b) and d) HAADF images of 5Å Au doped BN at higher magnification, d)BF image corresponding to c). Frame widths are 300, 30, 15, 15 nm in a), b), c) and d), respectively.

HAADF images of 1Å Ni doped BN and simultaneously taken corresponding BF images are shown in Figure 2. Figs 2a), c) and e) are subsequently acquired HAADF images from the same region with increasing magnification at 60 kV acceleration voltage; magnified areas are indicated by white squares in figs. 2a) and b). All HAADF images clearly show that there are no Ni atoms or cluster on clean BN surfaces. This is not obvious in the corresponding BF images, which reveal disturbances (Ni-islands) of the BN lattice; the impurities are in fact not sitting on the BN lattice itself, but in/on the contamination layer between Ni atoms and BN surface. This can only be revealed in the HAADF images, which is the reason why HAADF imaging is more powerful than phase contrast (BF) imaging in terms of revealing constellations and positions of impurities and ad-atoms. In particular, individual Ni atoms can be observed in the HAADF image in fig. 1c), on a very fine contamination layer, through which the hexagonal BN structure is still clearly visible. Without the guidance of the HAADF images it would not be possible to identify the sites of the ad-atoms on the honeycomb structure.

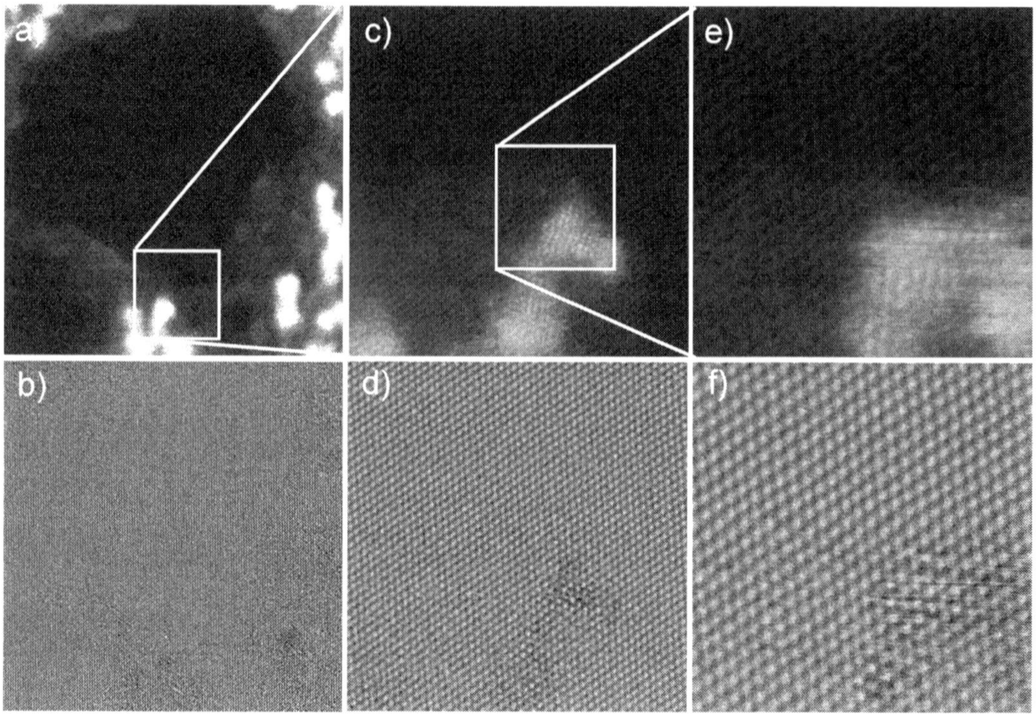

Figure 2. HAADF and BF images of 1Å Ni doped BN at different magnification, obtained at 60kV. a), c) and e) HAADF, b), d) and e) corresponding BF images. Frame widths are 25, 10, 5 nm in a) and b), c) and d), e) and f), respectively.

4. Conclusions

In conclusion, we used an aberration-corrected STEM operated at 60 and 100 kV, to investigate the behavior of metals (Au and Ni) on thin BN sheets in terms of coverage, interaction and also the quality of BN as a substrate. HAADF images reveal that both Au and Ni metals do not stick to the clean –free from contamination– surface of BN, however, they form ad-atoms and clusters on the contamination. This shows that the interaction between metals and BN is much weaker than the metal-metal interaction. It was also found that BN is a very good transparent support for the study of nanocrystals in the TEM.

References

[1] Novoselov, K. S.; Geim, A. K.; Morozov, S. V.; Jiang, D.; Zhang, Y.; Dubonos, S. V.; Grigorieva, I. V.; Firsov, A. A. **2004** *Science* 306 5696 666-669.
[2] Kubota, Y.; Watanabe, K.; Tsuda, O.; Taniguchi, T. **2007** *Science* 317 5840 932-934.
[3] Watanabe, K.; Taniguchi, T.; Kanda, H. **2004** *Nat Mater* 3 6 404-409.
[4] Xue, J.; Sanchez-Yamagishi, J.; Bulmash, D.; Jacquod, P.; Deshpande, A.; Watanabe, K.; Taniguchi, T.; Jarillo-Herrero, P.; LeRoy, B. J. **2011** *Nat Mater* 10 4 282-285.
[5] Yimin, A. W.; et al. **2011** *Nanotechnology* 22 19 195603.

[6] Yazyev, O. V.; Pasquarello, A. **2010** *Physical Review B* 82 4 045407.
[7] Zhou, Y. G.; Yang, P.; Sun, X.; Wang, Z. G.; Zu, X. T.; Gao, F. **2011** *J. Appl. Phys.* 109, 084308.
[8] Zan, R.; Bangert, U.; Ramasse, Q.; Novoselov, K. S. **2011** *Nano Letters* 11 3 1087-1092.
[9] Chan, K. T.; Neaton, J. B.; Cohen, M. L. **2008** *Physical Review B* **77** 23 235430.
[10] Giovannetti, G.; Khomyakov, P. A.; Brocks, G.; Karpan, V. M.; van den Brink, J.; Kelly, P. J. **2008** *Physical Review Letters* 101 2 026803.
[11] Gorbachev, R. V.; Riaz, I.; Nair, R. R.; Jalil, R.; Britnell, L.; Belle, B. D.; Hill, E. W.; Novoselov, K. S.; Watanabe, K.; Taniguchi, T.; Geim, A. K.; Blake, P. **2011** *Small* **7** 4, 465-468
[12] Jin, C.; Lin, F.; Suenaga, K.; Iijima, S. **2009** *Physical Review Letters* 102 19 195505.

Journal of Physics: Conference Series **371** (2012) 012051 doi:10.1088/1742-6596/371/1/012051

A nanomanipulation system for tomographic examination of nanostructures on demand

W. Guan, A. Lockwood, B.J. Inkson, G. Möbus

Department of Materials Science and Engineering, University of Sheffield, Mappin Street, Sheffield, S1 3JD, UK.

Email: w.guan@sheffield.ac.uk

Abstract. In this paper, we present a recently upgraded 7-degree-of-freedom nanomanipulation system inside a TEM with focus on applications in electron tomography of nanostructures. Previously it has been emphasised that the traditional way of mounting of nanoparticles on flat carbon films might not be the best way to achieve very high tilt angle series free of missing wedge for tomography. Mounting of particles ex-situ at the end of nanotips allows for higher tilt but poses challenges in the success rate of getting the object attached to the very tip. Our mixed translation & rotation nanomanipulator based on 3 coarse positioners in x, y, z by piezoelectric slip-stick actuation and 3 fine directions by a piezo-tube, as well as a rotational piezodrive, allows not only to rotate a cylindrical object by up to a full circle, but now also features a second specimen mount for a "fixed sample", which could host a reservoir of nanoparticles on carbon film, or nanoparticles inside nanopores. The translation system can then be used to pick up particles on demand and rotate them with the piezoelectric actuator. In particular, we describe how nanoparticles dispersed and transferred onto a normal TEM grid are picked up by a W tip for subsequent large tilt rotation.

1. Introduction

Electron tomography [1] is commonly used nowadays in materials research fields for the physical and chemical characterization of sub-micron sized materials with up to nanometer resolution in three dimensions. To reduce or eliminate missing wedge artifacts in tomography, methods include (i) wide angle single tilt (at best 180°) with special holders or TEM goniometers [2, 3] or (ii) double-axis tilt [4, 5], which is more challenging in alignment and higher in dose. Here we present a new nanomanipulation system [6] with 7 degrees of freedom inside a TEM with focus on applications in electron tomography of nanostructures.

The traditional way of mounting nanoparticles on a flat grid cannot achieve an 180° full tilt series due to the shadow effect of the grid even though the grid size can be smaller than the pole piece gap. Therefore, in order to avoid the missing wedge, it not only requires the holder/goniometer having the ability to rotate at least ±90°, but it also needs the objects not to be shadowed for a range more than 180°. The object could be mounted ex-situ at the end of a nanotip, but it is very challenging to succeed by chance mounting the object at the very end of the tip. Therefore we propose our 7-degree-of-freedom piezoelectric drive to lift out nanoparticles in-situ which can be controlled to attach to the end of the tip as much as possible to reduce the shadow effect from the tip, and a tomographic tilt series is collected afterwards for later tomographic reconstruction.

2. Materials and methods

The whole piezoelectric goniometer system [6] is composed of three sub devices: fine drive, coarse drive and the rotary drive. Three fine translational movements in X, Y and Z axes are achieved by a quartered piezo tube in the front of the whole system as the fine drive. In the middle of the system is the coarse drive containing three sliders which provide independent coarse translational movements along all three axes and is operated serially to the fine drive. At the back, the rotary drive is equipped with an angular sensor providing the rotational movement around the X-axis and accurate angular displacement feedback. The coarse X, Y, Z and fine x, y, z are independent coordinate systems, therefore we define them as different degrees of freedom, although strictly speaking in geometric terms, both coarse and fine drives move in a 3D Cartesian space. By adding the rotary drive, a rotary degree of freedom, seven degrees of freedom in total for the movements are achieved. Both the coarse drive and the rotary drive use a slip-stick inertial slider mechanism. For the rotary drive, the shaft on which the coarse and fine drives are mounted is centred inside a pair of piezos with opposite shearing directions along the Y-axis, and the rotation directions change accordingly when the polarity of the potential applied on the piezos is reversed. The translational drives are put ahead of the rotary drive on the specimen side, so that the specimen can be always centered after each angular movement. The whole piezoelectric goniometer is fitted inside a hollow standard specimen holder for JEOL 2000/3000 series microscopes, therefore there will be neither modification requirements to the TEM nor any compromise to the performance of the default TEM goniometer.

A tungsten tip is mounted on the fine drive to be used for picking up nanoparticles. A second specimen mount is fixed to the hollow holder shell, and can host a reservoir of nanoparticles on a cut grid, as shown in Figure 1 (a). The piezoelectric goniometer can work in parallel or in addition to the TEM goniometer, so we can use the TEM goniometer to move/rotate the grid to find interesting objects for lift-out. The W tip approaches the grid by using translational coarse and fine drive, then touches the desired particles and picks them up, shown in Fig 1 (b) and (c). The particle should be sitting at the end of the tip as much as possible to avoid the shadow effect for later rotation. Finally, the W tip with attached nano objects is retracted in a free space and rotated by the rotary drive for the tomographic experiment, which is shown in Fig 1 (d).

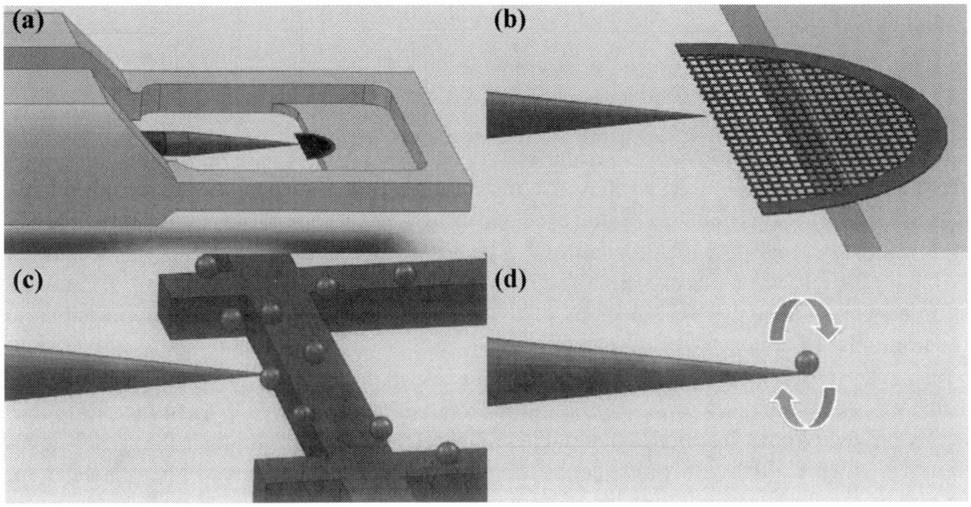

Figure 1. (a) A W tip is mounted on the fine drive of the piezoelectric system A 3mm grid cut in half is mounted on the other side fixed to the holder shell as the reservoir of nanoparticles. (b) The tip approaches to the grid holding nanoparticles. (c) Choosing and lift-out of particles (purple) on the grid. (d) The tip is retracted and a tilt series is conducted.

It is not necessary that the nanotip is made of W. For instance a Ni tip might be more helpful to pick up magnetic nanoparticles, while Al and porous Al_2O_3 tips have been tested for high transparency in the past, or Au tips would be most oxide-free. The fixed sample reservoir is not necessary to be a cut copper grid either. It could be another nanotip, metal wire, piece of Si, etc., whatever is holding the nano-objects of interest. In this paper, we choose different grids as the reservoir. As to the nanoparticles, in theory any particles, subject to suitable size range, could be picked up. For demonstration in this case, we choose hydroxyapatite (HA) nanoparticles (courtesy of nano4med TSB consortium Notthingham/Sheffield) for their wide and important applications in bio-engineering [7], medicine [8], and drug delivery [9].

3. Experimental Applications

Tungsten tips were prepared and mounted as reported in [10]. BF-TEM tilt series of the W tip with attached HA particles were acquired using a JEM 3010 TEM (JEOL, Japan) at 300kV.

The first attempt was to disperse HA particles on an ordinary copper grid with holey carbon film (Agar), then cut the grid in half and mount it onto the fixed end of the actuation holder, as shown in Figure 1 (a). The W tip was moved by the coarse piezo drive, and the step size can be controlled from about 20nm to 150nm by using the minimum and maximum driving voltage. The W tip carefully approached the carbon film containing HA particles (in plate shape) and then touched a bundle of them, which is shown in Figure 2 (a). Due to Van der Waals force or charging induced electrostatic force of the W tip, a couple of HA plates were attached to the W tip, one was on top left and the other was on the right, as shown in Figure 2 (b). Now the one on the top could be used for tomography.

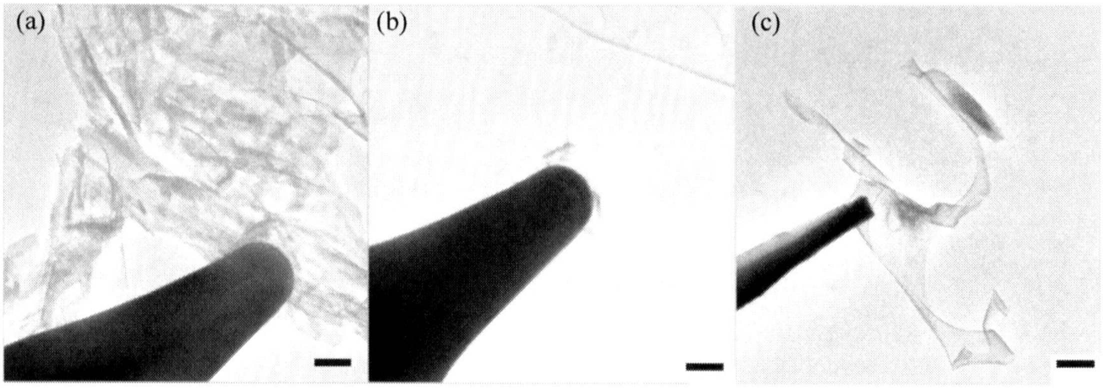

Figure 2. (a) The W tip is touching the carbon film containing HA platelets, scale bar is 200nm. (b) A couple of individual HA plate are picked up, scale bar is 200nm. (c) A piece of carbon film is lifted out by a W tip, scale bar is 500nm.

Not only particles, sometimes a piece of carbon film can also be lifted out by the W tip, which is shown in Figure 2 (c). The advantage of lifting out carbon film is that a reduction of the shadow effect is achieved as the objects on the film may be far away from the end of the tip. However, the size, shape and orientation of the carbon film fragment which can be picked up is totally uncontrollable, also the sticky carbon could crash/bend the tip before successful lift out. So a carbon film pick-up may not be very practical in reality.

When the HA nanoparticle size is very small, say less than 20nm, the W tip could not attract them from the thin film, as picking them up from a holey carbon film is equivalent to breaking the carbon film and lift out one piece of it. Therefore, for small HA particles, we dispersed them on a bare copper grid. The nanoparticle clusters, once loosely bonded on the copper wire, could then be easily broken and picked up by the W tip.

Figure 3 shows a lifted out HA particle cluster containing particles with size about 5nm on the top of a W tip. The clusters were rotated with the tip from 0° to 120° and images were taken with 5° intervals. A selection of images from this series is shown in Figure 3 with rotation angle at 0°, 25°, 50°, 75°, 100° and 120° respectively. Unfortunately the HA cluster was deformed by the irradiation damage, but from the combined shape of W tip and HA, a clear progress of tilt is still be seen.

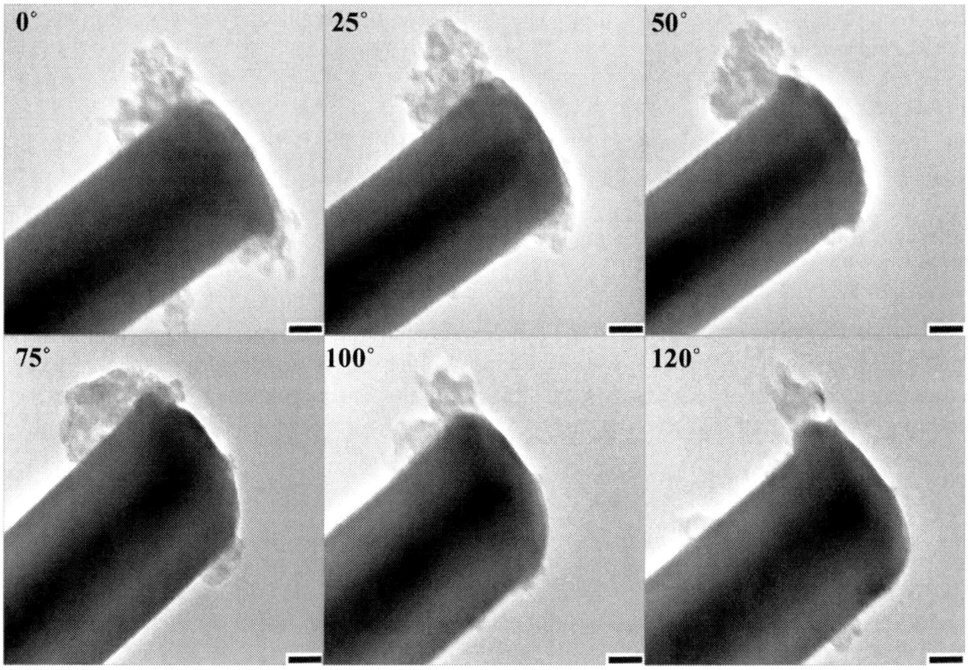

Figure 3. A selection of images from a tilt series with 5° intervals from 0° to 120°.Scale bar is 50nm.

4. Conclusion

In summary, a newly developed 7-degree-of-freedom piezoelectric drive inside TEM can be used to choose and pick-up a nanoobject from a particle reservoir, and subsequently electron tomography can be achieved via unlimited tilt series, as demonstrated in [6]. Artifacts related to the missing wedges can be eliminated and the main benefit is that picked-up nanostructures are sitting on the top of the tip to avoid any shadow effect during the rotation.

Acknowledgements

This work was supported in part by EPSRC, UK, GR/S85689/01 & EP/G036748/1, and via the TSB project Nano4Med.

References

[1] Frank J 2007 *Electron Tomography*, Plenum, New York, 2nd edition
[2] Xu X J, Lockwood A, Guan W, Inkson B J and Möbus G 2008 *Microsc. Res. Techniq.* **71** p 773
[3] Kamino K, Yaguchi T, Ohnishi T and Ishitani T 2008 *J. Electron. Microsc.* **53** p 583.
[4] Mastronarde D N 1997 *J. Struct. Biol.* **120** p 343
[5] Iancu C V, Wright E R, Heymann J B and Jensen G J 2005 *J. Struct. Biol.* **151** p 288
[6] Guan W, Lockwood A, Inkson B J and Möbus G 2011 *Microsc. And Microanal.* **17**
[7] Rusu V M, Ng C H, Wilke M, Tiersch B, Fratzl P and Peter M G 2005 *Biomaterials* **26** p 5414
[8] Cai Y R, Liu Y K, Tao J H, Zhang M, Shi Z L and Tang R K 2007 J. Mater. Chem. **17** p 3780
[9] Shinto Y, Uchida A, Korkusuz F, Araki N and Ono K 1992 *J. Bone. Joint Surg. [Br]* **74** p 600
[10] Xu X, Peng Y, Saghi Z, Gay R, Inkson B J and Möbus G 2007 *J. Phys.Conf. Ser.* **61** p 810

An *in-situ* TEM investigation of He bubble evolution in SiC

C J Pawley†[1,2], M F Beaufort[3], E Oliviero[4], J A Hinks[2], J F Barbot[3] and S E Donnelly[2]

[1] Centre for Materials and Physics, School of Computing, Science and Engineering, University of Salford, Salford M5 4WT, UK.

[2] Electron Microscopy and Materials Analysis Group, School of Computing and Engineering, University of Huddersfield, Huddersfield HD1 3DH, UK.

[3] Institut Pprime, Université de Poitiers, 86962 Futuroscope-Chasseneuil, France.

[4] Centre de Spectrométrie Nucléaire et de Spectrométrie de Masse, CNRS-IN2P3, Université Paris-Sud, Orsay Campus, France.

†Email: c.pawley@edu.salford.ac.uk

Abstract This paper presents work using the capabilities of two TEM with *in-situ* ion irradiation facilities: Microscope and Ion Accelerator for Materials Investigation (MIAMI) at the University of Huddersfield and Joint Accelerators for Nano-science and Nuclear Simulation JANNuS at Centre de Spectrométrie Nucléaire et de Spectrométrie de Masse (CSNSM), Orsay, France, to study the nucleation and growth of He bubbles in silicon carbide (SiC) and to carry out an investigation into bubble behaviour at high temperatures and under displacing irradiation. Preliminary results on bubble nucleation and growth during He irradiation of SiC are presented together with results from a simultaneous anneal and high-energy heavy-ion irradiation of samples containing He bubbles. The displacing irradiation is observed to impede He bubble growth resulting in smaller bubbles than those obtained from an anneal alone. A tentative interpretation of these observations is presented.

1. Introduction

Silicon carbide (SiC) is a candidate material for fusion reactor applications [1]. Current designs use SiC in plasma-facing components in fusion reactors because of its high-temperature stability and radiation hardness. Additionally it is attractive because it has the potential to be cost-effective and is suitable for structural applications. Other advantages include resistance to corrosion and to thermal fatigue and limited saturation swelling.

Expected conditions within a fusion reactor will give rise to significant material challenges. Materials within the reactor will be subject to irradiation by light elements through transmutation reactions, whilst being subjected to a flux of energetic neutrons which will cause significant displacements to the atomic structure of the SiC. Data from Zinkle [2] suggest that reactor lifetime concentrations of He will be of order 1 atomic-% with maximum neutron-caused damage levels of 150 displacements per atom (dpa). Operating temperatures are expected to be between 550 and 1000°C.

Whilst a significant amount of scientific research has focused on the performance of SiC in nuclear environments, there are still some critical questions (outlined by Riccardi [3]) that remain unanswered. We aim to address one of these in the current project – specifically the question of material stability under the combined effects of high temperature, He accumulation and displacing irradiation.

2. Experimental

Silicon carbide wafers of the 4H polytype were acquired from Cree Inc. The wafers were 3″ in diameter and were grown 8° off-axis from the [0001] direction. Samples were cut from the wafer using a diamond-wire saw and prepared for transmission electron microscopy (TEM) using the tripod method [4] and ion-beam thinning using a Gatan Model 691 PIPS with 4 keV Ar^+ ions incident at 4° to the sample surface. Samples were then irradiated within the Microscope and Ion Accelerator for Materials Investigations (MIAMI) facility at the University of Salford, UK (now at the University of Huddersfield), the details of which are documented elsewhere [5]. Samples were irradiated by 3.5 keV He^+ ions with a flux of approximately 5×10^{13} ions/cm^2/s to a final fluence of between 8×10^{16} ions/cm^2 and 10^{17} ions/cm^2. Calculations using the Monte-Carlo code SRIM [6] indicate that 3.5 keV He^+ ions have an average range of 25 nm in SiC – approximately half the thickness of the samples. The angle between the ion and electron beams was 25°. In order to improve bubble image quality, samples were tilted slightly to avoid aligning low-index directions with the electron beam, thus avoiding strong Bragg reflections. Bright-field off-zone-axis TEM micrographs were then taken using a JEOL JEM-2000FX operated at 200keV after each irradiation step to record the changes occurring within the sample. Irradiation and microscopy were performed at a constant temperature of 700°C.

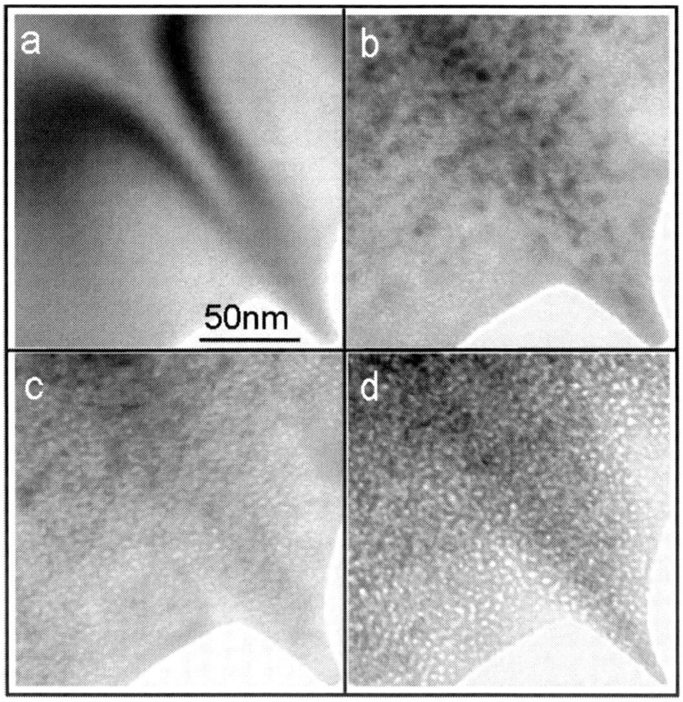

Figure 1. TEM off-zone-axis bright-field micrographs, 800 nm underfocus, of SiC irradiated with 5×10^{13} He ions/cm^2/s at 700°C, showing bubble distributions at fluences of: (a) 0; (b) 1.1×10^{16}; (c) 3.4×10^{16}; and (d) 10^{17} ions/cm^2. Scale marker applies to all panels.

Bubble-containing samples were irradiated during *in-situ* experiments at the Joint Accelerators for Nano-science and Nuclear Simulation (JANNuS) [7] facility in CSNSM, Orsay, France using 4 MeV Au^{2+} at a flux of 2.5×10^{11} ions/cm^2/s to a fluence of 2×10^{15} ions/cm^2. Samples were subjected to a temperature ramp from 700 to 1300°C at the same time as the irradiation. The angle between the ion and electron beams was 68°. Underfocussed bright-field TEM micrographs were recorded using a Technai G^2 20 operated at 200keV. At both MIAMI and JANNuS images were captured using 4 megapixel Gatan Orius SC200 digital cameras.

3. Results and Discussion

3.1. Bubble nucleation and growth during 3.5 keV He irradiation

Figure 1 shows a region of the sample before irradiation (Figure 1a) and following irradiation at 700°C to three increasingly high fluences. The dark mottled contrast in Figure 1b is typical of that due to interstitial clusters and the figure also shows contrast typical of small nanometer-sized bubbles (small white areas in this underfocused image). At the higher fluence in Figure 1c, the bubbles are now clearly visible and the interstitial contrast is less evident. Finally, Figure 1d shows a bubble distribution with a wider range of sizes.

Analysis of the images sequenced into a video provide a strong indication that, once nucleated, bubbles remain on fixed sites (no diffusive motion is observed) and continue to grow by accumulation of injected He and vacancies from the concomitant radiation damage. They eventually grow to a size where contact with a neighboring bubble may occur – at which point coalescence is observed followed by a change in the resulting elongated bubble to an assumed spherical shape (circular in projection). The coalescence gives rise to the wider range of bubble sizes visible in Figure 1d.

3.2. Effects of displacing irradiation and high temperature on He bubble distribution

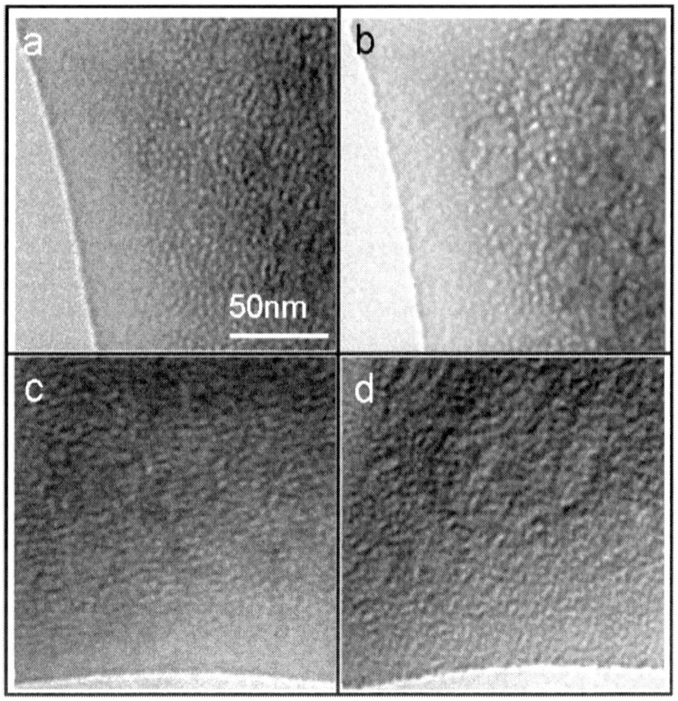

Figure 2. Bright-field micrographs of sample subjected to temperature ramp only (a and b) and subjected to identical temperature ramp and 4 MeV Au ion irradiation (flux: 2.5×10^{11} ions/cm^2/s. Images recorded at (a) 810 °C; (b) 1045 °C; (c) 770°C/1.75×10^{12} ions/cm^2 and (d) 1050°C/9.5×10^{14} ions/cm^2. Scale marker applies to all panels.

The images in the two left panels of Figure 2 show SiC containing He bubbles following irradiation to a fluence of 8×10^{16} ions/cm^2 as detailed in Section 2. Figure 2b shows the bubble morphology following a temperature ramp of 10°C/minute to a temperature of 1045°C. Figure 2d shows the bubble morphology following an identical temperature ramp to 1050°C with simultaneous irradiation with 4 MeV Au^{2+} ions. Inspection of the image reveals a mean bubble size significantly smaller than in Figure 2b. Specifically, the mean bubble radius in Figure 2d is 1.2 nm (1 nm at 700°C) compared with 1.5 nm (0.9 nm at 700°C) in Figure 2b. The mean radius of bubbles subjected to Au^{2+} irradiation during the temperature ramp grew by only 19% whereas those subjected to the temperature ramp alone exhibited a 60% increase in radius. In previous work on heavy-ion irradiation of He bubbles in Al

Birtcher *et al* [8] concluded that the relatively-dilute displacement cascades that occurred in the low-Z substrate resulted in a gradual displacement of He from the bubbles back into the lattice where it was able to diffuse and escape. The continuous production of Frenkel defects during irradiation allowed the bubbles to re-equilibrate by loss of vacancies resulting in a net reduction in mean bubble diameter. The reported shrinkage rates were 0.02 to 0.05 nm per dpa. In the current work, we measured a slightly lower mean shrinkage rate of 0.012 nm per dpa. Detailed modelling of this process is underway, but our tentative conclusion is that a similar process to that observed by Birtcher *et al* is operating in the He/SiC system in which similar dilute cascades occur to those in Al.

In the present work, measurements of bubble size distributions were rendered difficult at high Au fluences and high substrate temperatures by surface restructuring that occurred under these conditions. Future runs are scheduled at JANNuS in which similar experiments will be carried out on a population of larger, more clearly delineated bubbles and under conditions in which surface restructuring will be minimised.

4. Conclusions

4H-SiC TEM samples have been irradiated by 3.5 keV He^+ ions at a temperature of 700°C in order to investigate He bubble nucleation and growth. Aided by the *in-situ* nature of the experiments, we are able to conclude that bubble motion is minimal – final positions of bubbles are determined by the locations on which bubbles nucleate (or at least above the size at which they become resolvable by TEM). Bubbles subsequently grow until they overlap and coalesce. The mean bubble radius following irradiation to a fluence of 10^{17} ions/cm^2 was approximately 0.9 nm.

He bubbles in SiC were then subjected to a simultaneous temperature ramp and displacing irradiation in the form of 4 MeV Au^{2+} ions in a second set of *in-situ* experiments. Bubble growth was observed to be inhibited by the displacing irradiation as evidenced by reduced enlargement compared to the thermal ramp alone. The reduced growth rate is attributed to the displacement of He out of the bubbles by the collision cascades induced by the incident Au ions.

Acknowledgments

The authors wish to acknowledge the support of the Engineering and Physical Sciences Research Council (EPSRC) for providing the funding to construct the MIAMI facility (grant EP/E017266/1). This work is supported by the French network EMIR. The work was facilitated by a Royal Society / Centre National de la Recherche Scientifique (CNRS) International Joint Project grant between the Institut Pprime (Poitiers) and Centre for Materials and Physics (Salford). C.J.Pawley wishes to thank the University of Salford for partial funding of this research under the Graduate Teaching scheme.

References

[1] Najmabadi F, Conn R W, Bathka C G, Bromberg L and Cheng E T 1991 *Fusion Technol* **19** 783–790
[2] Zinkle S J 2005 *Phys Plasmas* **12** 058101-1 – 058101-8
[3] Riccardi B, Giancarli L, Hasegawa A, Katoh Y, Kohyama A, Jones R H and Snead L L 2004 *J Nucl Mater* **329-333(Part 1)** 56–65
[4] Klepeis S J, Benedict J P and Anderson R M 1987 *Specimen preparation for transmission electron microscopy: Proc. Mater Res Soc symp* vol 115 ed J C Bravman, R M Anderson and M L McDonald (Cambridge, UK: Cambridge Uni Press) pp179–184
[5] Hinks J A, van den Berg J A and Donnelly S E 2011 *J Vac Sci Technol A* **29(2)** 021003–6
[6] Ziegler J F 2004 *Nucl Instrum Meth B* **219-220** 1027–1036
[7] Serruys Y, Ruault M O, Trocellier P, Henry S, Katasov O and Trouslard P 2005 *Nucl Instrum Meth B* **240(1-2)** 124–7
[8] Birtcher R C, Donnelly S E, Templier C 1994 *Phys Rev B* **50** 764–9

Quantitative STEM: Experimental Methods and Applications

J M LeBeau[1], S D Findlay[2], L J Allen[3] and S Stemmer[4]

[1] Department of Materials Science and Engineering, North Carolina State University, Raleigh, NC 27606, U.S.A.
[2] School of Physics, Monash University, Victoria 3800, Australia.
[3] School of Physics, University of Melbourne, Victoria 3010, Australia.
[4] Materials Department, University of California Santa Barbara, Santa Barbara, CA 93106, U.S.A.

Abstract. High-angle annular dark-field (HAADF) imaging in scanning transmission electron microscopy (STEM) is highly sensitive to both the atomic number and the Debye-Waller factor of the atom columns. We will discuss the experimental requirements for a quantitative understanding of STEM image contrast, in particular the determination of the precise specimen thickness. We show that near perfect agreement can be achieved between theory and experiment and demonstrate that quantitative STEM allows for column-by-column counting of all the atoms in an arbitrarily shaped specimen.

1. Introduction

High-angle annular dark-field (HAADF) images acquired in scanning transmission electron microscopy (STEM) possess excellent atomic number (Z) sensitivity and have been widely applied to solve materials science problems such as determining interface atomic structures or the location and identity of individual atoms and clusters [1]. It has also been shown that image simulations are required to interpret HAADF image intensities quantitatively in terms of the elements present in the atom columns of a specimen [2]. With methods developed in the recent literature [3,4], experimental image intensities can be placed on the same absolute intensity scale as simulations, allowing for direct comparisons and interpretation of STEM images. It has been shown that experimental and simulated image intensities agree quantitatively, in both bright-field and HAADF STEM modes, provided that multiple thermal scattering and the effects of spatial incoherence are taken into account in the simulations [3,5]. Other inelastic scattering processes do not play a significant role in contrast formation for samples that are sufficiently thin [5].

An experimental challenge in quantitative STEM and for comparisons between experiment and theory is to accurately determine the TEM foil thickness. Two-beam convergent beam electron diffraction (CBED) is not compatible with atomic resolution STEM, because it requires specimen thicknesses that are too large, tilting the sample away from the zone axis conditions used for imaging and a small convergence semi-angle. The electron energy-loss spectroscopy (EELS) log-ratio approach requires knowledge of the total inelastic mean free path. In thin samples surface plasmons, amorphous or contamination layers reduce the accuracy and precision of this approach. A new STEM-based approach, position averaged CBED (PACBED) [6], is compatible with the thin specimens used in high-resolution STEM imaging. It is sufficiently sensitive to small thickness changes and does not require changes in the optical conditions and sample orientation away from those used for imaging or determination of additional specimen parameters. This approach is described in section 2. Taking the measurement of locale specimen thickness to its ultimate limit, we show in section 3 that truly quantitative atomic resolution HAADF-STEM can provide highly accurate, column-by-column counts of atoms without the need for a calibration standard or any a-priori knowledge of sample shape or thickness.

2. Position averaged convergent beam electron diffraction (PACBED)

Position averaged convergent beam electron diffraction (PACBED) can reliably measure sample thicknesses to better than 10%, specimen tilts to better than 1 mrad and sample polarity [7] for the same electron optical conditions and sample thicknesses as used in atomic resolution scanning transmission electron microscopy imaging.

Illuminating a zone-axis sample with a highly convergent electron probe, as used for STEM imaging, results in a CBED pattern containing overlapping orders of diffracted disks. These overlapping disks generate a complicated interference pattern that is highly dependent on the position of the finely focused (< 0.2 nm) electron probe within the unit cell [8]. Figure 1 shows, as an example, simulations of such patterns for an Ångstrom-size probe focused on different atom columns in a $SrTiO_3$ unit cell.

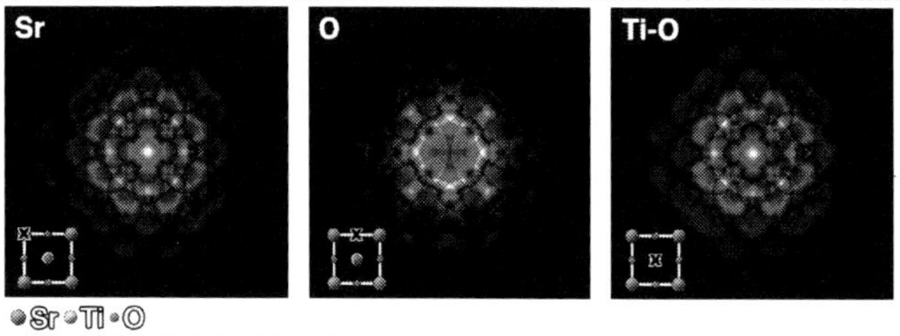

Figure 1. Simulations of coherent CBED patterns for $SrTiO_3$ along <100>, for different positions of the STEM probe within the unit cell (as indicated in the unit cell schematics in the lower left corner).

A PACBED pattern is formed through incoherent averaging of CBED patterns over many probe positions. The fine detail resulting from coherent interference is then lost, but better count rates and a reduced sensitivity to instability are obtained. The experimental PACBED pattern is highly sensitive to the sample structure, thickness and tilt, which can be quantitatively extracted through comparison with simulated PACBED patterns. The theory for simulating CBED patterns is well established and software packages are available [9]. An analytical depiction of PACBED has also been developed [6]. It has been shown that the PACBED pattern is independent of coherent aberrations of the probe-forming lens and of spatial incoherence [6]. Comparisons between PACBED experiments and simulations allow for determination of the local sample thickness, in regions as small as a few unit cells. Figure 2 shows a comparison between experimental and simulated PACBED patterns of a $SrTiO_3$ crystal. The thickness of the $SrTiO_3$ specimen can be determined with better than ± 1 nm precision.

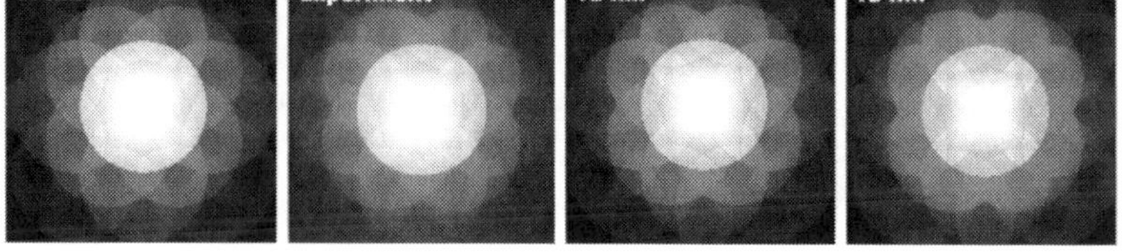

Figure 2. Comparison of an experimental PACBED pattern of $SrTiO_3$ with simulated patterns for different thicknesses, simulated using the absorptive multislice method [10].

3. Application of quantitative STEM: The electron microscope as an atom scale

Truly quantitative atomic resolution HAADF-STEM, i.e., the direct comparison between theory and experiment, can be used to take the determination of the specimen thickness to its ultimate limit: highly accurate, column-by-column counts of atoms without the need for a calibration standard or any a-priori knowledge of sample shape or thickness. The sample investigated in this example was a Au film evaporated onto a NaCl crystal [11]. To count the number of atoms in each column, an image analysis approach was developed to accurately determine the column location and corresponding column intensity [11]. This approach is schematically illustrated in Fig. 3. The maximum intensity at each column was extracted at the atom column centroid positions, determined as described in Fig. 3.

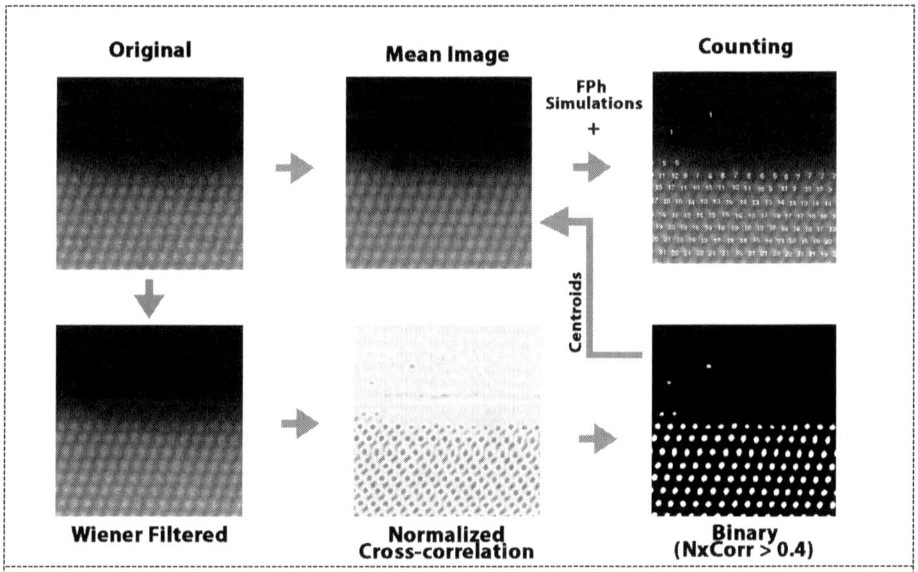

Figure 3. Overview of the image analysis workflow to determine the positions of gold atom columns and extract the image intensities [11]. FPh = frozen phonon.

The signal of the simulated atom column was extracted by averaging about the same circular region as the experimental data. Comparison between experimental and simulated atom column intensities was then used to determine the number of atoms in each column via linear interpolation and rounding to the nearest integer value (i.e. there are no fractional atoms). Figure 4 shows an experimental image of a wedge-shaped Au foil with a (111) surface observed along <110>. The labels indicate the number of Au atoms in each column. In the thicker region of the sample, the number of atoms in adjacent planes increases smoothly with thickness, generally by one atom.[11]

Figure 5 shows a histogram containing all the atom columns in Fig. 4, binned according the number of atoms they contain. The histogram shows that the atom columns in each bin are all self-similar, indicating precise counting. Visual inspection immediately shows that the precision is ± 1 atom. For example, the atom column intensities in the bin of columns containing 24 atoms are very distinct from those of bins 22 or 26.

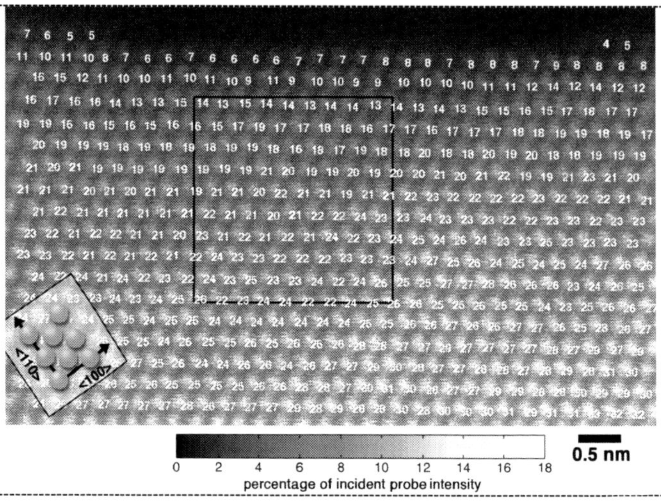

Figure 4. HAADF-STEM image of a wedge-shaped Au film viewed along <110>. The labels near the lower right of each atom column indicate the number of atoms contained in that column [11].

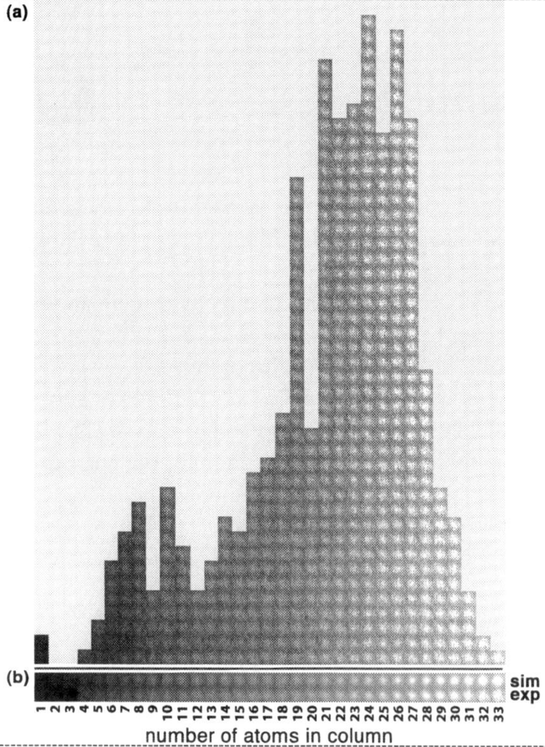

Figure 5. (a) Histogram of all the columns in Fig. 4 binned by the number of atoms. (b) Atom column images extracted from simulations (top) and those after averaging all the columns in each bin (bottom) [11].

Journal of Physics: Conference Series **371** (2012) 012053 doi:10.1088/1742-6596/371/1/012053

4. Summary

By placing STEM images on an absolute intensity scale, truly quantitative, direct comparisons between simulation and experiment are possible. A STEM diffraction technique, position averaged convergent beam electron diffraction, provides an independent measure of the local specimen thickness more accurately and precisely than other methods, and is essential for comparisons between simulation and theory. Applications of truly quantitative atomic resolution HAADF imaging include highly accurate, column-by-column counts of atoms in a three-dimensional volume of a sample without the need for a calibration standard. Future applications of the technique include obtaining accurate local concentrations of dopants and impurity atoms at defects, the shape and composition of nanostructures, reconstructions at interfaces and dislocation cores.

5. References

[1] *Scanning Transmission Electron Microscopy*, edited by Pennycook S J and Nellist P D (Springer, New York, 2011)
[2] LeBeau J M, Findlay S D, Wang X, Jacobson A J, Allen L J, and Stemmer S, *Phys. Rev. B* 2009 **79** 214110
[3] LeBeau J M, Findlay S D, Allen L J, and Stemmer S, *Phys. Rev. Lett.* 2008 **100** 206101
[4] LeBeau J M and Stemmer S, *Ultramicroscopy* 2008 **108** 1653
[5] LeBeau J M, D'Alfonso A J, Findlay S D, Stemmer S, and Allen L J, *Phys. Rev. B.* 2009 **80** 174106
[6] LeBeau J M, Findlay S D, Allen L J, and Stemmer S, *Ultramicroscopy* 2010 **110** 118
[7] LeBeau J M, D'Alfonso A J, Wright N J, Allen L J, and Stemmer S, *Appl. Phys. Lett.* 2011 **98** 052904
[8] Cowley J M, *Ultramicroscopy* 1979 **4** 435
[9] Kirkland E J, *Advanced Computing in Electron Microscopy* (Springer, New York, 2010)
[10] Allen L J and Rossouw C J, *Phys. Rev. B* 1989 **39** 8313
[11] LeBeau J M, Findlay S D, Allen L J, and Stemmer S, *Nano Letters* 2010 **10** 4405

Quantitative techniques for aberration corrected HAADF STEM of nano-materials

H E[1*], P D Nellist[1], A J D'Alfonso[2] and L J Allen[2]

[1] Department of Materials, Oxford University, Parks Road, Oxford OX1 3PH, UK
[2] School of Physics, University of Melbourne, Parkville, Victoria 3010, Australia

Email: haibo.e@materials.ox.ac.uk

Abstract. Electron microscopy is a powerful tool for directly visualising structure and with the advent of aberration correction, atomic columns and even individual atoms are now routinely imaged. HAADF STEM offers a great sensitivity to atomic number - the well known Z-dependence in the measured intensity. By calibrating the HAADF detector, putting images on an absolute scale, and using an absolute HAADF scattering cross section comparison, it is possible to compare experiment with simulations in a parameter robust manner. We show how the cross section method relieves the use of a fitting parameter for the probe and is insensitive to experimental parameters such as defocus. Finally, we test our analysis method using a 2-dimensional nano-structure to identify the composition of atomic columns in thin layered materials. We also discuss some of the potential sources of error in quantifying small, thin, low-scattering and beam sensitive samples.

1. Introduction

Using a high angle annular dark field (HAADF) detector in the scanning transmission electron microscope (STEM), it is possible to directly image the atomic configuration of nanostructures. This technique has often been chosen for heavy elements on light supports or structures with a combination of light and heavy elements for the strong correlation between atomic number and imaged intensity, known as Z contrast. The major advantage of HAADF STEM over conventional high resolution transmission electron microscopes (HRTEM) is its incoherence. In HAADF STEM there are no contrast reversals often seen in HRTEM due to the phase contrast nature of the imaging [1]. However, though it offers many advantages, meaningful quantitative compositional information is less than straight-forward to extract from HAADF images because intensities are dependent on both composition and thickness [1].

In this paper we show that by comparing detector calibrated experimental images with simulations, it is possible to compare the absolute scattering from a thin layered sample without fitting parameters or modelling of the whole structure. We show that this method is robust to experimental parameters such as defocus, which are often difficult to determine accurately. Finally, we discuss briefly the main sources of errors in routine HAADF quantification of beam sensitive samples.

2. Absolute HAADF Scattering Cross Section

"Cross sections" is a term often used in X-ray and neutron diffraction and also electron energy loss spectroscopy in electron microscopy, but not applied to image analysis. Earlier studies using HAADF-STEM showed interest in the idea of cross sections but without detailed description of the calibrations necessary to fully quantify them [2] and without the resolution needed to isolate individual atomic columns [3]. Since detector calibrations have been formalized to put imaging on an absolute scale and careful matching of experiment to simulations shown to be possible [4], other similar parameters have been used such as column mean-intensities to match experiment and theory for the heights of individual columns [5].

Since HAADF STEM is an incoherent technique by the nature of its detection method [1], we can consider the final image as a convolution in two dimensions of the probe $P(r_0 - r)$ and an object function $O(r)$:

$$I(r_0) = \int P(r_0 - r)\, O(r) d^2 r$$

At atomic resolution, so long as the columns are well resolved and the sample is sufficiently thin such that there is minimal "cross-talk" between columns, this can allow us to understand the image intensity intuitively: once the image intensity is calibrated as a function of the incident beam [5], we can then integrate around an isolated column contained in pixels i each of area A and intensity I such that:

$$\sum_i I_i A_i = \int \sum_i A_i P(r_0 - r)\, O(r)\, d^2 r$$

Due to the fact that the detector calibration expresses the intensity as a fraction of the incident probe, the summed probe function $\sum_i A_i P(r_0 - r)$ normalises to 1. Thus, the summed object function of that column is measured as:

$$A \sum_i I_i = \int O(r)\, d^2 r$$

What remains is an integration of the object function which has no dependence on the probe and has dimensions of $[L^2]$. We call this parameter the absolute HAADF scattering cross section [5] and can use this to compare simulations to experiments without concerns about the probe.

3. Parameter Robustness

When comparing experimental images to simulations quantitatively, some experimental parameters that affect the imaged intensity must be known accurately and thus their identification and calibration is vital. Some of the most influential and difficult to determine experimental parameters are defocus and source effects of the finite probe [7]. Optimum focus is usually found by manual adjustment of experimental conditions or choosing the image which maximises contrast in a through focal series. However, the latter often requires prolonged exposure and issues of damage in beam sensitive samples are paramount, and the former is prone to human error. Thus it is essential to have a metric for comparing experiments to simulation which is robust to this experimental parameter.

This work was originally born out of investigation of catalyst nanoparticles of Pt/Pd. Thus we used an absorptive potential multislice simulation which takes into account phonons and TDS and is ideally suited for thin and small nano-materials. Simulations show that the peak value for a single atom drops significantly (Figure 1) with defocus, as is expected when imaging with a broadened defocused beam. However, the cross section is insensitive to defocus so long as the area of column integration is sufficiently such that the entire column is included.

Such large variations in peak intensity over a small range of defocii means that accurate comparison with simulations is limited unless an independent measure of that parameter is done. Cross-sections are more robust.

However, other parameters such as local sample mistilt, accurate knowledge of the Debye-Waller thermal factor still remain as concerns in the object function. We note also that the incoherence is only transverse and this approach is no longer valid for thicker crystals, especially where cross-talk and probe tails mean that imaged intensity cannot be clearly attributed to one atomic column.

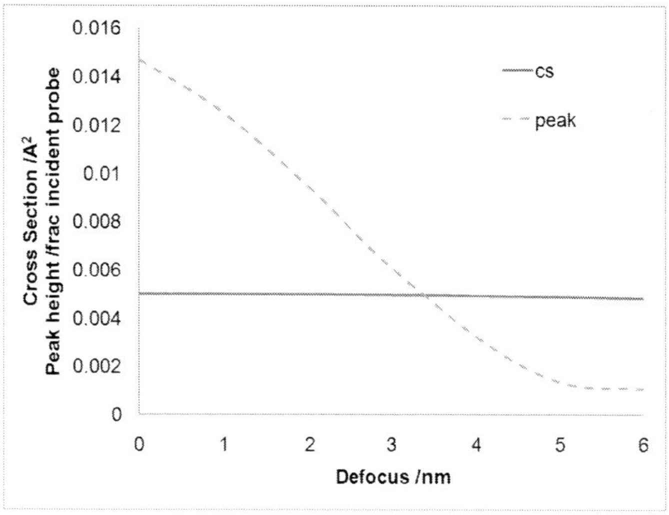

Figure 1: The cross section (blue solid line) of a single Pt atom, a heavy element which scatters strongly to the HAADF detector, does not vary significantly with defocus. However, the peak intensity (green dashed line) decreases significantly from the value at focus, even within a few nm of defocus, which is often a reasonable experimental error. Simulation is of a 200kV probe with convergence angle of 25mrad.

4. Thin 2D Layered Material

MoS_2 is a thin 2D layered material and thus offers the ideal test-bed for elemental identification, as shown previously [8]. We apply our method to show that this parameter robust method allows for identification of elements by matching experiment to theory. The sample was prepared by liquid exfoliation [9] and imaged using a C_s-probe-corrected JEOL ARM-200F at 80kV with inner and outer detector angles of 55mrad and 180mrad respectively. The small probe of the C_s-corrected instrument ensures that individual columns can be well resolved, eliminating issues of cross-talk. The detector was calibrated for quantitative imaging using different probe currents to allow the nonlinear digital to analogue converter gain to be kept constant between detector maps and imaging while also allowing a good dynamic range to be used for the imaging of such a low scattering sample [10]. We used 2.6pA for the detector map and 33.8pA for imaging, giving a ratio of 1:13. The probe current was measured immediately prior to and post imaging both the detector and sample, with a Faraday cage, to ensure that there was no significant probe current decay from the cold FEG.

Careful background subtraction was necessary since the sample showed contamination in some areas (Figure 2, bottom) which would greatly affect the measured cross section value. Four line profiles were taken from the vacuum into the monolayer (Figure 2, dashed lines) and the vacuum "black level" was subtracted. The intensity between the columns is attributed to probe tails and thus included in the final cross section value. Only atoms far from edges and steps were considered as areas close to the edges of thin 2D sheets often have ad-atoms [8]. Only those in the top half of the image were investigated, where the sample was cleaner.

The measured values from 42 atoms gave a mean cross section of $0.0098A^2$. The main sources of calculable errors were due to the accuracy of the pico-ammeter in the measuring of the small 2.6 ± 0.1pA current used to map the detector and also the large variance of integrated intensities due to noisy data. The standard deviation of the cross section was ±0.0031A so a large sample size of Mo atoms was taken and the measured cross sections averaged. Thus the error on the mean was much lower: $\pm 0.0006A^2$.

Comparison with theoretical cross section values revealed that the mid area atoms were single Mo (calculated cross section of $0.0092A^2$), confirming that this is a monolayer. It was not possible to identify the S_2 as the columns of S_2 were not well resolved above the background. They would have had a much smaller cross section of $0.0022A^2$.

Cross sections of other columns (Figure 2, right) showed that these were either two S on top of one Mo or one Mo on top of two S, confirmed by the open hexagon structure of multilayer MoS_2. It is not possible to distinguish between the two stacking configurations as in such thin materials, channelling of the probe has not been established. Thus the scattering from both stacking orientations is identical.

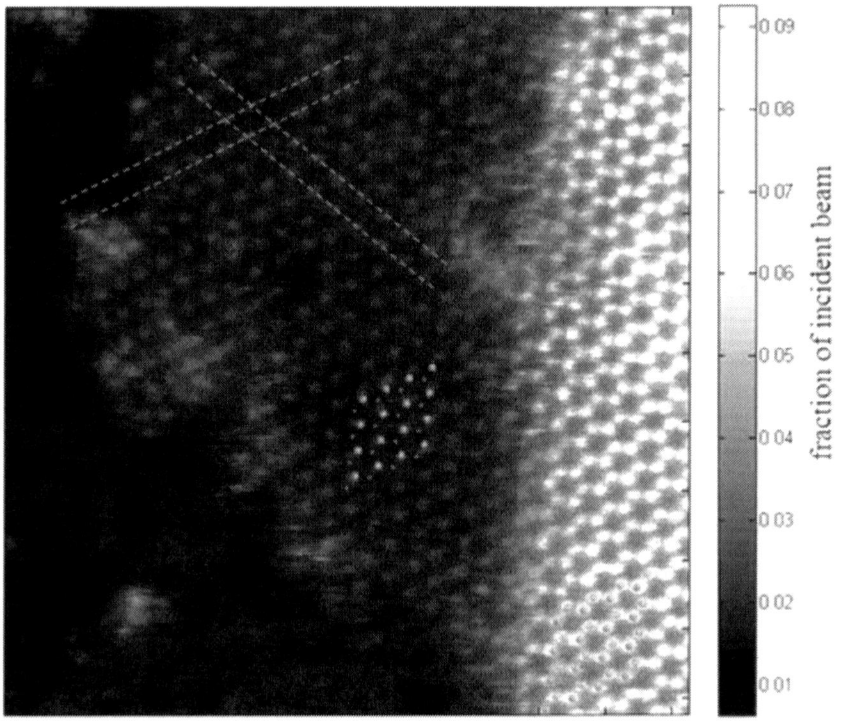

Figure 2: Calibrated HAADF STEM image of the edge of a MoS_2 flake showing monolayer (centre) and bilayers (right). Vacuum is seen in the top left corner, as well as carbon contamination in the bottom left. The image is calibrated for the incident probe current and image saturated for display so that the monolayers can be clearly seen. The image has not been altered in any other way.

5. Conclusions

We have shown that analysis of quantified HAADF STEM images can be carried out by summing around well resolved columns and that this can be understood very intuitively in terms of the cross-sections of scattering. We have shown that this approach is more robust to imaging parameters, such as defocus, which may be experimentally difficult to determine. This is especially useful when imaging conditions are limited such as for beam sensitive specimens so rapid acquisition of images is necessary before beam damage occurs.

Applying this analysis to an ideal sample of a thin 2D layered material of MoS_2 prepared by liquid exfoliation; we show that it is possible to match with excellent agreement the cross sections of Mo in experimental images with simulations of single atoms, removing the need to do large, time consuming calculations of the whole image. We highlight some of potential sources of error in quantifying images, especially in detector calibrations and background subtraction.

Acknowledgements

This work has been funded by the EPSRC and Johnson Matthey. L.J. Allen acknowledges support from the Australian Research Council. The authors would like to thank JEOL for their help with the experimental facilities and A. Shmeliov for the MoS_2 sample.

References

[1] Pennycook SJ, Jesson DE, McGibbon AJ and Nellist PD 1996 *J. Electron Microscopy* **45** 36-43
[2] Singhal A, Yang J and Gibson J 1997 *Ultramicroscopy*, *67*(1-4), 191-206
[3] Menard LD, Gao SP, Xu H, Twesten RD, et al. 2006 *J Phys Chem. B*, *110*(26), 12874-83.
[4] LeBeau JM, Findlay SD, Allen LJ, Stemmer S, 2008 *PRL.* **100** 1-4
[5] LeBeau JM, Findlay SD, Allen LJ, Stemmer S, 2010 *NanoLett.* **10** 4405-8
[6] E H, Nellist PD, D'Alfonso AJ, Allen L *in preparation*
[7] Dwyer C, Erni R, and Etheridge J. 2010 *Ultramicroscopy* 1-6
[8] Krivanek OL, Chisholm MF, Nicolosi V, et al. 2010 *Nature*, *464*(7288), 571-4.
[9] Nicolosi V, McCarthy D, et al. 2007 *Eur.J. AppPhys, 37, 149-159*
[10] E H, Nellist PD, Lozano-Perez S, Ozkaya D 2010 *J. Phys.: Conf. Ser.* **241** 012067

On the visibility of "heavy" atoms in dark-field STEM

M D Shannon[1], P D Nellist[2], V Nicolosi[2], G B Shannon[3], A Shmeliov[2]

[1]SuperSTEM (EPSRC National STEM Facility), Daresbury Laboratory, Keckwick Lane, Daresbury WA4 4AD (and School of Engineering, University of Liverpool);
[2]University of Oxford, Department of Materials, Parks Road, Oxford OX1 3PH;
[3]School of Pharmacy, University of Nottingham, Nottingham NG7 2RD

mervyn.shannon@liv.ac.uk

Abstract. One of the most useful and apparently straightforward attributes of annular dark-field (ADF) STEM imaging is the ability to image heavy atoms on relatively light substrates using Z-contrast (High-Angle ADF) imaging. From multislice calculations, however, some isolated heavy atoms, e.g. Re, can go undetected in a thin Co matrix in standard aberration-corrected HAADF-STEM imaging of catalyst nanoparticles oriented accurately on a zone-axis [2]. More recently simultaneous Medium-Angle and High-Angle ADF imaging of a few layers of BN has shown experimentally that some adatoms are invisible in MAADF whilst exhibiting very strong contrast in HAADF. Simulations confirm that this can occur for atoms such as sodium. In both situations HOLZ ring imaging can be used to extract additional information.

1. Introduction

High-Angle Annular Dark-Field (HAADF) or Z-contrast STEM imaging has an elegant simplicity when it comes to detecting and imaging heavy atoms on or in a relatively light substrate. It is a potential pitfall to assume that all such heavy atoms are visible in thin samples (< 3nm). It is important that we appreciate and understand the circumstances in which this assumption breaks down.

Such images of Au atoms within crystalline Si nanowires can be directly interpreted [1]. Voyles et al [3] discussed the more complex case of Sb doping of crystalline silicon. They showed that in a zone-axis orientation the contrast ratio I_{Sb}/I_{Si} between a column of atoms containing just one Sb atom in a substitutional site and a column of pure Si is a function of the depth of the dopant atom, whereas the ratio for an interstitial Sb atom is close to one and nearly independent of depth (in a 4.5nm thick slab). For the former configuration the intensity ratio is low for a dopant near the entrance surface and rises approximately linearly to a thickness of about 10nm after which it exhibits damped oscillation.

To maximize the activity of the catalyst, additives such as Pt, Re, Ir are commonly used to "promote" Co catalysts used in the Fischer-Tropsch process which synthesizes liquid paraffins from carbon monoxide and hydrogen. In the manufacture of the catalyst these platinum group metals (PGM) lower the temperatures at which chemical reduction from Co_3O_4 to CoO and then Co metal occur and hence inhibit sintering and maximize surface area of the active metal. However in a typical preparation about 20 such heavy atoms are used per 3-4 nm Co metal particle. HAADF imaging of the location of the PGM atoms at various stages is one approach to understanding how this number might be lowered to reduce cost. In samples reduced and passivated such high numbers were not observed [2] and image simulations were performed to understand whether this indicated that the PGM were interstitial or that electron channelling rendered them less visible than expected.

Journal of Physics: Conference Series **371** (2012) 012055 doi:10.1088/1742-6596/371/1/012055

The properties of graphene and related sheet materials such as BN are modified by the presence of adatoms. Recently the use of Medium-Angle ADF (MAADF) imaging at 60kV has been advocated for the imaging of BN single sheets with discrimination of B, C, N and O atoms by absolute intensity [4]. However simultaneous MAADF and HAADF imaging has shown that on occasion "heavy" atoms are visible in the HAADF signal but not MAADF in regions with only 3 or 4 layers of BN.

2. Experimental and computational methods

The catalyst samples were imaged at 100kV in a modified VG-HB501 STEM with a second generation 3rd order Cs-corrector from Nion. HAADF images were recorded with a convergence half-angle of 24mrad and 70-210mrad collection angles and these parameters were used for the calculations reported here. The BN samples were imaged at 60kV in a Nion UltraSTEM 100 5th order corrected instrument with 24mrad convergence angle. MAADF and HAADF images were recorded simultaneously with 60-100 and 100-200mrad collection angles respectively.

Frozen phonon multislice image simulations were performed using either Kirkland's TEMSIM code [5] or Koch's QSTEM code [6] with 30 iterations to represent thermal diffuse scattering (TDS).

3. Results

The common theme here is that imaging using HOLZ rings can be useful in elucidating structure in very thin crystals.

3.1. PGM atoms substituted in the surface of Co nanoparticles

The results of simulations for a 3.5nm particle with 22 Re atoms randomly substituted in the surface are shown in Fig 1. There are 15 Co atoms in the central column and 8 Co/Re in the edge-on facets.

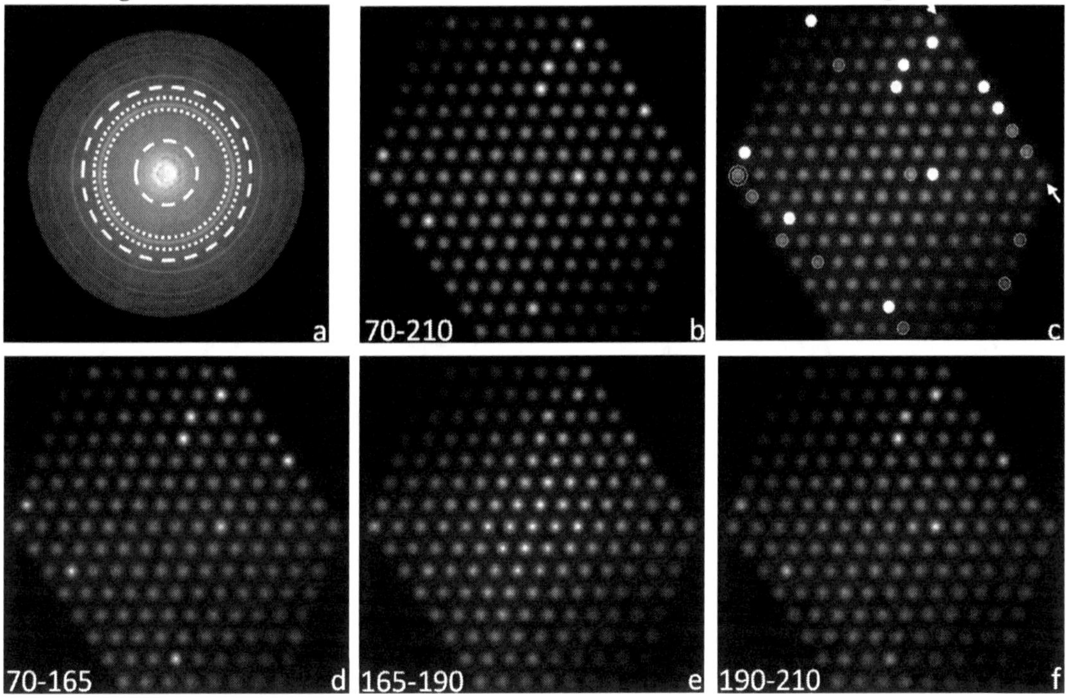

Fig 1. Images calculated for [110] projection of a 1415 atom cuboctohedral particle using QSTEM [6] with TDS, C3 = -10 μm, C5 = 10mm, defocus = -10 Å, source size 0Å. The inner and outer radii of the annuli of the detectors used to calculate (b),(d)-(f) are superimposed on the diffraction pattern in (a) calculated with the probe placed at the centre of the particle. In (c) circles indicate a surface Re atom in the bottom half (the e⁻ beam enters from below) of the crystal (open) and top half (solid).

Images calculated precisely along the [110] zone-axis with the standard HAADF detector (70-210mrad) show that, due to electron channeling effects, Re atoms near the exit surface of the crystal are readily seen with enhanced contrast whilst those at or near the entrance surface are difficult to detect by eye. Indeed anomalies arise such as the 2 atoms of Re in the leftmost column (2 circles in Fig. 1(c)) have lower intensity than the single Re in the column next to it. The 2 are 1st and 2nd in the column of 8 atoms whilst the single Re is 7th of 7 atoms.

The calculated contrast is very similar to the standard HAADF image for annuli 70-165 and 190-210mrad. However if an annulus containing just the first-order Laue zone (Fig. 1(e)) is chosen the contrast is more as one would anticipate in the absence of channeling. The column of 8 atoms with 2 Re atoms is brighter than its neighbours with 1 Re. Also in the row of atom columns (each containing 8 atoms) indicated between the arrows in Fig. 1(c), the intensities of the four columns with 1 Re atom each are almost equal whereas in the standard HAADF image the 2 columns with Re near the beam exit are much more intense than the two with Re near the beam entrance [2]. The combination of HAADF and HOLZ ring imaging could be exploited to derive the dopant depth.

3.2. Atoms invisible in Medium-Angle Annular Dark-Field (MAADF) images of BN

Whilst studying a delaminated BN sample a number of simultaneously acquired MAADF/HAADF image pairs showed strong contrast due to the presence of adatoms in the HAADF image but not the MAADF. Two such adatoms are shown in Fig. 2. These were typically in regions of 3 or 4 BN layers thick, and were positioned on lattice sites.

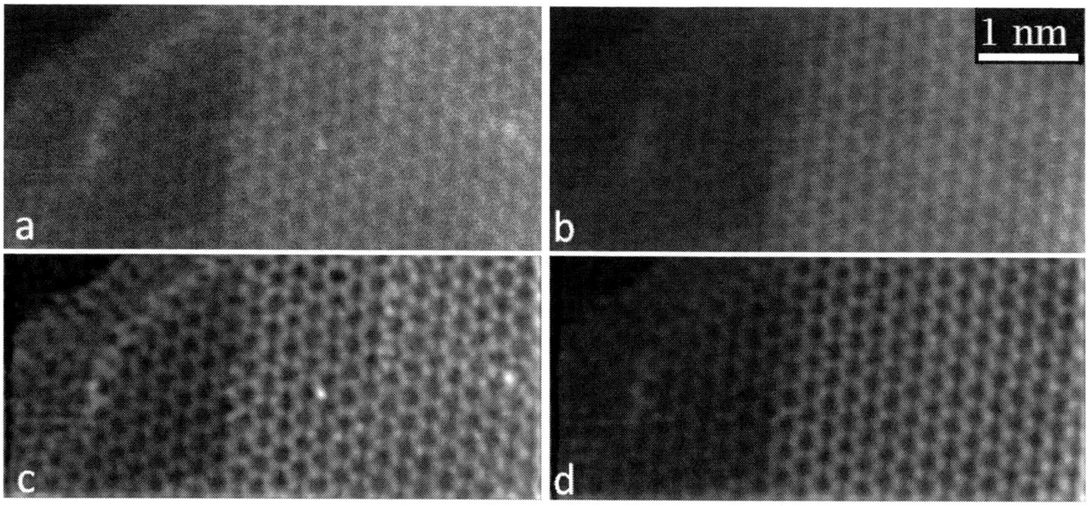

Fig. 2. Simultaneously acquired raw HAADF (a) and MAADF (b) images of the edge of a BN flake and as processed by the filter described in [4] (c) and (d).

A number of simulations have been performed placing different adatoms above or below B or N atoms in a wedge ranging from 1 to 5 layers thick. An example with Na as the adatom placed 0.33nm above or below a B or N atom is given in Fig. 3. All of the adatoms are readily seen in the HAADF simulation; they are also visible in the MAADF simulation for 1 or 2 BN layers but not for 3, 4 or 5 BN layers. Increasing the atomic number of the adatom from 11 to 14 (Si) the adatoms become visible in the MAADF image if they are on the beam exit side of the (up to 5) BN layers and for 1, 2 and 3 BN layers but not 4 or 5 BN layers if on the beam entrance side of the wedge. If the Na adatoms are slightly displaced along an in-plane BN bond direction they become mostly visible in the MAADF image with the same restrictions as for Si above. The height above the BN layer is significant. 0.33nm is equivalent to the interlayer spacing in BN. However if the Na adatom is placed 0.165nm above or below B or N the adatoms become visible, though those before 4 or 5 BN layers are again difficult to detect (Fig 3(c)).

For a model placing Na atoms midway between layers without relaxation of the structure (Fig. 3(d)-(f)), the result is even stronger contrast in the simulated MAADF image. A simulated image formed using just a narrow annulus at the SOLZ ring position, which corresponds to the interlayer spacing in BN, is unaware of the intercalated atoms. This may be useful in distinguishing between surface adatoms and intercalated atoms. Of course the intensity of this signal is very weak. These simulations using QSTEM assume a reasonable source brightness of $5x10^8$ A/cm^2 and a pixel dwell time of 30μs.

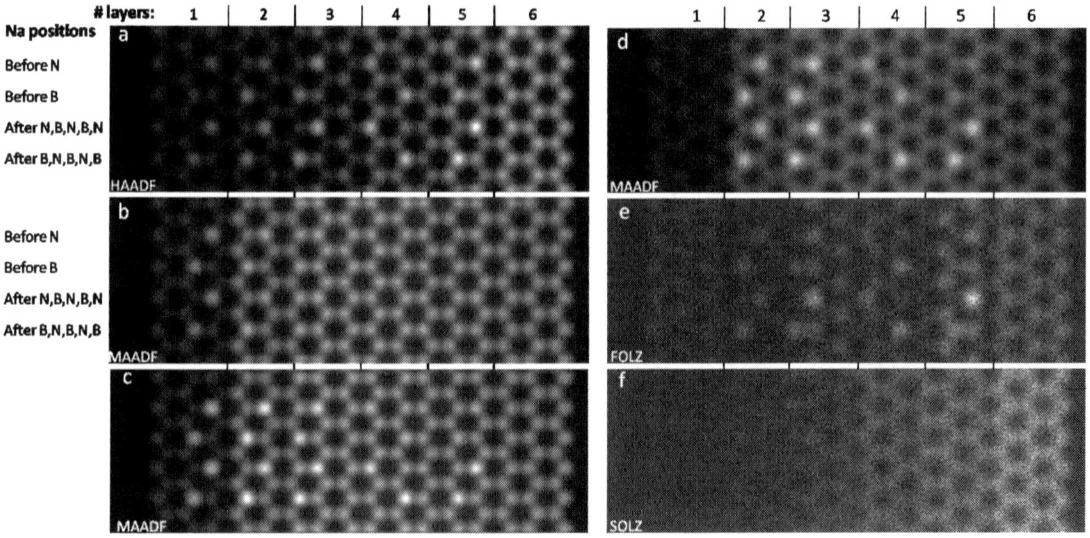

Fig 3. HAADF (a) and MAADF (b) simulations of Na adatoms placed 0.33nm above or below BN layers compared with the MAADF image (c) for Na placed 0.165nm above or below the BN layers. The simulations (d)-(f) are for Na intercalated between the last-but-one and last BN layer.

4. Conclusions

Even for very thin samples (<2nm) heavy atoms may not be visible in standard HAADF or MAADF images along accurately aligned zone axes. Careful choice of angular ranges, such as a HOLZ ring, can help to resolve the position in *z* of the dopant or adatom. A position sensitive detector of high sensitivity and high dynamic range is needed to facilitate such studies.

Acknowledgements

MDS thanks the Royal Society for an Industrial Fellowship whilst studying PGM dopant atom visibility and EPSRC under grant EP/D040396/1 for the BN image simulation study.

References

[1] Allen J E, et al 2008 *Nature Nanotech.* ***3*** 168
[2] Shannon M D, Lok C M and Casci J L 2007 *J. Catal.* ***249*** 41
Shannon M D and Shannon G B 2010 *Microsc. Microanal.* ***16*** (Suppl 2) 99
[3] Voyles P M, Muller D A, Grazul J L, Citrin P H and Gossmann H-J L 2002 *Nature* ***416*** 826
Voyles P M, Grazul J L, Muller D A 2003 *Ultramicroscopy* ***96*** 251
[4] Krivanek O L, Chisholm M F, Nicolosi V, Pennycook T J, Corbin G J, Dellby N, Murfitt M F, Own C S, Szilagy Z S, Oxley M P, Pantelides S T and Pennycook S J 2010 *Nature* ***464*** 571
[5] Kirkland E J 1998 *Advanced Computing in Electron Microscopy*, Plenum, New York
[6] Koch C T 2002, *Ph.D. Thesis*, Arizona State University

Quantitative HAADF-STEM on heterostructured GaAs nanowires

H Kauko, R Bjørge, R Holmestad and A T J van Helvoort
Norwegian University of Science and Technology (NTNU), Trondheim, Norway
E-mail: a.helvoort@ntnu.no

Abstract. GaAs nanowires are a highly promising candidate for future optoelectronic applications. Here quantitative high-angle annular dark field scanning transmission electron microscopy (HAADF-STEM) has been used for studying the Sb content x in zinc blende (ZB) $GaAs_{1-x}Sb_x$ axial inserts in otherwise pure wurtzite (WZ) GaAs nanowires. Direct comparison of the normalized experimental HAADF-STEM image intensities with simulated intensities displayed a good match for the pure structure. It was discovered that the Sb content in a 20 nm long insert is not uniform but increases from 13 ± 3 at.% at the lower to 21 ± 3 at.% at the upper WZ-ZB interface. Non-compositional effects on HAADF intensity have been identified.

1. Introduction

GaAs nanowires are a promising candidate for future optoelectronic applications, such as solar cells, light-emitting diodes (LEDs) and nanolasers [1]. Pure GaAs nanowires usually have the wurtzite (WZ) phase, but alloying with Sb can change the phase into zinc blende (ZB). It is important for the applications to understand the effects of alloying and to measure the actual composition. With high-angle annular dark field scanning transmission electron microscopy (HAADF-STEM), compositional information can be extracted directly from the images with a high spatial resolution.

Here we have applied quantitative HAADF-STEM for characterizing heterostructured GaAs nanowires. More precisely, the Sb content x and distribution in ZB $GaAs_{1-x}Sb_x$ axial inserts of otherwise pure WZ GaAs nanowires have been studied by directly comparing experimental image intensities with simulated intensities. Prerequisite for direct comparison is that the experimental images are normalized relative to the incident beam intensity, which was obtained by scanning the ADF detector prior to the measurements [2]. The work was performed on a non-corrected TEM/STEM at a relatively low magnification.

2. Experimental details

The nanowires were grown on a ZB GaAs(111)B substrate in a Varial Gen II Modular MBE system, using Au particles as catalysts. The ZB GaAsSb insert was grown for 30 s after 20 min of WZ GaAs nanowire growth, and followed by 5 min of WZ GaAs nanowire growth. The targeted Sb concentration for the insert was 15 %. The growth directions were [0001] and [111] for the WZ and ZB segments, respectively.

A schematic diagram of the insert area and its dimensions is shown in figure 1(a). As illustrated in the figure, the nanowires are hexagonal in cross-section. This specific geometry

– the thickness profile can be deduced exactly from the projected width – and the presence of two crystal phases make the nanowires an interesting system for quantitative HAADF studies. Figure 1(b) shows a HAADF-STEM image of an entire nanowire with insert.

For TEM-work, the nanowires were scraped off the substrate, dispersed in isopropanol and transferred to a graphene-coated Cu-grid (Graphene Laboratories Inc.). The microscopy was performed on a JEOL 2010F (200 kV, $C_s = 1$ mm). The probe-forming aperture semi-angle was 8.4 mrad, the probe size approximately 0.2 nm and the ADF detector range was 29-114 mrad.

In order to directly compare the experimental and simulated intensities, they have to be normalized relative to the incident probe intensity. Therefore a detector scan was performed by adjusting the intermediate and projector lenses so that the probe was scanning directly over the detector. Using the resulting detector image (figure 1(c)), the experimental images were normalized according to [3]

$$I_{norm} = \frac{I_{raw} - I_{vac}}{I_{det} - I_{vac}} \tag{1}$$

where I_{raw} is the raw image intensity, I_{vac} is the vacuum intensity and I_{det} is the incident probe intensity directly on the detector. I_{vac} and I_{det} were determined from the detector image as the mean intensities over the vacuum and the detector regions, respectively. While scanning the detector, care was taken that the gain and offset (contrast and brightness) were such that the detector was not saturated. The same gain and offset settings were used for image collection.

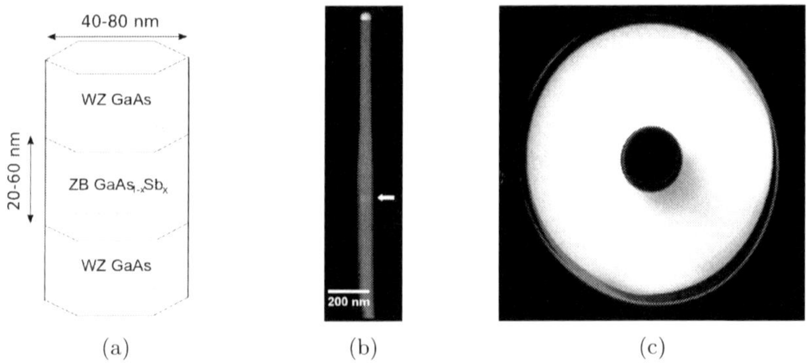

Figure 1. (a) A schematic diagram of a nanowire segment with insert, (b) HAADF-STEM image of an entire nanowire with the insert indicated by an arrow and (c) intensity map obtained by scanning the probe directly over the ADF detector.

3. Simulations

For the simulations, a frozen lattice multislice simulation software from Dwyer was employed [4], using 8 phonon configurations. Both WZ GaAs and ZB $GaAs_{1-x}Sb_x$ were simulated for a thickness range of 0-100 nm, the latter with Sb concentration x varying from 0 to 25 at.% with an interval of 1 at.%. Sb takes the As position in the lattice, and was included in the simulations by randomly substituting a certain percentage of As atoms by Sb atoms.

For ZB GaAs(Sb) a 4x6 supercell in [110] orientation and for WZ GaAs a 5x5 supercell in $[11\bar{2}0]$ orientation was employed. The microscope parameters were as given in section 2. The Debye-Waller factors for both the structures were based on theoretical calculations by Schowalter [5]. The changes in lattice parameters and Debye-Waller factors of Ga due to change in composition were taken into account by calculating the parameters for each composition as a linear combination of the parameters for the pure structures (GaAs and GaSb) [3].

Using the average supercell image intensities for ZB $GaAs_{1-x}Sb_x$, a matrix for intensity as a function of thickness and concentration (CT-matrix) was generated [6]. The CT-matrix was used for analyzing the Sb-content directly from the normalized experimental intensities. The WZ parts in the images were analysed similarly by using average simulated image intensity as a function of thickness.

4. Results and discussion

Figure 2(a) shows a high-resolution TEM (HRTEM) image of the insert area of a nanowire. In the image stacking faults above the insert are clearly visible; lattice defects are often observed at the upper ZB-WZ interface for these heterostructured nanowires. Another prominent feature in the image is the dark area in the lower left corner of the insert, which is probably due to local strain of a growth-related origin. The strained area shows considerable intensity variations in the corresponding HAADF-STEM image (figure 2(b)) as well. The crystal phases were confirmed by electron diffraction and the presence of Sb only in the ZB region by EDX (not shown).

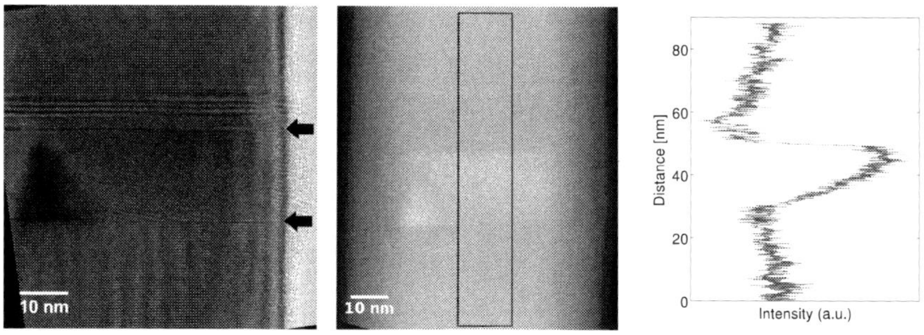

(a) (b) (c)

Figure 2. (a) A HRTEM image of an insert, with the WZ-ZB interfaces marked by arrows; (b) HAADF image of the same area and (c) intensity profile integrated along the area indicated in the HAADF image.

Figure 2(c) shows the intensity profile over the area indicated in the HAADF image (figure 2(b)). Several observations can be made from this graph. Firstly, the intensity is clearly higher at the insert than at the WZ parts, showing clear Z-contrast. Secondly, the intensity is not uniform over the insert but rather increasing until reaching a plateau, indicating that the Sb content is increasing from the lower to the upper interface. This trend has been confirmed by EDX. Ideally, the growth should be optimized to produce an even Sb concentration within the insert, yielding a well-defined change in the band structure. Furthermore, there is a clear drop in the WZ intensity at the upper ZB-WZ interface, probably due to the presence of the stacking faults. Finally, the intensities of the WZ areas below and above the insert are approximately equal, indicating that the wire was lying on zone across the imaged area.

In figure 3, normalized experimental intensity profiles across the nanowire are plotted together with the corresponding simulated profiles. Figure 3(a) is for the WZ part below the insert, and figures 3(b) and 3(c) are for the lower and upper parts of the insert, respectively. The simulated ZB $GaAs_{1-x}Sb_x$ intensity profiles were constructed using the calculated CT-matrix, and the experimental intensity profiles were constructed by averaging over narrow areas across the nanowire. For the lower part of the insert an approximate Sb concentration of 13 at.% was found and for the upper part 21 at.%. By considering the sensitivity of simulated HAADF intensity to changes in thickness and composition, as well as the accuracy of the experimental intensity levels and the thickness determination, the accuracy for the deduced concentration was estimated to be approximately 3 at.%.

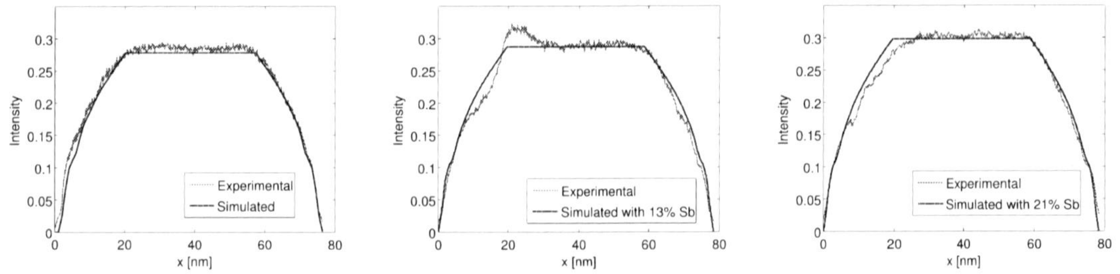

(a) (b) (c)

Figure 3. Experimental intensity profiles across the (a) WZ area below the insert, (b) lower part and (c) the top part of the ZB insert together with the corresponding simulated profiles.

The simulated intensity profiles match the experimental profiles rather well across the whole nanowire, except for the local distinct variation in experimental intensity in figure 3(b) due to the strained area (see figure 2). The resulting Sb concentrations are however higher than expected. One explanation for this can be that static atomic displacements (SAD), i.e., shifts in atomic positions due to the alloying Sb atoms [7], were not taken into account in the simulations. A recent study shows that the SADs increase the simulated intensity significantly, thereby lower the resulting alloy concentration [6]. The Debye-Waller factors have a significant effect on the intensity as well. Initial simulations with different Debye-Waller factors (from [8]) produced clearly higher intensities, thus lower Sb concentrations (15 % instead of 21 %).

5. Conclusions

We have used quantitative HAADF-STEM for studying GaAs nanowires with axial $GaAs_{1-x}Sb_x$ inserts. The work involved normalization of the experimental images from the basis of a detector scan and simulations with a multislice simulation software. The simulated intensities were in good agreement with the normalized experimental intensities for WZ GaAs, and the comparisons indicated that the Sb content in a 20 nm long ZB insert increases from approximately 13 ± 3 at.% to 21 ± 3 at.%. Non-compositional factors affecting the experimental and simulated intensities, such as strain and static atomic displacements, were discussed.

Acknowledgements

The authors would like to thank H Weman's group at the NTNU for the nanowire material, and K Müller and A Rosenauer at the University of Bremen for advice in HAADF image simulations and quantification.

References

[1] Yang P, Yan R and Fardy M 2010 *Nano Letters* **10** 1529–1536
[2] LeBeau J M and Stemmer S 2008 *Ultramicroscopy* **108** 1653–1658 ISSN 0304-3991
[3] Rosenauer A, Gries K, Mueller K, Pretorius A, Schowalter M, Avramescu A, Engl K and Lutgen S 2009 *Ultramicroscopy* **109** 1171–1182
[4] Dwyer C 2010 *Ultramicroscopy* **110** 195–198
[5] Schowalter M, Rosenauer A, Titantah J T and Lamoen D 2009 *Acta Crystallographica Section A* **65** 5–17
[6] Rosenauer A, Mehrtens T, Müller K, Gries K, Schowalter M, Satyam P V, Bley S, Tessarek C, Hommel D, Sebald K, Seyfried M, Gutowski J, Avramescu A, Engl K and Lutgen S 2011 *Ultramicroscopy* **111** 1316–1327
[7] Grillo V, Carlino E and Glas F 2008 *Physical Review B* **77**
[8] Reid J S 1983 *Acta Crystallographica Section A* **39** 1–13

Toward 3D structural information from quantitative electron exit wave analysis

Konstantin B. Borisenko,[1] Grigore Moldovan,[1] Angus I. Kirkland,[1] Amy Wang,[2] Dirk Van Dyck,[2] and Fu-Rong Chen[3]

[1]Department of Materials, University of Oxford, Oxford, OX1 3PH, UK; [2]Department of Physics, University of Antwerp, Antwerp, Belgium; [3]Department of Engineering and System Science, National Tsing Hua University, Hsin Chu, Taiwan

Corresponding authors: konstantin.borisenko@materials.ox.ac.uk (K.B. Borisenko), angus.kirkland@materials.ox.ac.uk (A. I. Kirkland)

Abstract. Simulations show that using a new direct imaging detector and accurate exit wave restoration algorithms allows nearly quantitative restoration of electron exit wave phase, which can be regarded as only qualitative for conventional indirect imaging cameras. This opens up a possibility of extracting accurate information on 3D atomic structure of the sample even from a single projection.

1. Introduction

Electron exit wave restoration in High Resolution Transmission Electron Microscopy (HRTEM) is a well established technique for quantitative investigation of structure at the atomic level. It requires a series of images recorded at different defocus levels, or a combined defocus series / tilt series, and is used widely to improve image interpretation and to increase resolution [1,2]. Recent developments include quantitative analysis of electron exit waves to obtain three-dimensional information about the sample at the atomic level [3]. Although images obtained in HRTEM are two-dimensional projections, the electron exit wave reconstructed from these images does contain encoded three-dimensional information about the sample due to the specimen thickness, relative position of the atomic layers and multiple scattering events as the electrons pass through the material.

The theoretical work presented here examines if it is possible to extract three-dimensional information about a crystalline Pt model, including the position of a single Co substituent atom, using realistic HRTEM image simulations and accurate electron exit wave restoration techniques. The influence of electron optics and detector parameters on the restoration is also studied and the optimal conditions for a successful experiment are discussed.

2. Experimental

Two cubic models of crystalline (001) Pt with 1.57 nm edge with a substituent Co atom at the top or bottom of the central column in the cell were prepared, as shown in Figure 1. The electron exit wave was simulated at 200 kV using a multislice method and weak phase-object approximation as implemented in E. Kirkland [4]. The simulated waves were used for subsequent simulations of a series of defocused images (as shown in Fig. 2). Typical microscope parameters for aberration corrected TEM were used, including a 2×10^{-5} electron beam energy spread, $\Delta E/E$, and a 0.001 mm spherical aberration coefficient, *C*s. The sampling interval was 0.0123 nm/pixel for 128 x 128 pixels array. Two different chromatic aberration coefficients were tested in the calculations, *C*c = 0.5 and *C*c = 0.1 mm. The chromatic aberration coefficient has an effect of reducing the information limit of the microscope as an incoherence envelope function.

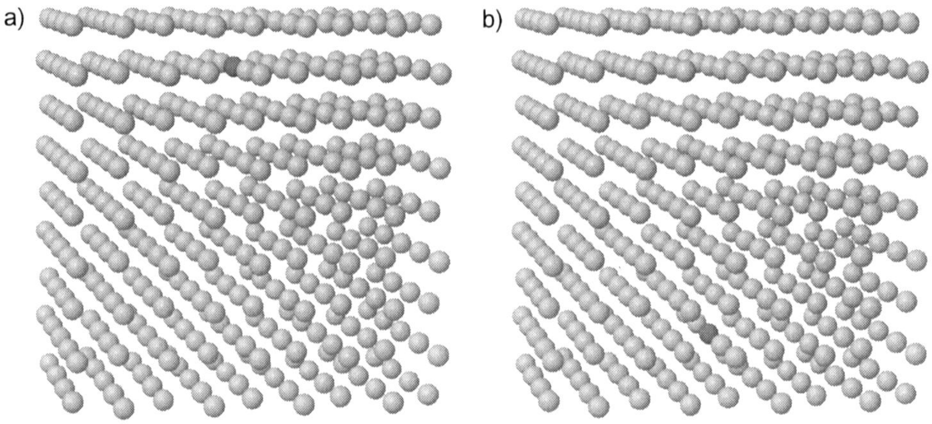

Figure 1. Models of Pt cell with a substituent Co atom in high (a) and low (b) positions.

Accurate simulations of detector contributions to the simulated images were performed to account for Poisson noise, Detective Quantum Efficiency (DQE), Modulation Transfer Function (MTF) and any further noise introduced by the detector. These calculations included the contribution of each individual electron to the final image, where the input position into the detector corresponded to a probability function defined by the original intensity, which also provides the Poisson contribution to noise. The output pixel signals for each of these electrons were calculated using Monte Carlo simulations of their interaction with the sensor, following the standard approach with added relativistic corrections and generation of secondary electrons [5]. Two detector models for TEM were tested, an indirect scintillator-coupled charge-coupled device (CCD) camera [6] and a direct monolithic active pixel sensor (MAPS) [7]. CCD configuration is typical of conventional cameras. MAPS illustrate the much improved MTF and DQE of thinned direct detectors. Complete detector parameters, including detector dark noise, readout noise and dynamic range limitations, were set at typical values for such commercial cameras. To reflect typical experimental conditions, an average electron dose of 100 electrons/pixel and an integration time of 1 sec were used for both detector models. Examples of simulated defocused images, including the detector contribution, are presented in Figure 2.

The exit wave restorations were performed utilising two algorithms, a modified Fienup input-output algorithm [8] and an unconstrained trust region optimization algorithm [9]. Refinements in the wave restoration routine were carried out until the change in the sum of squared deviations between the input and the output images in two consecutive refinement steps was less than 0.1.

3. Results and discussion

The simulated electron exit wave phase shift shows clear differences between the two models, as illustrated by the line scan through the centre of exit waves in Figure 3a. There is also a considerable difference in the phase shift for the column containing the Co atom relative to the similar columns without the substituent atom (Figure 3a), as well as for the columns containing different number of Pt atoms.

As the restored exit wave phase can only be obtained up to the information limit of the microscope, the information limit plays an important role in determining the differences in phase shifts for the two tested models. It is found that a *C*c of 0.5 mm reduces these differences considerably (Figure 3c), and therefore only results with *C*c=0.1 are used for further discussion.

Figure 2. Simulated images in a focal series of a PtCo model, with defocus (overfocus) ranging from 0 to 5 nm with step of 1 nm from left to right (*C*c = 0.1 mm). Image intensity before detection (a); image intensity with indirect CCD (b) and direct MAPS (c) detector calculations. The substituent Co atom is located in the atomic column in the centre of the images.

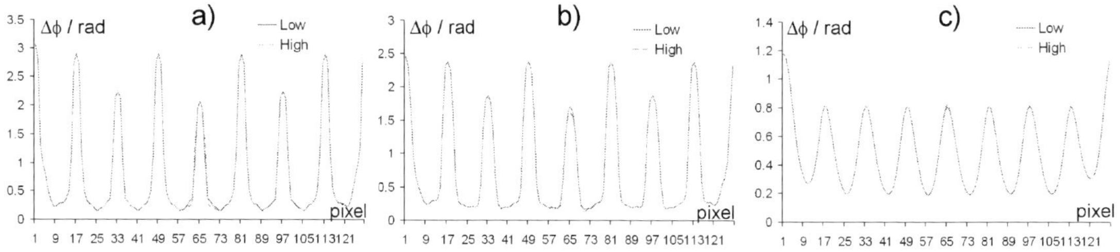

Figure 3. Line scans of phase shifts in the simulated exit waves for models with low and high Co atom positions. Original exit wave (a); information transfer-limited exit wave with *C*c=0.1 mm (b) and *C*c=0.5 mm (c), showing the influence of the information limit.

Figure 4 presents a comparison between restorations using Fienup and trust region optimization algorithms, employing full derivatives of the exit wave and the performance of CCD and MAPS detectors. By comparison with the original exit wave in Figure 3b, it is found that the Fienup algorithm (Figures 4a,b), although fast, results in agreement only for the positions of peaks. The relative peak heights are not reproduced. This may be due to long periods of stagnation of this algorithm, which leads to early termination of the refinements. The trust region optimization algorithm with CCD detector shows a better performance, which however remains only qualitative (Figure 4c). The MAPS detector and the trust region optimization algorithm result in a considerably better agreement with the original exit wave phase shifts. For this case all columns of atoms show correct relative phase shifts and their absolute values are close to the original exit wave.

It is apparent, however, that all of the methods within the tested parameters failed to distinguish between different positions of Co substituent in the sample, because the difference in position results in a relatively small phase shift difference, beyond the detection limit. It is anticipated that such detection will become possible with parameters or techniques that allow higher information transfer.

4. Conclusions

It is found that direct detectors will allow quantitative restoration of electron exit waves and extraction of accurate 3D structural information. This shows that realistic modelling of the detector contribution to recorded scattering intensities is essential in assessing the applicability of novel approaches for electron exit wave analysis and for informing further development of such techniques.

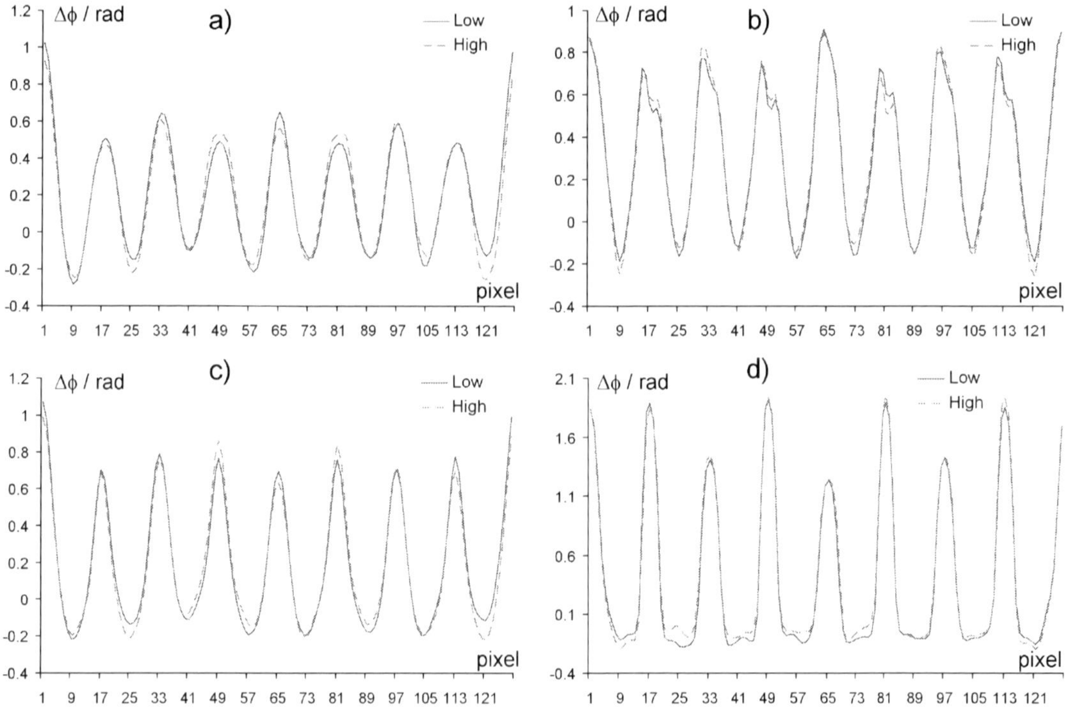

Figure 4. Line scans of phase shifts in restored exit waves (*Cc*=0.1 mm) for low and high Co atom position, including detector contribution for CCD camera (a,c) and MAPS detector (b,d) models. Restorations were performed using Fienup (a,b) and trust region optimization (c,d) algorithms.

For the sample model considered here, the use of a direct imaging camera together with an accurate phase restoration algorithm will allow close to quantitative determination of phase shifts in differently substituted columns. However, it was not possible to distinguish the position of a single Co substituent atom within a column from the restored phase shifts in a single projection condition. This task, nevertheless, seems amenable for the focal series/tilt series technique which allows higher information transfer for increased sensitivity in the direction of projection. It is envisaged that accuracy of this method will improve with further improvement in electron microscope lenses and imaging detectors.

References

[1] H. W. Zandbergen and D. Van Dyck, Microscopy Research and Technique 49 (2000) 301.
[2] A. I. Kirkland and R. R. Meyer, Microsc. Microanal. 10 (2004) 401.
[3] A. Wang, F. R. Chen, S. Van Aert, and D. Van Dyck, Ultramicroscopy 110 (2010) 527.
[4] E. J. Kirkland, *Advanced Computing in Electron Microscopy*. New York: Plenum Press, 1998.
[5] D. C. Joy, *Monte Carlo Modelling for Electron Microscopy and Microanalysis*. Oxford University Press, 1995.
[6] R. R. Meyer and A. I. Kirkland, Microscopy Research and Technique 49 (2000) 269.
[7] G. Moldovan and A. I. Kirkland, J. Phys.: Conf. Ser. 241 (2010) 012009.
[8] J. R. Fienup, Applied Optics 21 (1982) 2758.
[9] M. Celis, J. E. Dennis and R. A. Tapia "A trust region strategy for nonlinear equality constrained optimization", in *Numerical Optimization 1984* (P. Boggs, R. Byrd and R. Schnabel, eds.), Philadelphia: SIAM, 1985, pp. 71–82.

Journal of Physics: Conference Series **371** (2012) 012058 doi:10.1088/1742-6596/371/1/012058

Reduced density function analysis of titanium dioxide doped tantalum pentoxide

K Evans[1], **R Bassiri**[1,3], **I Maclaren**[1], **S Rowan**[1], **I Martin**[1], **J Hough**[1] and **K B Borisenko**[2]

[1] SUPA, School of Physics and Astronomy, Glasgow University, Kelvin building, Glasgow, G12 8QQ, UK
[2] Department of Materials, University of Oxford, Oxford, OX1 3PH, UK.
[3] Ginzton, E.L. Laboratory, Stanford University, Stanford, California 94605, USA

E-mail: k.evans@physics.gla.ac.uk

Abstract. Future advanced gravitational wave detectors will need to be constructed using ultra low loss materials to achieve the desired sensitivity levels. Coating thermal noise which is related to mechanical loss has been determined to be a limiting factor for the sensitivity of these detectors and must be reduced. To achieve this goal, work is on going to identify the causes of coating mechanical loss. Recent experiments suggest this is micro-structural in origin, thus determining the structural features of coatings via appropriate modelling is of interest. It will be shown that through Transmission Electron Microscopy and reduced density function analysis, the local structural changes due to doping of coating materials can be identified. Furthermore by using this data as an empirical constraint for combined reverse Monte Carlo and Density Functional Theory simulations the trends seen in the data can be replicated in the modelling.

1. Introduction

Einstein's general theory of relativity (GR) predicts the existence of wave-like distortions in space-time responsible for the transmission of gravitational information, which propagate at the speed of light [1]. These Gravitational Waves (GWs), when detectable will provide a new and unique way of studying some of the most exotic and violent objects in our Universe.

Current GW detectors are based on Michelson-type laser interferometers. The principle of laser interferometer GW detectors is that a GW would slightly change the arm length of the interferometer, which would alter the interference pattern created by the recombination of the two beams [2]. Due to the extremely small expected effects of GW signals from astrophysical sources, detectors must be carefully constructed to minimise any noise sources affecting their detection capabilities [3]. For the next generation of GW detectors, research suggests that thermal noise present in the mirror coatings will be a limiting factor within the most sensitive operational frequency range [4].

Two forms of thermal noise which are of concern for GW detectors are Brownian thermal noise [4] and thermoelastic noise [5]. Brownian thermal noise arises due to random thermal motion of atoms, and thermoelastic noise results from the statistical fluctuations in temperature of the materials coupling to displacements of the material surface via coupling through the thermal expansion coefficient. Although thermal noise can be difficult to measure directly at the frequencies of interest,

its effects can be determined through the measurement of mechanical loss, to which it is directly related [4].
The current material of choice for the high refractive index layers in multilayer mirror coatings is tantala (Ta_2O_5) due to its optical properties and relatively low loss at room temperature. The optical properties of tantala can be improved by heat-treating the coating up to 600°C [6] without unduly affecting the loss at room temperature [7]. However, heat-treatment has also been shown to increase the loss at low temperature [7], which is of interest for possible future cryogenic detectors [8]. Furthermore, it has previously been shown that the room temperature loss of tantala can be reduced by doping with titania (TiO_2) and that this effect peaks at ~30 cation % Ti [7]. It is not currently understood why titania reduces loss at room temperature nor why heat-treatment increases it at low temperature. Through work on silica, which shares similar low temperature loss properties with tantala, it has been postulated that observed loss peaks at low temperature may be due to specific structural features [9,10]. It is currently believed that the variations in loss seen in tantala due to heat-treatment and doping are caused by associated structural changes in the material micro-structure[9,10]. In order to determine what structural changes or features are leading to loss, we have studied the structural variations in a range of TiO_2 doped Ta_2O_5 coatings with Ti contents from 0 – 50 cation % Ti using a combination of electron diffraction measurements and atomistic modelling.

2. Method

The coatings studied in this work were deposited using dual ion beam assisted sputtering by LMA (Lyon, France) and were deposited onto 1 inch diameter by ¼ inch thick fused silica substrates held at a temperature of 90°C. Coated substrates were then heat-treated in air at 600°C for 24 hours to relax internal stresses and to allow the structure to equilibrate [11].

Specimens were prepared for Transmission Electron Microscopy (TEM) using a standard cross sectional method based on mechanical thinning, dimpling and ion beam thinning. The primary experimental constraint for the modelling process was electron diffraction data, which was recorded on a FEI Tecnai T20 TEM operated at 200 keV with a beam convergence semi-angle (α) of 1.89 mrad. To ensure accurate calibration of the TEM diffraction data, diffraction patterns were recorded from a single crystal silicon specimen under exactly identical conditions just prior to recording the electron diffraction data from the glasses. The typical experimental discrepancy in the lattice parameters is below 1%, and given an accurate calibration, the value dq is also calculated. This value dq is a measure of the q resolution of the CCD under the experimental conditions (where q, the diffraction vector, is defined as $q = 4\pi\sin\theta/\lambda$) and is typically of the order of 0.2 $Å^{-1}pixel^{-1}$.

The diffraction data is then used to calculate the Reduced Density Function (RDF), which is a statistical representation of where the nearest neighbours (NNs) lie in real space [12]. This is done by radially averaging the diffraction pattern, removing the intensity from individually scattering atoms and Fourier transforming the result. RDFs can be directly interpretable; quantities such as peak height, width and position provide information about the NN relationships in the structure.

The modelling process consists of using a Reverse Monte Carlo (RMC) simulation to create a structure which matches the experimental RDF [13]; Electron Energy Loss Spectroscopy (EELS) and X-ray reflectometry are used to constrain the stoichiometry [13] and density [14], respectively. Density Functional Theory is then used to simulate a liquid quench procedure, in order to relax the structure to an energetically lower state [13] using the simulation software CASTEP [15]. Studies of the energy evolution with time have allowed for a reduction in simulation time compared to previous work [13] by reducing the equilibration time from 5 to 2ps per temperature step; the one possible cost of reducing the equilibration time in this way is to reduce the probability of slower processes such as tunnelling occurring. The model is then refined further using a second RMC step on a larger model, constructed from the DFT model, to ensure both an energetically stable configuration and an acceptable fit to the experimental RDF [13].

3. Results and Discussion

It is known from previous work on tantala that the first two peaks in the RDF correspond to the Ta-O and Ta-Ta NNs [13]. This is assumed to be similar for titania doped tantala, since Ti-O bonds have similar lengths to Ta-O bonds in published oxide crystal structures [16]. Thus, the first peak will also contain Ti-O bonds, and the second will also contain Ti-X NN information, where X can be Ta or Ti. Clear trends are observed as a result of the increase in Ti content. Firstly, the X-O peak increases in height, becomes narrower, and moves to slightly larger distances (figure 1), suggesting that that the addition of Ti increases the ordering in the first coordination shell, as well as slightly increasing bond lengths. Secondly, the X-X peak develops a prominent shoulder indicating an increase in abundance of a NN distance slightly shorter than the Ta...O distance responsible for the 2nd peak in pure Ta_2O_5

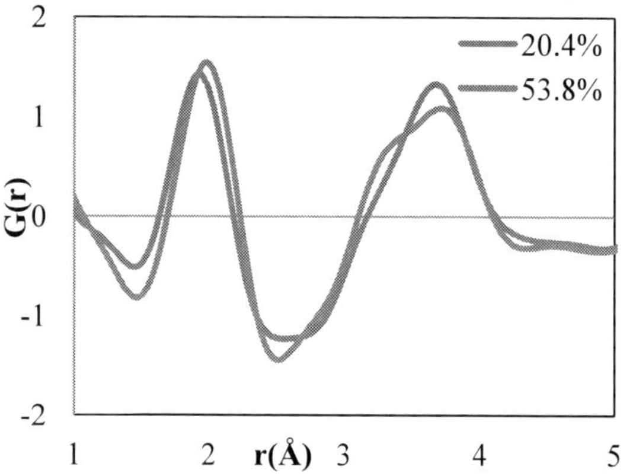

Figure 1: Experimental(?) RDFs of 20.4 cat. % Ti and 53.8 cat. % Ti TiO_2 doped Ta_2O_5 showing the appearance of a "shoulder" in the 2nd peak at high TiO_2 percentage

Preliminary modelling suggests that titanium preferentially bonds to the oxygen and tantalum, rather than forming separate Ti-rich clusters in the structure. This is seen in Figure 2 where the abundances of Ti-O and Ti-Ta are greater than Ti-Ti. This suggests that the titanium atoms are substituting for tantalum atoms in the structure and not changing it fundamentally. This is illustrated in Figure 3, which shows the typical Ta_2O_2 planar fragment, which is one of the key building blocks of amorphous Ta_2O_5 (see supplementary material to ref [13], http://dx.doi.org/10.1063/1.3535982), and its modification as a result of Ti addition to hybrid $TaTiO_2$ and pure Ti_2O_2 rings.

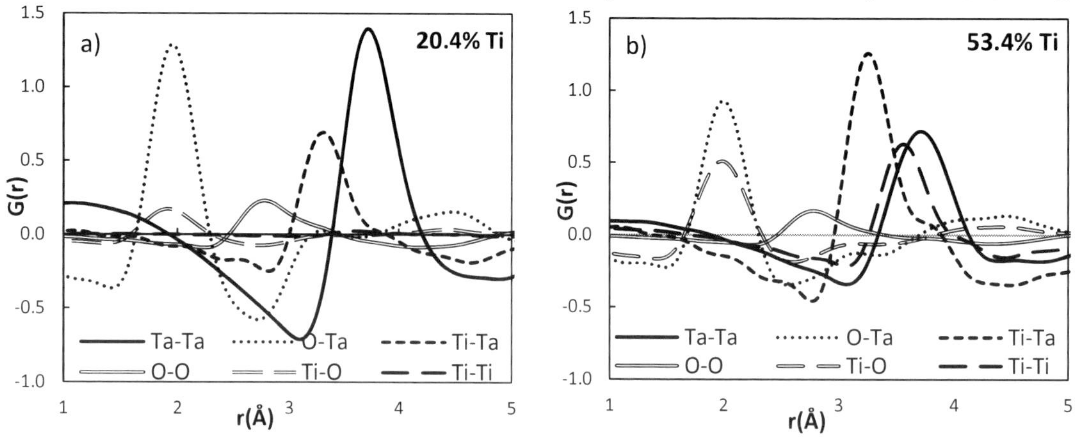

Figure 2: Partial RDFs for (a) a 20.4% TiO_2 doped sample and (b) a 53.8% TiO_2 doped sample

Analysis of the partial RDFs resulting from the fitted(?) models reveals that the noticeable shoulder on the 2nd peak in Figure 1 for the 53.8 cat. % Ti sample arises from the high abundance of Ti-Ta distances at this doping level. Thus, the hybrid $TiTaO_2$ ring structure is far more abundant than either of the other ring structures as is evident from the large Ti-Ta peak height in Figure 2b, and is probably preferred at this composition. This distance is also shorter than Ta-Ta or Ti-Ti distances, as can also be seen from Figure 2b, which could indicate some form of interactions between Ti and Ta in these planar fragments. It should also be noted that the pure Ti_2O_2 ring structure does not appear to have a Ti-Ti link unlike the other two variations. This happens since the O-Ti-O typical bond angle is more acute than the O-Ta-O bond and as such the Ti atoms are further apart and outside the normal NN criteria used by ourselves or others of the sum of covalent radii plus a small percentage (e.g. 10%).

Figure 3: Three variations of the Ta_2O_5 crystalline building block found in TiO_2 doped Ta_2O_5: a) The pure Ta-Ta case b) the hybrid Ti-Ta case and c) pure Ti-Ti case. The metal atoms are at the top and bottom of each structure, and the oxygen atoms at the side.

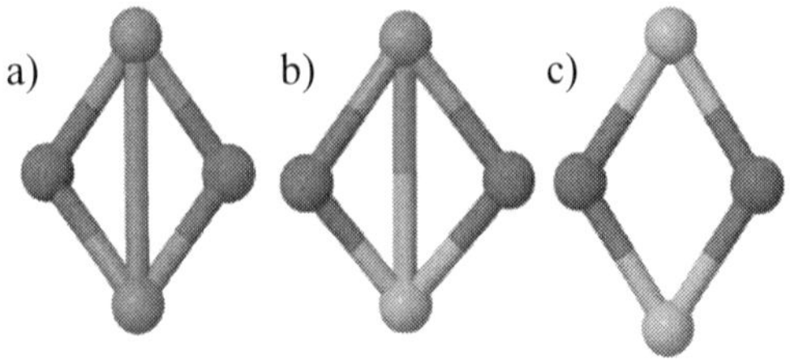

4. Conclusions

In conclusion, diffraction data from transmission electron microscopy combined with reduced density function analysis has been able to detect a significant structural change due to doping in an amorphous mixed tantalum titanium oxide. This detectable change allows the development of unique structural atomic models for each dopant percentage. Initial structural models for samples containing about 20 and 50 cation % titanium show a strong preference towards Ti-Ta bonding over Ta-Ta or Ti-Ti, suggesting that titanium substitutes for tantalum in the structure, modifying it subtly as it does so. Understanding these structural changes in detail is the first step to understanding the processes which causes mechanical loss and thermal noise, ultimately allowing the design of lower loss materials.

Acknowledgements

The Science and Technology Facilities Council is gratefully acknowledged for the provision of a studentship to KE, and for funding of the Institute for Gravitational Research and of Scotgrid.

References

[1] Einstein A, 1916 *Sitzungsber. K. Preuss. Akad. Wiss., Part 1* 688
[2] Saulson P 1994 *Cosmology and Particle Physics* **427** 233
[3] Martin I 2008 *Title,* Ph.D. Thesis (University of Glasgow, UK)
[4] Saulson P R 1990 *Phys. Rev. D* **42** 2437
[5] Fejer M M, Rowan S, Cagnoli G, Crooks D R M, Gretarsson A, Harry G M, Hough J, Penn S and Sneddon P H and Vyatchanin S P 2004 *Phys. Rev. D* **70**, 082003
[6] Netterfield R P *et al.* 2005 *Proc. SPIE* 5870 58700H
[7] Harry G *et al.* 2007 *Class. Quantum Grav.* **24** 405–415
[8] Einstein Gravitational Wave Telescope Conceptual Design Study 2011 https://tds.ego-gw.it/ql/?c=7954
[9] Levin Y 1998 *Phys. Rev. D* **57** 659–63
[10] Harry G M, Gretarsson A M, Saulson P R, Kittelberger S E, Penn S D, Startin W J, Rowan S, Fejer M M, Crooks D R M, Cagnoli G and Hough J and Nakagawa N 2002 *Class. Quantum Grav.* **19** 87–917
[11] Mackowski J M, Morgado N and Pinard L and Remillieux A 2004 *Proc. SPIE: 5250 (Bellingham, WA: SPIE)*
[12] Cockayne D J H 2007 *Annu. Rev. Mater. Res.* **37** 159–87
[13] Bassiri R, Borisenko K B, Cockayne D J H, Hough J and Maclaren I and Rowan S 2001 *Appl. Phys. Lett.* **98** 031904
[14] Parratt L G 1954 *Phys. Rev.* **95** 359.
[15] Clark S J, Segall M D, Pickard C J, Hasnip P J, Probert M J, Refson K and Payne M C 2005 *Z. Kristallogr.* **220** 567
[16] Zhang R Z, Wang C L, Li J C, Zhang J L, Yang K and Zhang C and Chen H M 2007 *Appl. Phys. Lett.* **91** 092909

Electron Microscopy and Analysis Group Conference 2011 (EMAG 2011) IOP Publishing
Journal of Physics: Conference Series **371** (2012) 012059 doi:10.1088/1742-6596/371/1/012059

Ab-initio modelling, polarity and energetics of clean rutile surfaces in vacuum and comparison with water environment

T.P. Hardcastle[1], R.M.D. Brydson[1], K.J.T. Livi[2], C.R. Seabourne[1] and A.J. Scott[1]

1. Inst. for Materials Research, SPEME, University of Leeds, Leeds LS2 9JT, UK

2. HRAEM/IIC Facility, Johns Hopkins University, Baltimore, MD 21218 USA

pmth@leeds.ac.uk

Abstract. All terminations of the (1x1) rutile (110), (101), (001), (100) and (111) surfaces are classified according to their electrostatic polarity. Six are found to be non-polar. The plane-wave density functional theory code CASTEP is used with a GGA-PBE exchange-correlation functional and a vacuum/material slab supercell method to calculate the surface energy density of symmetric thin rutile films with the six non-polar terminations in vacuum. The ratio of the surface energy densities of a rutile crystal with {111} and {110} facets in water is deduced using Lagrange multipliers and found to be consistent with the DFT vacuum results.

1. Introduction

The currently active research area of metal oxide surfaces has the curious and unusual attribute of producing fertile developments on a fundamental theoretical level by means of application of physics developed 150 years ago. Maxwell's theory of classical electrostatics was applied by Tasker in 1979 [1] to prototype model oxide surfaces providing a simple recipe for their classification on the basis of energetic stability. Subsequent developments regarding the polarity of oxide surfaces were elegantly summarised with classical arguments by Noguera in 2000 [2] and Goniakowski et al. in 2008 [3]. Alongside this, a modern quantum theory of dielectric polarisation emerged in the 1990's [4, 5, 6] in which polarisation is interpreted as a displacement of the centres of charge of Wannier functions. It is suggested [3] that the success of a classical approach based on point charges for insulating materials with highly localised orbitals is because Wannier functions can be mapped to a lattice of effective point charges as shown by Marzari and Vanderbilt in 1997 [7]. The merit of this suggestion is not investigated here, but the general success [8] of Tasker's system makes for convincing testimony.

In Goniakowski's 2008 paper, Maxwell's microscopic equations are assumed to apply at the surface of a uniform semi-infinite isotropic dielectric material. It is shown that such a material existing in the ground state cannot sustain a macroscopic polarisation in the absence of an external electric field:

$$\langle \bar{P}_x \rangle = {\mu_B(x_s)}/{aS} - \sigma_s(x_s) = 0, \quad (1)$$

where $\langle \bar{P}_x \rangle$ is the laterally-averaged polarisation density of the surface cell in the surface-normal direction x, μ_B is the x-component of the electric dipole moment of the bulk cell, σ_s is the total charge in the surface cell divided by its surface area S, a is the surface-normal lattice parameter of the bulk cell and x_s is the x coordinate of a Gibbs dividing surface (on which the bulk and surface cell definitions depend) chosen to denote the location of the boundary shared by the surface cell and the bulk cell directly below it. Equation (1) is useful since it can be interpreted as a stability condition. Its violation implies non-zero polarisation and consequent instability of the termination in question. It is also independent of the chemical identity of the charge distribution, allowing for the stabilising (or otherwise) effect of charged adsorbates placed in the surface cell to be predicted. Furthermore, it is just a more general way of expressing Tasker's well-known surface instability condition, $\mu_S \neq 0$, where μ_S is the surface cell dipole. By defining the bulk and surface cells identically using $\mu_B \equiv \mu_S$, we obtain $\sigma_S = 0$ on account of bulk cell neutrality and hence recover $\mu_S = 0$ for stability and $\mu_S \neq 0$ for instability as expected. Tasker's classification does not depend on exact ionic charge allocations and applies *prior to cleaving* (Indeed, it can be used to *predict* surface reconstructions required to stabilise polar surfaces.) so a robust, canonical indication of stability can be obtained using frozen-bulk termination atomic positions and formal ionic point charge values in equation (1) by merely identifying $\langle \bar{P}_x \rangle \neq 0$ as the necessary and sufficient condition for surface termination instability.

For Miller indices $h, k, l = 0, 1$ there are 14 unique frozen bulk terminations of the rutile structure, having eliminated equivalent terminations by virtue of space group symmetries. Summing over atomic species labels j and same-species atom labels i, with $\mu_B(x_s) = \sum_{bulk\ cell} q_i^j (x_i^j - x)$ and $\sigma_s(x_s) = \frac{1}{S}\sum_{surface\ cell} q_i^j$ with formal charges $q_i^{Ti} = +4$, $q_i^{O} = -2$ and using bulk atomic positions, equation (1) can be used to classify these 14 terminations by polarity as shown in Table 1.

Surface plane	Termination	$\langle \bar{P}_x \rangle$ / eÅ^{-1}	Surface plane	Termination	$\langle \bar{P}_x \rangle$ / eÅ^{-1}
(111)	TiO_2/O/Ti/O/...	0	(101)	O_2/Ti_2/O_2/...	0
	O/Ti/O/TiO_2/...	0		Ti_2/O_2/O_2/...	- 0.16
	Ti/O/TiO_2/O/...	-0.07		O_2/O_2/Ti_2/...	0.16
	O/TiO_2/O/Ti/...	0.07	(100)	O/Ti/O/...	0
(110)	O/Ti_2O_2/O/...	0		Ti/O/O/...	- 0.15
	Ti_2O_2/O/O/...	-0.10		O/O/Ti/...	0.15
	O/O/Ti_2O_2/...	0.10	(001)	TiO_2/ ...	0

Table 1. *The formal polarity of low-index rutile frozen bulk terminations in vacuum. The termination columns list the repeat sequence of atomic planes lying parallel to the surface, where the leftmost entry denotes the surface-terminating plane. 6 terminations have exactly zero (by inversion symmetry) polarisation. The remaining 8 are polar. All arithmetic used bulk atomic positions obtained via the ICSD. Precision errors were negligible.* $\langle \bar{P}_x \rangle \neq 0$ *values are shown to 2 significant figures.*

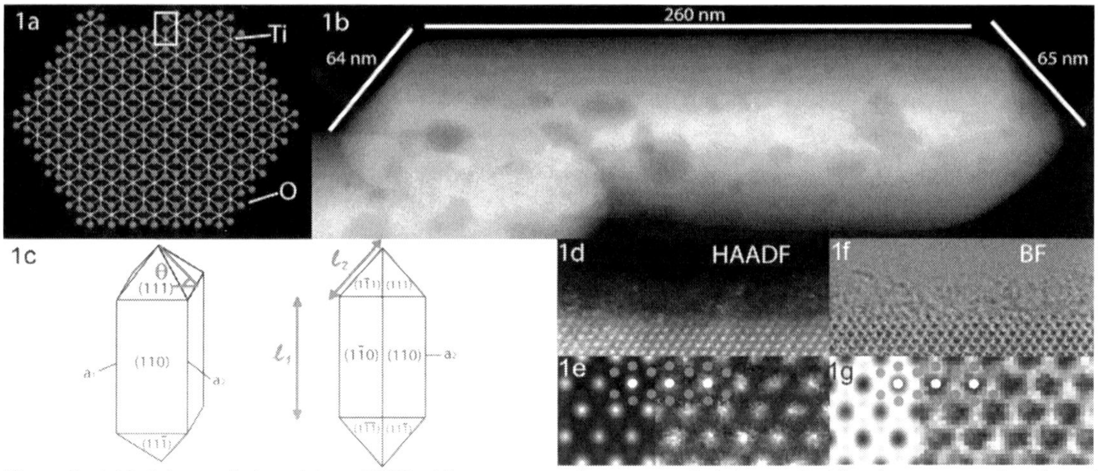

Figure 1. *a) Model crystal viewed down [100] with steps at top. Steps currently being studied for abstract quoted at the end of this caption. Unit cell in white. b) A double-terminated crystal viewed down [100]. The length ratio of the {110} and {111} edges is 4:1. c) Crystal habit with edge lengths l_1, l_2 and θ =(90 - 47.67)° = 42.33°. d) Aberration-corrected HAADF image of {110} side viewed down [100] e) Simulated HAADF image with Ti atoms highlighted in white and O atoms in red. f) BF of {110} side viewed down [100] direction g) Simulated BF image with atoms highlighted as in 1e. Intensity at the edge in 1d and 1f is organic matter.* ***Images taken and adapted from abstract entitled "Imaging the surface of Rutile by STEM and its implication for organic molecule bonding" for MC2011 Conference Kiel, Germany, with permission of K. J. T. Livi.***

The vacuum polarity classification in Table 1 filters out polar surfaces, whose surface energies diverge [1], leaving the 6 non-polar surfaces for which convergent vacuum surface energy calculations can be carried out using the periodic supercell method, and comparisons drawn with crystals in non-vacuum environments. One current area of research ripe for such comparison is that of rutile surfaces in an aqueous environment with organic adsorbates. Livi et al. used aberration-corrected STEM HAADF and BF imaging at SuperSTEM in Daresbury, UK, to study rutile crystals which crystallised with a square prism and double-terminated square pyramid habit with {110} and {111} facets, different from the well-known vacuum equilibrium habit [12], exposed to water and organic matter as shown in Fig. 1. Using an approach like that of Table 1 for the real surface in the presence of a non-trivial chemical environment would require a much more realistic approach to surface chemistry. The adsorption modes of the organic material can not be determined from the images owing to its amorphous appearence as seen in Fig. 1d and 1f. However, the surface energy density *ratio* of the aqueous-exposed crystal facets, $\frac{\gamma_{111(aqueous)}}{\gamma_{110(aqueous)}}$, can be inferred from the images, without specific knowledge of atomic-scale surface phenomena, by appealing to the Gibbs-Wulff theorem which states that a macroscopic crystal in equilibrium exists in a state of minimum total surface energy. An inverse Wulff construction can not be performed directly on Fig. 1b because of the [100] viewing direction; the crystal facets are not lying perpendicular to the plane of the page. Instead, elementary geometry can be applied to express the total crystal surface energy E_s and volume V in terms of the STEM-observed edge lengths l_1 and l_2, and the (fixed) angle θ as per Fig. 1c, to give the expressions $E_s(l_1, l_2, \theta) = 8(1 + sec^2\theta)^{-1/2}(l_1 l_2 \gamma_{110(aqueous)} + l_2^2 \gamma_{111(aqueous)} sec\theta)$ and $V(l_1, l_2, \theta) = 4(1 + sec^2\theta)^{-1}\left(l_1 l_2^2 + \frac{2}{3}(1 + sec^2\theta)^{-1/2} tan\theta\, l_2^3\right)$. Then, E_s can then be minimised with respect to l_1 and l_2 whilst imposing the constraint $V > 0$ by using the method of Lagrange multipliers, to relate the quantity $\frac{\gamma_{111(aqueous)}}{\gamma_{110(aqueous)}}$ to the observables l_1 and l_2. This method is just an algebraically explicit inverse Wulff construction and is entirely equivalent. The result is equation (2a). The $l_1 = 4l_2$, $\theta = 42.33°$ case, pertaining to the crystal in Fig. 1b, is evaluated to 3 significant figures in equation (2b).

$$\frac{\gamma_{111(aqueous)}}{\gamma_{110(aqueous)}} = \frac{l_1}{l_2}\left(\frac{\sqrt{1 + sec^2\theta}}{2sec\theta}\right) + \frac{tan\,\theta}{sec\theta} \quad (2a) \qquad \left.\frac{\gamma_{111(aqueous)}}{\gamma_{110(aqueous)}}\right|_{\substack{l_1 = 4l_2 \\ \theta = 42.33°}} = 3.16 \quad (2b)$$

2. Method

The density functional theory code CASTEP [9] was used to calculate the geometry-optimised ground state enthalpy of symmetric (same termination on each side) thin rutile films of increasing thickness in vacuum with the 6 non-polar terminations of surfaces (110), (101), (001), (100) and (111). Supercells containing a slab of rutile material and a vacuum slab were used along with the GGA-PBE exchange-correlation functional picked on the basis of a bulk lattice parameter validation test. The vacuum thickness c_{vac}, kinetic energy cut off E_k and k-point spacings $s^{i=1,2,3}$, where i denotes correspondence to the reciprocal space axis b_i, were set at $c_{vac} = 10Å$, $E_k = 400\ eV$ and $s^{i=1,2,3} < 0.055\ Å^{-1}$ for $i = 1, 2, 3$, based on numerical convergence tests. Slabs were constructed by stacking single unit cells with a starting thickness of 1 Ti_2O_4 unit cell for the (110), (101), (001), (100) films and 1.5 Ti_2O_4 unit cells for the two (111) films. Distinct calculations for each slab were performed up to a thickness of ~ 20 Å. All atoms in the slabs were allowed to relax, with the exception of atoms located further than 8 Å in the surface normal direction from both surfaces for slabs of total thickness > 16 Å. The positions of these atoms were fixed to reduce computational cost. The surface energy density γ in each case was calculated using the ground state enthalpy output values $E_{Ti_nO_{2n}}$ of the slab/vacuum supercells of formula Ti_nO_{2n} and total surface area $2S$, and a Ti_2O_4 bulk cell enthalpy:

$$\gamma = {}^{1}/_{2S}\left(E_{Ti_nO_{2n}(supercell)} - {}^{n}/_{2}\, E_{Ti_2O_4(bulk)}\right) \quad (3)$$

3. Results and Discussion

The CASTEP output is shown in Fig. 2. The plots were extrapolated with a 4-point weighted mean to give the γ values in Table 2. No absolute measurements of termination-specific rutile surface energies yet exist, but these results are consistent with previous ab-initio studies of rutile surfaces in vacuum [10, 11, 12]. The (111)(O) slabs underwent relaxations with a non-trivial relationship with thickness. Slab energy convergence behaviour appears to have been investigated only for the (110) surface [13].

Non-polar termination type	(110)	(100)	(101)	(001)	(111)(TiO_2)	(111)(O)
$\gamma_{(termination\ type)(vac)}$ / Jm^{-2}	0.42 ± 0.08	0.62 ± 0.08	0.98 ± 0.08	1.15 ± 0.08	1.33 ± 0.08	1.35 ± 0.08

Table 2. *CASTEP results for γ in vacuum to 2 significant figures. Precision errors from basis set and extrapolation errors.*

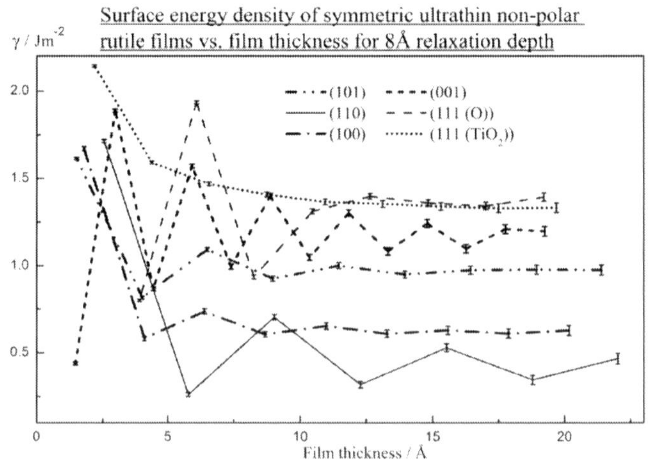

Figure 2. *Symmetric non-polar ultrathin film surface behaviour converging to that of semi-infinite surfaces in vacuum with increasing thickness. Termination type indicated in brackets for (111) plots.*

The importance of surface chemistry on crystal growth has been demonstrated elsewhere [14]. The CASTEP results obtained in this paper for clean surfaces in vacuum should be regarded as a basis for comparison with non-vacuum environments. The crystalline rutile studied by Livi et al., the example studied here, in the water/organic environment shown in Fig. 1b shows a length ratio of 4:1 for the water/organic-exposed {110} and {111} facet edge lengths l_1 and l_2, corresponding to a surface energy density ratio $\frac{\gamma_{111(aqueous)}}{\gamma_{110(aqueous)}} = 3.16$ as shown in equation (2b). In order to gain some insight into the significance of the aqueous/organic environment on the relative surface energy densities of the (111) and (110) surfaces this ratio can be compared to the vacuum case by using the CASTEP values from Table 2 to evaluate the two ratios $\frac{\gamma_{(111\,(TiO_2))(vac)}}{\gamma_{(110)(vac)}}$ and $\frac{\gamma_{(111\,(O))(vac)}}{\gamma_{(110)(vac)}}$. These values, calculated to 3 significant figures, are shown in equations (4a) and (4b). The precision errors arise primarily from the extrapolation method used.

$$\frac{\gamma_{(111\,(TiO_2))(vac)}}{\gamma_{(110)(vac)}} = 3.17 \pm 0.63 \quad (4a) \qquad\qquad \frac{\gamma_{(111\,(O))(vac)}}{\gamma_{(110)(vac)}} = 3.21 \pm 0.64 \quad (4b)$$

It can be seen that $\frac{\gamma_{(111\,(TiO_2))(vac)}}{\gamma_{(110)(vac)}} \approx \frac{\gamma_{(111\,(O))(vac)}}{\gamma_{(110)(vac)}} \approx \frac{\gamma_{111(aqueous)}}{\gamma_{110(aqueous)}}$, suggesting that *the aqueous environment has no significant effect on the relative surface energy densities of the (110) and (111) surfaces*. This result predicts nothing about absolute changes in the (110) and (111) surface energies.

4. Conclusions

The surface energy densities of the rutile (111)(O) and (111)(TiO_2) surfaces in vacuum were predicted for the first time at 1.35 ± 0.08 Jm^{-2} and 1.33 ± 0.08 Jm^{-2} respectively. The surface energy density ratio of (110) and (111) rutile surfaces in vacuum does not differ significantly in an aqueous environment.

5. References

[1] Tasker P 1979 *Journal of Physics C Solid State Physics* **12** 4977
[2] Noguera C *2000 Journal of Physics: Condensed Matter* **12** R367
[3] Goniakowski J et al. 2008 *Reports on Progress in Physics* **71** 016501
[4] Vanderbilt D 1993 *Physical Review B.* **48** 4442–4455
[5] King-Smith R and Vanderbilt D 1993 *Physical Review B.* **47** 1651–1654
[6] Resta R 1994 *Reviews of Modern Physics* **66** 899–915
[7] Marzari N and Vanderbilt D 1997 *Physical Review B.* **56** 12847-12865
[8] Diebold U 2003 *Surf. Sci. Reports* **48** 53-229
[9] Clark S J et al. 2005 *Zeitschrift fuer Kristallographie* **220** 567-570
[10] Perron H et al. 2007 *Theor. Chem. Acc.* **117** 565-574
[11] Kiejna A et al. 2006 *Journal of Physics: Condensed Matter* **18** 4207-4217
[12] Ramamoorthy M and Vanderbilt D 1994 *Physical Review B.* **49** 16721–16727
[13] Bredow T et al. 2004 *Physical Review B.* **70** 035419
[14] Barnard A S and Curtiss L A 2005 *Nano Letters* **5** 1261-1266

Low-loss EELS of 2D boron nitride

R.J. Nicholls[1], J.M. Perkins[2], V. Nicolosi[1], D.W. McComb[2], P.D. Nellist[1] and J.R. Yates[1]

[1]Department of Materials, University of Oxford, Parks Road, Oxford OX1 3PH, UK
[2]Department of Materials, Imperial College London, London SW7 2AZ, UK

rebecca.nicholls@materials.ox.ac.uk

Abstract. Electron energy loss spectroscopy provides a probe of the dielectric function of a material which can be affected by the size and morphology of a sample. In this paper, the effect on the loss function when hexagonal boron nitride is reduced to a single layer is investigated. The previously predicted red-shift of the spectrum is seen in experiment and reproduced by modelling using density functional theory. The dielectric function shows that, for a single layer, the lack of screening causes the real part of the dielectric function to tend to 1 so that the loss function resembles the imaginary part of the dielectric function.

1. Introduction

Many layered structures can be exfoliated down to 2D materials in a similar way to which graphene can be produced from graphite [1]. As in the case of graphene and graphite, these 2D materials can possess different properties to the related bulk materials and this has the potential to be exploited in a new generation of nanoelectronics. In order to reach this goal, understanding and characterization of these new 2D materials is vital. Low-loss electron energy loss spectroscopy (EELS) has already been used successfully to characterize graphene and to map the number of layers [2]. The EELS spectrum is related to the dielectric function of a sample, which in turn can be affected by size and morphology. Graphene and all 2D materials have been predicted to have a distinctive EELS spectrum [3]. In this paper we explore the low-loss spectrum of 2D hexagonal boron nitride and in order to understand the changes in the spectrum, we compare experimental data to simulated low-loss EELS spectra.

2. Experiment

Flakes of hexagonal BN were produced using liquid-phase exfoliation [1]. EELS spectra were then obtained using an FEI Titan 80/300 TEM/STEM operating in STEM mode. As BN suffers from beam irradiation in the same way as carbon materials [4], the microscope was operated at 80kV in order to minimize the damage. The energy resolution, defined by the full width half maximum of the zero loss peak, was 0.7eV.

A typical high angle annular dark field image taken from the BN sample is shown in Figure 1a. From the image, it can be seen that the thickness of the flakes vary. EELS spectra taken from a thick and a thin part of a BN flake are shown in Figure 1b. It can be seen that the reduction in thickness of the sample has a large effect on the low-loss EELS spectrum. The spectrum taken from the thin region shows a decrease in the peak at 26.5eV when compared to the spectrum from the thicker sample, as well as a shift of the peak at 8.5eV. This behaviour is similar to that seen for graphene [2].

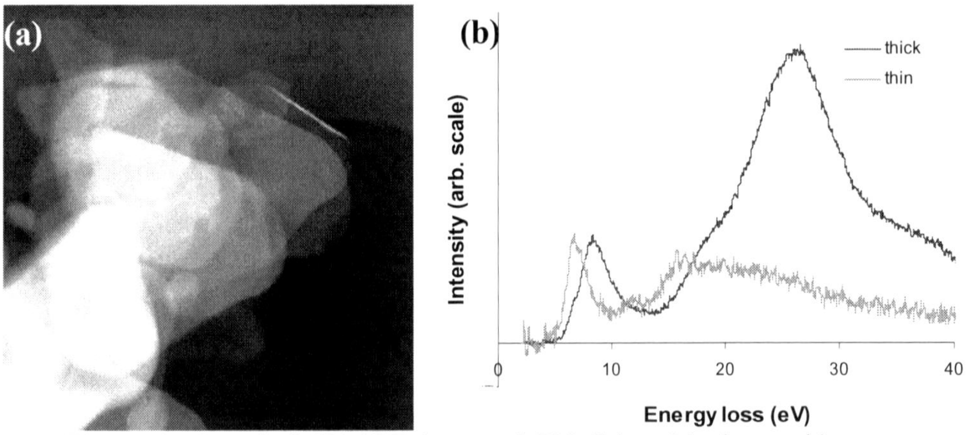

Figure 1: A typical HAADF image of BN flakes (a) along with spectra obtained from thick and thin regions of a flake (b)

3. Loss function simulations

The low-loss part of the EELS spectrum is described by the loss function, which is related to the dielectric function. There are several possible approaches to modelling the dielectric function of a composite material and the method used here is to calculate the dielectric function of the composite system using density functional theory (DFT). An alternative approach, which is also very popular, is to use an analytical expression to relate the bulk dielectric function to a specific geometry. Although that approach has been used to model the loss function of multi-walled carbon nanotubes [5], we have found that it is less successful for single BN layers.

DFT allows calculation of the ground state electronic structure. The dielectric function describes the response of a material to an external field, and is not a ground state property. The DFT ground state can be used as a starting point to calculate the dielectric function and some level of approximation must be made. All the simulations presented here have been carried out using the pseudopotential DFT code CASTEP [6] which uses the random phase approximation (RPA) to calculate the dielectric function. RPA has been shown to be adequate for EELS simulations due to cancellations between the quasiparticle shift and long range term near the plasmon energy [7]. Pervious calculations on graphite [3] and BN [8] layered structures have shown that local field effects are negligible when the polarisation is in the plane of the material, but are required to accurately reproduce the spectrum from the out of plane polarisation. In this paper we are looking to explain the general trends, and so local field effects have not been included.

The simulations in this paper have been carried out using the PBE exchange-correlation functional and ultrasoft pseudopotentials. Planewaves up to a maximum of 340eV where included in the basis set, and the Brillouin zone was sampled using a regular grid with a maximum spacing of 0.03Å^{-1}. These parameters were checked so that any increases in cut-off energy or decrease in k-point spacing had no visible effect on the spectrum. The simulations have had a Gaussian broadening of 1eV applied.

Figure 2a shows the simulated loss function for bulk BN along with the experimental data from the thick region of the flake. There are several potential ways to stack the BN layers to produce bulk BN. Five different stacking geometries were simulated, and the results were all found to be very similar so only simulations from the AA' structure (i.e. the atom type alternates along the *c*-axis) are shown here. As BN has an anisotropic structure, the dielectric function is also anisotropic. As a result the loss function varies with the direction of **q**. Figure 2a shows spectra simulated for **q** parallel to the *c*-axis (q||c), **q** perpendicular to the *c*-axis (q-c) and **q** averaged over all directions (poly). The experimental data in Figure 2a is reproduced relatively well by the spectrum simulated for an averaged **q** which tells us that the thick areas of the flake are bulk like.

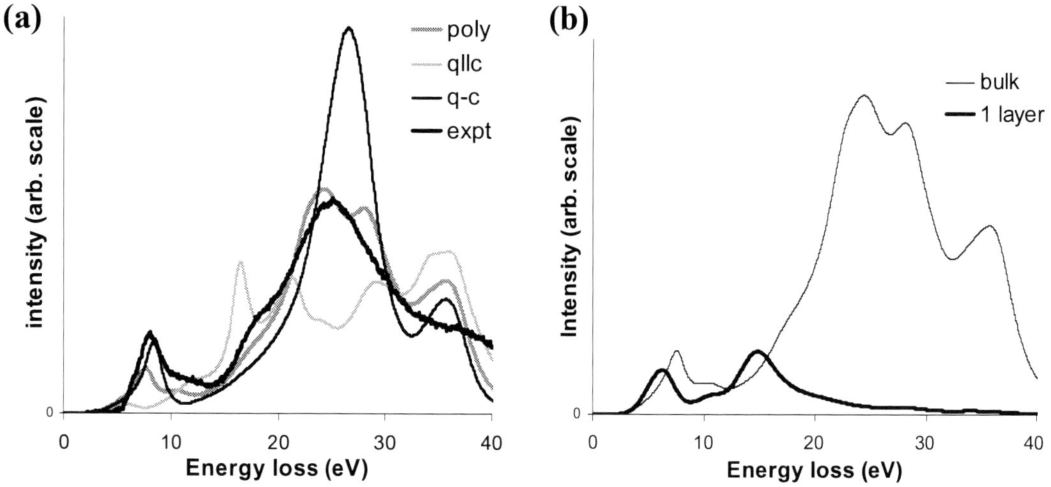

Figure 2: Comparison of average **q**, **q**llc and **q**-c bulk BN simulations (a) and comparison of a single layer with bulk (b)

In order to simulate single layers of BN, the *c* parameter was increased to separate the layers. As the layers do not interact, the stacking is unimportant and an AA stacking is used for this calculation. The *c* parameter was increased until the changes in the spectrum became very small and it was found that a spacing of 4.5nm was required. This is a larger value than would have been required to converge core-loss EELS spectra because in that case the initial state is very localised. Figure 2b shows spectra simulated for a single layer along with the previous bulk calculation. Only data for an average value of **q** is shown.

4. Discussion

The simulated data shows that the main plasmon peak and the peak around 8eV move downwards in energy and that the plasmon peak decreases in intensity when the structure is reduced to a single layer. The differences between the single layer and bulk are in agreement with the generic changes predicted for 2D layered systems by Marinopoulos *et al.* [3].

The experimental data from the thick section of the BN flake has peaks at 8.4eV, 12eV, 18eV and 26eV. Figures 3a and c show the real (ε_1) and imaginary (ε_2) parts of the dielectric function for bulk BN with q-c (a) and q||c (c) along with the corresponding loss function. If ε_1 is zero and ε_2 is small, the loss function will show a peak, which is generally interpreted as a plasmon. The peak at 26eV in the experimental data corresponds to a zero-crossing of ε_1 for q-c and is the collective excitation of all the electrons (i.e. the bulk plasmon). Figures 3b and 3d show ε_1 and ε_2 for the layers with a spacing of 4.5nm (single layers). For q-c, the zero-crossing of ε_1 has disappeared which indicates a strong damping of the bulk plasmon. As seen by Marinopoulos *et al.* for graphite → graphene, as the interplanar distance in BN is increased ε_1 tends to 1 and the loss function tends towards ε_2. The plasmons have been damped to the extent that the loss function is dominated by single electron transitions. The situation for q||c is more complicated due to a strong coupling between plasmons and interband transitions. For a single layer, however, the damping of the plasmons again means that the loss function follows the form of ε_2. Comparison of the simulated data for a layer and the spectrum from the thin region of the flake suggests that the flake is thicker than a single layer.

5. Conclusions

When hexagonal BN is reduced to a single sheet, the low-loss EELS spectrum changes in the way predicted by Marinopoulos *et al.* [3]. DFT is able to model these changes and explain the behaviour of the loss function. As the layer separation is increased, the screening within the sample is reduced such that the real part of the dielectric function tends to 1 and the loss function resembles the imaginary part of the dielectric function.

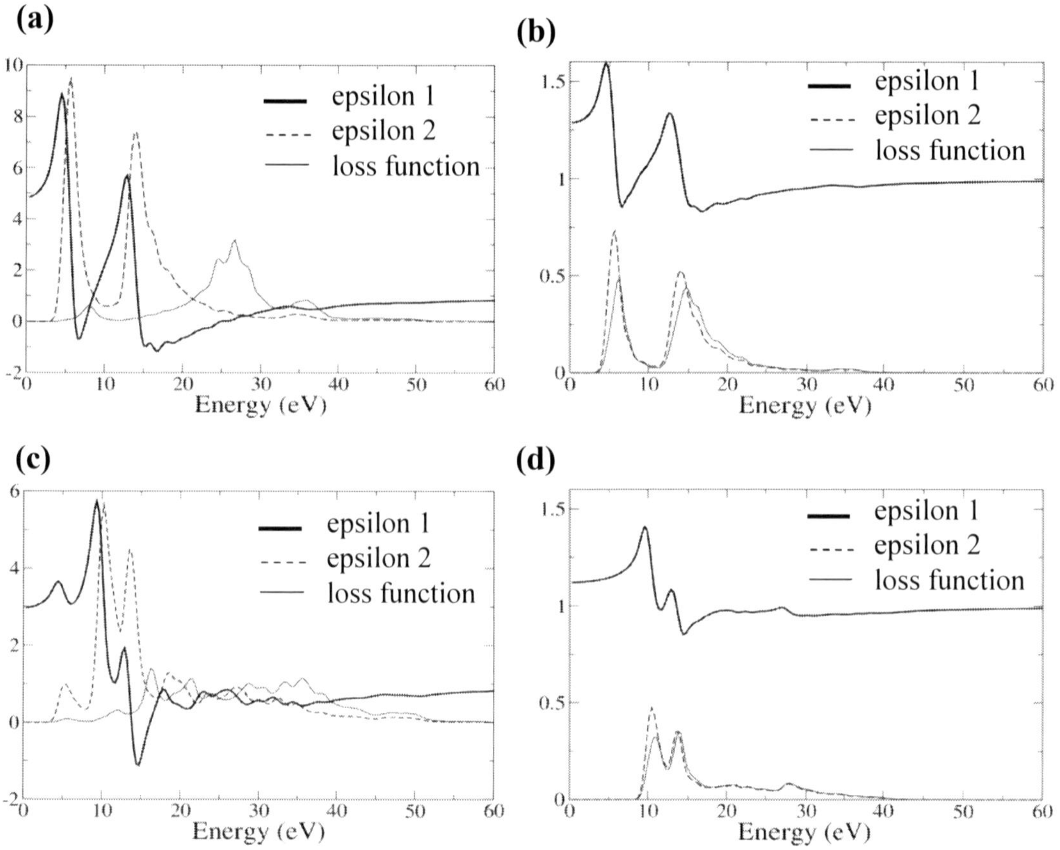

Figure 3: Real and imaginary parts of the dielectric function and loss function for bulk (a and c) and layer (b and d) BN with **q**-c (a and d) and **q**||c (c and d).

Acknowledgements
RJN gratefully acknowledges financial support from EPSRC grant EP/H046550.

References

[1] Coleman JN *et al.*, 2011 *Science* **331** 568-571
[2] Eberlein T, Bangert U, Nair RR, Jones R, Gass M, Bleloch AL, Novoselov KS, Geim A, Briddon PR 2008 *Phys. Rev.* B **77** 233406 -1-4
[3] Marinopoulos AG, Reining L, Olevano V, Rubio A, Pichler T, Liu X, Knupfer M, Fink J 2002 *Phys. Rev. Lett.* **89** 076402- 1- 4
[4] Smith BW, Luzzi DE 2001 *J. Appl.Phys.* **90** 3509-15
[5] Kociak M, Henrad L, Stèphan O, Suenaga K, Colliex C 2000 *Phys. Rev. B* **61** 13936-44
[6] Clark SJ, Segall MD, Pickard CJ, Hasnip PJ, Probert MJ, Refson K, Payne MC 2005 *Zeitschrift für Kristallographie* **220** 567-70
[7] Dash LK, Bruneval F, Trinitè, Vast N, Reining L 2007 *Computational Materials Science* **38** 482-93
[8] Marinopoulos AG, Wirtz L, Marini A, Olevano V, Rubio A, Reining L 2004 *Appl. Phys.* A **78** 1157-67

Electron microscopy of nuclear graphite: A modelling approach

C.R. Seabourne[1], R. Brydson[1], M.I. Heggie[2], C.D. Latham[2] and A.J. Scott[1]
[1]Institute for Materials Research, SPEME, University of Leeds, Leeds, LS2 9JT, United Kingdom
[2]Chemistry Subject Group, School of Life Sciences, University of Sussex, Falmer, Brighton, East Sussex, BN1 9QJ, United Kingdom
c.r.seabourne@leeds.ac.uk

Abstract. Graphite acts as both a major structural component and moderator in nuclear reactors. Upon neutron irradiation, various structural and property changes occur. Property changes include; coefficient of thermal expansion (CTE), Young's modulus and thermal resistivity. This work focuses on the characterisation of irradiated graphite models using both electron energy-loss spectroscopy (EELS) and imaging techniques. A number of models of irradiated graphite have been found, whereby one or more interstitial atoms form between the hexagonal layers. In this work, density functional theory (DFT) modelling is used to predict EEL spectra (carbon K edges) at the spiro-interstitial position, and contrast those to bulk predictions. We observe that for a 'bulk-like' position in the spiro-interstitial model, the carbon K edge shape is similar to that of true bulk, thus confirming the validity of the model used. For the carbon K edge prediction at the spiro-interstitial position, although peaks in the π^* region remain in approximately the same energy positions, there is considerable broadening suggesting the presence of strained sp^2 bonding character. The σ^* peak is significantly altered, both in energy position and intensity relative to the π^* region. These observations are arguably consistent with a spiro-interstitial strained from the ideal bonding angles observed in spiro-pentane. Simulations of TEM and HAADF images of the spiro-interstitial model were also performed. These suggested that in a typical TEM, for the (100) orientation, even at a thickness of ~15 Å the interstitial would be difficult to observe. In the case of (S)TEM, a similar situation exists.

1. Introduction

Graphite is a major structural component as well as a moderator in nuclear reactors. It is likely to continue to be used in high temperature reactors (HTRs), which are currently being planned. These are projected to have an operating lifetime of between 50 and 100 years [1]. Over the course of a reactor's lifetime, neutron irradiation of graphite causes various structural and property changes, including for example Young's modulus and the thermal resistivity [2–3]. A new EPSRC consortium, 'Fundamentals of Nuclear Graphite' has been established, involving the authors of this work. Nanoscale characterisation techniques have been markedly improved since the original theories regarding irradiation damage were developed. A major aim therefore of the consortium is to improve the mechanistic understanding of the damage processes.

On the atomic level, a number of models of irradiated graphite have been developed, and the overwhelming majority invoke Frenkel pairs as the principal defects arising [4]. Here we examine a single interstitial (di-interstitials are also possible) to see if it could be observable in the electron microscope. The ground state for a single interstitial is the 'spiro-interstitial' structure, so-called due to its resemblance to spiro-pentane [5–7].

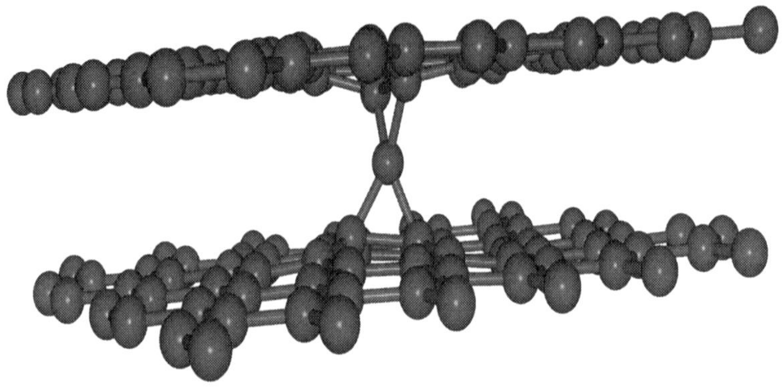

Figure 1. The 'spiro-interstitial' structure of irradiated graphite. This is the ground state structure for a single interstitial, the formation energy relative to pure graphite is 5.85 eV [6].

The CASTEP density functional theory (DFT) code is used to predict EEL spectra [8]. Image simulation in the electron microscope is also investigated, using a multislice methodology for TEM simulations, and the QSTEM code for (S)TEM imaging [9–11].

2. Methodology

In terms of EELS simulations, the employed methodology was similar to previous work involving some of the authors of this paper [12-13]. Throughout the DFT calculations presented, the local density approximation (LDA) was used, consistent with previous work in this field, which was used to form the basis of the structural models in this paper [6]. In terms of other parameters, there are two major considerations; the basis set size (defined by the kinetic energy cut-off value) and the density of *k*-point sampling. These values were converged specifically against the predicted carbon K edge. The kinetic energy cut-off value was increased in steps of 100 eV from 200 eV. The *k*-point mesh density was doubled from $4\times4\times4$ upwards. A Monkhorst-Pack grid of *k*-points was used, with equal numbers of points along each reciprocal lattice direction, thus increasing the sampling density in the *c* direction. The successive predicted K edge spectra (ground state predictions) were analysed numerically. For example, upon doubling the *k*-point mesh density, the percentage change in intensity value was found for each energy-axis value, these then being averaged across a range of −5 eV to 50 eV above the Fermi level. Convergence was achieved when this average percentage was less than 15%. Practically speaking, once broadening is applied much less rigorous parameters would lead to highly similar results.

Table 1. Chosen parameters for bulk graphite carbon K edge modelling.

Parameter	Value
Kinetic energy cut-off	500 eV
EELS calculation *k*-point mesh	$32\times32\times32$

High-resolution TEM (HRTEM) modelling was carried out using the Cerius2 code, according to a multislice methodology [9-10]. The module allows the customisation of the microscope parameters being simulated. In this work, they were set to match the FEI CM200 device at Leeds (200 keV source voltage, C_s 1.2 mm), with thickness-defocus plots then being found. All (S)TEM results were simulated using Koch's QSTEM code [11], customised to use the following parameters throughout the simulations (based on those for SuperSTEM 1; 100 keV source voltage, −2.7 mm defocus, 0.005 mm spherical aberration (C_s), C_c 1.0 mm, 32 mrad convergence angle, dE 0.4 eV. The bright-field (BF) detector range was set as 0 – 4.3 mrad and for high-angle annular dark-field (HAADF) a range of 101 – 185 mrad was used.

3. Results and Discussion

The spiro-interstitial structure shown in Figure 1 was investigated. Initially, modelling was carried out using the full 289 atom structure found previously [6]. In order for these calculations to be tractable in terms of time and computational memory, a less dense *k*-point mesh was used than given in Table 1. EELS is a highly localised property, and the 289 atom structure can be 'cut-down' to the region of interest if a ground state calculation is utilised.

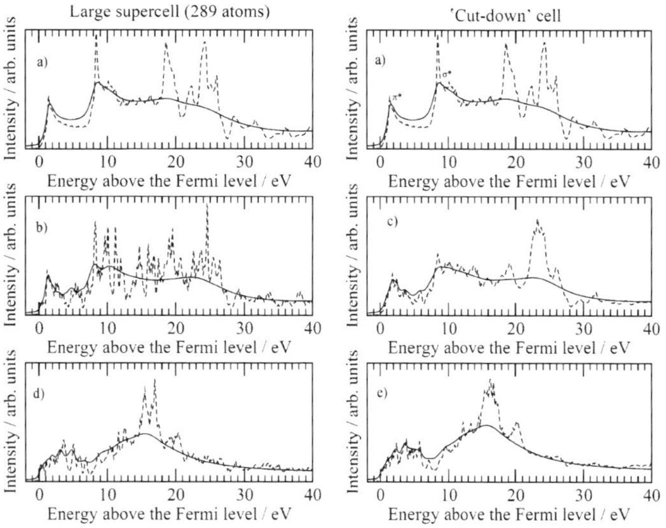

Figure 2. All plots shown are predicted carbon K edges. The dashed lines represent effectively unbroadened spectra, the solid lines show the spectra with the addition of energy-dependent final state lifetime broadening [14]. a) is shown twice. This plot is a bulk graphite carbon K edge prediction in the ground state. b) and c) Represent an edge in the spiro-interstitial model structures at a 'bulk-like' position remote from the interstitial, for the full 289 atom structure, and 'cut-down' structure respectively. d) and e) are the carbon K edge predictions for the spiro-interstitial position for the full 289 atom structure and 'cut-down' structure respectively.

For a position in either the large or 'cut-down' models that was remote from the spiro-interstitial, the prediction closely matches that of bulk. Therefore, changes we observe at the spiro-interstitial position itself can be considered valid. Comparing for example results a) and c) to e), there is broadening in the π^* region, suggesting strained sp^2 bonding, rather than all sp^2 character being eliminated. The relative intensity of the σ^* transition peak increases, and the energy position of the peak is shifted. These observations are arguably consistent with the spiro-interstitial structure, which is strained from ideal sp^3 bonding character as would be observed in spiro-pentane.

Bulk graphite was investigated. For a (100) projection, several phase reversals occur over typical TEM thicknesses or defocus ranges. For example, for a thickness of 49.2 nm, a phase reversal occurs approximately every 40 nm of sequential defocus change, approximately true of all thickness values that were investigated. For a given value of defocus, for example −50 nm, a phase reversal occurs approximately every 10 nm of thickness change. In terms of the spiro-interstitial, consider the results presented in Figure 3:

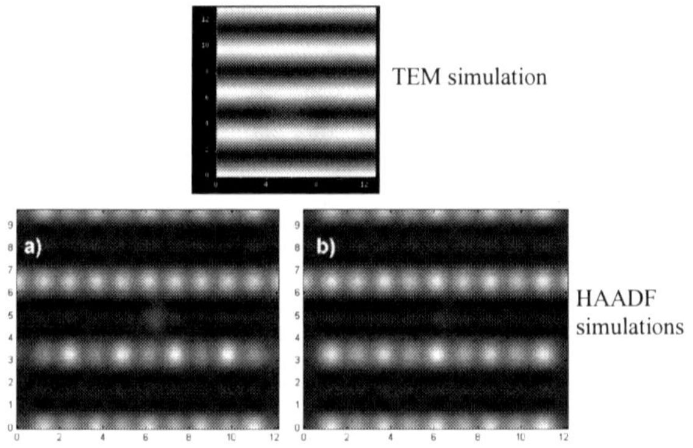

Figure 3. Various image simulations. The TEM simulation uses parameters as described in the methodology section for a CM200 device. The image is for zero defocus. In this case the spiro-interstitial was placed at the centre of a 14.67 Å thick 'sample'. a) and b) are both (S)TEM simulations carried out using QSTEM, according to parameters discussed in the methodology section of this proceeding. In a), the interstitial is at the edge of a 5.8 Å thick 'sample', and in b) it is at the centre of a 12.5 Å thick 'sample'.

4. Conclusions

Early-stage EELS modelling suggests that for a spiro-interstitial as compared to the bulk graphite carbon K edge, the π^* peak is broadened, and the σ^* peak shifted with an intensity increase relative to the π^*. These observations are consistent with the formation of a strained spiro-interstital. There is evidence for cross-linking defects in double-walled nanotubes (DWNT) that could reasonably be assumed to be due to spiro-interstitial formation or intimate Frenkel pairs [15], DWNT TEM imaging (side-on) having been achieved [16]. This simulation work further suggests that graphene-like ultrathin samples should be used if interstitials are to be observed in the TEM / (S)TEM.

References

[1] Zinkle S J, Busby J T 2009 *Mater. Today* **12** 12.
[2] Kelly B T, Brocklehurst J E 1971 *Carbon* **9** 783.
[3] Agranovich V M, Semenov L P 1964 *Journal of Nuclear Energy, Parts A/B Reactor Science Technology* **18** 141.
[4] Kelly B T. *Physics of Graphite*. Applied Science Publishers, London 1981.
[5] Heggie M I, Eggen B R, Ewels C P, Leary P, Ali S, Jungnickel G, Jones R and Briddon P R 1998 in *Fullerenes Vol. 6: Recent Advances in the Chemistry and Physics of Fullerenes and Related Materials (11th International Symposium),* eds. Kadish K M, Ruoff R S, Electrochemical Society: Pennington, N.J., 60-67.
[6] Latham C D, Heggie M I; Gámez J A; Suárez-Martínez I, Ewels C P and Briddon P R 2008 *J. Phys.: Condens. Matter* **20** 395220.
[7] Telling R H, Ewels C P, El-Barbary A A and Heggie M I 2003 *Nat. Mater.* **2** 333.
[8] Clark S J, Segall M D, Pickard C J, Hasnip P J, Probert M I J, Refson K and Payne M C 2005 *Zeitschrift für Kristallographie* **220** 567.
[9] Cowley J M, Moodie A F 1957 *Acta Cryst.* **10** 609.
[10] Saxton W O, O'Keefe M A, Cockayne D J H and Wilkens M 1983 *Ultramicros.* **12** 75.
[11] Koch C 2002 PhD Thesis, Arizona State University. *Determination of Core Structure Periodicity and Point Defect Density Along Dislocations.*
[12] Seabourne C R, Scott A J, Brydson R and Nicholls R J 2009 *Ultramicros.* **109** 1374.
[13] Seabourne C R, Scott A J, Vaughan G, Brydson R, Wang S G, Ward R C C, Wang C, Kohn A, Mendis B and Petford-Long A 2010 *Ultramicros.* **110** 1059.
[14] Moreau P, Boucher F, Goglio G, Foy D, Mauchamp V and Ouvrard G 2006 *Phys. Rev. B* **73**, 195111.
[15] Ewels C P, Telling R H, El-Barbary A, Heggie M I and Briddon P R 2003 *Phys. Rev. Lett.* **91** 025505.
[16] Urita K, Suenaga K, Sugai T, Shinohara H and Iijima S 2005 *Phys. Rev. Lett.* **94** 155502.

Acknowledgements

EP/I002588/1, EP/I003312/1 and related grants. CRS thanks the HPC team at the University of Leeds.

OptaDOS - a new tool for EELS calculations

R J Nicholls[1], A J Morris[2], C J Pickard[2], J R Yates[1]

[1]Department of Materials, University of Oxford, Parks Road, Oxford OX1 3PH, UK
[2]Department of Physics and Astronomy, University College London, Gower Street, London WC1E 6BT, UK

E-mail: `rebecca.nicholls@materials.ox.ac.uk`

Abstract. Many modern (Scanning) Transmission Electron Microscopes ((S)TEMs) are equipped with an energy loss spectrometer. The Electron Energy Loss (EEL) spectra collected provide an experimental method of probing the bonding within a material. With the extra addition of monochromators, the energy resolution obtainable means that even more information is revealed within the fine structure of the spectra. Interpreting the fine structure can often be aided by simulation. Density-functional theory (DFT) is one method of simulating EEL spectra. DFT allows us to simulate DOS and EEL spectra from different structures. This comparison between simulation and experiment enables us to explore how changes in the spectrum are related to changes within the sample. CASTEP is a pseudopotential DFT code which can simulate both low-loss and core-loss EEL spectra. Recent developments have resulted in a separate analysis tool, OptaDOS. This package computes various spectral properties including DOS, projected DOS, joint DOS, core-loss and low-loss EEL spectra and optical spectra. One of the important aspects of the code is the way in which the DOS is calculated. This is done via linear extrapolation or adaptive smearing, methods which are not currently available within CASTEP (or indeed any other code) and which allow detailed analysis of spectral properties. This paper summarises these developments and what they mean for the interpretation of EEL spectra.

1. Introduction

Spectrometers which allow electron energy loss (EEL) spectroscopy to be performed are now quite common additions to (Scanning) Transmission Electron Microscopes ((S)TEMs). The EEL spectra they produce contain information about the bonding within the sample and with the addition of monochromators, it is possible for the fine structure to contain a large amount of detail. Comparison of the experimental data with simulated spectra can help aid interpretation. Several computer codes exist which are able to simulate spectra, for example CASTEP [1], Feff9 [2] and WIEN2k [3].

In this paper, we will outline a new analysis tool, OptaDOS, which calculates the density of states (DOS) and spectral properties from the output of an electronic structure code, such as CASTEP, and discuss how it can help in the interpretation of EEL spectra.

2. OptaDOS

OptaDOS (www.optados.org) is an analysis tool which is run after an electronic structure calculation and uses the computed energy bands to calculate DOS and joint DOS (jDOS). Spectral properties, such as core-loss and low-loss EEL spectra, can be thought of as weighted

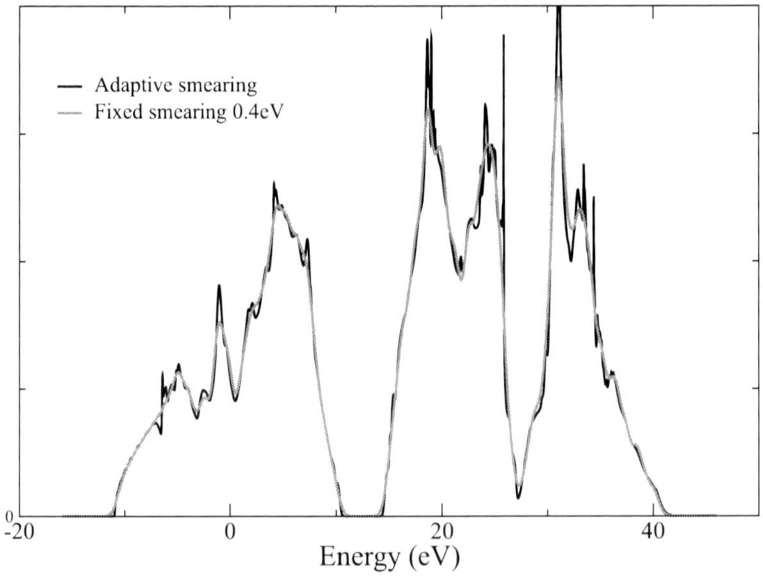

Figure 1. DOS of diamond calculated with both fixed and adaptive broadenings.

DOS or jDOS. These can be produced by OptaDOS using appropriate matrix elements as calculated by the electronic structure program. OptaDOS was developed for use with the pseudopotential code CASTEP, but there is no reason why it cannot also be used as a tool with other electronic structure programs. Further details about the technical aspects of OptaDOS can be found in Morris *et al.* [4].

2.1. Density of States (DOS)

The DOS per unit energy (E) is given by [5]

$$\rho(E) = \frac{1}{N_k} \sum_{i,\mathbf{k}} \delta(\epsilon_{i\mathbf{k}} - E) \qquad (1)$$

where N_k is the number of k-points, and $\epsilon_{i\mathbf{k}}$ is the energy of the ith state at k-point $\mathbf{k}$. In order to plot the DOS, the delta functions can be replaced by Gaussians. The width of the Gaussian has an effect on how smooth the DOS looks, and whilst a narrow Gaussian will wash out fewer of the details, it will take more k-points to obtain a converged DOS. OptaDOS uses two methods based on Gaussian broadening to plot DOS, the first is a fixed Gaussian broadening, which shall be referred to as fixed broadening, and the second method, which shall be referred to as adaptive smearing, uses the gradient of the band to determine the width of the Gaussian [6].

Figure 1 shows the DOS of diamond calculated using both fixed and adaptive broadenings for a 12x12x12 k-point grid. The fixed smearing width is 0.4eV, which is the smallest that can be used for this k-point grid without the spectrum showing unphysical oscillations due to under-sampling. It can be seen that the fine features of the DOS which are captured by the adaptive smearing are washed out by the fixed smearing. If a finer k-point grid is used, the smearing width at which there are no artificial oscillations for the fixed smearing will be reduced and will need to be re-determined. For a particular k-point grid, the adaptive smearing method produces a DOS closer to the 'true' density of states. It also has the additional benefit of a single parameter (k-points) controlling the quality of the DOS rather than the two parameters (k-points and smearing width) needed for the fixed smearing.

2.2. Partial DOS, joint DOS and spectral properties

The partial DOS (pDOS) is obtained by projecting the DOS onto different atomic orbitals

$$pDOS(E) = \frac{1}{N_k} \sum_{i,\mathbf{k}} | < \psi_{i\mathbf{k}} | \psi_{atomic,\mathbf{k}} > |^2 \delta(\epsilon_{i\mathbf{k}} - E) \tag{2}$$

By comparing equations 1 and 2, it can be seen that the pDOS is a DOS weighted by a matrix element. Core-loss spectra, which are calculated by OptaDOS using the approach of Pickard [7, 8, 9]

$$I(E) = \frac{4\pi e^2}{\Omega} \sum_{i,\mathbf{k}} | < \psi_{i\mathbf{k}} | \hat{\mathbf{u}}.\mathbf{r} | \psi_{core,\mathbf{k}} > |^2 \delta(\epsilon_{i\mathbf{k}} - E) \tag{3}$$

can also be thought of as weighted DOS. Both pDOS and core-level spectrum calculations rely on an accurate calculation of the DOS.

The jDOS is given by

$$jDOS(E) = \frac{1}{N_k} \sum_{i,j,\mathbf{k}} \delta(\epsilon_{i\mathbf{k}} - \epsilon_{j\mathbf{k}} - E) \tag{4}$$

and is related to the energy difference between states i and j. It is also possible to calculate the jDOS using the two approaches outlined earlier. Figure 2 shows the jDOS of diamond calculated using both methods for a k-point grid of 12x12x12. The value of the fixed smearing is 0.4eV, which is the smallest smearing it is possible to use for this case without getting oscillations due to the under-sampling. The jDOS calculated with fixed smearing has the main features seen in the adaptive smearing jDOS, but, as with the DOS, the finer details are not present.

OptaDOS uses the approach outlined by Pickard [8] to calculate the imaginary part of the dielectric function which is given by

$$\epsilon_2(E) = \frac{2\pi e^2}{\Omega \epsilon_0} \sum_{i,j,\mathbf{k}} | < \psi_{i\mathbf{k}} | \hat{\mathbf{u}}.\mathbf{r} | \psi_{j\mathbf{k}} > |^2 \delta(\epsilon_{i\mathbf{k}} - \epsilon_{j\mathbf{k}} - E) \tag{5}$$

By comparing equations 4 and 5, it can be seen that ϵ_2 takes the form of a weighted jDOS. Once ϵ_2 has been calculated, ϵ_1 is obtained using the Kramers Kronig relation. The loss function (which is what is obtained in a low-loss EEL spectroscopy experiment) is related to the dielectric function by $Im[-1/\epsilon]$.

3. Core-loss spectra

Good DOS calculations are vital for accurate calculation of theoretical spectra. OptaDOS allows the calculation of a spectrum based on the DOS produced with adaptive smearing. The advantage of this is that it is possible to compare the spectrum with the pDOS (also produced by OptaDOS) which can help assign the origin of peaks in the spectra. Once the theoretical spectrum has been understood, OptaDOS can then be used to convolve the spectrum with appropriate broadening representing lifetime and instrumentation effects.

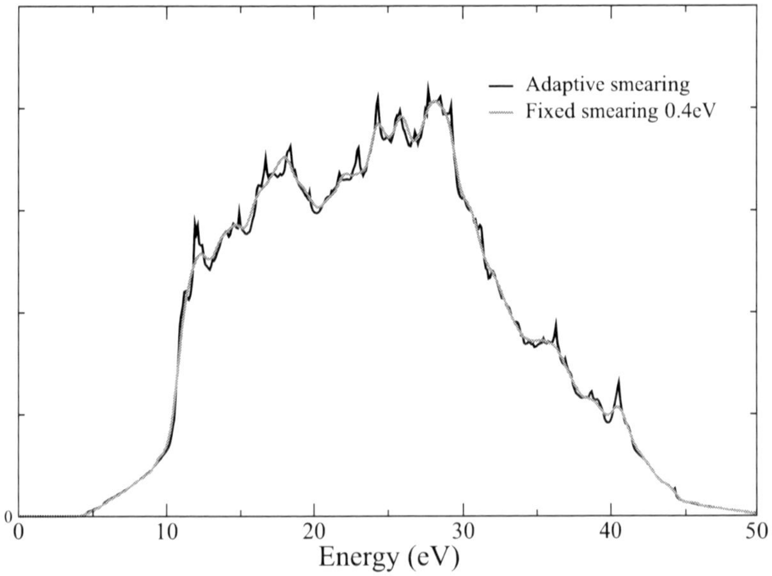

Figure 2. JDOS of diamond calculated with both fixed and adaptive broadenings.

4. Low-loss spectra

Low-loss spectra can be calculated using the OptaDOS 'optics' task, which calculates the dielectric function ($\epsilon(\omega)$) using the random phase approximation without local field effects. In many cases neglecting local field effects gives qualitative agreement with experiment. The precise importance of local field effects depends on the material and the spectral features of interest [10]. As well as the real and imaginary parts of $\epsilon(\omega)$ and the loss function, the optics task also calculates the absorption coefficient, conductivity, reflection coefficient and refractive index. All of these properties can be calculated for a polycrystalline material or for q in a specific direction. It is also possible to calculate the average response in a plane perpendicular to a specific direction.

Acknowledgments

RJN gratefully acknowledges financial support from EPSRC grant EP/H046550.

References

[1] Clark SJ, Segall MD, Pickard CJ, Hasnip PJ, Probert MJ, Refson K and Payne MC 2005 *Z. Kristall.* **220** 567-70
[2] Rehr JJ, Kas JJ, Prange MP, Sorini AP, Takimoto Y and Vila F 2009 *Comptes Rendus Physique* **10** 548-59
[3] Blaha P, Schwarz K, Madsen GKH, Kvasnicka and D Luitz J 2001 *WIEN2k, An Augmented Plane Wave + Local Orbitals Program for Calculating Crystal Properties* (Karlheinz Schwarz, Techn. Universität Wien, Austria) ISBN 3-9501031-1-2
[4] Morris AJ, Nicholls RJ, Pickard CJ and Yates JR *in preparation*
[5] Martin RM 2004 *Electronic Structure* (Cambridge: Cambridge University Press)
[6] Yates JR, Wang X, Vanderbilt D and Souza I, 2007 *Phys. Rev.* B **75** 195121-1-11
[7] Pickard CJ and Payne MC 1997 *Inst. Phys. Conf. Ser.* **153** 637-40
[8] Pickard CJ 1997 *PhD Thesis* University of Cambridge
[9] Gao S-P, Pickard CP, Perlov A and Milman V 2009 *J. Phys.: Condens. Matter* **21** 104203-1-12
[10] Dash LK, Bruneval F, Trinité V, Vast N and Reining L 2007 *Comp. Mat. Sci.* **38** 482-493

Self-consistent absorption corrections for low-energy X-ray lines in energy-dispersive X-ray spectroscopy

MC Parri and T Walther

Department of Electronic & Electrical Engineering, University of Sheffield, Mappin Street, Sheffield S1 3JD, UK

M.Parri@sheffield.ac.uk

Quantification of energy-dispersive X-ray spectra in transmission electron microscopy is often unreliable for X-ray lines of low energy, as they are subject to strong absorption and fluorescence corrections which will depend critically on both the sample thickness and the detector properties (take-off angle, detector material and type, thickness and cleanliness of any detector window). By using the method to vary the take-off angle between repeated measurements in energy-dispersive X-ray spectroscopy from the same region, an attempt has been made to study our ability to determine the depth of a buried epitaxial thin layer within a plan-view sample and to correctly assess the composition of specimens containing lighter elements. By using several different Monte Carlo simulation programs, the attenuation of the X-rays along their path towards the detector can be calculated and the predicted results can be compared to experimental measurements. Comparison of two different programs to experimental results has been made, for an InAs quantum well buried beneath a GaAs cap layer and a NiO thin film specimen.

1. Introduction

Attempting to use energy dispersive X-ray spectroscopy (EDXS) for quantification using low energy X-rays is often unreliable. Since the intensity of such low energy radiation is strongly susceptible to various factors, such as the type of the detector and in particular its entrance window, acquiring data for a useful absorption correction is experimentally difficult. As such, Monte Carlo simulation programs have been used in an attempt to calculate this information.

Further, a method for calculating the depth of a layer within another material or the thickness along the beam direction of a homogenous sample has been described. If expanded, this method could be used as verification of other techniques, such as depth-profiling by secondary ion-mass spectroscopy or scanning Auger spectroscopy, allowing for greater accuracy in their depth calibration.

2. Monte Carlo program comparison for absorption

Building upon previous work [1], four Monte Carlo simulation programs will be compared for their ability to accurately model X-ray absorption. This has been done by simulating X-ray generation and absorption in a gallium arsenide sample of varying thickness and calculating the ratio of the As K-line to the As L-line intensity.

Figure 1 is a plot of these curves on a logarithmic scale, along with experimental measurements from a cleaved GaAs wedge sample. All four programs show similar curves (though EDAX Electron Flight Simulator [2] has a hard coded limitation prohibiting it from calculating values for t>3μm), however, there is considerable shift between these curves.

Recalling that no detector is perfectly efficient, it is clear that taking the ratio of two X-ray intensities of different energies would have to include a correction factor, since both lines, depending on the precise energies involved, may not be detected with the same efficiency. As a first order approximation, a typical efficiency graph [3] implies that a Si:Li detector with atmospheric thin window would detect As_L at between 60 and 80% while As_K at ~100% efficiency. As such, the experimental results should be corrected by this factor so as to be comparable to the simulated data, however this correction is negligible on a logarithmic scale. As an approximate compromise, this efficiency has been presumed to be 70%, and the experimental data in figure 1 adjusted accordingly.

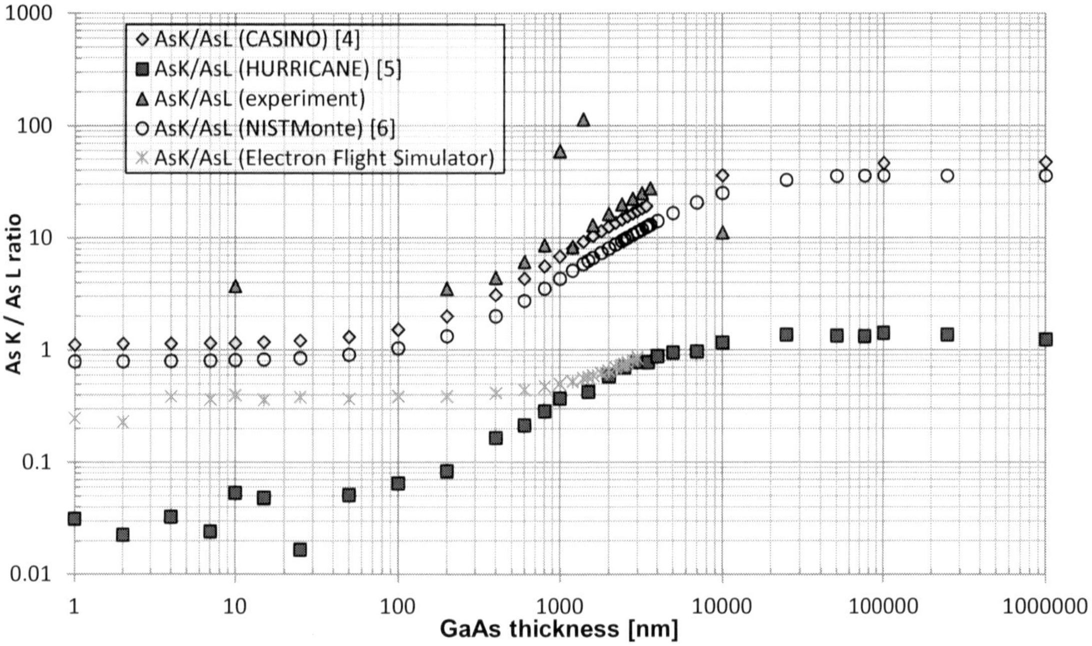

Figure 1: As_K to As_L ratio comparison for four programs and experimental data from a cleaved GaAs wedge for primary energy of 197kV and an approximate take-off angle for flat samples of 20°.

3. Depth of an InGaAs single quantum well buried in an InAs/GaAs specimen

The existing method of calculating the depth of a layer in a thin foil [7] has proved to be very sensitive to noise and inconsistencies in the sample position and thickness. An alternative method has been developed, that is somewhat more insensitive to noise.

Recalling that X-rays of different energies are absorbed differently, normalising an X-ray intensity of a certain energy with respect to another at a different energy will provide information about the sample if the take-off angle (and therefore the effective thickness) is varied. This can then be compared to previously calculated standards.

Using indium in gallium arsenide as an example, taking the intensity ratio of In_L (3.287keV) over As_L (1.282keV) and In_L over As_K (10.543keV) gives two ratios with different dependences on sample thickness and hence take-off angle. The ratio In_L / As_L, since As_L is more heavily absorbed even considering self-fluorescence, will decrease as the take-off angle increases, while In_L / As_K will increase with take-off angle. In theory, these two ratio curves can be matched to data from simulations. The data presented below was gathered with a nominal take-off angle of θ=25° and a solid-angle of detection of 0.12srad and specimen tilt α where α>0 means tilt towards the detector.

Figure 3 shows these ratios for two regions of nominally identical GaAs samples with an 8nm thick $In_{0.24}Ga_{0.76}As$ layer under a GaAs cap of 95nm, in agreement with annular dark-field scanning TEM in cross-section, as shown in figure 2. Using the As_K / As_L ratio at α=0° specimen tilt gave 2.60 and 2.99 for the first and second sample, respectively. While this data was collected with a take-off angle of 25° instead of 20° the difference seemed small enough to vanish into the error. These values were then

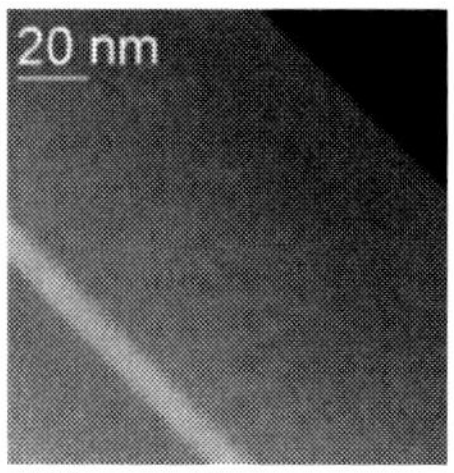

Figure 2: Annular dark-field STEM of cross-section of GaAs sample with InGaAs quantum well.

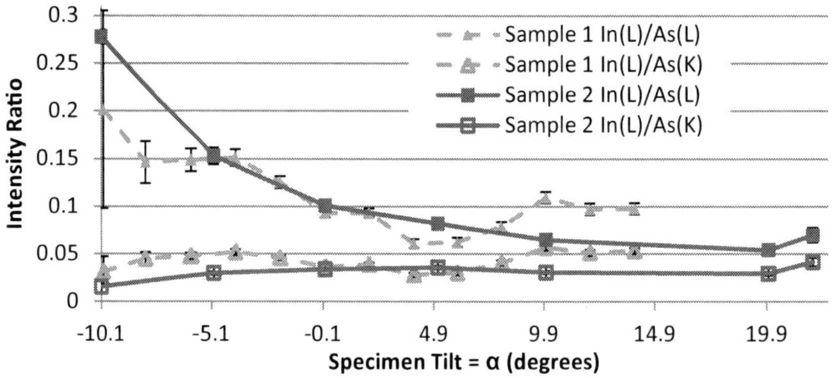

Figure 3: Experimental results of intensity ratios In_L / As_L and In_L /As_K for two different samples of nominally same structure and composition. Error bars are standard deviations. Take-off angle is $\theta = \alpha + 25°$.

compared to the CASINO and NISTMonte plots from figure 1 and the calculated total thickness was then read off the x axis. This resulted in a thickness estimate of 236nm (Casino) to 424nm (NISTMonte) for the first sample and 276 to 546nm for the second, respectively. The degree of scatter is considerable, but a clear trend in agreement with the above described expectation is observed in the In_L / As_L ratio curve for the second sample. The two In_L / As_K ratio plots (open data symbols) are flatter because both In_L and As_K are much less absorbed than As_L.

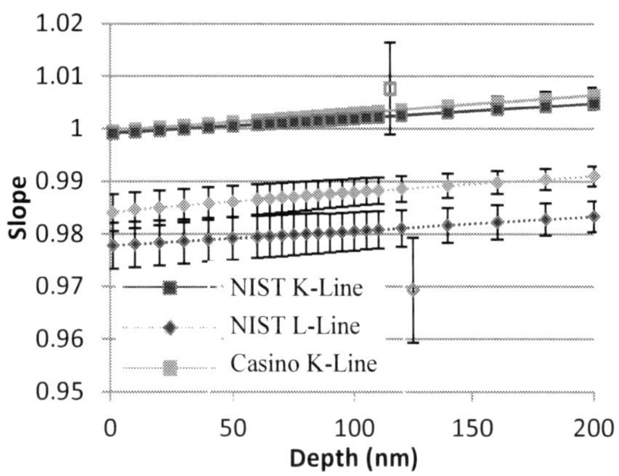

Figure 4: Comparison of calculated slope to capping layer thickness. Legend is shared.

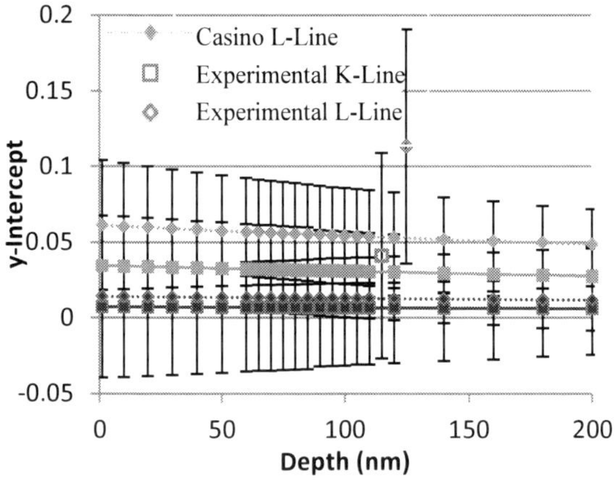

Figure 5: Comparison of calculated y-intercept vs. capping layer thickness. Legend is shared.

For a depth d below the surface an X-ray intensity decays exponentially with an absorption length λ_i such that $I_i(d) = I_{oi} \exp(-d^*/\lambda_i)$, where $d^* = d / \sin(\theta + \alpha)$. Then the logarithm of a ratio of two X-ray lines $i = 1,2$ is given by $\log I_1/I_2 = -d^* / (\lambda_1 + \lambda_2) + \log I_{o1}/I_{o2}$, which is a linear function of d^* with slope $m = -1/(\lambda_1 + \lambda_2)$ and y-intercept $b = \log I_{o1}/I_{o2}$.

Calculating the slopes and y-intercepts of In_L / As_L and In_L / As_K ratios, as provided by MS Excel 2007's *LOGEST* function, and comparing them for equivalent data from simulations conducted by NISTMonte and CASINO, which figure 1 indicated to be the simulation programs that seem to be able to predict K/L line ratios most reliably, does not give readily interpretable results, as seen in figures 4 and 5. These results are for the first sample, with total sample thickness of 236nm from CASINO and 424nm from NISTMonte, as described above. The second sample gives similar data. The y-intercept is not strongly dependent on the InGaAs layer's depth, and the size of the error bars precludes using it for an accurate comparison. The slope evaluation of the NISTMonte In_L / As_L plot from figure 4 would indicate that the layer must be buried at a depth somewhere between 1 and 160nm, while the CASINO data do not match at all.

The reasons for the discrepancy between the experimental and the simulated results are not clear. Most obviously, the simulations included a perfect detector, without any loss of efficiency along the energy spectrum, whereas the experiment uses a Si (Li) detector with a polymer thin window capable of withstanding the pressure from the evacuated column. Such reduced detector efficiency would cause a change in the absolute values but not of the curvature, and should not be of great concern.

4. Chemical composition of a NiO bulk test specimen

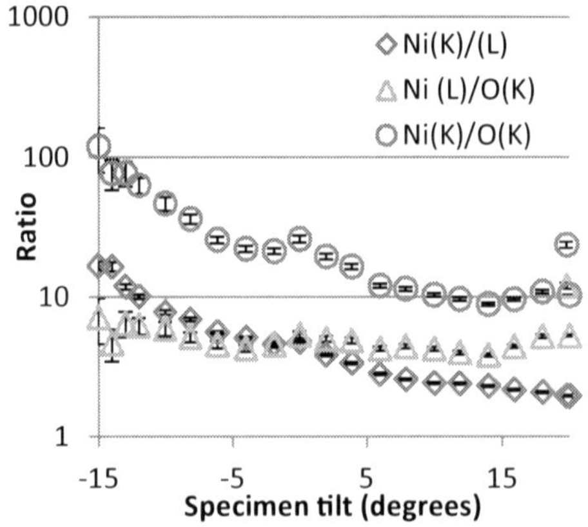

Figure 6: Experimental NiO results.

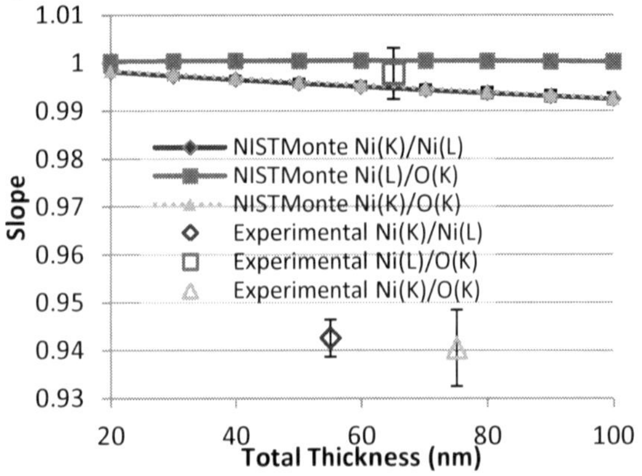

Figure 7: NiO slope vs. thickness along beam direction

A comparison between a nominal 58nm thick NiO bulk sample and NISTMonte simulations of different thicknesses was carried out. This comparison involved the Ni_K (7.478keV), Ni_L (0.851keV) and O_K (0.525keV) lines. Figure 6 displays the experimental results. The Ni_K / Ni_L and Ni_K / O_K intensity ratios decrease with increasing tilt, as expected, while the Ni_L / O_K ratio remains almost constant. Both of these trends agree with theory.

Figure 7 shows fits to the simulated and experimental results. Both the Ni_K / Ni_L and the Ni_K / O_K lines show increasing negative slope with increasing thickness, while the Ni_L / O_K ratio stays relatively constant: given their similar line energies, this is not a surprise. The numerical discrepancy between experiment and simulation is obvious: the O K-counts decrease faster than the NISTMonte simulations suggest. This difference is likely due to incorrect absorption correction in NISTMonte, rather than e.g. build-up of ice on the detector window, which would have decreased the Ni L-edge intensity similarly. This was not observed.

5. Conclusions

The Casino and NISTMonte codes seem to reproduce experimental ratios of K/L line intensities more accurately than the Hurricane and EDAX Electron Flight Simulator codes.

The qualitative agreement between these codes and theory is good and the depth of a buried thin layer within a sample and the average chemical composition of a thinned bulk sample can be determined to some degree, however, for highly precise measurements almost noiseless experimental spectra will be needed.

References

[1] Walther T, 2010, Proc. MSMXVI, Oxford, *J. Phys.: Conf. Ser.* **209**, 012029
[2] SmallWorld, 2011, LLC: http://www.small-world.net/efs.htm
[3] Williams DB, Carter CB, 1996, *Transmission electron microscopy: a textbook for materials science*, New York and London, Plenum Press, Figure 32.6
[4] Hovington P, Drouin D, Gauvin R, 1997 *Scanning* **19**, 1
[5] Rickerby DG, Thiot JF, 1994 *Mikrochmica* Acta **114**, 421
[6] Ritchie N W M, 2005, *Surf. Interface Anal.* **37**, 1006
[7] Walther T, 2008, Proc. EMAG 2007, Glasgow, *J. Phys.: Conf. Ser.* **126**, 012090

Electron backscatter diffraction of MgB_2

P. Mikheenko
School of Metallurgy and Materials, University of Birmingham, Edgbaston, Birmingham B15 2TT, UK

E-mail: p.mikheenko@bham.ac.uk

Abstract. MgB_2 is one of important new functional materials, which can play significant role in emerging Hydrogen Economy. If adopted, Hydrogen Economy would be a low-CO_2 fossil-fuels-free answer to the current and future energy demands. To contribute to this economy, advanced methods of MgB_2 preparation and an insight into its properties are required. This paper reports electron backscatter diffraction (EBSD) of dense polycrystalline MgB_2 prepared by Hot Isostatic Pressing and Resistive Sintering. The EBSD study was performed in combination with polarised optical microscopy, scanning electron microscopy, transmission electron microscopy, atomic force microscopy and scanning conductivity probe microscopy. The superconducting properties were measured in order to select specimens with highest critical current density to associate this density with structural features of the samples. The investigation shows importance of twist grain boundaries in increasing critical current density in MgB_2.

1. Introduction

MgB_2 is a relatively new superconducting material promising to radically change the area of applied superconductivity. As the amount of available of fossil fuels steadily decreases while demand for energy continues to grow, superconductivity is posed to play an increasing role in the energy economy. This role is directly linked to the use of hydrogen as a clean fuel. Hydrogen could be derived in abundance from the renewable energy resources such as solar, wind, marine, hydro, geothermal, biomass and also produced in nuclear and thermonuclear power stations. The convenient way of hydrogen transport and storage is in its liquid form and the liquid hydrogen would enable wide use of superconductivity. There are many superconducting materials capable of working effectively at temperature of liquid hydrogen. MgB_2 has the far largest potential for these applications.

The mechanical and superconducting properties of brittle and low-critical-field MgB_2 are still far from ideal for power applications. Significant efforts are required in order to improve its properties with the imaging playing crucial role in supporting these efforts. In this paper, several imaging techniques are applied to MgB_2 to clarify its structural properties on micro and nano scale in order to link them with superconducting properties. The electron backscatter diffraction (EBSD) appeared to be a central imaging technique. It allowed to obtain crucial information about grains, grain boundaries and crystal lattice of MgB_2. It gave access to a wealth of information not possible to reach with other imaging techniques and was guiding those techniques to obtain additional information.

2. Experimental

MgB_2 was prepared using hot isostatic pressing (HIP) and resistive sintering (RS). HIP is a standard technique successfully applied to MgB_2 [1]. RS is an original technique, its closest analogue being spark plasma sintering [2]. Details of HIP and RS in application to MgB_2 are described in [3,4].

Critical current density (J_c) of MgB_2 samples was derived from magnetisation measurements performed on a Quantum Design Magnetic Properties Measurement System. The samples were imaged by polarised optical microscopy (POM), scanning electron microscopy (SEM), transmission electron microscopy (TEM), electron backscatter diffraction (EBSD), atomic force microscopy (AFM) and scanning conductivity probe microscopy (SCPM). Moving the sample from one instrument to another, selected areas were visualised by all these techniques except for TEM, which requires special preparation of the sample. In this study, EBSD emerged as a central technique. It has been run on an electron microscope JEOL 7000. The crystallographic MgB_2 file for INCA software was generated by Oxford Instruments Ltd.

3. Results and discussion

POM, EBSD and AFM images of a selected area of MgB_2 are shown in figure 1. The grains in the material are clearly distinguished by the colour in POM and EBSD and by the topography in AFM. An important feature is that some grains, which are seen in POM as one, are split in EBSD in two or more grains by straight lines.

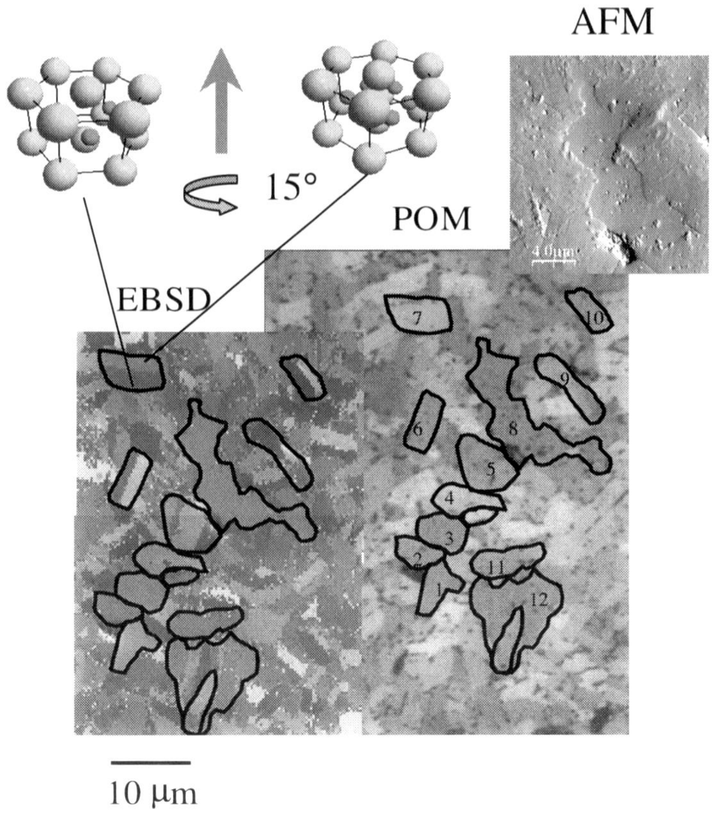

Figure 1. EBSD, POM and AFM (left to right) images of an area of MgB_2. The grains 5-7, 10 and 12 seen in POM as one, are split in EBSD by straight lines in two or more grains. The AFM image is taken around the grain 8. Top left shows the crystal orientation of the parts of the grain 7, which are separated by a 15°-twist grain boundary.

Analysis of the crystal lattice orientation in the grains separated by straight lines, like that shown for the grain 7 in the top left part of the plot, reveals that the lines are twist grain boundaries with rotations around *c*-axis of MgB_2. It is natural to expect the appearance of such twist boundaries in the pressed MgB_2 since it is a highly anisotropic layered material with weak forces between *ab*-planes. In HIP and RS, small grains are pressed into each other at an arbitrary angle creating large number of such boundaries on a micro and nano-scale. These boundaries coexist with other high-angle boundaries as shown in figure 2. In this figure, small square area is selected on a large EBSD map on the left. It is re-mapped in details by EBSD (not shown) and imaged by AFM (centre of the plot). The EBSD gives full description of the crystal orientation in the grains as is shown by the unit-cells around the AFM image, where green atoms are for magnesium and red are for boron. The grain boundary misorientation angles are also shown in the plot. The same square area was also mapped by conductivity scanning probe microscopy as described in [4].

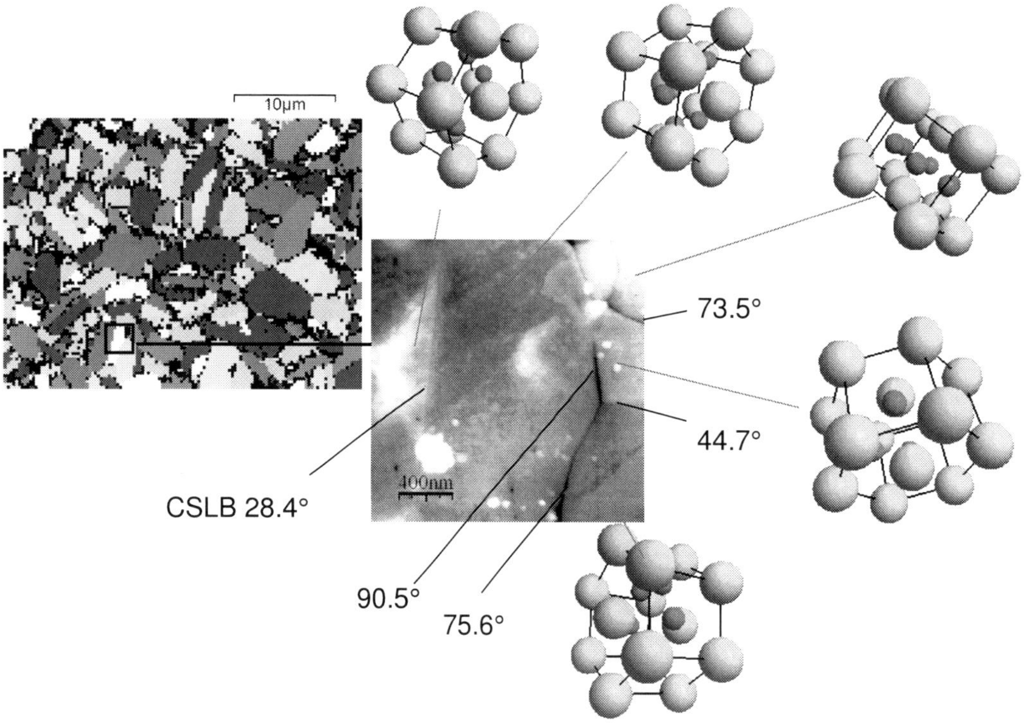

Figure 2. EBSD map (left) and AFM image (centre) of a selected square area of MgB_2. Differently directed unit-cells around the AFM image show crystal lattice orientation in the grains. The grain boundary misorientation angles are also shown in the plot. In the AFM image, the grain boundary on the left is a low-energy coincidence site lattice boundary (CSLB).

Most of the grain boundaries seen in AFM image in figure 2 are of high angle. The grain boundary on the left is low-energy coincidence site lattice boundary (CSLB) [3]. It is also a twist boundary with 28.4°-rotation around *c*-axis. CSLB boundaries are most frequent in MgB_2. Their high angles provide large density of dislocations that effectively pin magnetic flux and support high superconducting current in the sample [3]. The CSPM study [4] gives evidence that grain boundaries with relatively high angle are not detrimental for transport properties. This is further confirmed by J_c measurements in a large number of samples with different grain size D as shown in figure 3.

In this figure, open red circles are for the pure MgB_2 samples and black filled squares are for the samples with nanoparticles of external materials. The latter were added to further deform MgB_2 grains and increase the number of twist boundaries. The scattering of the points is lower for the samples with nanoparticles and J_c is inversely proportional to the average grain size determined from the EBSD maps. The reciprocal $J_c(D)$ dependence is confirmed by the double-logarithmic presentation of the data in the inset. The experimental points fit well a model describing pinning on grain boundaries [3].

The clarification of the nature of pinning in MgB_2 became possible due to the use of EBSD. In addition to POM, EBSD detected twist boundaries, guided AFM, CSPM and TEM in selection of most important areas for imaging, allowed to determine grain size and visualised effect of external nanoparticles on micro and nano-structure of the samples.

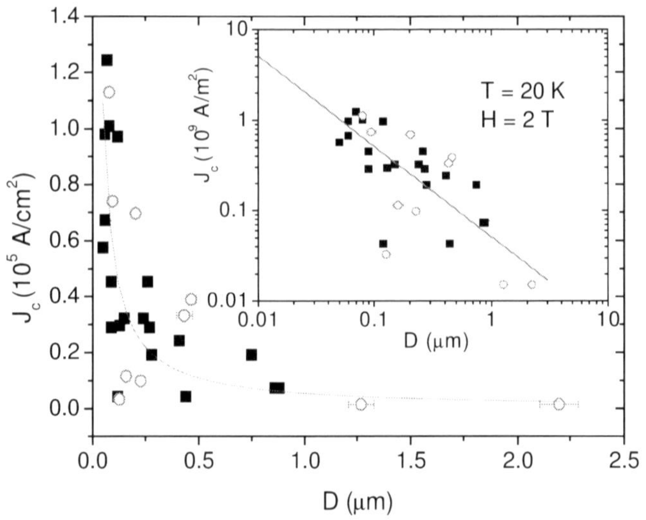

Figure 3. Grain size dependence of critical current density in MgB_2 at 20 K and magnetic field of 2 T. Inset shows the data in double-logarithmic scale. Black filled squares are for the samples with nanoparticles of external phases and open red circles are for pure MgB_2. The red lines in the plot show reciprocal J_c dependence on D.

4. Conclusions

Electron backscatter diffraction was successfully applied to MgB_2 produced by the advanced methods of resistive sintering and hot isostatic pressing. EBSD allowed to obtain important information about the micro and nano-structure of the samples and clarified nature of the pinning in the material. The EBSD findings can be useful in designing new processing techniques that would be able to increase critical current density in MgB_2. These techniques could be based on the treatment of pure MgB_2 or, alternatively, on the nano-additions of external phases.

References

1 Shields T C, Kawano K, Holdom D and Abell J S 2002 *Supercond. Sci. Technol.* **15** 202
2 Dancer C E J, Mikheenko P, Bevan A, Abell J S, Todd R I, Grovenor C R M 2009 *J. Europ. Ceramic Soc.* **29** 1817
3 Mikheenko P, Martínez E, Bevan A, Abell J S and MacManus-Driscoll J L 2007 *Supercond. Sci. Technol.* **20** S26
4 Mikheenko P 2011 *Journal of Physics: Conference Series* **286** 012014

FIB Patterning of Stainless Steel for the Development of Nano-Structured Stent Surfaces for Cardiovascular Applications

M Schmidt[1], F Nazneen[2], Y Georgiev[1], G Herzog[2], P Galvin[2], N Petkov,

Electron Microscopy and Analysis Facility (EMAF)[1] and and Life Sciences Interface (LSI)[2] group, Tyndall National Institute, University College Cork, Cork, Ireland

Corresponding author: michael.schmidt@tyndall.ie; nikolay.petkov@tyndall.ie

Abstract. Coronary artery disease is a major problem worldwide. Stent implantation is a percutaneous interventional procedure that mitigates vessel stenosis, providing mechanical support within the artery. However, stenting causes physical damage to the arterial wall. The research presented here develops novel nano-structured features on stent surfaces to promote rapid endothelial cell adhesion to reduce in-stent re-stenosis. Nano-structured features (concaves) ordered in rectangular arrays were patterned on 316L Stainless Steel (SS) surfaces using focused ion beam (FIB) milling after electropolishing using linear sweep voltammetry and chronoamperometry. Various dose test experiments were performed, aiming at an array of 120 nm diameter holes with pitch of 240 nm and depth of 50-100 nm on sample area of 400 μm × 400 μm. Studies on FIB milling rates were carried out to optimise the time and to create a uniform array of holes. Based on the SEM examination of the hole arrays (plane view and cross section) it can be concluded that a low ion beam current created well shaped uniform concave structures (Gaussian shape) with good depth profile, while a high current resulted in an array of holes with sine profile. Further, a higher current created larger diameter holes with less defined depth profile and deviation from Gaussian shape. We demonstrate that the ability to perform nano-structuring with FIB milling is greatly affected by the polycrystalline nature of SS.

1. Introduction

Coronary artery disease causes 7.2 million deaths worldwide annually [1], resulting from vascular occlusion, myocardial infarction and its complications. Stent implantation is a percutaneous interventional procedure that mitigates vessel stenosis, providing mechanical support within the artery. However, stenting causes physical damage to the arterial wall, leading to inflammation, thrombosis (clotting), and neo-intimal hyperplasia (re-stenosis) [2]. Further improvements in next generation stent coating are needed to enhance the attachment and survival of endothelial cells on the stent surface. The research presented here develops novel nano-structured features on 316L Stainless Steel (SS) stent surfaces to promote rapid endothelial cell adhesion to reduce instent re-stenosis. Based on a literature survey, promising cell responses to nano-structured surfaces were identified including nano-pit features [3].

Conventional techniques for topographical surface modification or surface patterning at the nanoscale are lithography based techniques. In technologies typically derived from microelectronics, such as in X-ray lithography or UV lithography, patterns are formed by using a mask [4]. Such lithography processes are time-consuming with many steps, and inherently inappropriate for prototype designs and processes. Electron beam lithography, and lithography based on scanning tunnelling microscopy or atomic force microscopy are high-resolution mask less procedures, but with a very low throughput and unsuitable for wide surface nano-patterning [5]. Imprint lithography is a high resolution technique for nano-patterning of large surfaces, but it requires moulds and is restricted to polymeric materials which could then be used as etch masks [6]. Patterning by Focus Ion-Beam (FIB) milling offers several advantages for flexible prototyping (i) practically any substrate material that is able to withstand high vacuum conditions of the microscope chamber can be used, (ii) there is high flexibility in the obtainable shapes and geometries by modulating the ion beam current and the patterning conditions, (iii) reduced complexity of the pattering process e.g. it is a single-step process with a possibility of real-time monitoring of the milling progression. Thus for any particulate type of substrate various depths as well as lateral dimensions including the optimal feature size can be obtained at minimum number of processing steps.

In this report we describe the development and the limitations of a new methodology for surface nano-patterning of 316L (SS) stent surfaces based on the use of F. The optimal FIB patterning conditions for achieving reasonably high throughput (patterned rate of about 0.03 mm^2 per hour) and nano-size accuracy in dimensions and shapes of the features, are discussed. Additionally we describe a characterisation protocol for analysis of such structures by combination of SEM, AFM and cross-sectional imaging.

2. Experimental

2.1. Focussed Ion Beam (FIB) patterning

The FIB system used in the current study is the FEI Helios NanoLab 600i, which is a Dual beam FIB for localized milling and deposition, transmitting a 30 keV beam of Ga+ ions combined with a high resolution scanning electron microscope (SEM). The resolution of FIB milling is approximately the same as that of the "spot" size (i.e., the diameter of the beam at the sample surface). In our process, tuning the working current from 0.28 nA up to 0.92 nA, the feature size on the surface of the 316L SS stent surfaces changes from 120 nm up to 180 nm.

Electropolished 316L steel substrates were used for this study. Prior to nano-structuring, the polished specimens were cleaned in acetone, ethanol, and finally in ultrapure water via an ultrasonic treatment for 10 min. Nano-structured features (concaves) ordered in rectangular arrays were patterned on 316L steel surfaces using FEI Helios NanoLab 600i system. This system was used because of high beam quality and stage stability. Concave structures of 120 nm and 180 nm diameters with pitch of 240 nm and 360 nm, and targeted depths of 50 - 100 nm were fabricated on three electropolished steel surfaces on areas of 400 $\mu m \times 400$ μm . Two types of concave arrays were fabricated at 30 keV at two different beam currents and writing times e.g. (i) 120 nm diameter at 0.28 nA beam current and 12 h patterning time and (ii) 180 nm diameter were at 0.92 nA and 6 h writing time.

2.2. Plan view and cross-sectional SEM imaging.

Scanning electron microscope (SEM) was used to analyze the topography of nano textured surfaces of the polished and patterned stainless steel. The SEM images presented were obtained using the SEM at the FEI Helios NanoLab 600i at an electron beam current of 5 kV and 86 pA beam current. Samples for cross-sectional imaging were developed by a standard preparation protocol after depositing protective carbon and Pt layers. The electron beam induced (EBID) carbon deposition supplied necessary contrast difference between the protective Pt and the SS stent surface, hence enabling accurate determination of the concave's shape and depth. Final polishing was performed at 30 kV and ion current of 93 pA .

2.3. Atomic Force Microscopy (AFM) imaging.

AFM examinations were performed in ambient air with a commercial microscope (Dimension 3100 controlled by a Nanoscope IIIa controller equipped with a phase imaging extender, Digital Instruments, Santa-Barbara, CA, USA) in tapping mode. The silicon cantilevers (Windsor Scientific Ltd, UK) had a <10 nm radius of curvature and a 40N m^{-1} spring constant (nominal values). Topographic images were recorded at a scan rate of 0.5 Hz and a resonance frequency of 300 kHz.

3. Results and Discussion

Before attempting prototyping on large 400x400µm areas used for the biological tests we have performed optimisation tests on relatively small test patterns, one such area is shown in Figure 1 (a). From the known polycrystalline nature of the SS stent material one can assume that when subjected to ion milling or imaging it will show pronounced channelling contrast. It is well known for Si or Cu that it etches and mills faster in preferred directions [7], [8], [9]. Similarly polycrystalline metals will show milling rates that are varying by the different orientation of grains towards the incoming beam.

Figure 1 illustrates how much this anisotropic milling affects the desired outcome of uniform concaves. Shown in Figure 1 are examples from the pre-tests on 10µm x 10µm areas with 120 nm diameter concaves at 240 nm pitch. The structures that appear with the brightest contrast (Figure 1(a), marked as region 1) showed deeper and sharper edges than the structures that appear darker in contrast. The FIB/SEM cross sections across a 10 µm width as depicted in Figure 1 (b) reveal that the difference in depth varies in the range from 20 - 85nm. This variation of more than 4 times indicates that an exact match of the desired depth of 50 - 100nm can only be approximated. Nevertheless the correlation of FIB/SEM cross-sectional imaging with the top-down appearance of known patterned area (grain) was essential in establishing accurate size and shape distribution of the formed concaves.

Figure 1. a) SEM image of 10µm x 10µm patterned area with 120nm holes. b) Depth profile across a length of 10µm (stretched in height to pronounce region of interest). Clearly to see the difference in depth for different crystal orientations.

Figure 2 demonstrates another problem that is pronounced only for the extremely large patterning areas of 400x400 µm. In the overview SEM image (Figure 2 (a)) a central triangular shape becomes visible in the patterned square. Closer inspection reveals that the concave structures within this triangular region are circular as targeted for (AFM image Figure 2b). The corners however exhibit structures with deformed circular shapes. Figures 2 (c) and (d) show typical examples of those deformed structures that were examined by AFM in order to study the surface topography, uniformity and dimensions of the nano-holes. From the stigmated nature one can conclude that there are problems with the focusing and astigmatism of the caused by the FIB optics. This is not surprising as the 200 times magnification is close to the lower limit of the FIB Helios column. Smaller areas of about 150x150 µm showed circular shapes when imaged in top-down direction in the whole patterned area.

4. Summary

FIB has compelling advantages for flexible prototyping compared to other traditional techniques, however the milling rates and the corresponding shape and size of the formed structures is largely affected by the grain size of the polycrystalline stainless steel. Moreover this method is limited to 120 nm resolution for the desired concave depth and uniform scan size of 200μm x 200μm. Nevertheless formed structures show large variation of concave depths and shapes and as such might serve as a resourceful platform for screening large variations of cell/pattern SS stent interactions.

a) b) c) d)

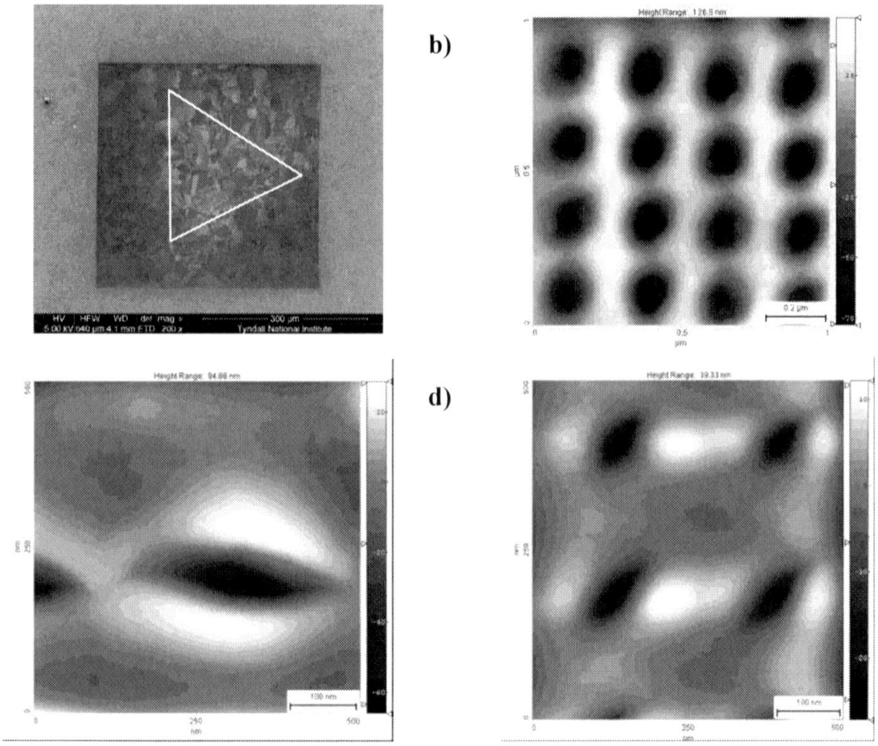

Figure 2. a)Triangular area covering more than half of the 400 μm × 400 μm square reproduced b) perfect holes, while c-d) shallower partially stigmated holes were reproduced at the corners.

Acknowledgements
This work was supported through 09/SIRG/I1621, NBIPI, and INSPIRE initiatives funded by the Irish Government PRTLI 4, NDP 2007-2013.

References

[1] http://www.who.int/mediacentre/factsheets/fs317/en
[2] Lowe H C, Oesterle S N, and Khachigian L M 2002 *J. Am. Coll. Cardiol* **39** 183-193
[3] Dalby M J, Gadegaard N, and Wilkinson, C D W 2008 *J Biomed Mater Res Part A* **4** 973-979
[4] Spille E and Feder R 1977 *Topics in Applied Physics* Volume 22 (Berlin: Springer) pp35-92
[5] Vieu C, Carcenac F, Pépin A, Chen Y, Mejias M, Lebib A, Manin-Ferlazzo L, Couraud L, Launois H 2000 *Appl. Surface Sci.* **164** 111
[6] Chou Y S, Krauss P R, and Renstrom P J 1995 *Appl. Phys. Lett.* **67** 3114
[7] Bean K E 1978 *IEEE Transactions on Electronic Devices* **25** 1185-1193
[8] Seidel H, Csepregi L, Heuberger A, Baumgärtel H 1990 *J Electrochem Soc* **137** 3612
[9] Kempshall B W, Schwarz S M, Prenitzer B I, Giannuzzi L A, Irwin R B, Stevie F A 2001 J *Vac Sci Technol B* **19** 749-754

Journal of Physics: Conference Series **371** (2012) 012066 doi:10.1088/1742-6596/371/1/012066

Analysis of few-nm sized metal nanoparticles on carbon nanostructures

M W Fay[1], A La Torre[2], A N Khlobystov[2] and P D Brown[3]
[1]Nottingham Nanotechnology and Nanoscience Centre, University of Nottingham, University Park, Nottingham, NG7 2RD, UK.
[2]School of Chemistry, University of Nottingham, University Park, Nottingham, NG7 2RD, UK.
[3] Division of Materials, Mechanics and Structures, Department of Mechanical, Materials and Manufacturing Engineering, Faculty of Engineering, University of Nottingham, University Park, Nottingham, NG7 2RD, UK.
michael.fay@nottingham.ac.uk

Abstract. HAADF-STEM tomography has been used to analyse the size and distribution of encapsulated gold nanoparticles (NPs) in hollow graphitised carbon nanofibres. HAADF combined with a rolling-ball filter allows few-nm sized metallic NPs to be imaged with sufficient contrast against the higher volume carbon support structures to enable tomographic reconstruction.

1. Introduction

There has been increasing interest in hybrid metal nanoparticle (NP)/carbon nanostructures over the past decade [1-3]. For such materials, information on the distribution of the NPs on, or encapsulated within, the carbon structure is of particular importance. In most cases, the metal NPs are formed *in situ*, e.g. during arc discharge growth of the carbon nanostructure, complicating control of the size, shape and dispersion of the NPs. Where pre-formed NPs are encapsulated [4,5], the relative orientations of NPs within the nanostructure and the distances between the NPs become the main factors of interest.

Tilt-series TEM is the most suitable analytical technique for investigating the location of metal NPs adsorbed on the exterior or encapsulated within a three-dimensional (3D) carbon nanostructure. For particle size distribution analysis and tomographic reconstruction, it is particularly important to quickly and reproducibly separate the contributions to the signal from the metal NPs and the carbon support structure. While elementally sensitive techniques such as EDX mapping may produce a lower contribution to the signal from the carbon content, the additional acquisition time per frame required makes this approach inconvenient in terms of total time requirements, whilst increasing the potential to alter the sample, e.g. by the beam induced coalescence and ripening of metastable metal NPs.

Traditional Bragg diffraction contrast bright-field and dark-field TEM analysis of such structures can be difficult due to the significantly large volume of the carbon structure compared to the metal NPs, particularly for few-nm sized NPs. Conversely, high angle annular dark field (HAADF) STEM provides for atomic number contrast, and increases the metal NPs' relative contribution to the signal.

In this work, we apply HAADF-STEM imaging to Au-NPs encapsulated in hollow graphitised carbon nanofibres (GNF). These nanofibres comprise graphene layers, forming a herringbone stack at an angle to the main axis, and thus have a corrugated interior [6]. Consequently, the interaction between encapsulated NPs and the GNF is potentially more significant than the NP-NP interaction, in terms of positioning and aggregation [7].

2. Experimental

Samples consisting of dodecanethiolate-stabilised AuNPs were inserted into GNFs in supercritical CO_2 [7]. TEM analysis was performed using a JEOL 2100F FEGTEM, equipped with a Gatan Orius camera for conventional imaging and a JEOL digital STEM system for annular dark field imaging. TEM specimens were deposited onto copper grid mounted holey carbon films from methanolic solution and dried under a stream of nitrogen. Acquisition of HAADF tilt series was performed using a Gatan 916 tomography holder, with a tilt range of -55 to +55 degrees, with one image taken per degree. Post acquisition filtering of the tilt series was performed using ImageJ software. Tomographic reconstructions were produced using IMOD 3.13.2 software.

3. Results and discussion

With conventional diffraction contrast imaging (figure 1), 'large' (i.e. > 5 nm) NPs are clearly discernible, but smaller NPs, particularly when viewed along the edge of the inside of the hollow GNF, are difficult to identify.

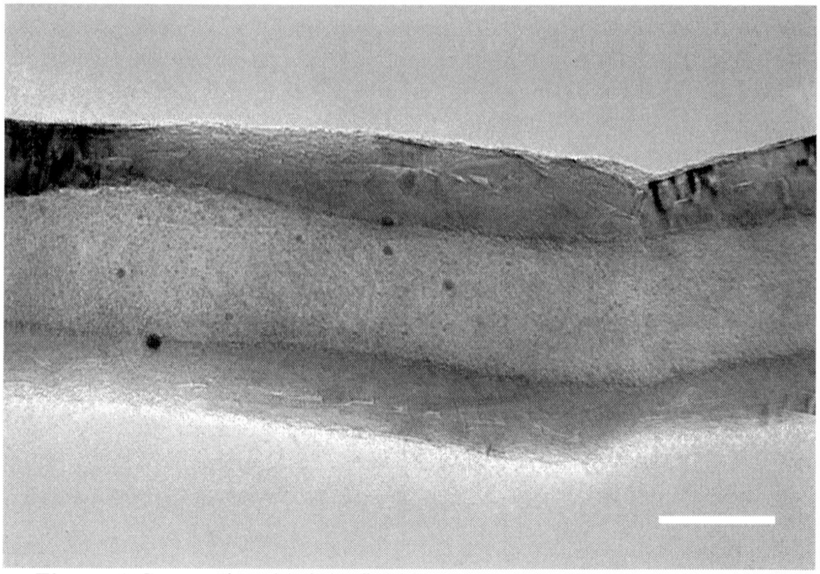

Figure 1 Conventional bright field image of Au NPs encapsulated in a graphitic nanofibre (scale bar = 50 nm).

The contrast between the metal NPs and the carbon of the nanofibre is greatly improved by using the Z-contrast sensitivity of HAADF imaging (figure 2). Such images allow the locations of NPs to be identified unambiguously, relative to the structure of the GNF, particularly as part of a tilt series. Comparison of the original zero-tilt image before tilt series acquisition with that at the end indicated that no discernable ripening of the NPs had occurred during tilt-series acquisition. However, even though the NPs are now clear enough to be identified, the contrast levels between the metal NPs and carbon structure remains problematic for tomographic reconstruction.

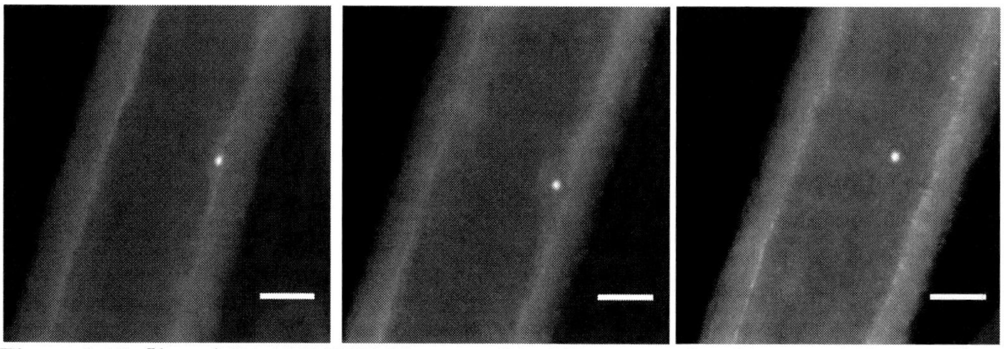

Figure 2 Unfiltered HAADF STEM images of Au NPs encapsulated in a graphitic nanofibre, tilted at -42, 0, +42 degrees. (Scale bars = 20 nm.)

To further enhance the contrast for reconstruction from such tilt series, a number of filters were tested, to remove the large scale carbon background from the image. It was found that a 50 pixel rolling ball filter produced clear results (figure 3), although some contrast from the graphitic nanofibre remained at positions where the walls were viewed edge on. The size of the rolling ball filter in this case was chosen to avoid losing signal from any of the NPs, rather than to minimise the contribution from the carbon at the expense of losing NP information.

As some contribution from the carbon material is still clearly visible, careful thresholding of the subsequent reconstruction was required to optimise the final 3D model, in terms of identifying the location of NPs on the inside and outside surfaces of the GNF. Tomographic reconstruction of this sample (figure 4) confirmed the presence of metal NPs on both the exterior and interior surfaces of the hollow GNF in this instance.

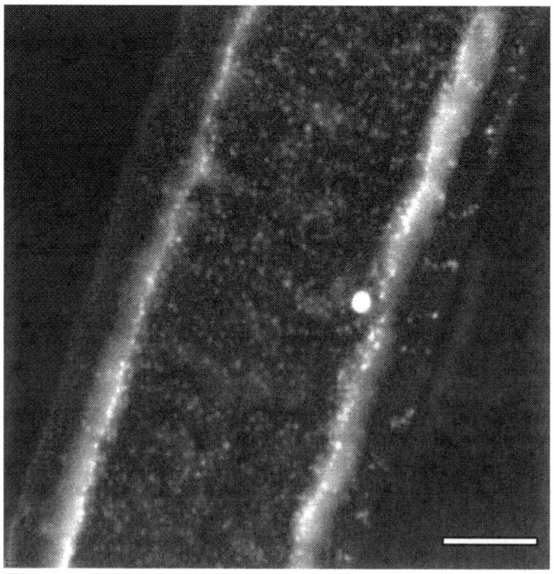

Figure 3 HAADF STEM image after applying a 50 pixel rolling ball filtered image (1 pixel = 0.113nm). Scale bar = 20nm

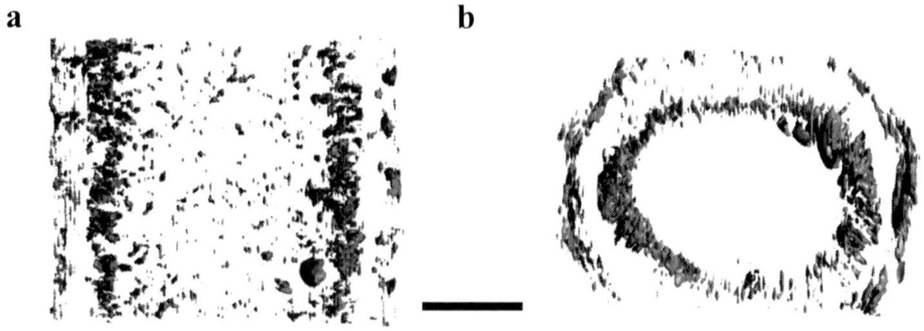

Figure 4 Tomographic reconstruction from rolling ball filtered HAADF STEM images of Au NPs on the inner and outer surfaces of a hollow graphitic nanofibre, viewed (a) edge on and (b) down the long axis. Scale bar = 20nm

4. Conclusion

The Z-contrast sensitivity of HAADF imaging combined with post-acquisition image rolling-ball filtering has been used to extract information on the localisation of metal NPs encapsulated in graphitic nanofibres, with minimised signal contribution from the carbon support.

References

[1] Sloan J, Wright D M, Woo H G, Bailey S, Brown G, York A P E, Coleman K S, Hutchison J L, Green M L H Capillarity and silver nanowire formation observed in single walled carbon nanotubes *Chem. Commun.* 1999, 699-700.

[2] Day T M, Unwin P R, Macpherson J V Factors controlling the electrodeposition of metal on pristine single walled carbon nanotubes. *Nanoletters* 2007, 7, 51-57.

[3] Costa P M F J, Coleman K S, Green M H L, Influence of catalyst metal particles on the hydrogen sorption of single-walled carbon nanotube materials *Nanotechnology* 2005, 16, 512-517.

[4] Castillejos E, Debouttiere P J, Roiban L, Solhy A, Martinez V, Kihn Y, Ersen O, Philippot K, Chaudret B, Serp P An efficient strategy to drive nanoparticles into carbon nanotubes and the remarkable effect of confinement on their catalytic performance. *Angew. Chem. Int. Ed.* 2009, 48, 2529-2533.

[5] La Torre A, Rance G A, El Harfi J, Li J, Irvine D J, Brown P D, Khlobystov A N Transport and encapsulation of gold nanoparticles in multiwalled carbon nanotubes. *Nanoscale* 2010, 2, 1006-1010.

[6] Endo M, Kim A Y, Hayashi T, Fukai Y, Oshida K, Terrones M, Yanagisawa T, Higaki S, Dresselhaus M S Structural characterization of cup-stacked-type nanofibers with an entirely hollow core. *Appl. Phys. Lett.* 2002, 80, 1267-1269.

[7] La Torre, A, Fay M W, Rance G A, Gimenez-Lopez M C, Solomonsz W A, Brown P D, Khlobystov A N Interactions of Gold Nanoparticles with the Interior of Hollow Graphitised Carbon Nanofibres, submitted to *ACS Nano*

Three-dimensional structure of Au nanoparticles supported on amorphous silica and carbon substrates

A Bruma, Z Y Li[1]

Nanoscale Physics Research Laboratory, School of Physics and Astronomy, University of Birmingham, B15 2TT, UK

Abstract. Scanning Transmission Electron Microscope (STEM) has been employed to study the three-dimensional structure of gold (Au) nanoparticles deposited by means of thermal evaporation in high vacuum on amorphous silica (a-SiO_2) and amorphous carbon (a-C) supports. By performing quantitative analysis on the evolution of the high angle annular dark field (HAADF) images, we studied the influence of the nature and the temperature of support on the growth mode of gold nanoparticles.

1. Introduction

Gold nanoparticles in the size regime of a few nanometers, supported on insulating oxides, have attracted much research efforts recently due to a variety of applications such as nanoelectronics and catalysis [1,2]. Silica (SiO_2) is one of such supports [3]. Transmission Electron Microscopy (TEM) have been employed to establish the influence of the environment on the size and structure of Au nanoparticles on/in SiO_2 [4, 5], yet the information about the growth modes of the nanoparticles has been lacking. In this study, we address this issue by application of scanning transmission electron microscopy (STEM).

Recently, we have demonstrated that quantitative analysis of high angle annular dark field (HAADF) imaging in a STEM can be employed to explore the three-dimensional shapes of nanoparticles when combined with simple geometric models [6]. Using this approach, we can gain insights into growth modes of nanoparticles and the interaction of nanoparticles with substrate. The growth of Au on amorphous carbon (a-C) via physical vapor deposition has been well documented [6, 7]. It is generally accepted that the bonding of Au nanoparticles with a-C substrate is weak [8] and the 3D shape of the Au nanoparticles can be approximated as hemispherical-like. However, this is not the case when using silica as supports for Au nanoparticles. The shape and thermal stability seems to vary, depending on how the supports are prepared [9]. In the present study, we conduct a comparative investigation on Au nanoparticles grown on amorphous SiO_2 substrate (Au/a-SiO_2) with Au grown on amorphous carbon (Au/a-C) to emphasize the influence of the substrate on the size and shape of nanoparticle formation. On addition, we investigated effects of substrate temperature on the growth mode of the Au nanoparticles.

2. Experimental details

The Au nanoparticles are produced via thermal vapour deposition on amorphous carbon and SiO_2 thin film-covered TEM grids (Agar Inc.), under high vacuum conditions (~10^{-7} mbar). During the deposition of Au, the substrate temperature was maintained at either 300 K or 473 K. A JEM 2100F STEM, with an acceleration voltage of 200 kV and fitted with a CEOS probe corrector, was employed

[1] To whom any correspondence should be addressed. E-mail address: Z.Li@bham.ac.uk

to image supported nanoparticles. The typical number of nanoparticles examined in each experimental series was 100-150. All the HAADF-STEM images within the series are recorded using the same brightness and contrast setting. A camera length of 10 cm was used, which corresponds to the inner-angle of the HAADF detector of 62 mrad. The emission current was noted stable throughout the investigation.

3. Results and discussions

Figure 1 shows four representative HAADF-STEM images of Au nanoparticles deposited on SiO_2 (a, b) and carbon (c, d) maintained at 300 and 473 K, respectively, during the deposition process. It is clear that the deposited Au forms separated nano-islands on both substrates. Most nanoparticles are circular in projection, except for a few closely spaced clusters that tend to be touching. The histograms of diameters of the nanoparticles are presented in Figure 1 (e-h).

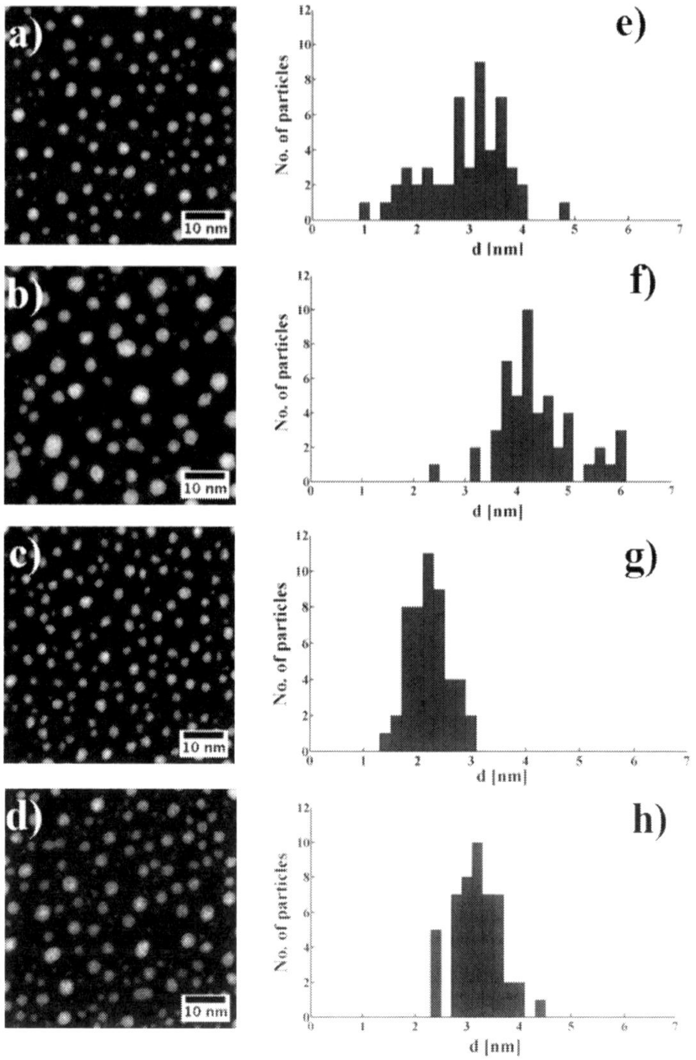

Figure 1. Representative HAADF-STEM images and corresponding size distribution of evaporated Au nanoparticles on a-SiO_2 at 300K (a) and 473K (b) respectively; Deposition on a-C at 300K (c) and 473K (d) respectively. Histograms of the corresponding diameter distribution are shown in (e)-(h), placed directly besides the representative figures.

As the substrate temperature increases from 300 K to 475 K, the mean value of the diameters increases from 3.45 nm to 4.39 nm for Au/a-SiO_2, and from 2.22 nm to 3.18 nm for Au/a-C. It can be seen that the higher the temperature is, the larger the particles are, indicating that the nano-islands are formed as a result of kinetic limited aggregation. The larger particles formed on a-SiO_2 may be associated with the longer diffusion length of Au on a-SiO_2 than that on a-C. It is also interesting to note that the standard deviation of the size distribution is independent on the substrate temperature. The larger standard deviation for nanoparticles on SiO_2 support suggests a more random distribution of defects on the SiO_2 substrate, which act as nuclear center for Au.

The integrated HAADF-STEM intensities over each particle are analyzed and plotted as a function of diameter in Figure 2. Here, the intensities are determined for each particle after local background subtraction and the diameters are estimated by taking the average of the long and short axis of the projection of each particle. HAADF-STEM is an incoherent imaging technique and, as a consequence, the integrated intensity of small Au nanoparticles has been shown to be proportional to their volume [6, 10]. If we take a simple geometrical consideration by assuming that the particles can be modeled by either a 3D sphere or hemisphere or a 2D circular raft-like disk, then the HAADF-STEM intensities would be proportional to d^3 or d^2 in each cases, where d is the diameter of the nanoparticles.

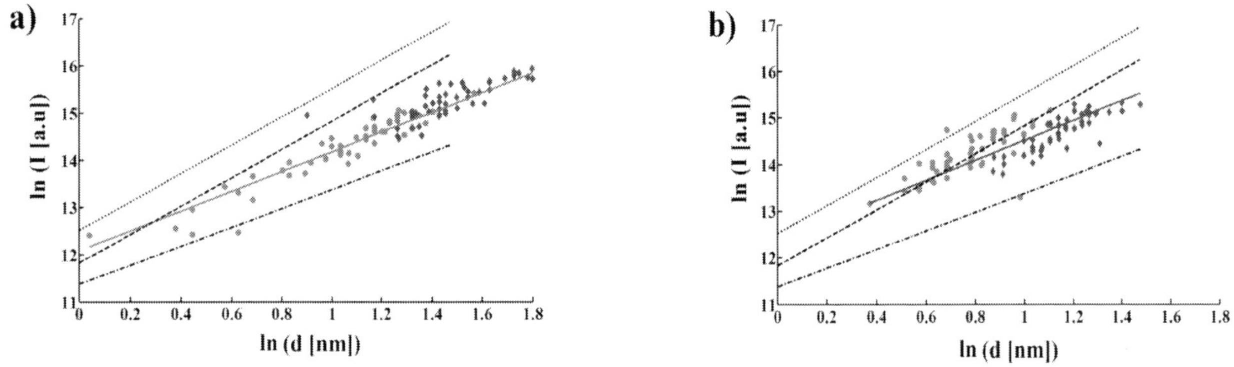

Figure 2. Double logarithmic plot of the evolution of integrated intensity as a function of nanoparticle diameters: (a) Deposition on a-SiO_2 at 300K (•) and 473K (♦). (b) Deposition on a-C at 300K (•) and 473K (♦). The solid lines are the best fit for the data for SiO_2 (R=0.93) and carbon (R=0.87) (R is the fitting residual). The 3D spherical and hemispherical model are shown in dashed lines together with the 2D model in dashed dotted line.

In Figure 2, a double logarithmic plot is shown for a-SiO_2 (a) and a-C (b) together with 2D disk and 3D spherical and hemi-spherical models. The solid lines show the least square fitting of the data, which give the slope 2.09 in the case of SiO_2 and 2.14 in the case of carbon. Both are clearly shown to be away from the 3D spherical model, which has been described well for colloidal nanoparticles as well as for soft-landing nanoclusters formed in the gas phase [6]. The plots suggest that the particle shape is more close to the hemispherical or other 2D models than to the spherical growth model, meaning Au particles are wetting on both substrates. What is the most interesting observation from the Figure 2 is that the scaling exponents are unchanged, i.e. the growth mode kept the same, irrespective of the substrate temperature change from 300 to 473 K. This phenomenon again confirms the conclusion drawn from Figure 1 above that the nucleation of Au nanoparticles is controlled by kinetics within the temperature range. The almost identical scaling exponent for Au on amorphous SiO_2 and carbon may be interpreted as the similar size-dependent 3D shape of Au nanoparticles on these two supports, which indicates the similar binding energy for Au/a-SiO_2 and Au/a-C. It is interesting to compare our results to that reported by Veith *et al.* on Au/SiO_2, where the high thermal stability of Au nanoparticles of size ~2.5 nm was shown, up to at least 773 K in an oxygen containing environment [11]. The authors attributed the stability of Au to the substrate surface defects where protons or surface hydroxyl groups have been removed during the Au deposition process using the magnetron sputtering technique. The mechanism of Au nanoparticles growth and the role of the supports on the shape of Au nanoparticles is clearly a topic of further investigation.

4. Conclusions

In summary, we have shown in this study that quantitative analysis of HAADF-STEM intensity is invaluable to nanoparticle characterization. We have established a correlation between integrated HAADF-STEM intensity over each Au nanoparticle with the diameter of nanoparticles. A quantitative analysis of the scaling exponent together with the simple geometrical models allows us to obtain characteristics of the nanoparticles' three-dimensional morphology. The growth via thermal vapor deposition is intermediate between 3D and 2D mode. It is shown that the scaling exponent is not influenced by the temperature of the substrates. The broad-brush approach described in this paper may be applied to similar systems to gain a quick and efficient assessment of nanoparticle's shape. The method would be particularly useful for industrial catalysts, where performance is dependent on particles having complex shapes with large surface areas. Combined with aberration-corrected STEM imaging, one may gain such information at atomic-details.

Acknowledgements

We are grateful to Dr. Y. Han for the provision of technical assistance and microscope training and Drs. Q. Guo and L. Tang for the guidance regarding the evaporation process. We acknowledge financial support from UK Engineering and Physical Science Research Council. The JEM 2100F STEM used in this work was obtained through Birmingham Science City project, supported by Advantage West Midlands and part-funded by the European Regional Development Fund.

References

[1] Lee J S 2010 *Gold Bull.* **43** 189-199
[2] Zhen X, Veith G M, Redekop E, Lo C S, Yablonsky G S, Gleaves J T 2010 *Ind. Eng. Chem. Res.* **49** 10428-10437
[3] Zhu K, Hu J, Richards R 2005 *Catal. Lett.* **100** 195-199
[4] Ruffino F, Bongiorno C, Giannazzo F, Roccaforte F, Raineri V, Grimaldi M G 2007 *Nanosc. Res. Lett.* **2** 240-247
[5] Ruffino F, Grimaldi M G, Giannazzo F, Roccaforte F, Raineri V 2008 *Nanosc. Res. Lett.* **3** 454-460
[6] Young N P, Li Z Y, Chen Y, Palomba S, Di Vece M, Palmer R E 2008 *Phys. Rev. Lett.* **101** 246103
[7] Koga K, Sugawara K 2003 *Surf.Sci.* **529** 23-25
[8] Mason G M 1983 *Phys. Rev. B* **27** 748-762
[9] Martra L, Prati L, Manfredotti C, Biella S, Rossi M, Coluccia S 2003 *J. Phys. Chem. B* **107** 5453-5459
[10] Treacy M M J, Rice S B 1989 *J. Microsc.* **156** 211-234
[11] Veith G M, Lupini A R, Rashkeev S, Pennycook S J, Mullins D, Schwartz V, Bridges C A, Dudney N 2009 *J. Catal.* **262** 92-101

'Ex-Situ' Annealing and Structural Transformations in Gold Nanoparticles

K.E. MacArthur, N.P. Young, J.W. Critchell, A.I. Kirkland

Department of Materials, University of Oxford, Parks Road, Oxford OX1 3PH, UK

E-mail: katherine.macarthur@materials.ox.ac.uk

Abstract. The atomic structure and morphology of gold nanoparticles have been studied using High-Resolution Transmission Electron Microscopy (HRTEM). The structural transformations of these particles have been probed through ex-situ annealing experiments, yielding population statistics on structure as a function of particle size and annealing temperature. The results show that for an original structurally heterogeneous particle population, there is a rise in the percentage of particles with the decahedral morphology as annealing temperature is increased. This can be correlated with a reduction in the number of icosahedral particles as a function of annealing temperature. HRTEM imaging following low temperature annealing reveals a range of modified and distorted decahedral particles, giving insight into the temperature threshold of the solid state transformation and the microstructural processes occurring.

1. Introduction

The structure of metallic nanoparticles continues to attract considerable interest due to its fundamental importance in relation to their physical and chemical properties and the application of such properties to technologies such as catalysis. A detailed understanding of the thermodynamics of specific nanoparticle systems is now becoming possible through combined in-situ experiments and detailed structural calculations [1]. However, the behavior of real catalysts under industrially relevant conditions continues to be an area of interest. Factors such as high-temperature annealing, reactive gas environments and contaminants differentiate industrial scenarios from high-vacuum experimental studies on model catalysts or the modeling of idealized free-standing nanoparticles. In the present study we report initial results of variable temperature experiments on gold nanoparticles using ex-situ annealing under argon followed by HRTEM characterisation. This study aims to compare structural transformations initialed through ex-situ annealing with those observed using micro-heater experiments in the TEM [1,2]. Ex-situ experiments also yield better statistics on particle transitions.

2. Experimental Method

Our experiments used colloidal gold particles in the size range 6-20 nm. Colloidal suspensions prepared as described previously [1] were initially size separated into three classes and subsequently studied independently. Particles were drop-cast from aqueous solution onto holey carbon coated copper TEM grids. Ex-situ particle heating was carried out using a tube furnace. Specimen grids were placed in a ceramic boat and inside an alumina furnace tube. The ends of the tube were sealed and the system evacuated using a rotary pump to ~10^{-2} psi, prior to the admission of argon gas and

Published under licence by IOP Publishing Ltd

Journal of Physics: Conference Series **371** (2012) 012068 doi:10.1088/1742-6596/371/1/012068

controlled heating. HRTEM was performed using a JEOL JEM-3000F operating at 300kV. Particle structures were initially identified through inspection of the particle outline shape and symmetry. Multislice simulations (using the JEMS software) of particles of different structures in various orientations were also performed to enable comparison with the experimental HRTEM data as an aid to structural classification. There is not space to include the figures here however previously publish multislice simulations can be found in [3].

3. Results and Discussion

The sizes of particles from the different colloidal suspensions were first determined from HRTEM images as detailed in the table (1). Size measurements were taken directly from HRTEM images with an average diameter being estimated from multiple measurements of individual particles. Examples of the equilibrium structure of nanoparticles at room temperature are shown in figure 1. All specimens showed a mixture of icosahedral (Ic), decahedral (Dh) and FCC crystalline particles in various orientations. Many of the individual FCC particles were twinned, others were single crystals and some displayed a trigonal lamellar morphology.

	Mean Particle Size (nm)	Standard Deviation (nm)	Minimum Size (nm)	Maximum Size (nm)
Suspension (A)	10.8	1.7	6.0	16.3
Suspension (B)	14.2	1.6	10.4	21.2
Suspension (C)	15.8	2.7	7.4	22.8

Table 1: Summary of the particle sizes witnessed for each colloidal suspension.

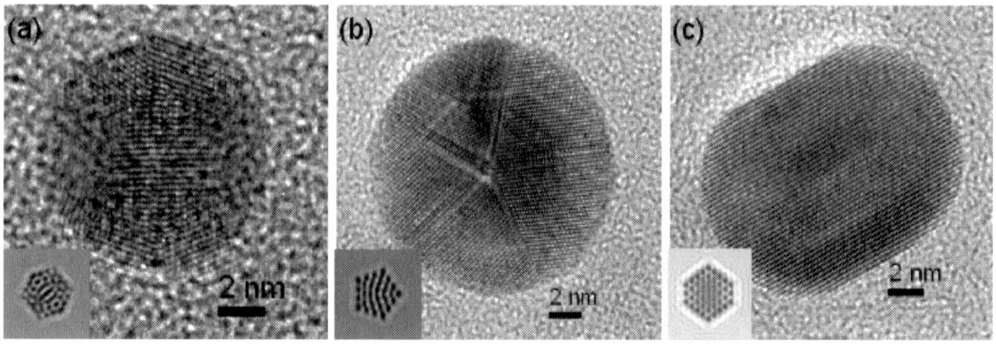

Figure 1: Examples of images of gold nanoparticles obtained during the HRTEM study with multislice simulation image of the same particle orientation (view direction given relative to orthogonal coordinates) inset. All images were taken at room temperature. (a) An icosahedral particle viewed along the <112> axis. (b) A decahedral particle orientated close to the <110> axis. (c) A single crystalline FCC particle oriented along <110>.

The percentage of particles in each main structural category was calculated for the three different colloidal preparations by analysis of single particle HRTEM images. A sample of images of 100 particles prior to heating and after each annealing temperature was recorded. Preliminary investigations of the particle dispersion on the holey carbon support membranes ensured that single particles were well separated. This is particularly important to avoid particle coalescence, interaction and ripening during the annealing stages. Each specimen was then heated in the tube furnace to 100°C under an atmosphere of argon. The total annealing time was 250 minutes after which specimens were removed and new images recorded. This process was also repeated for an annealing temperature of 250°C for 350 minutes. Figure 2 shows plots of the percentage of particles from each colloidal sample for particular morphology as a function of annealing temperature. A clear increase in the occurrence of decahedral particles is evident in figure 2(a) and an increase in the number of particles adopting this structure as a function of temperature was identified for all three sets of colloids. Figure 2(b) shows a decrease in the occurrence of icosahedral particles across all three sets of colloids during the same heating program. The percentage increase in decahedral particles and decrease in the percentage of icosahedral particles observed are comparable in magnitude (within statistical errors) for any one colloidal preparation across the temperature range studied. Figure 2(c) shows a subset of decahedral particles taken from figure 2(a) and discriminates between those decahedral particles which display a clear, ordered decahedral structure and those which are modified or distorted. Modification, in this study, is defined as a significant deviation from the accepted conventional decahedral structure of five distorted FCC tetrahedral units sharing a common vertex. The more commonly observed distortions and modifications included a five-fold axis shifted from the centre of the particle towards one edge, additional twining of a decahedral particle, the occurrence of bi-decahedral particles with two separate axes of five-fold symmetry, and particles showing co-existing icosahedral and decahedral symmetry. Examples of modified decahedral particles are shown in figure 3.

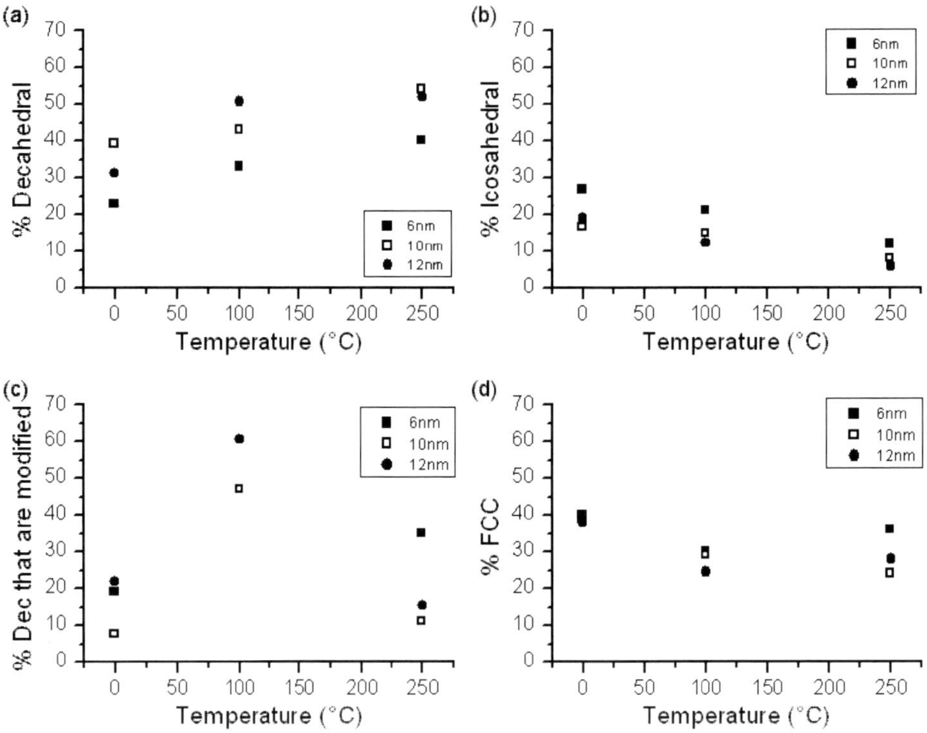

Figure 2: Percentage of the particle population studied which take a particular crystallographic structure as function of annealing temperature. (a) Decahedral particles, (b) Icosahedral particles, (c) percentage of the decahedral subset which display modified structures, (d) FCC particles.

The increase in the percentage of decahedral particles in the size and temperature range studied is consistent with previous in-situ experiments and the calculated phase diagram for nanogold [1]. Our ex-situ annealing studies also show that the decahedral morphology is energetically favorable in an argon atmosphere and the larger population statistics act as confirmation of the previous limited single-particle in-situ studies. Further annealing at discrete temperatures and subsequent cooling back to room temperature for HRTEM observation has identified some of the processes occurring during the transformation from icosahedral to decahedral structures. Annealing temperatures up to 250°C are below the melting point for gold particles of this size and thus, suggests that any transformation must occur in the solid state. A plausible microscopic mechanism has been suggested by Koga [4] based on cooperative slip of the {111} planes in the icosahedral structure, allowing realignment of crystallographic planes and twisting of the 5-fold axis to form a decahedron. However, the energy barriers for such a process are poorly defined and thus the work presented here provides an approximation for the annealing temperature required to overcome such energy barriers.

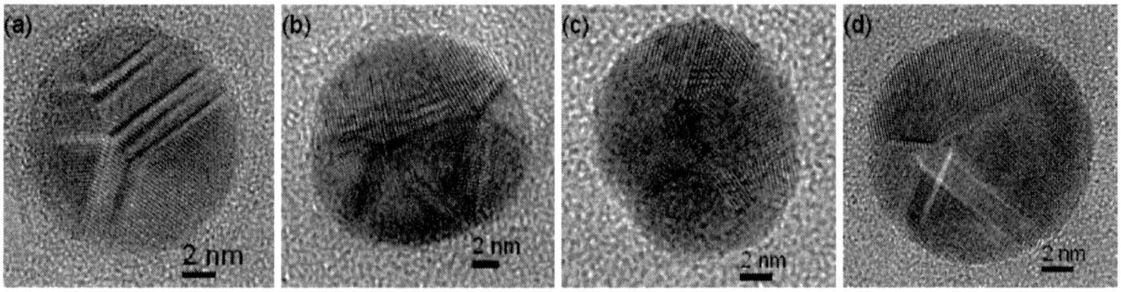

Figure 3: HRTEM images of modified decahedral particles. (a) Decahedral particle with additional parallel {111} twinning. (b) Bi-decahedral particle sharing a common {111} twin plane. (c) Particle displaying both decahedral and icosahedral symmetry with a common {111} twin plane. (d) Decahedral particle with a non-central 5-fold symmetry axis. Such structures are often formed after incomplete transformations from icosahedra.

Close inspection of decahedral particles present after annealing at 100°C reveals that many of these decahedra adopt a modified structure in comparison to the surveys at room temperature and at 250°C. At 100°C incomplete Ic to Dh transformations are evident as shown by an increased number of decahedral particles showing additional {111} twinning, shifted 5-fold axes, or bi-decahedral [5] or combined Dh-Ic symmetry. Figure 2(d) also shows a relatively constant percentage of FCC particles indicating that transformation of the FCC to Dh structure is not favorable at the temperatures used in this study.

Conclusions

The structure of gold nanoparticles in the size range 6 – 20 nm has been investigated through HRTEM and ex-situ annealing under an argon atmosphere. Annealing of particles at 250°C revealed an increase in the percentage of decahedral particles and a correlated reduction in the number of icosahedral particles across all particle sizes studied. This indicates a solid-state structural transformation from the icosahedral to the decahedral structure. Particle annealing at 100°C revealed many decahedral particles with a modified structure including increased {111} twinning, shifting of the 5-fold axis to the periphery of the particle and the formation of bi-decahedral particles. This suggests the onset of structural transformation occurs at 100°C or lower, but is not completed until higher temperatures are reached.

References

[1] A.S. Barnard, N.P. Young, A.I. Kirkland, M.A. van Huis, H. Xu, *ACS Nano*, **3** (2009) 1431
[2] N.P. Young et al. *Journal of Physics: Conference Series*, **241** (2010) 012095
[3] J.A. Ascencio et al. *Surface Science*, **396** (1998) 349-368
[4] K. Koga, T. Ikeshoji, K. Sugawara, *Physical Review Letters*, **92** (2004) 115507
[5] K.Koga, *Physical Review Letters*, **96** (2006) 115501

Acknowledgements

Colloidal specimens were supplied by Dr. H. Xu of the University of Wisconsin-Madison. K.E.M. thanks Prof. Roy Johnston and Mr Andrew Logsdail of the School of Chemistry, University of Birmingham for providing nanoparticle coordinate files used in the multislice simulations in this work. N.P.Y. acknowledges JEOL (UK) Ltd and EPSRC for financial support.

Scanning Transmission Electron Microscopy of Metal-Graphene Interaction

R Zan[1,2], U Bangert[2], Q Ramasse[3] and K S Novoselov[1]

[1]School of Physics and Astronomy, The University of Manchester, Manchester, UK
[2]School of Materials, The University of Manchester, Manchester, UK
[3]SuperSTEM, STFC Daresbury Laboratory, Warrington, UK

recep.zan@postgrad.manchester.ac.uk

Abstract. Metal impurities, gold, chromium and titanium, have been deliberately introduced into graphene. The structural and topographic properties of doped graphene have then been studied by using conventional transmission electron and aberration corrected scanning transmission electron microscopy. Analysis revealed that metal atoms cluster preferentially in/on contaminated areas. Contrarily to observations that gold atoms do not adhere to clean patches of monolayer graphene, chromium and titanium were found to be more reactive with clean monolayer graphene, and the coverage was higher than for gold for the same evaporated amount.

1. Introduction

Graphene has spurred an intense research activity since 2004 [1]. The interaction between metal atoms and graphene is of great interest because metals, in particular gold, are used for a range of graphene applications, including making contacts to circuits [1] electrochemical catalysis [2] and biosensors [3].

This interaction has been studied theoretically by means of Density Functional Theory (DFT) [4], however, it has not been experimentally confirmed yet, since transmission electron microscopy observations are practically none existent [5]. So it would be important to observe, in particular via TEM, metal behavior on graphene to be able to understand the nature of the interaction, the effect of contacts (contact improvement) on electrical transport measurement, and to prove the theoretically predicted positions of metal atoms in the hexagonal graphene structure.

Here we present a (S)TEM study of metal-graphene interactions.

2. Experimental

We employ two different techniques for graphene preparation, exfoliation from HOPG graphite [1] and the CVD method, in which graphene is grown on copper [6]. Graphene layer numbers have been identified from electron diffraction patterns by comparing the first and second ring intensities, and by high angle annular dark field (HAADF) imaging in conjunction with electron energy loss spectroscopy (EELS). Metal impurities have been introduced onto mono- and few- layer graphene using electron beam and thermal evaporators. For accurate comparison, the same amount of all three metals (2Å) has been evaporated on pristine graphene sample at room temperature.

High resolution imaging of metal atoms and clusters on the graphene surface has been carried out in stationary transmission electron microscopy mode (TEM) as well as in scanning mode (STEM), the former in a Tecnai F30, the latter in an aberration corrected dedicated STEM (Daresbury SuperSTEM). Atomic sites of singular metal atoms and clusters on graphene have been directly revealed in HAADF imaging mode by their approximate Z^2 dependence, and by relating HAADF with simultaneously obtained BF images.

3. Results and Discussion

Figure 1 shows results of 2 Å gold deposited on mono-layer graphene. We never observed gold atoms on single or de-coupled (turbostratic) layers, however, gold clusters and nano-crystals form in the hydrocarbon contamination, which indicates the extreme mobility of gold on the pristine (clean) graphene surface (figs 1a), b) and c)). Single gold atoms can only be observed on few-layer graphene films, where the bonding contribution from sub-surface layers appears to prohibit their diffusion and enhance the sticking probability to the graphene surface. The gold atoms, which have appeared on few-layer graphene (fig. 1d: some of them are numbered), were dissociated from the bigger gold cluster (left, fig. 1d) as a result of repeated scanning in the STEM. The atoms stay for a few scans in their positions. By comparing locations of gold atoms in the HAADF image (fig, 1d) with identical positions in the BF image, (fig. 1e), the sites can be identified as on top of carbon atoms (T-sites) in the graphene honeycomb structure. The position of metal impurities with respect to the hexagon structure depends on the metal's electron affinity to, and the electron density on the hexagon structure, e.g., on top of carbon atoms, bridging on top of between two carbon atoms or centered at the middle of the hexagon.

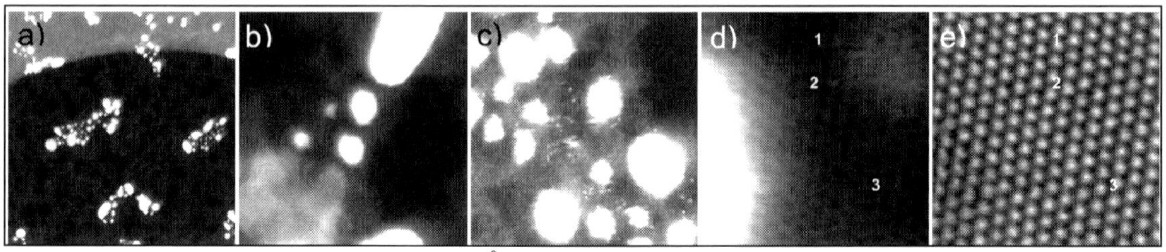

Figure 1. a)-c) HAADF images of 2Å Au doped mono-layer graphene shown at different magnifications and obtained at 60kV. d) and e) 5Å Au doped few-layer graphene: HAADF and corresponding BF image (the gold atoms are right of the numbers). Frame widths are 150, 25, 15, 3 nm in a), b), c), d) and e), respectively.

Chromium, similar to gold, was found mainly on hydrocarbon deposits (fig. 2a). However, the fractional coverage was larger (fig. 2a); this indicates that the Cr-clusters are flatter or less dense than the Au clusters. The crystallinity of the Cr clusters is lower than that of the Au clusters, suggesting that the Cr deposits in molecular form, possibly partially as Cr-oxide (fig. 2b); indeed EEL spectrum images have proven that high Cr levels coincide with oxygen signals. Interestingly, single Cr atoms do appear in some images on mono-layer graphene. These atoms stay in place over several image scans and then disappear, leaving di-vacancies behind. In subsequent scans further impurity atoms move to the edge of the vacancies, disappear upon scanning whilst enlarging the vacancy clusters: hole-drilling in graphene, as shown in fig. 2c), takes place. The hole keeps increasing upon repeated scanning. This is strong indication for the fact that Cr is more reactive with graphene than Au, and that Cr-mediated

C-C dissociation and vacancy formation takes place. This mechanism has been derived from observations of many image scans in different areas. Figure 2c) shows Cr-atoms decorating the edges of a hole in graphene (indicated by white arrow).

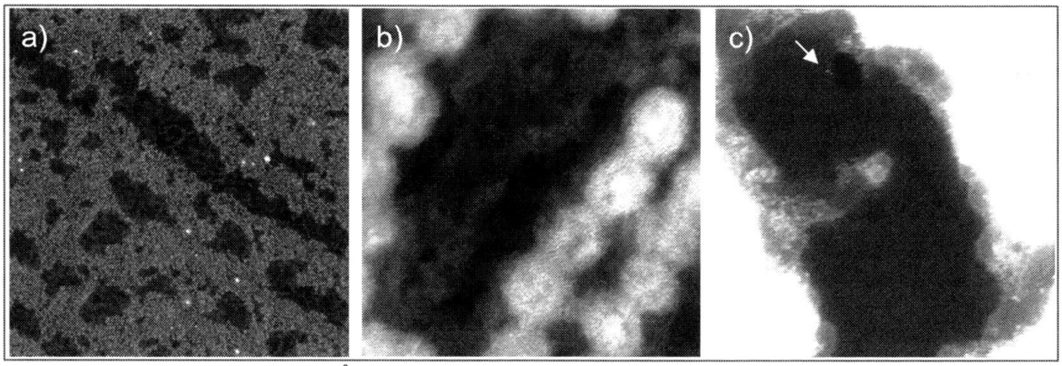

Figure 2. HAADF images of 2Å Chromium doped mono-layer graphene at different magnification, taken at 60kV. Frame widths are 200, 50, 20 nm in a), b), and c), respectively.

Titanium, in terms of coverage, is similar to Cr, and like Au and Cr it does not sit on the clean pristine graphene surface, as shown in low magnification images in fig. 3. Similar to Cr, Ti mediates vacancy formation in graphene at 60 kV, which is much lower than the threshold for carbon atom (as well as for Au, Cr and Ti atom) displacements by the electron beam [7]. This means that both, Cr and Ti have lowered the damage threshold for graphene as theoretically predicted [8]. Figure 3b) shows drilling of graphene: as soon as a hole appears it gets bigger after each repeat scan. Again the process is mediated by impurity atoms, which decorate and move along the hole's edge (fig. 3c). Observation of relatively strong Ti-graphene interaction has been predicted in recent DFT calculations [9]. The Ti binding energy on graphene is calculated to be much higher than the gold, which also explain the reason for weak interaction between gold and graphene and thus why gold is mobile on clean graphene [10]. Furthermore, energy transferred to Au ad-atom from the beam is the lowest among the studied metals [11].

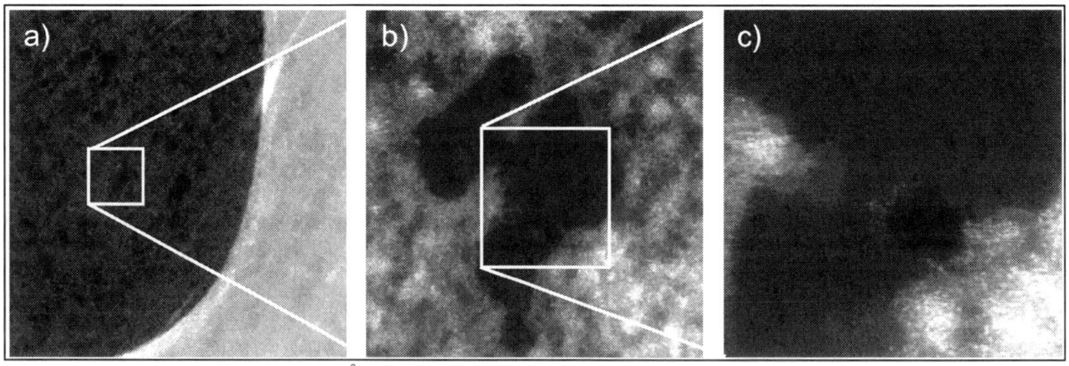

Figure 3. HAADF images of 2Å Titanium doped mono-layer graphene at different magnification, taken at 60kV. Frame widths are 250, 25, 10 nm in a), b), and c), respectively.

4. Conclusions

In conclusion, it appears that individual gold atoms do not adhere to clean single-layer graphene, whereas they are found to reside on top of carbon atoms on few-layer graphene. Cr and Ti appear to interact more strongly with mono-layer graphene and were observed to catalyse the dissociation of C-C and bonds under the electron beam, even at 60kV acceleration voltage. Bonding of occasional, singular Cr and Ti atoms to graphene is believed to initiate this process; it could also be a precursor for the frequent nucleation of clusters, which provide more continuous contact throughout the membrane (in contrast to Au). Thus the importance of Cr and Ti, which are used as sublayer underneath Au for improving electrical contacts on graphene, can be understood.

References

[1] Novoselov, K. S.; et al. A. *Science* **2004,** 306, (5696), 666-669.
[2] Muszynski, R.; Seger, B.; Kamat, P. V. *The Journal of Physical Chemistry C* **2008,** 112, (14), 5263-5266.
[3] Hong, W.; Bai, H.; Xu, Y.; Yao, Z.; Gu, Z.; Shi, G. *The Journal of Physical Chemistry C* **2010,** 114, (4), 1822-1826.
[4] Chan, K. T.; Neaton, J. B.; Cohen, M. L. *Physical Review B* **2008,** 77, (23), 235430.
[5] Zan, R.; Bangert, U.; Ramasse, Q.; Novoselov, K. S. *Nano Letters* **2011,** 11, (3), 1087-1092.
[6] Li, X.; et al. *Science* **2009,** 324, (5932), 1312-1314.
[7] Girit, C. O.; et al., A. *Science* **2009,** 323, (5922), 1705-1708.
[8] Boukhvalov, D. W.; Katsnelson, M. I. *Applied Physics Letters* **2009,** 95, (2), 023109.
[9] Yazyev, O. V.; Pasquarello, A. *Physical Review B* **2010,** 82, (4), 045407.
[10] Zan, R.; Bangert, U.; Ramasse, Q.; Novoselov, K. S. *Small* **2011,** 7, (20), 2868-2872.
[11] Egerton, R. F.; McLeod, R.; Wang, F.; Malac, M. Ultramicroscopy **2010,** 110, 991-997.

Scanning Tunnelling Microscopy of Suspended Graphene

R Zan[1,2], U Bangert[2], C Muryn[3], P. Mattocks[4], B Hamilton[4] and K S Novoselov[1]

[1]School of Physics and Astronomy, The University of Manchester, Manchester, UK
[2]School of Materials, The University of Manchester, Manchester, UK
[3]School of Chemistry, The University of Manchester, Manchester, UK
[4]School of Electrical and Electronic Engineering, The University of Manchester, Manchester, UK

recep.zan@postgrad.manchester.ac.uk

Abstract. We have obtained for the first time atomic resolution STM images of free-standing graphene, and have found UHV high temperature annealing conditions to controllably achieve atomically clean surfaces. This is an important premise for fundamental studies of the interaction of pristine graphene with foreign atom species, in particular metals.

1. Introduction

Graphene, just one atom layer thick, has spurred a flurry of investigations into its structural properties, morphology and chemistry via number of different experimental techniques and instruments [1-3]. One of them is scanning tunnelling microscopy (STM), which has been used to investigate the atomic-scale topography of graphene. However, the graphene used in all such studies is supported by substrates [4, 5] and no observations for suspended graphene have been made [6]. Our experience of suspended graphene is based on transmission electron microscopy (TEM) studies and invasiveness of the electron beam in TEM has prompted us to investigate suspended membranes via STM as an alternative method [7, 8]. We have obtained for the first time atomic resolution STM images of free-standing graphene, and have found UHV high temperature annealing conditions to controllably achieve atomically clean surfaces. This is particularly important for probing the interaction of foreign species, e.g., metal with the pristine graphene surface, since TEM revealed poor adhesion. All reports of graphene-impurity interaction, graphene dosing, doping and functionalisation so far have been made without proof of the presence of a pristine surface- all graphene in previous studies either has an abundance of contaminants or is presented on substrates. The interaction between metal atoms and graphene is of great interest because metals, e.g. gold, chromium etc. are used in macroscopic electrical transport measurements of graphene as electrodes. We will give an account of new cleaning procedures and of our progress in observing atomic structure of suspended graphene at room temperature.

2. Experimental

Two different techniques for the preparation of graphene membranes have been used, which are micromechanical cleavage from HOPG (highly ordered pyrolytic graphite) [3] and CVD growth of films on Cu-substrates [9]. The graphene membranes are transferred to a TEM grid (400 mesh lacey carbon copper) after several chemical treatments [10] and finally dried in a critical point dryer to avoid possible damage. Identification of the graphene layer number has been done via electron diffraction patterns by comparing first and second ring intensities and via high angle annular dark field (HAADF) imaging in conjugation with ultra-high spatially resolved electron energy loss spectroscopy (EELS) [8]. Later, the graphene on the TEM grid is transferred to the STM chamber for analyses.

Initial verification of the existence of monolayer graphene and its coverage on the grid has been obtained in a TEM, a Tecnai F-30. An Omicron low-temperature UHV system (residual pressure 10^{-10} mbar) and electrochemically etched tungsten tips were used for the STM experiment. Standard tip cleaning and forming procedures were carried out in UHV. All STM data were taken at room temperature (300 K). Sample-tip bias voltages used were typically in the range of 0.5-1.5V and tunneling currents were between 0.1 nA and 1.0 nA. The tip performance has been checked on gold on mica and graphite (HOPG) samples as measurement references. To make sure that we are tunneling on suspended parts and not on grid bars, we have systematically taken images at small spatial separation across entire grid squares (each square ~40 micron), with all high quality images yielding similar results.

3. Results and Discussion

We have started the STM measurements to establish annealing condition for suspended graphene (by stepwise increase of the temperature) because TEM studies of suspended graphene have revealed that only few tens of nm^2 clean surface (free from the contamination) can be obtained; these clean patches are surrounded by contamination (hydrocarbon), shown in figure 1a). Perfectly clean surface is required to be able to obtain stable tunneling from graphene in the STM. Our first attempt to image suspended graphene has been after degassing the sample at ca. 50 °C for a few hours, shown in figure 1b). This resulted in unstable tunneling behavior, characterized by diffuse images with poor reproducibility, which indicates that the surface was not clean enough for imaging with the STM tip. After annealing at ca. 100 °C for 1 h, the images showed improved stability although large amounts of contamination remained. Annealing at ca. 270 °C for 4 hours, and subsequently to ca. 300 °C for 18 hours further improved the image stability; however residual surface contamination still remained. An additional annealing step at ca. 550 °C for 15 minutes produced a clean surface, shown in figure 1c), and showed stable tunneling with smooth line profiles.

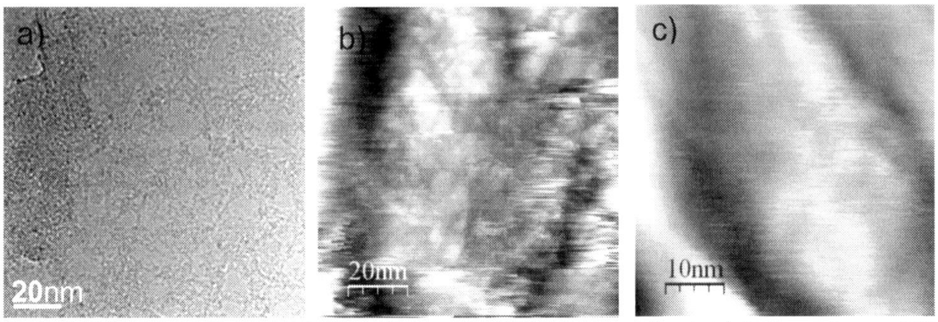

Figure 1. a) TEM image of suspended graphene. (b) STM image after initial degassing at 50°C for 4 hours. (c) Annealing at 550°C for 15 minutes. All images are raw.

By further exploring the annealing regimes in terms of time and temperature it was found that annealing the pristine sample for 24 hours ~400 °C is usually sufficient to remove almost all contamination, and with a well formed tip we were able to obtain atomic resolution images, shown in figure 2. A clear honeycomb monolayer structure was observed and the average bond length of 1.42 Å is consistent with literature values. This cleaning procedure allowed us to acquire atomic resolution graphene images at almost every probing position on the suspended layer.

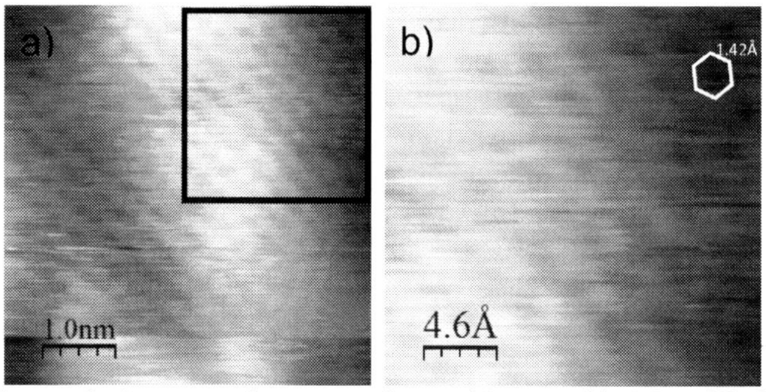

Figure 2. STM high resolution monolayer graphene (a) raw (b) Magnified area of square in a). Hexagonal monolayer structure is observed and one hexagon ring is overlaid in white in (b). The images represent raw data.

In some areas of the sample both hexagonal and triangular structures were observed in a single image frame, corresponding to monolayer and bilayer graphene respectively. Figure 3 shows such mono- and bi-layer regions. Magnified mono- and bi-layer region in squares 1 and 2 in figure 3a) are shown in figures 3b and 3c, respectively.

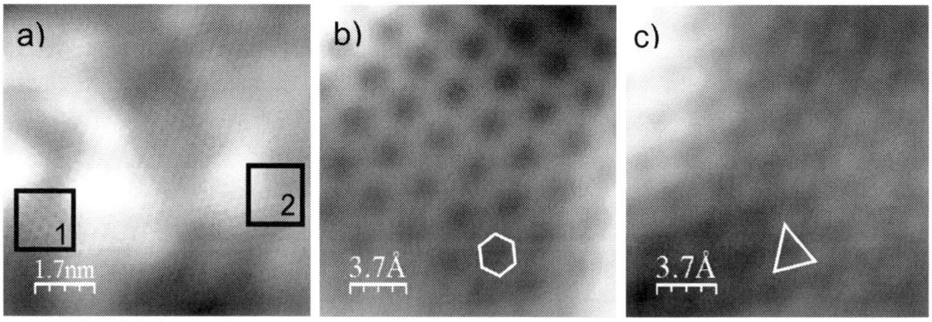

Figure 3. (a) STM high resolution mono- and bi-layer regions (b) magnified monolayer region, square 1 in a). (c) magnified bi-layer region, square 2 in a). One hexagon ring is overlaid in white in (b), and a white triangle in (c). The images represent filtered data.

4. Conclusions

Suspended, i.e., free-standing graphene has been imaged at atomic resolution in the STM for the first time. Cleaning and annealing conditions have been established that enabled us to observe large monolayer (hundreds of nm^2) and also some bilayer regions (tens of nm^2). The cleaning and imaging conditions reported here should make further development and application of UHV STM imaging and spectroscopy of suspended graphene possible.

References

[1] Berger, C.; Song, Z.; Li, X.; Wu, X.; Brown, N.; Naud, C. c.; Mayou, D.; Li, T.; Hass, J.; Marchenkov, A. N.; Conrad, E. H.; First, P. N.; de Heer, W. A. *Science* **2006,** 312, (5777), 1191-1196.

[2] Meyer, J. C.; Geim, A. K.; Katsnelson, M. I.; Novoselov, K. S.; Booth, T. J.; Roth, S. *Nature* **2007,** 446, (7131), 60-63.

[3] Novoselov, K. S.; Geim, A. K.; Morozov, S. V.; Jiang, D.; Zhang, Y.; Dubonos, S. V.; Grigorieva, I. V.; Firsov, A. A. *Science* **2004,** 306, (5696), 666-669.).

[4] Ishigami, M.; Chen, J. H.; Cullen, W. G.; Fuhrer, M. S.; Williams, E. D. *Nano Letters* **2007,** 7, (6), 1643-1648.

[5] Stolyarova, E.; Rim, K. T.; Ryu, S.; Maultzsch, J.; Kim, P.; Brus, L. E.; Heinz, T. F.; Hybertsen, M. S.; Flynn, G. W. *Proceedings of the National Academy of Sciences* **2007,** 104, (22), 9209-9212.

[6] Zan, R.; Bangert, U.; Muryn, C.; Mattocks, P.; Hamilton, B.; Novoselov, K. S. **submitted**.

[7] Bangert, U.; Gass, M. H.; Bleloch, A. L.; Nair, R. R.; Eccles, J. P*hysica Status Solidi (a)* **2009,** 206, (9), 2115-2119.

[8] Gass, M. H.; Bangert, U.; Bleloch, A. L.; Wang, P.; Nair, R. R.; Geim, A. K. *Nat Nano* **2008,** 3, (11), 676-681.

[9] Li, X.; Cai, W.; An, J.; Kim, S.; Nah, J.; Yang, D.; Piner, R.; Velamakanni, A.; Jung, I.; Tutuc, E.; Banerjee, S. K.; Colombo, L.; Ruoff, R. S. *Science* **2009,** 324, (5932), 1312-1314.

[10] Zan, R.; Bangert, U.; Ramasse, Q.; Novoselov, K. S. *Nano Letters* **2011,** 11, (3), 1087-1092.

Imaging and diffraction characterisation of 2D inorganic nanostructures

A. Shmeliov[1], M. Shannon[2], P. Wang[1], H. E[1], P. D. Nellist[1] and V. Nicolosi[1]

[1]University of Oxford, Department of Materials, Parks Road, OX1 3PH, Oxford, UK
[2]SuperSTEM, STFC Daresbury Laboratories, Keckwick Lane, Warrington, WA4 4AD, UK

Email: aleksey.shmeliov@materials.ox.ac.uk

Abstract. 2D nanomaterials are a novel class of materials, noted for their potential in a range of high impact applications, such as sensing catalysis, and composite reinforcement. However, their structural features have not yet been fully characterised. In this work we have shown that 2D nanostructures of hBN, MoS_2, and WS_2, prepared by liquid exfoliation, can stack in sequences different from the bulk stacking. In 2D hBN non-sequential stacking was observed. 2D MoS_2 and WS_2 appear to have stacking where Mo/W atoms are located on the top of each other which differs from naturally occurring 2H and 3R polytypes. The majority of the hBN and MoS_2 2D nanostructures retain the stacking of their bulk counterparts, while in 2D WS_2 non-bulk stacking dominates. The conclusions are based primarily on the atomically resolved ADF STEM images with supporting evidence from electron diffraction.

1. Introduction

Initially, two-dimensional (2D) nanostructures were considered as the fundamental building blocks for layered materials [1] and later for nanotubes [2]. Following the discovery of graphene [3], 2D nanostructures became a separate class of materials, and were noted for their potential in a range of high-impact applications, such as sensing [4], catalysis [5], and composite reinforcement [6]. Liquid exfoliation [6] and chemical vapour deposition [7] demonstrated the feasibility of mass-scale production. With the advent of mass-produced 2D nanostructures a new focus of the research is to characterise these materials.

Structural properties of any material play a particularly important role as they determine its intrinsic properties and performance in applications. Furthermore, for the case of 2D nanomaterials knowledge of the topological defects and edge structure might offer insight towards functionalisation approaches.

The primary focus of this work is stacking variations in 2D nanostructures of hexagonal boron nitride - hBN, molybdenum disulfide - MoS_2, and tungsten disulfide - WS_2, investigated using aberration corrected annular dark field (ADF) scanning transmission electron microscopy (STEM) with supporting evidence from electron diffraction. The particular advantage of this study is the choice of ADF STEM imaging as the main characterisation technique. The acquired images show intensities that are strongly dependent on the atomic number of the species present in the sample. All of the samples were produced by liquid exfoliation of the bulk material in either N-methyl-2-pyrrolidone or isopropanol (more details on the sample preparation and analysis of the exfoliated material can be found here [6]). Suitability of this production method for inorganic layered materials has been demonstrated only recently. Out of 2D nanostructures of hBN, MoS_2, and WS_2 produced by liquid exfoliation, only hBN has been studied before [8][9].

2. Results and Discussion

2.1 2D Hexagonal Boron Nitride

In this work, initial experimental evidence of the stacking variation came from electron diffraction patterns. The ratio of the spot intensities in the electron diffraction pattern can be related to the corresponding squared modulus ratio of the structure. Depending on the stacking this ratio will change. Here we consider 4 different stacking sequences of hBN (Fig. 1). Experimentally, two different ratios of the (100) and (110) spot intensities, I_{100}/I_{110}, were observed (Fig. 2). In Figs. 2b and 2e ratios are 9.3±2.5 and 0.29±0.09, correspondingly. While the first ratio doesn't agree with any of the calculated values (Table 1), the second corresponds to AB. Tilt of the sample combined with stacking faults may be the reason for the discrepancy between theoretical prediction and experimental values. Furthermore, the observed differences cannot be explained by considering fractional unit cells. Overall, for this case electron diffraction doesn't enable reliable determination of the stacking sequence. However, it suggests that stacking variations can exist in 2D hBN nanostructures.

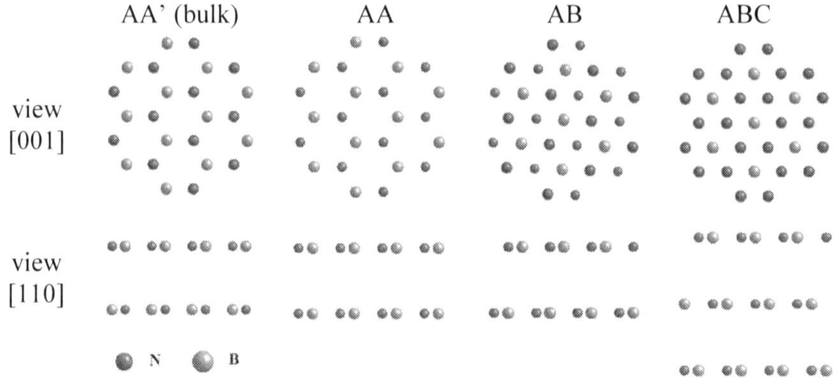

Figure 1: hBN structural models. View along [001] shows pattern expected for atomically resolved ADF image. View along [110] unambiguously shows stacking sequence. The interlayer distance is shown to scale.

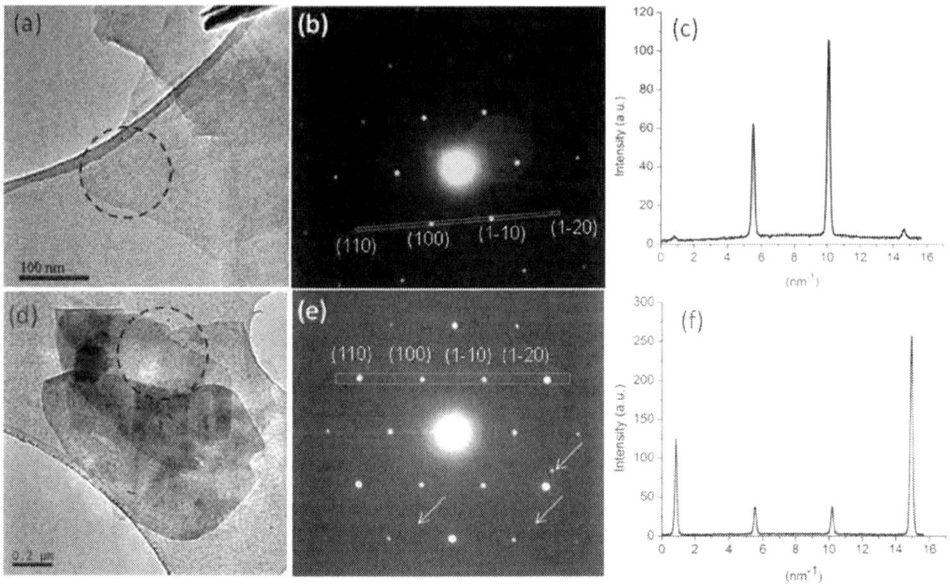

Figure 2: (a), (d) Low resolution TEM images with corresponding (b), (e) diffraction patterns and (c), (f) intensity profiles. Additional spots marked with arrows in (e) are likely to correspond to the presence of in-plane rotated layers.

Table 1: Calculated ratios of intensities, I_{100}/I_{110}, for the different stacking sequences described in Fig. 1. Ratios were calculated using atomic scattering amplitudes for neutral atoms. The electron beam direction is considered perpendicular to the hBN basal plane.

	Ratio (I_{100}/I_{110})					
No. of layers:	1	2	3	5	11	bulk
AA', AA	1.1	1.1	1.1	1.1	1.1	1.1
AB	1.1	0.28	0.37	0.31	0.28	0.28
ABC	1.1	0.28	0	0.04	0.01	0

Fig. 3a shows characteristic atomic resolution ADF STEM image of a typical hBN nanosheet. An accurate comparison between the ADF STEM projected pattern in Fig. 3a with the models in Fig. 1 shows that the stacking sequence corresponds either to AA' or AA. Theoretical calculations [10] indicate that AA stacking is the least energetically favourable; therefore one can assume that the structure in Fig. 3a should correspond to bulk.

Two out of approximately 20 high resolution ADF images did not show the honeycomb pattern observed in Fig. 3a (Fig. 3b). Fig. 3c shows region where the transition between the honeycomb and the new pattern can be seen. This non-honeycomb pattern is similar to patterns formed by AB and ABC structures. Another feature of Figs. 3b,c, which is not present in Fig. 3a, is the large number of bright spots, corresponding to the presence of impurity atoms. One possibility is that high concentration of impurities can stabilise stacking disorder.

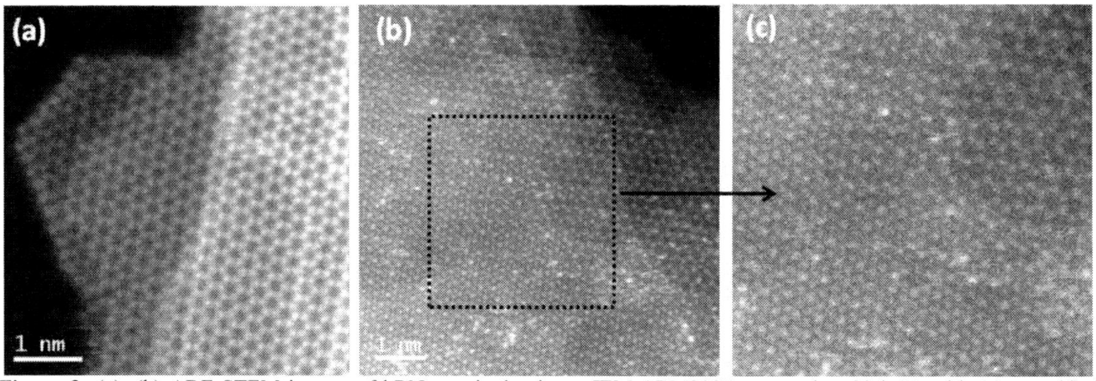

Figure 3: (a), (b) ADF STEM images of hBN acquired using a JEM-ARM200F operated at 80 keV, with (c) magnified region of the image in (b).

2.2 2D Molybdenum Disulfide and Tungsten Disulfide

The structure of the 2H polytype, which is dominant for MoS_2 and WS_2, is shown in Fig. 4a. Since the atomic number of Mo/W is much higher than S, in MoS_2/WS_2 ADF images the signal from S atom is swamped by the neighbouring W/Mo signal. Indeed, experimental images of MoS_2 in Fig. 5 do not show S signal. As a result, the pattern created by the Mo/W atoms for the 2H polytype is different for the mono- and the multi-layer regions (Fig. 4b). Therefore, it follows that in Fig. 5a the 2D nanostructure has stacking similar to 2H polytype. In Fig. 5b the pattern is the same for the entire field of view of the image and the presence of a step can be clearly seen. It means that Mo atoms must be stacked on the top of each other. Such stacking sequence exists in neither 2H nor 3R polytypes. Statistical analysis on the intensities of the possible S sites might allow determination of the complete crystallographic structure. Out of approximately 20 MoS_2 nanostructures, only one of them had the structure with Mo atoms stacking on the top of each other, while the rest of them were consistent with 2H polytype. For WS_2 nanostructures, similar results were observed. However, unlike the case of 2D MoS_2, stacking of W on top of each other is dominant.

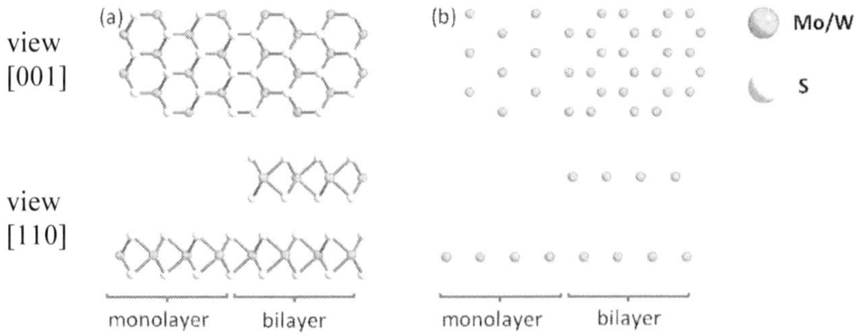

Figure 4: (a) The structure of MoS_2 2H polytype. (b) The structure of MoS_2 2H polytype with S atoms hidden.

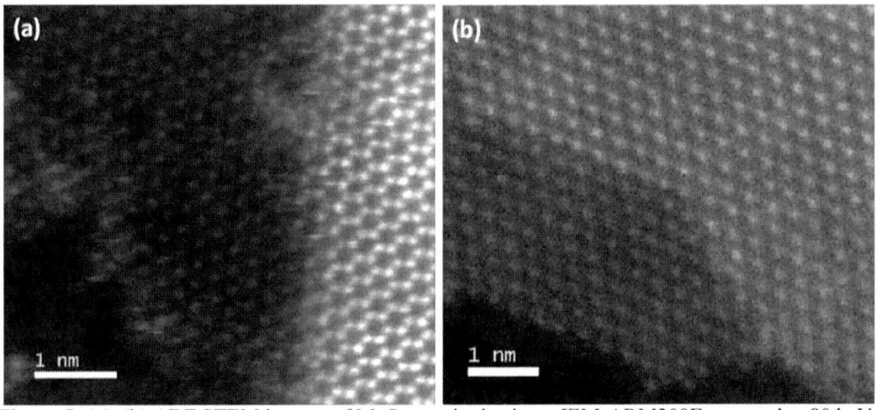

Figure 5: (a), (b) ADF STEM images of MoS_2 acquired using a JEM-ARM200F operated at 80 keV.

3. Conclusions

In this work imaging and structural characterisation of 2D inorganic nanostructures, produced by liquid exfoliation, was carried out. The primary focus of the structural analysis was stacking. On the basis of electron diffraction and ADF STEM non-bulk stacking was reliably determined in 2D nanostructures of hBN, MoS_2, and WS_2. However, based on the set of acquired images, with the exception of WS_2, the majority of the 2D nanostructures retain the stacking of their bulk counterparts. Observed variations in stacking sequences are expected to affect the electronic properties of 2D nanostructures, having considerable impact for their potential applications.

References

[1] Wallace P, Physical Review **71**, 622–634 (1947).
[2] Iijima S, Nature **354**, 56–58 (1991).
[3] Novoselov K S, Geim A K, Morozov S V, Jiang D, Zhang Y, Dubonos S V, Grigorieva I V, and Firsov A A, Science (New York, N.Y.) **306**, 666-9 (2004).
[4] Schedin F, Geim A K, Morozov S V, Hill E W, Blake P, Katsnelson M I, and Novoselov K S, Nature Materials **6**, 652-5 (2007).
[5] Chianelli R, Siadati M, De la Rosa M, Berhault G, Wilcoxon J, Bearden R, and Abrams B, Catalysis Reviews **48**, 1-41 (2006).
[6] Coleman J N et al, Science **331**, 568-571 (2011).
[7] Li X, Cai W, An J, Kim S, Nah J, Yang D, Piner R, Velamakanni A, Jung I, Tutuc E, Banerjee S K, Colombo L, and Ruoff R S, Science **1312**, 1312-14 (2009).
[8] Krivanek O L et al, Nature **464**, 571-4 (2010).
[9] Kim J S, Borisenko K B, Nicolosi V, and Kirkland A I, ACS Nano **5**, 3977-86 (2011).
[10] Ribeiro R and Peres N, Physical Review B **83**, 1-6 (2011).

Direct Measurement of the Crystallographically Sensitive Atomic Termination of Nanophase Cerium Dioxide

N.P. Young[1], S.J. Haigh[2], H. Sawada[3], K. Takayanagi[4], A.I. Kirkland[1]

1. Department of Materials, University of Oxford, Parks Road, Oxford OX1 3PH, UK
2. School of Materials, University of Manchester, Grosvenor Street, M1 7HS, UK
3. JEOL Ltd. 1-2 Mushasino 3-Chome, Akishima, Tokyo 196, JAPAN
4. Department of Physics, Tokyo Institute of Technology, 2-12-1-H-51 Oh-okayam, Meguro-ku, Tokyo 152, & CREST, Kawaguchi, Saitama 332-0012, JAPAN

E-mail: neil.young@materials.ox.ac.uk

Abstract. Aberration-corrected high-resolution transmission electron microscopy (HRTEM) was used to study the surface termination of nanophase cerium dioxide particles. Application of computational exit-wave restoration and comparison with simulations confirmed the {111} surfaces to have an oxygen termination, and while structurally unstable at room temperature the {100} facets exhibited a metal terminating plane. This crystallographically sensitive atomic termination was found to be maintained under the reducing conditions of the electron beam. During electron beam irradiation, Electron Energy Loss Spectroscopy (EELS) analysis revealed that a change from Ce^{4+} to Ce^{3+} originates at the particle surface, and gradually extends into the particle as reduction continues. The results aid understanding of the oxygen buffering mechanism in cerium dioxide, and provide valuable insight into the use of CeO_2 nanoparticles in catalysis.

1. Introduction

Optimisation of activity and selectivity in catalysis requires deep understanding of the atomic and molecular-level processes occurring during reactions. A first step towards this goal is to obtain robust information on the structure of nanophase catalyst materials at the atomic level, particularly with respect to nanocrystal surfaces and the immediate sub-surface region. At the nanoscale, morphological constraints upon thermodynamically stable particle structures dictate the type and extent of different crystallographic surface facets. In such situations bulk stoichiometry may be modified and/or variations in atomic termination of different surface facets may be observed. The ability of cerium dioxide to catalyse redox reactions and its application in solid oxide fuel cell anodes is due to the multiple valency of cerium in both 3^+ and the 4^+ oxidation states, and the mobility of oxygen anions. A better understanding of the mechanism of oxygen buffering in this material is sought and experimental studies of cerium dioxide nanoparticle surfaces will provide key insights. In the present study, advanced electron microscopy based techniques are used to experimentally deduce the physical and chemical structure of cerium dioxide nanoparticle surfaces.

2. Experimental Method

Our methodology employs commercially available cerium dioxide nanophase powder (average particle size 15 nm, Pi-KEM Ltd.). Particles were suspended in isopropanol, ultrasonicated for 10 minutes and then dropcast directly onto holey carbon coated copper TEM grids. Electron microscopy was performed using a JEOL JEM-2200MCO aberration-corrected (S)TEM, operating at 200kV. The electron energy spread was 0.6eV, spherical aberration corrected to a value of -5μm, focal spread 3.8nm and beam divergence was 0.1eV. High Angle Annular Dark Field (HAADF) Scanning Transmission Electron Microscope (STEM) imaging and EELS was performed on a JEOL R005 (S)TEM, also operating at 200kV, with EEL dispersion of 0.2eV/channel and collection angle 50mrad. Focal series reconstruction was performed using the FTSR plug-in for Digital Micrograph (HREM Research) and multislice HRTEM image simulation.

3. Results and Discussion

A selection of cerium dioxide nanoparticles were imaged by aberration-corrected HRTEM. Figure 1 shows typical examples of nanoparticles originating from the nanophase powder. These images are the phase of the specimen electron exit-wavefunction, each is reconstructed from a focal series of 20 images, with half taken below Gaussian focus and half over. In figure 1(a) part of a single nanoparticle orientated close to the <110> direction with respect to the electron beam is shown. Typically particles would flocculate such that only part of their structure was clearly visible in HRTEM. This particle shows clear {111} facets and also a {100} facet. It should be noted that under electron irradiation the surface atomic structure of {111} type facets was very stable over the course of the experiment, however the {100} type showed significant mobility of the Ce ions for almost all particles surveyed. From individual HRTEM images and certainly the exit-wavefunctions presented here, features corresponding to both cation and anion species may be tentatively identified within the images. Particles orientated in the <110> direction have {111} planes that alternate between complete Ce and O planes. Interpretation in the <110> direction is further simplified by each individual column having either a sole Ce or O occupancy along the beam direction. In figure 1(a), large bright spots corresponding to Ce columns and smaller satellite spots corresponding to the oxygen columns may be seen. Figure 1(b) shows a cerium dioxide particle orientated near to the <001> direction, particles orientated in this direction were much less common and also less suitable for surface termination studies since facets contain both Ce and O atoms along their length.

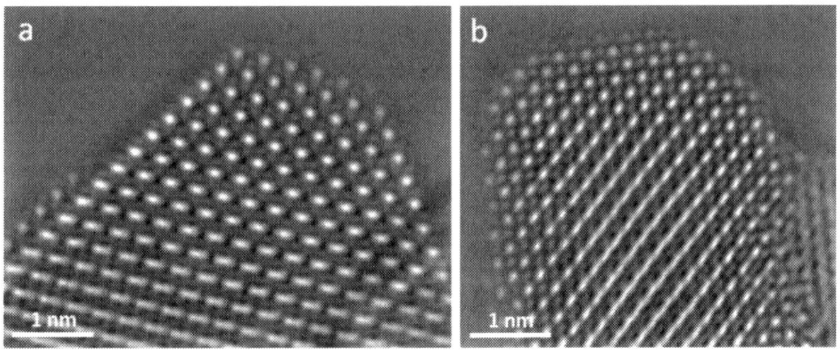

Figure 1: Phase images of electron exit-wavefunctions, from focal series restorations of 20 individual images, centred around Gaussian defocus and with a focal step of 3.6nm. (a) Nanoparticle orientated near to the <110> direction, (b) nanoparticle orientated near to <001> with respect to the beam.

Figure 2(a) shows the phase of the restored specimen exit surface wavefunction of another cerium dioxide nanoparticle viewed close to a <110> direction. This particle was chosen for detailed analysis since it is attached to other particles at only one side, leaving a number of free surfaces to be studied in isolation. The surfaces in the restored phase image parallel to the electron beam have been characterised by extracting average line profiles and matching these to equivalent profiles extracted from simulated structural data. One set of profiles bisects a {111} facet (profile 1) and the other bisects a {100} facet (profile 2). Exit-wavefunctions are generally free from significant aberrations and imaging artefacts so greatly simplify comparison with calculated data, relative to single conventional HRTEM images. Figure 2(b) plots these average line profiles, where the vacuum level is represented on the far left of the profile. The main peaks in both profiles represent the terminal white spot at either surface; that is the final Ce column position. The main difference between the two profiles is that profile 1 exhibits a smaller secondary peak to the left of the main peak. This smaller peak (indicated by an arrow in figure 2(b)) at the very surface of the {111} facet was found for all {111} facets surveyed for all particles in this orientation, and corresponds to the small spots visible in the image. We attribute these to an oxygen surface termination of the {111} facets.

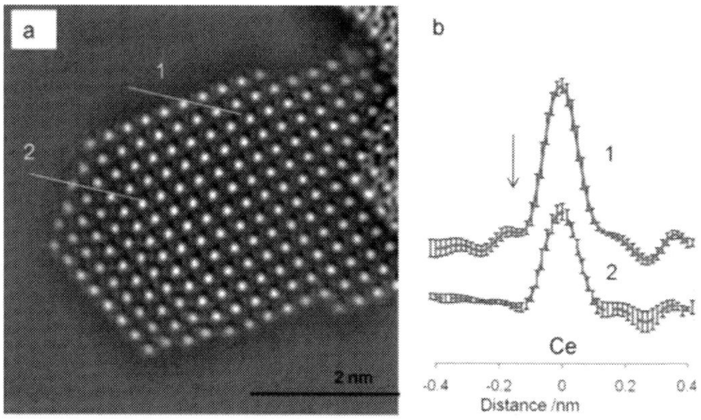

Figure 2: (a) Phase of the specimen exit surface wave-function for a small cerium dioxide nanoparticle imaged close to a <110> direction. (b) Average line profiles across surfaces along a <001> direction as indicated in (a). Profile 1 shows the average obtained for a {111} surface with a peak closest to the surface corresponding to oxygen. This peak is absent in Profile 2 for a {100} surface. In each profile the solid line plots the mean value of the experimentally measured phase shift and the error bars show the standard deviation in this measurement for equivalent surface sites as a function of pixel position.

Multislice simulations of the phase of the exit wave-function were performed, where the termination of the cerium dioxide particle was varied between Ce and O for both the {111} and {100} surfaces. Figure 3 demonstrates that when the surface is oxygen terminated a line profile across the restored phase exhibits a shoulder with a maximum phase shift of ~20% of the maximum at the cerium metal position (indicated by the arrows in Figures 3(a) and 3(b)). Comparison of these simulations with the experimentally restored phase images leads to the conclusion that the {111} surfaces are oxygen terminated whereas the {001} surfaces are metal terminated, in agreement with previously reported theoretical calculations [1]. The valency of the cerium ions was deduced using Electron Energy Loss Spectroscopy (EELS) via observation of the ratio of the Ce M_4/M_5 peaks. Figure 4 shows EEL spectra recorded as a function of distance from a {111} surface in order to monitor the cerium oxidation state during beam induced reduction. Figure 4(c) shows that the perpendicular distance from the {111} surface at which a Ce^{3+} species is present may be determined from local measurements of the M_4/M_5 peak ratio. This distance varied with increased exposure to the electron beam and demonstrated a change from Ce^{4+} to Ce^{3+} beginning at the surface of the particle. This result provides direct experimental verification that oxygen vacancy diffusion into the particle is favored until the whole particle is eventually transformed. Sputtering of oxygen ions via the focused electron beam is also a factor in the observed change in Ce ion valency, however given the low electron dose in the initial exposure this is thought to have negligible influence upon the results. Further details are given in another article [3].

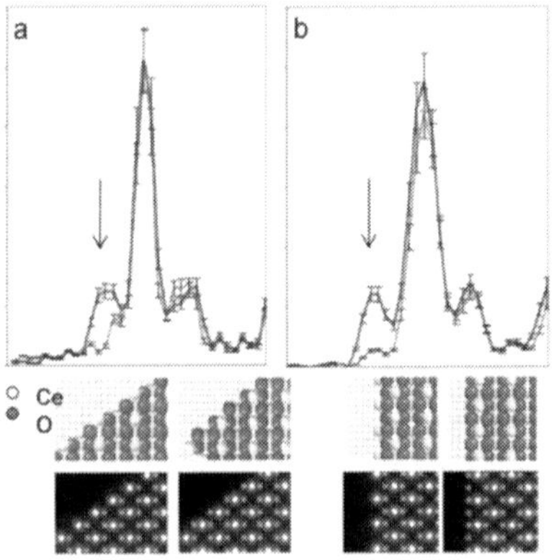

Figure 3: Simulated average line profiles in a <100> direction across (a) a {111} surface, and (b) a {100} surface, when cerium dioxide is imaged along a <110> direction. Metallic (fine lines) and oxygenated (bold lines) surfaces are compared, with oxygen termination giving rise to a shoulder as indicated by the arrows in (a) and (b). The atomic models used to perform the multislice calculations and simulated phase images from which the profiles were extracted are also shown.

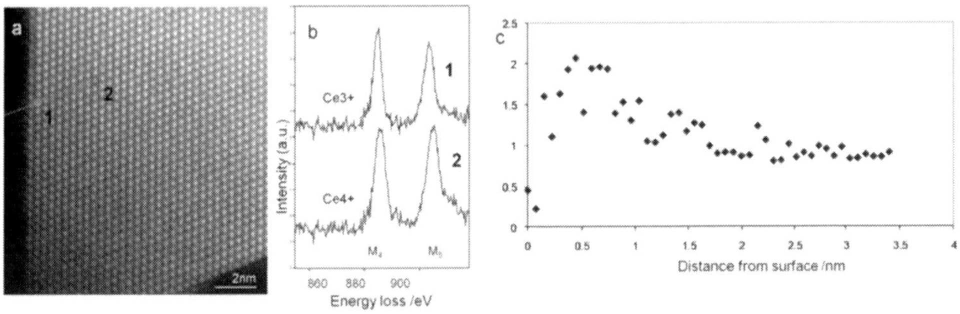

Figure 4: (a) HAADF STEM image of a cerium dioxide nanoparticle. (b) EEL spectra of the Ce-$M_{4,5}$ edge the positions marked in (a). Spectra were acquired sequentially along a <111> direction across a {111} surface as marked in (a). (c) Ratio of the intensity of the Ce-M_4 and Ce-M_5 white lines as a function of the distance from the surface.

Conclusions

The surface structure of cerium dioxide nanoparticles was investigated through aberration-corrected HRTEM and exit-wavefunction restoration. The {111} surfaces of nanoparticles viewed in the <110> direction were found to have an oxygen termination, while the apical {100} surfaces were seen to have a cerium termination. EELS was used to measure the valency of cerium ions on the surface and sub-surface layers of the nanoparticles. A trend from Ce^{3+} at the surface to Ce^{4+} within the interior was noted and showed that particle reduction proceeds from the surface inwards.

References

[1] A.S. Barnard, A. I. Kirkland. *Chem. Mater.* **2008**,20, 5460
[2] C. Binet, A. Badri, J. C.Lavalley, *J. Phys. Chem.* **1994**, 98, 6392
[3] S.J. Haigh, N.P.Young, H.Sawada, K.Takayanagi, A.I. Kirkland, *ChemPhysChem*, **2011**,12, 2396

Acknowledgements
This work was supported by EPSRC (grant EP/F048009/1), JEOL (UK) Ltd and the Japan Science and Technology Agency (JST) under the CREST project.

Journal of Physics: Conference Series **371** (2012) 012073 doi:10.1088/1742-6596/371/1/012073

Aberration corrected and 3D cryo-tomography HAADF-STEM surface studies of ZnO tetrapods

M R Ward[1,2], H Sugiura[3], K Yoshida[2,4], N Tanaka[3], P L Gai[1,2,5] and E D Boyes[1,2,6]

[1] Department of Physics, University of York, York, YO10 5DD, UK
[2] York JEOL Nanocentre, University of York, York, YO10 5DD, UK
[3] Ecotopia Science Institute, Nagoya University, Chikusa-ku, Nagoya, 464-8603, Japan
[4] Japan Fine Ceramics Centre, 2-4-1 Mutsuno, Atsuta-ku, Nagoya, 456-8587, Japan
[5] Department of Chemistry, University of York, York, YO10 5DD, UK
[6] Department of Electronics, University of York, York, YO10 5DD, UK

E-mail: mrw502@york.ac.uk

Abstract. We present a morphology study of ZnO tetrapods using aberration corrected TEM, HAADF-STEM and 3D HAADF-STEM cryotomography as an alternative to more conventional TEM and SEM techniques. We use 3D IMOD reconstructions to show that the $\{1\bar{1}0\}$ facets dominate the total surface area of uniform hexagonal prism tetrapods. Using HRTEM we show that the small tetrapods have a zincblende phase core from which the four legs extend. The facets and the edges of these legs were found to be atomically clean and flat with the potential for ZnO tetrapods as model substrates. We deposited ultrafine Pt/Pd nanoparticles onto the tetrapods and investigated the resulting morphologies. We found using HAADF-STEM cryotomography and reconstruction techniques that the nanoparticle coverage gave separate nanoparticles and overall uniform coverage. We believe these techniques and the results from them could be useful for the development of nanoparticle-ZnO tetrapod composite systems with applications in optoelectronics, gas sensing and catalysis.

1. Introduction

ZnO is a semiconductor with a large direct band gap of 3.37 eV [1]. At the nanoscale, it forms many different structures that range from simple rods to complex 3D formations such as flowers. The simplest 3D structure is the tetrapod which consists of four tetrahedrally coordinated arms that grow from a central nucleus [1]. The arms are usually single crystal wurtzite with lattice parameters a = 0.25 nm and c = 0.52 nm. The structure of the nucleus is of some debate because for different tetrapod sizes and morphologies different conclusions about its core structure have been made [2].

Nanoscale ZnO is a candidate material for use in UV optoelectronics, field emission displays, solar cells and gas sensors but the branched nature of ZnO tetrapods enable it to have a more diverse range of applications compared to its non-branched equivalents (e.g. rods). For example, branching enables multiple interfaces to be made and as a result entire devices have been demonstrated with single tetrapods [3].

The applications of ZnO have been further expanded by adding metals to ZnO structures to form metal-metal oxide composite structures. The addition of metallic nanoparticles, usually larger than 5 nm in diameter leads to a profound change in the band structure and the chemical properties of the structure as a whole [4]. The metal-metal oxide composite systems have also been found to increase the catalytic activity of both metal and the substrate [5].

Although ZnO tetrapods and related composite systems are abundant in the literature, most of the data revolves around conventional SEM and TEM data. Here we present an alternative approach to characterising the morphology of such structures using a combination of aberration corrected HRTEM and HAADF-STEM, and 3D HAADF-STEM cryotomography. Using these techniques we have indicated that small tetrapod structures can have a zincblende nucleus and that the tetrapod legs are formed from atomically clean surfaces. A mixture of 1.5 nm Pt/Pd nanoparticles were sputtered onto the surfaces and with HAADF-STEM cryotomography a detailed analysis of the resulting morphology was performed which showed that the nanoparticle coating was uniform. We believe these techniques and the results could be useful for the development of tetrapod and metal particle composite systems.

2. Experimental Procedure

The ZnO tetrapods were synthesised by rapidly heating pure Zn pellets in air. The pellets were placed into a tungsten wire cup and a large current was fed through to rapidly heat the pellets. The resulting vapour oxidised in air and a substrate was transported through the vapour to make TEM specimens. Controlling the current through the wire allowed some control over the growth rate. A mixture of 1.5 nm Pt/Pd nanoparticles was used to coat the ZnO TEM grids using a JEOL JFC-2300HR depositing machine. The overall coating thickness was 0.2 nm.

HAADF-STEM was preferred for the cryotomography series because it suffers less from diffraction contrast and interference effects that can sometimes appear in TEM images [6]. Furthermore, the focus here was the morphology of the tetrapods before and after the coating process hence the Z dependant intensities of HAADF-STEM allowed for more quantitative analysis of the morphologies. HRTEM however was used primarily for examining the overall structure of the tetrapods.

Cryotomography was achieved using an FEI Tecnai Polara fitted with a large objective polepeice gap to give a high tilt range of $\pm$ 70° using a liquid nitrogen cooled specimen holder. The low temperature ensured that the specimen would be more resilient to beam damage during the extended data acquisition time. The specimens were tilted between the two tilt extremes in 1° increments; so 140 frames were recorded in total. Using frames from the tilt series, the 3D reconstructions were generated with IMOD using the general back-projection method. An in-house program was used to evaluate the intensities of the Pt/Pd coated tetrapods to distinguish Pt/Pd intensities from the ZnO and vacuum intensities.

Aberration corrected TEM and HAADF-STEM was carried out using a double aberration corrected JEOL 2200FS with better than 0.1 nm resolution. Correcting the aberrations in the STEM probe ensured that the probe was as sharp as possible to ensure maximum intensity which is necessary for examining small structures such as ultrafine nanoparticles.

3. Results and Discussion

Figure 1 shows three different views using the HAADF-STEM. The low magnification image (a) shows a typical tetrapod leg which has hexagonal prism morphology although other structures can also be seen. We believe this to be a consequence of variations in the growth rate. The aberration corrected image (b) shows a section of a tetrapod leg orientated in the $[\bar{1}\,0\,1]$ zone axis. The uniform contrast at the edge implies that the edge formed by the $(\bar{1}\,1\,0)$ atomic planes are atomically clean and the $(1\,\bar{1}\,0)$ facets are flat. The image in (c) is a frame from an IMOD reconstruction of a HAADF-STEM cryotomography series (not shown here). This technique allows straightforward analysis of the morphologies of the tetrapods. The tetrapod shown in (c) has legs that differ from one another but all generally show a hexagonal prism

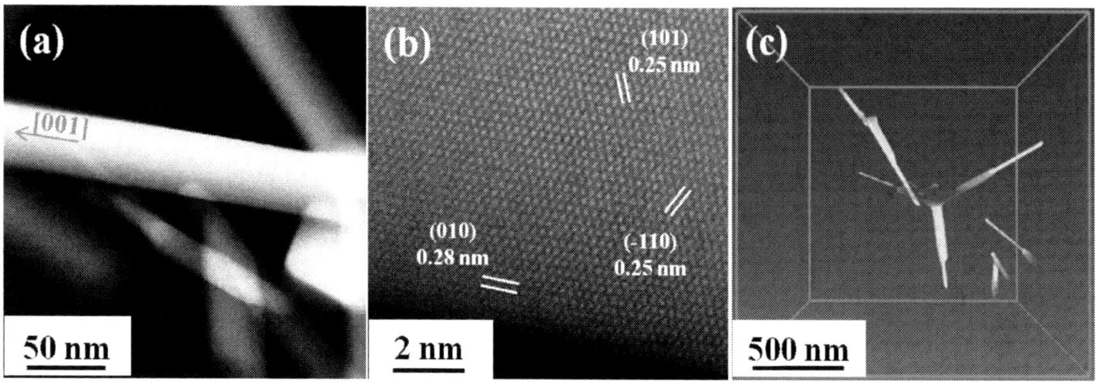

Figure 1. HAADF-STEM images of ZnO tetrapods with (a) showing a low-mag image of a tetrapod leg, (b) an aberration corrected HAADF-STEM image and (c) showing a frame from a 3D IMOD reconstruction built using a HAADF-STEM tilt series

with hierarchical step. Figure 2(a) shows an HRTEM image of a thin tetrapod suspended over a hole. The angle between the two legs is approximately 112°. As indicated, each leg grows out in the [0 0 1] direction from the core. It can be seen that near the core there is a high density of defects which appear to be stacking faults and via these stacking faults, the wurtzite arms grow from the core [7]. For small tetrapods there have been suggestions of a zincblende core. The appearance of the atomic columns in the core in (b) looks much like cubic [1 1 0]. To confirm this, the FFT of 2(a) is shown in 2(b) which identifies the zincblende spatial frequencies assuming zincblende ZnO has lattice parameter a = 0.462 nm [8]. The FFT can be seen as an overlap of three diffraction patterns, two WZ[1 1 0] and a ZB[1 1 0]. The three crystal directions are identified in (a) and (b). Unlike the HAADF-STEM images in figure 1, the HRTEM images do not quantitatively suggest that the facets are flat. The HRTEM images show what appear to be atomic steps on the facet edges but we believe these may be a result of electron beam damage as their appearance varied between different images of the same area. Separated nanoparticles offer higher surface areas and such a coverage would represent a more realistic catalyst. Figure 3 shows a HAADF-STEM image of the Pt/Pd coated tetrapods and two alternate 3D views from an IMOD reconstruction. The difference in Z between the Pt and Pd is enough to clearly identify the two. The contrast from such particles on an HRTEM image such as figure 2 (b) would make it difficult to distinguish Pt and Pd from ZnO and would suffer from interference effects (e.g. Moire fringes). Overall, the nanoparticles did not seem to prefer a particular $\{1\,\bar{1}\,0\}$ facet with the coated surfaces showing individual nanoparticles and no evidence of coalescence. We believe such a system could be useful for model catalyst studies.

4. Conclusions

We have demonstrated quantitatively that the facets of hexagonal prism ZnO tetrapod legs are flat using aberration corrected HAADF-STEM and elucidated with 3D HAADF-STEM cryotomography that the $\{1\,\bar{1}\,0\}$ facets dominate the tetrapod surface area. To investigate the morphology of nanoparticle-ZnO composite systems we sputtered 1.5 nm particles onto the tetrapods. The resulting 3D HAADF-STEM cryotomography demonstrated that the coverage on the coated sides is uniform with minimal coalescence under normal observations. HRTEM showed directly that smaller tetrapods have a ZnO zincblende phase core with the wurtzite legs growing in the WZ[0 0 1] from the ZB{1 1 1} faces via dislocations. We believe the technique and the results could be useful for similar systems in catalysis and technological applications.

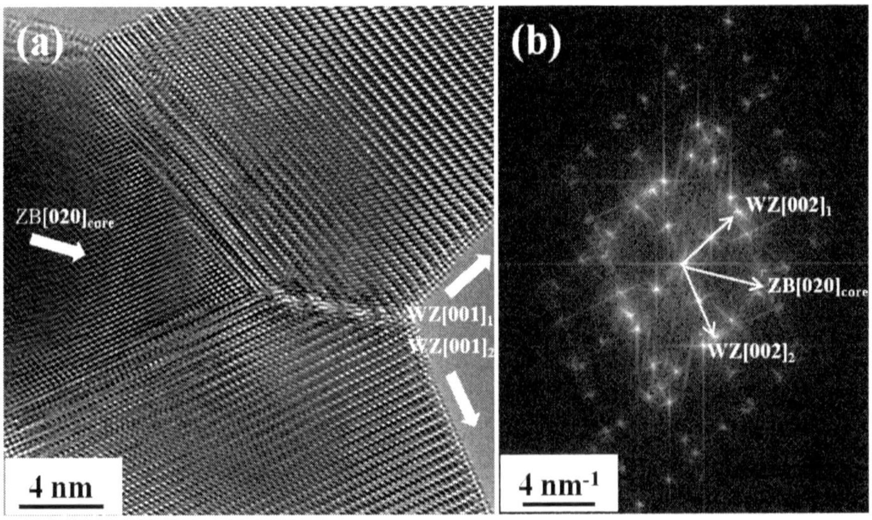

Figure 2. HRTEM image of a small tetrapod (a) with the two leg and core crystal directions labelled and its corrresponding FFT (b)

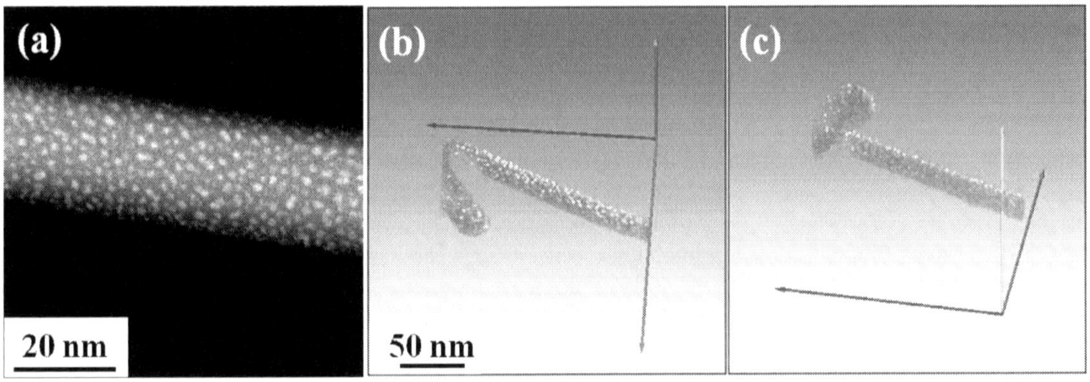

Figure 3. Pt/Pd nanoparticle coated ZnO tetrapods with (a) showing a HAADF-STEM image of a coated tetrapod leg and (b) and (c) showing top and bottom views from within a 3D IMOD reconstruction from a HAADF-STEM tilt series. The brighter areas are the Pt/Pd nanoparticles

Acknowledgments
M R Ward thanks the EPSRC for PhD funding.

References
[1] Newton M C and Warburton P A 2007 *Mater. Today* **10** 50–4
[2] Ronning C, Shang N G, Gerhards I, Hofsass H and Seibt M 2005 *J. Appl. Phys.* **98** 034307
[3] Lupan O, Chow L and Chai G 2009 *Sensor. Actuat. B - Chem.* **141** 511–7
[4] Lin J M, Lin H Y, Cheng C L and Chen Y F 2006 *Nanotech.* **17** 4391–4
[5] Daniel M-C and Astruc D 2004 *Chem. Rev.* **104** 293–346
[6] Midgley P A and Weyland M 2003 *Ultramicroscopy* **96** 413–431
[7] Shiojiri M and Kaito C 1981 *J. Cryst. Growth.* **52** 173–7
[8] Ashrafi A and Jagadish C 2007 *J. App. Phys.* **102** 071101

Hydrothermal synthesis of mixed cobalt-nickel ferrite nanoparticles

Trevor P. Almeida[1], Michael W. Fay[2], Yanqiu Zhu[3] and Paul D. Brown[1]

[1]Division of Materials, Mechanics and Structures, Department of Mechanical, Materials and Manufacturing Engineering, Faculty of Engineering; &
[2]Nottingham Nanotechnology and Nanoscience Centre;
University of Nottingham, University Park, Nottingham, NG7 2RD, UK.
[3]College of Engineering, Mathematics and Physical Sciences, University of Exeter, Streatham Campus, Exeter, EX4 4QF, UK.

Abstract. The hydrothermal synthesis of $Co_xNi_{1-x}Fe_2O_4$ nanoparticles (NPs) from mixed $FeCl_3$ / $CoCl_2$ / $NiCl_2$ precursor solution has been investigated as a function of reaction temperature, using the complementary characterisation techniques of X-ray diffractometry, transmission electron microscopy, energy dispersive X-ray analysis and electron energy loss spectroscopy. At pH ~ 8, the HS of $Co_{0.5}Ni_{0.5}Fe_2O_4$ NPs, up to ~ 25 nm in size, was found to proceed through the formation and dissolution of intermediate $Fe(OH)_3$ and $[FeCoNi_2(OH)_8]^+.[Cl^-.H_2O]^-$ phases.

1. Introduction

Spinel ferrites have attracted considerable attention in recent years because of their applicability in high-density magnetic recording devices, microwave devices, magnetic fluids and sensors [1,2].The cubic inverse spinel ferrite structure comprises $(metal)^{2+}$ ions, occupying octahedric B-sites, and Fe^{3+} ions, which occupy both tetrahedral A-sites and octahedral B-sites [3]. Cobalt ferrite is a well-known hard magnetic material with relatively high coercivity, whilst nickel ferrite is a soft magnetic material with low coercivity. $CoFe_2O_4$ and $NiFe_2O_4$ both exhibit ferrimagnetism, originating from the magnetic moment of anti-parallel spins between Fe^{3+} ions at tetrahedral sites and Co^{2+} or Ni^{2+} ions at octahedral sites [4]. Nanosized magnetic ferrites have been prepared by various methods such as sol–gel processing, hydrothermal processing, forced hydrolysis, ball milling and microemulsion. The approach of hydrothermal synthesis (HS), in particular, offers effective control over the size and shape of $NiFe_2O_4$ and $CoFe_2O_4$ nanoparticles (NPs) at relatively low reaction temperatures and short reaction times, providing for well-crystallised reaction products with high homogeneity and definite composition. Iron (III), cobalt (II) and nickel (II) chlorides act as simple precursors for the formation of these spinel ferrites. However, not much has been reported on the HS of mixed nickel-cobalt ferrites and improved understanding of the mechanistics of $Co_xNi_{1-x}Fe_2O_4$ NP growth is required in order to control fully their magnetic properties. In this context, an investigation of the development of $Co_xNi_{1-x}Fe_2O_4$ NPs, at pH ~ 8, as a function of reaction temperature, is presented.

2. Experimental

For the HS of $Co_xNi_{1-x}Fe_2O_4$, 0.05 ml $FeCl_3$ aqueous solution (45% pure $FeCl_3$; Riedel-de Haen, Germany), 12.5 mg $CoCl_2$ powder (98.0% $CoCl_2$; Sigma-Aldrich, UK) and 12.5 mg $NiCl_2$ powder (98.0% $NiCl_2$; Sigma-Aldrich, UK) were further diluted in 40 ml distilled water and mechanically stirred in a 125 ml Teflon-lined steel autoclave. The precursor solution was adjusted to pH ~ 8 with the drop-

*Author to whom correspondence should be addressed: paul.brown@nottingham.ac.uk

Sample	Temperature / °C	Phase	Morphology	Dimensions / nm
S100	100	$Fe(OH)_3$ $[FeCoNi_2(OH)_8]^+.[Cl^-.H_2O]^-$ $Co_xNi_{1-x}Fe_2O_4$	Amorphous Sheets NPs	- - < 5
S120	120	$[FeCoNi_2(OH)_8]^+.[Cl^-.H_2O]^-$ $Co_xNi_{1-x}Fe_2O_4$	Sheets NPs	- < 5
S140	140	$[FeCoNi_2(OH)_8]^+.[Cl^-.H_2O]^-$ $Co_xNi_{1-x}Fe_2O_4$	Sheets NPs	- < 20
S160	160	$[FeCoNi_2(OH)_8]^+.[Cl^-.H_2O]^-$ $Co_xNi_{1-x}Fe_2O_4$	Sheets NPs	- < 20
S180	180	$Co_xNi_{1-x}Fe_2O_4$	NPs	< 25
S200	200	$Co_xNi_{1-x}Fe_2O_4$	NPs	< 25

Table 1 HS processing conditions and a summary of the product phases, morphologies and dimensions.

wise addition of aqueous NaOH solution, then sealed and inserted into a temperature controlled furnace at the reaction temperature (summarised in Table 1) for 2 hours. The autoclave, once removed from the furnace, was allowed to cool down to room temperature naturally.

The synthesised reaction products were centrifuged for 6 minutes at 6000 rpm, cleaned with acetone and deposited onto single crystal silicon substrates for the purpose of structural characterisation using a Bruker D8 Advance X-ray diffractometer (Cu Kα radiation; θ/2θ diffraction geometry). The crystalline reaction products were identified using DIFFRACplus software. For the purpose of transmission electron microscopy (TEM) investigation, the acetone cleaned HS suspensions were dispersed using an ultrasonic bath before deposition onto lacey carbon / copper mesh support grids (Agar Scientific Ltd, UK). Conventional diffraction contrast bright field (BF) and phase contrast imaging were performed using Jeol 2000fx and 2100f electron microscopes. Energy dispersive X-ray (EDX) analysis (ISIS, Oxford Instruments) and electron energy loss spectroscopy (EELS) (Gatan Tridiem GIF) provided information on the elemental constituents, whilst selected area electron diffraction (SAED) patterns allowed for phase identification.

3. Results

The hydrothermal processing of mixed $FeCl_3$ / $CoCl_2$ / $NiCl_2$ precursor solution led to the synthesis of intermediate or primary reaction products (Table 1), depending on the reaction temperature; the evidence for which is now presented in detail.

The XRD patterns of Figure 1 illustrate the effect of increasing reaction temperature (Samples S100 – S200). The XRD pattern of Sample S100 synthesised at 100°C is consistent with the presence of amorphous $Fe(OH)_3$ (JCPDS 22-346), whilst peaks in the range of 11° to 22° are attributed to a variant of 'hydroxychloride green rust 1' (GR1(Cl^-)): $Fe_{3.6}Fe_{0.9}(O, OH, Cl)_9$ (JCPDS 13-88), whilst it is noted that the $FeCl_3$ precursor only contains Fe^{3+} cations and Sugimoto *et al* [5] suggest that the Fe^{2+} sites in GR1(Cl^-) can be substituted with Co^{2+} or Ni^{2+} to form a compound with the same crystal structure, *i.e.* $[FeCo_xNi_{3-x}(OH)_8]^+.[Cl^-.H_2O]^-$. The XRD patterns of Samples S120 – S200, consistent with the presence of both $NiFe_2O_4$ (JCPDS 10-325) and $CoFe_2O_4$ (JCPDS 1-1121), are assigned to the development of crystalline $Co_xNi_{1-x}Fe_2O_4$.

The BF TEM images of Figures 2a-f present the HS reaction products of Samples S100 – S200, as summarised in Table 1. Investigation of Sample S100 (Figure 2a) showed this suspension to comprise amorphous $Fe(OH)_3$, as identified by SAED (inset), along with GR1(Cl^-) sheets (arrowed) and small, < 10 nm, $Co_xNi_{1-x}Fe_2O_4$ NPs. Samples S120 – S160 (Figures 2b-d) consisted of GR1(Cl^-) sheets (arrowed) and $Co_xNi_{1-x}Fe_2O_4$ NPs (< 20 nm), whilst Samples S180 and S200 comprised < 25 nm $Co_xNi_{1-x}Fe_2O_4$ NPs (Figures 2e,f).

The BF and phase contrast images of Figure 3 provide a more detailed view of the nanostructures in Sample S160. Lattice fringes consistent with the development of crystalline GR1(Cl^-) sheets were identified (Figure 3b, inset), confirmed by associated Fast Fourier Transform (FFT) patterns (also

Journal of Physics: Conference Series **371** (2012) 012074 doi:10.1088/1742-6596/371/1/012074

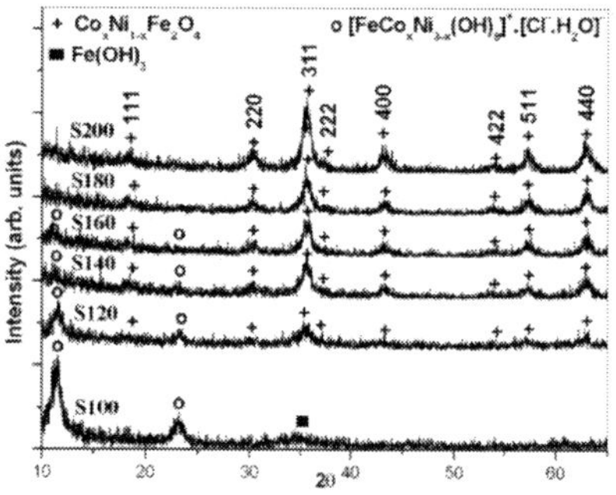

Figure 1 XRD patterns of the reaction products from hydrothermally synthesised $FeCl_3$ / $CoCl_2$ / $NiCl_2$ precursor solution as a function of increasing temperature.

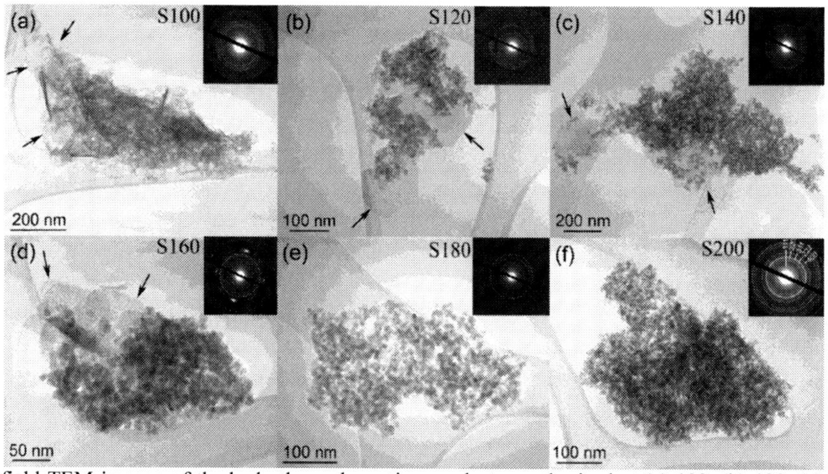

Figure 2 Bright field TEM images of the hydrothermal reaction products synthesised at (a) 100, (b) 120, (c) 140, (d) 160, (e) 180 and (f) 200°C, respectively (Samples S100 to S200), with corresponding SAED patterns (inset – (f) is indexed to both $CoFe_2O_4$ and $NiFe_2O_4$).

inset). Small $Co_xNi_{1-x}Fe_2O_4$ NPs were identified by characteristic lattice fringes (Figure 3c, inset). Figure 3d shows a group of GR1(Cl^-) sheets, whilst Figure 3e shows an HAADF image of an individual GR1(Cl^-) sheet (outlined in Figure 3d). The EDX chemical maps of Figures 3f-h present the elemental distribution of Ni, Co and Fe, respectively, within the GR1(Cl^-) sheet shown in Figure 3e. Further, the atomic composition of an individual $Co_xNi_{1-x}Fe_2O_4$ NP (Sample S200), as calculated by EELS analysis, is summarised in Table 2.

4. Discussion

This combined XRD and TEM investigation of the hydrothermal synthesis of mixed $FeCl_3$ / $CoCl_2$ / $NiCl_2$ precursor solution, at pH ~ 8, as a function of processing temperature, has provided detailed evidence on the development of $Co_xNi_{1-x}Fe_2O_4$ NPs via intermediate phases. HS at 100°C (Sample S100) resulted in the formation of amorphous $Fe(OH)_3$, GR1(Cl^-) sheets and embryonic $Co_xNi_{1-x}Fe_2O_4$ NPs. An increase in reaction temperature to 120°C and above promoted the growth of $Co_xNi_{1-x}Fe_2O_4$ NPs and apparent dissolution of the $Fe(OH)_3$ (Sample S120 – S160). At 180 and 200°C, $Co_xNi_{1-x}Fe_2O_4$ NPs was the only reaction product observed in the TEM data. Hence, it is evident that an increase in reaction temperature drives the development of $Co_xNi_{1-x}Fe_2O_4$ NPs, through the formation and dissolution of intermediate $Fe(OH)_3$ and GR1(Cl^-) phases.

Complementary EDX and EELS investigation of the GR1(Cl^-) and $Co_xNi_{1-x}Fe_2O_4$ nanostructures provided information on their atomic compositions. For example, EDX mapping of an individual GR1(Cl^-) sheet (Figures 3f-h) suggested Ni as the predominant elemental constituent, whilst Fe and Co showed similar proportions. Accordingly, the GR1(Cl^-) sheets are considered to be Ni-rich variants of $[FeCoNi_2(OH)_8]^+.[Cl^-.H_2O]^-$. Conversely, EELS investigation of the atomic composition of individual $Co_xNi_{1-x}Fe_2O_4$ NPs, indicated the presence of equal proportions of Ni and Co, hence assigned as stoichiometric $Co_{0.5}Ni_{0.5}Fe_2O_4$.

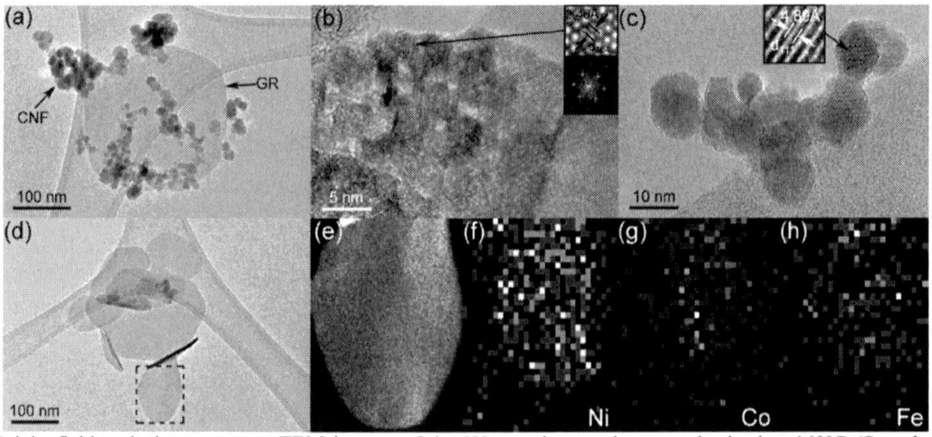

Figure 3 Bright field and phase contrast TEM images of the HS reaction products synthesised at 160°C (Sample S160). (a) Large GR1(Cl^-) sheet (labelled GR) and small $Co_xNi_{1-x}Fe_2O_4$ NPs (labelled CNF); (b) large GR1(Cl^-) sheet identified by lattice fringes and corresponding FFT (inset); (c) small $Co_xNi_{1-x}Fe_2O_4$ NPs identified by lattice fringes (inset); and (d) GR1(Cl^-) sheets. (e) HAADF image of the GR1(Cl^-) sheet outlined in (d) and EDX maps showing (f) nickel; (g) cobalt; and (h) iron, content.

Element	Atomic ratio (/ O)	Content / at%
O	1± 0.00	42
Fe	0.85 ± 0.12	35
Co	0.27 ± 0.04	11
Ni	0.28 ± 0.04	12

Table 2 Atomic composition calculated from a cumulative EELS spectrum of an individual $Co_xNi_{1-x}Fe_2O_4$ NP (Sample S200).

The kinematically accessible amorphous $Fe(OH)_3$ and poorly crystalline GR1(Cl^-) sheets exhibit high surface areas and their stability decreases with increasing HS reaction temperature. Both $Fe(OH)_3$ and GR1(Cl^-) are considered to act as precursors for formation of the more thermodynamically stable $Co_{0.5}Ni_{0.5}Fe_2O_4$. The Ostwald rule of stages concerns the competition between kinetics, irreversible thermodynamics and equilibrium thermodynamics [6]. A decrease in total surface energy is considered to be the driving force for the crystal growth kinetics of $Fe(OH)_3$, GR1(Cl^-) and $Co_{0.5}Ni_{0.5}Fe_2O_4$. As the more thermodynamically stable $Co_{0.5}Ni_{0.5}Fe_2O_4$ NPs nucleate and grow with increasing hydrothermal temperature, the equilibrium Fe^{3+}, Co^{2+} and Ni^{2+} concentration at the surface of $Co_xNi_{1-x}Fe_2O_4$ becomes lower than that of $Fe(OH)_3$ and GR1(Cl^-). The resulting concentration gradient leads to Fe^{3+}, Co^{2+} and Ni^{2+} ions flowing from $Fe(OH)_3$ and GR1(Cl^-) to the newly formed $Co_{0.5}Ni_{0.5}Fe_2O_4$ NPs, prompting $Fe(OH)_3$ and GR1(Cl^-) dissolution to restore the equilibrium Fe^{3+}, Co^{2+} and Ni^{2+} concentration at their surfaces. The growth of the intermediate phases and $Co_{0.5}Ni_{0.5}Fe_2O_4$NPs can be summarised by the generalised equations:

$$\mathbf{Fe^{3+}_{(aq)} + 3OH^-_{(aq)} \rightarrow Fe(OH)_{3(s)}} \quad (1)$$
$$\mathbf{Fe^{3+}_{(aq)} + Co^{2+}_{(aq)} + 2Ni^{2+}_{(aq)} + Cl^-_{(aq)} + 8OH^-_{(aq)} + H_2O \rightarrow [FeCoNi_2(OH)_8]^+.[Cl.H_2O]^-_{(s)}} \quad (2)$$
$$\mathbf{2Fe^{3+}_{(aq)} + {}^1/_2Co^{2+}_{(aq)} + {}^1/_2Ni^{2+}_{(aq)} + 4H_2O \rightarrow Co_{0.5}Ni_{0.5}Fe_2O_4 + 8H^+} \quad (3)$$

In summary, the growth of $Co_{0.5}Ni_{0.5}Fe_2O_4$ NPs at pH ~ 8 has been investigated as a function of HS temperature and proceeds through the formation and dissolution of intermediate $Fe(OH)_3$ and $[FeCoNi_2(OH)_8]^+.[Cl.H_2O]^-$ phases.

References

[1] Pallai V and Shah D O, 1995 *J. Magnetism Magnetic Mater.* **163,** 243.
[2] Skomski R, 2003 *J. Phys. Condens. Matter* **15,** R1.
[3] Singhala S, Singha J, Barthwalb S K and Chandraa K, 2005 *J. Solid State Chem.* **178,** 3183-3189.
[4] Kambale R C, Shaikh P A, Kamble S S and Kolekar Y D, 2009 *J Alloys Compd.* **478,** 599-603.
[5] Sugimoto T, Shimotsuma Y and Itoh H, 1998 *Powder Technol.* **96,** 85-89.
[6] Hurle D T J, editor. 1993 *Handbook of crystal growth, 1a Fundamentals: Thermodynamics and Kinetics*, Mutaftschiev B. *Chapter 4: Nucleation Theory,* (Amsterdam: Elsevier).

Electron Microscopy and Analysis Group Conference 2011 (EMAG 2011) IOP Publishing
Journal of Physics: Conference Series **371** (2012) 012075 doi:10.1088/1742-6596/371/1/012075

Investigating the optimum conditions for the formation of calcium oxide, used for CO_2 sequestration, by thermal decomposition of calcium acetate

M Bilton, A P Brown, S J Milne

Institute for Materials Research, SPEME, University of Leeds, Leeds, LS2 9JT, UK
E-mail: pre5mb@leeds.ac.uk

Abstract. Calcium oxide, CaO, has long been recognized as an effective, regenerative sorbent for the sequestration of CO_2. Calcium acetate hydrate, CaAc, ($Ca(CH_3COO)_2 \cdot xH_2O$) has been highlighted as one of the most efficient precursors for CaO production with regards to CO_2 sequestration. In this study, thermal decomposition of CaAc to CaO, via an intermediate phase of calcium carbonate ($CaCO_3$), has been examined at different heating rates using thermal and powder diffraction analysis. The decomposition from CaAc to CaO has yet to be analysed in detail at the particulate level, however, knowledge of how the transformation proceeds at this level should lead to control over the size and morphology of the final CaO product. Bulk characterisation using X-ray diffraction (XRD) and thermogravimetric analysis (TGA) has been used to benchmark the transformation at different heating rates and final calcination temperatures. TEM imaging is used to show how the chemical changes proceed particularly at the particle level.

1. Introduction

The drive for low carbon technologies, particularly carbon capture and storage (CCS), is an area of significant interest due to the rise in emissions of carbon dioxide and the widely recognised effects of global warming. Such effects have led to a proposed minimum worldwide target of a 50% reduction in CO_2 emissions (relative to 1990 levels) by the year 2050 for which CCS will play a pivotal role in helping to reach [4, 5]. One method for sequestration of CO_2 gas, prior to its exposure to the atmosphere, is to capture it using a calcium-based sorbent such as calcium oxide, CaO (e.g. calcined limestone). CaO is known to absorb CO_2 in a flue gas to produce calcium carbonate $CaCO_3$ (carbonation, equation 1), a process which is reversible. Once the metal oxide has reached its ultimate conversion, thermal regeneration can be utilised whereby $CaCO_3$ is heated beyond its calcination temperature (~750 °C) to reform CaO and near pure CO_2 gas which can be collected and stored (calcination, equation 2) [3, 11].

$$CaO + CO_2 \rightarrow CaCO_3 \qquad (1)$$

$$CaCO_3 \rightarrow CaO + CO_2 \qquad (2)$$

Calcium acetate hydrate, CaAc, ($Ca(CH_3COO)_2 \cdot xH_2O$) has been recognised as an effective precursor for the production of CaO [6, 9]. The cyclic capture of CO_2 using CaO is an area of great interest [1, 7, 2], however sintering during calcination leads to a decline in the adsorption capacity of CaO [10]. By carefully monitoring the decomposition process of CaAc to CaO, it is hoped that sintering of the first CaO product can be minimised, subsequently allowing for a larger surface area to be available for CO_2 pick-up.

2. Experiment

The decomposition of commercial CaAc (Acros Organics, 99%, extra pure) was firstly monitored using thermogravimetric analysis (Shimadzu TGA-50). A sample was heated to 1000 °C at a rate of 20 °C min^{-1}, in air. For X-ray diffraction (XRD), a CaAc sample was analysed with a temperature controlled heating stage (up to 800 °C) using a Philips PanAnalytical X'pert Diffractometer with Cu Kα radiation source. Each temperature interval was reached at a rate of 20 °C min^{-1}, and then held for

~ 1 hour. Samples for TEM imaging were produced by firstly directly heating CaAc samples in air, by direct insertion into preheated furnaces for 10 minutes at 500 °C and 800 °C respectively. Three further samples were prepared by a slow decomposition from room temperature to 200, 500 and 800 °C (at 20 °C min^{-1}); each held at the respective temperatures for 1 hour. Both sets of samples were immediately removed from the furnace after calcination, and air cooled. TEM analysis was performed using an FEI Tecnai F20 FEG-TEM operated at 200 kV and fitted with a Gatan Orius SC600A CCD camera.

3. Results & Discussion

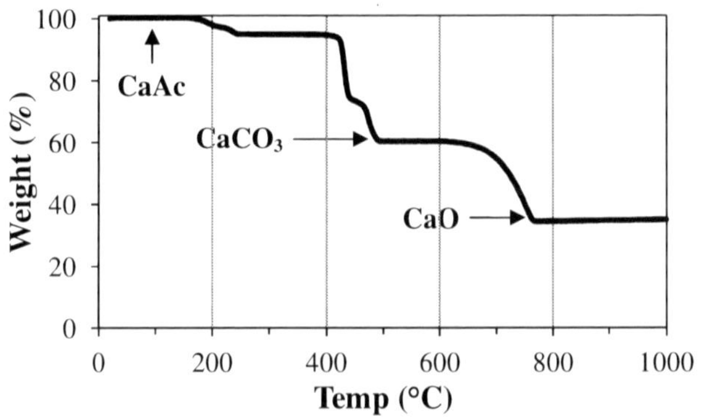

Figure 1: TGA curve showing that the thermal decomposition of calcium acetate to CaO is completed by 800 °C.

Temp (°C)	Mass Loss (%)
~150-240	**5.0**
~400-520	**37.2**
~650-765	**26.6**
Total Loss =	**65.8**

Table 1: Percentage mass loss for thermal decomposition of CaAc.

TGA data displays CaAc decomposed at a constant rate of 20 °C min^{-1} to 1000 °C. The thermal decomposition of CaAc measured here (Fig. 1) is in good agreement with previously reported data [8, 9]. Table 1 highlights the percentage mass loss at three given temperature ranges. There is an initial two-stage mass loss of water molecules, the first at ~150-200 °C, and the second at ~200-240 °C , with a total percentage mass loss of 5% (calculated xH_2O molar loss = 2.3) [8]. Between ~400-520 °C, literature indicates that the dehydrated CaAc breaks down to form acetone and $CaCO_3$, seen here as a combined percentage mass loss of 37.2% [8]. The loss observed at ~440 °C is likely to be attributable to the eventual loss of acetone [8]. The calcium carbonate then decomposes to form calcium oxide from ~650 °C to completion at ~765 °C [8].

XRD data (Fig. 2) shows how CaAc decomposes when heated (at 20 °C min^{-1}) and held at 100 °C intervals for 1 hour. The breakdown of hydrated CaAc to dehydrated CaAc is seen to occur via an apparent amorphous compound at 200 °C; this leads to dehydrated CaAc at 300 °C. This subsequently breaks down further to form $CaCO_3$ at 400 °C. At 600 °C traces of CaO are evident, indicating the start of the breakdown of $CaCO_3$. Decomposition to CaO is complete at 700 °C.

Due to the slower decomposition rate of CaAc, using hot-stage XRD, decomposed phases ($CaCO_3$ and CaO) are allowed a greater time for formation, and subsequently are formed at slightly lower temperatures than those produced with a quick heating rate, using TGA.

TEM images (Figs. 3 & 4) highlight the transformation from CaAc to CaO at a particle level. As received CaAc particles (Fig. 3a) display long lath-like structures, (typical length = 1-10 μm). Figures 3b-c display CaAc directly placed, in a pre-heated furnace, at 500 and 800 °C respectively for 10 minutes, to simulate a rapid decomposition, relatable to TGA data. CaAc heated by direct insertion into a preheated furnace at 500 °C provides $CaCO_3$ (Fig. 3b) with particles which vary from rounded and faceted shapes, which are slightly sintered, typically 100-200 nm in length. CaO (Fig. 3c), produced from direct insertion into at 800 °C preheated furnace, displays highly sintered agglomerates of fine platelet-like particles (<100 nm).

Figures 4a-c display a slower rate of decomposition, with powders decomposed from room temperature to 200, 500 and 800 °C respectively (at 20 °C min^{-1}) and held at that temperature for 1h, simulating the XRD analysis. At 200 °C, observed particles are likely to be broken laths (typical length = 1-5 μm) of CaAc; future experiments will examine this apparent amorphous phase further. The $CaCO_3$ formed at 500 °C after 1h (Fig. 4b) appears as much finer, sintered agglomerates in comparison to the rounded particles formed upon direct heating, this may suggest progression to the eventual decomposition of $CaCO_3$. The CaO formed at 800 °C after 1h (Fig. 4c) displays larger and heavily sintered agglomerates in relation to CaO formed by direct heating. There is also a significant amount of porosity evident, which is not observed in the CaO formed by direct heating. This may be due to a fast rate of sintering entrapping CO_2 or air between the particles; future experiments will investigate this further.

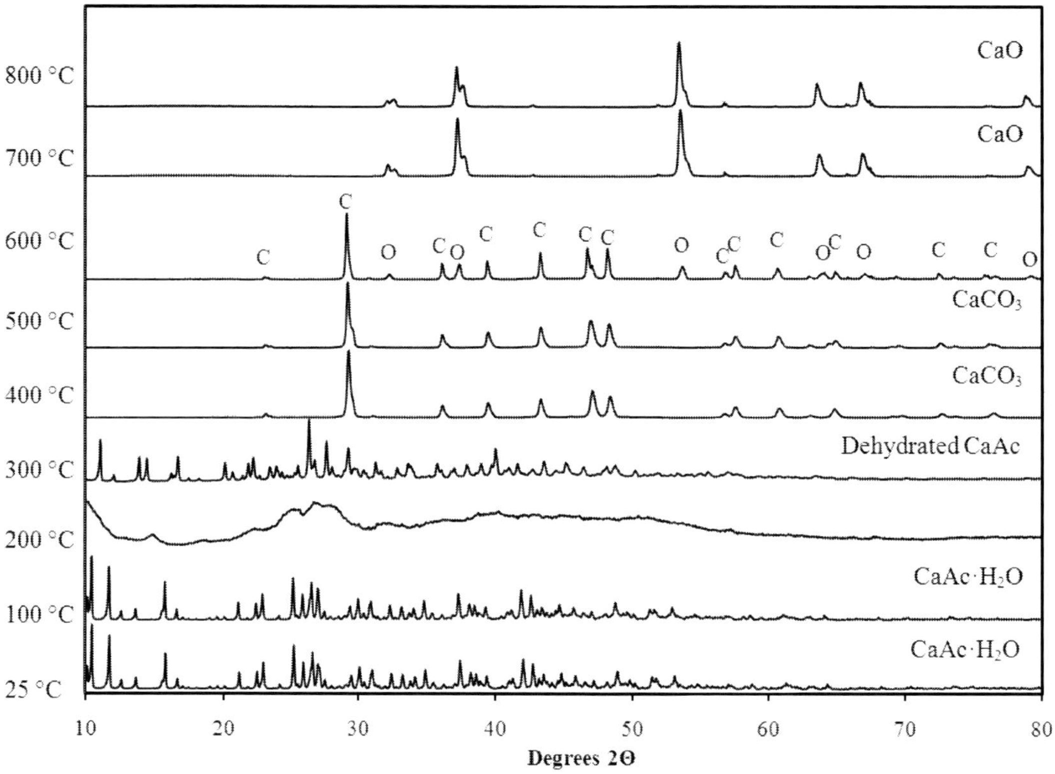

Figure 2: XRD data for the decomposition of CaAc from room temp (25 °C) up to 800 °C at intervals of 100 °C, held for 1h. CaAc patterns at 25 and 100 °C are in close agreement with JCPDS file number 019-0199. The pattern at 300 °C is in reasonable agreement with dehydrated CaAc: JCPDS file number 00-019-0198. Peaks labelled 'C' corresponds to $CaCO_3$: JCPDS file number 00-05-0586, and peaks labelled 'O' denote CaO: JCPDS file number 04-006-5940

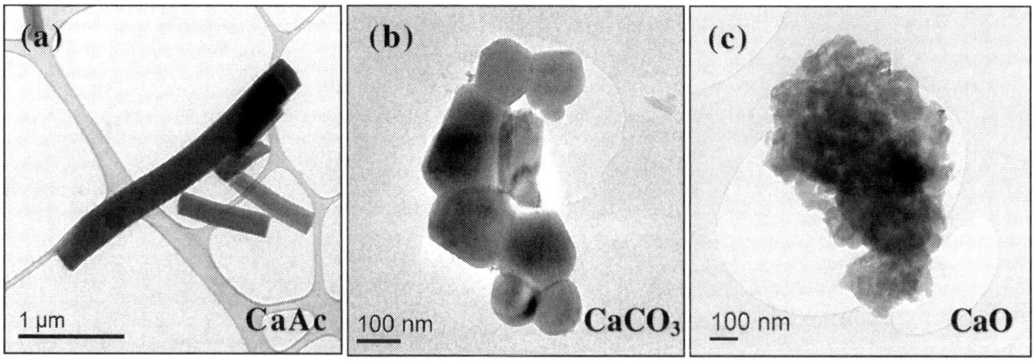

Figure 3: TEM images of (a) as received CaAc, with (b) $CaCO_3$ and (c) CaO formed from CaAc direct insertion into preheated furnaces at 500 and 800 °C respectively.

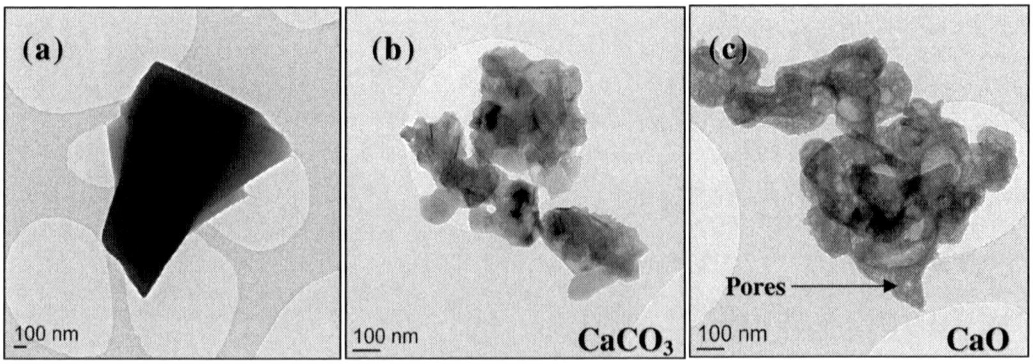

Figure 4: TEM images of (a) CaAc heated to 200 °C and held for 1h (b) CaAc heated to 500 °C ($CaCO_3$) and held for 1h, and (b) CaAc heated to 800 °C (CaO) and held for 1h.

4. Conclusion

CaAc has been decomposed at different heating rates using TGA and hot-stage XRD, to investigate its transformation to CaO. TGA and XRD data suggest the heating rate significantly affects the temperature of formation of CaO, with a slower heating rate, with 100 °C intervals of 1h temperature holds (using hot-stage XRD), showing a full conversion to CaO at 700 °C; and at a faster rate (20° C min^{-1} without hold, using TGA) at just below 800 °C. TEM images confirm this and display a high level of sintering of the final CaO products, a problem likely to affect sequestration of CO_2 to be carried out in future work; however porosity in the CaO product held for 1h at 800 °C may prove beneficial towards CO_2 uptake. Further experiments investigating careful control of final calcination temperature, heating rate and addition of particulate spacers will hopefully allow for a reduction in CaO sintering. Future work will also aim to confirm the cause of the porosity observed in CaO, produced by 1h decomposition of CaAc at 800 °C.

Acknowledgements
This work has been supported by a DTA scholarship courtesy of the EPSRC.

5. References

[1] Abanades J C and Alvarez D 2003 *Energy & Fuels* **17** 308-15
[2] Fennell P S, Pacciani R, Dennis J S, Davidson J F and Hayhurst A N 2007 *Energy & Fuels* **21** 2072-81
[3] Gupta H and Fan L-S 2002 *Industrial & Engineering Chemistry Research* **41** 4035-42
[4] Hester R E and Harrison R M 2009 *Carbon Capture - Sequestration and Storage*: RSC Publishing)
[5] Lackner K S 2003 *Science* **300** 1677-8
[6] Lu H, Reddy E P and Smirniotis P G 2006 *Industrial & Engineering Chemistry Research* **45** 3944-9
[7] Manovic V and Anthony E J 2011 *Fuel* **90** 233-9
[8] Niu S, Han K, Lu C and Sun R 2010 *Applied Energy* **87** 2237-42
[9] Silaban A, Narcida M and Harrison D P 1992 *Resources, Conservation and Recycling* **7** 139-53
[10] Sun P, Grace J R, Lim C J and Anthony E J 2007 *AIChE Journal* **53** 2432-42
[11] Wang Y, Lin S Y and Suzuki Y 2010 *Fuel Processing Technology* **91** 958-63

Investigation of anomalous grains in polynanocrystalline nickel

J E Darnbrough[1,2], S Mahalingam[1], P E J Flewitt[1,2]

[1]Interface Analysis Centre, University of Bristol, Oldbury House, 121 St. Michael's Hill Bristol, BS2 8BS, UK

[2]School of Physics, University of Bristol, H. H. Wills Physics Laboratory, Tyndall Avenue, Bristol, BS8 1TL, UK

E-mail: J.E.Darnbrough@bris.ac.uk

Abstract. Polynanocrystalline nickel, with an as received grain size distribution in the range 30 to 80 nm, when heat treated at a temperature of 500°C causes overall coarsening of the grains and anomalous growth of some grains. This leads to a bimodal grain size distribution. The anomalous large grains, typically 75-100 μ m, have an angular morphology that is not typical of the polyhedral grains within the overall distribution. Using a combination of focus ion beam milling and imaging, and electron backscatter diffraction it has been established that the interfacial planes forming the overall grain boundaries for these anomalous large grains include 110, 210, 112 and 310 planes. In addition a substructure is produced within the large grains interrelated in size with the subsumed nano-scale grains. The results are discussed briefly with respect to the more energetically favourable 111 boundary planes and the overall mobility.

1. Introduction

Polynanocrystalline materials are those with a structure of grains with a sub 100 nm size. In general such materials have improved mechanical properties compared with the corresponding microscale materials. These materials have high yield strength and hardness but are less ductile than their poly-macrocrystalline cohorts [1]. Whilst the materials are important for potentially a range of applications, the presence of high energy grain boundaries promotes coarsening to reduce the overall energy of the system [1]. Previous work has shown that a bimodal grain distribution with a polynanocrystalline structure embedded with larger grains retained the hardness associated with the smaller grains and improved the ductility [2]. Previous investigation showed that heat treatment of polynanocrystalline nickel causes a bimodal grain structure to develop; a coarsened matrix of smaller grains and large anomalous grains [3]. This paper considers polynanocrystalline nickel grain coarsening following heat treatment at temperatures of 420°C and 500°C for various times and the resulting bimodal grain structure. Primary interest comes from the angular anomalous grains with planar grain boundaries.

2. Experimental Procedure

The polynanocrystalline nickel was obtained from Integran Ltd and is proposed to be produced by pulsed electrodeposition to a thickness of 2 mm. This method of creation can introduce impurities into the material, predominantly sulphur. Energy dispersive x-ray spectroscopy (EDX) of a large area of the sample shows that there are traces of sulphur present in the material. As received material has no texture and when studied by x-rays diffraction (XRD) gives a grain size of 30-80 nm. Coarsening of the nano-scale nickel has been considered following heat treatment at temperature of 420°C and 500°C in vacuum, table 1. The microstructure changes have been characterised using a combination of transmission electron microscopy (TEM) using a Philips EM430 transmission electron microscope working at 200kV and focus ion beam (FIB) milling and imaging using a FEI Helios dual beam workstation and electron backscattered electron diffraction (EBSD) in a Zeiss Evo scanning electron microscope(SEM).

3. Results

3.1. Grain Growth

The grain structure of the as received polynanocrystalline nickel can be seen in Figure 1. The TEM image of the material, with the corresponding electron diffraction pattern, displays the grain size of 30 to 80 nm and containing the FCC crystal structure. This image is diffuse due to overlapping contrast caused by the number of grains that are contained in the 150nm thick foil section. An initial matrix of heat treatments lead to coarsening of the nano-grains and in one case a bimodal distribution including both the coarsened grains and large anomalous grains, figure 2. This was by achieved heat treating a sample at 500°C for 3 hours; table 1 shows the heat treatments and the average grain size measured by linear intercept. As such the average grain size for 3 hours at 500°C has been split for the bi-modal nature of the specimen, denoted in table 1. Specimens which have a bimodal grain size distribution developed the larger anomalous grain size of the order of 30-50 μ m where the size of the coarsened matrix grains surrounding them are 500-600nm. In order to find a temperature dependence at a heat treatment time of 3 hours further work was then done using a known constant temperature gradient held over the sample. This isothermal heat treatment allows investigation of the relative temperatures that cause both grain coarsening and the anomalous grain growth. EBSD mapping, shown in figure 2, reveals the microscructure of coarsened grains and anomalous grains. As table 1 shows, the size of the coarsened matrix does not increase at longer time however anomalous grains appear. From investigation of the sample which was held with a temperature gradient across it for 3 hours it was deduced that the anomalous grain growth occurs from 485°C, and matrix grain coarsening from 400°C [4].

Table 1: Average grain size measured by linear intercept method, including the bi-modal system developed at 500°C for 3 hours.

Sample	Average Grain Size (nm)	Standard Deviation	
virgin	95	65	
420°C 10 mins	580	150	
420°C 3 hrs	510	70	
420°C 12 hrs	540	90	
500°C 10 mins	630	160	
500°C 20 mins	680	400	
500°C 3 hrs	520	340	coarsened matrix
500°C 3 hrs	40.4×10^3	2.9×10^3	anomalous grains

3.2. Anomalous Grains

To gain knowledge about the planar faces of anomalous grains requires two surface analysis. A combination of SEM and EBSD is used to identify, and find the orientation of, large anomalous grains. With this knowledge of the grain surface and with FIB milling to reveal the relative angle of the second face extending into the material, figure 3, allows for two surface analysis and calculation of the orientation of the grown plane. Hence, given an angle between planes, the orientation of one and the fact that the two planes have a common edge direction there is unique solution for the grain boundary relative to the anomalous grain. Further ESBD investigation addressing miss-orientation within the large anomalous grains has discovered small (°5 degrees) variations in orientation, figure 4. This shows between 0 (blue) and 5 (red) degrees, green 1 degree, small regions of miss-orientaion in the anomalous grain and thus a substructure with cells typically ~500nm which are of a similar size to the surrounding matrix grains. The colours shown in the EBSD image, figure 2, represent the orientation of the grains determined by the position of the kikuchi bands relative to the pattern that would be produced by a 100 plane. The FIB milled section, figure 3, shows a large grain and the angle this plane has with respect to the top surface and milled surface. Given this geometric information and the orientation of the grain from the EBSD these planar boundaries were evaluated, table 2. This shows that the majority of planes are 211 together with other planes and gives a 94% confidence that there are no low index 111 planes present.

Table 2: All measured growth planes associated with anomalous grains

Number	Plane
8	211
4	210
3	221
2	110, 310, 320
1	311, 322, 430, 432

4. Discussion

The abnormally large grains observed as part of the coarsening of the nano-scale nickel have an unusual morphology with planar faces. These grain faces give a range of planes, table 2, that do not include the low energy, close packed 111 plane. These planar faces of the anomalous grains are formed and advance through the grain coarsened matrix. As such the driving force will be a balance between more energetically stable and crucially more mobile grain boundaries. The mobility of a grain boundary may be considered with respect to the ease of adding atoms to the boundary plane. For example, a 211 plane will contain stepped ledges that facilitate growth. Clearly the presence of a substructure within the larger grains of a dimension equal to that of the smaller surrounding grains implies some local interactions occur. However, since the grain boundaries are overall planar suggests that they act as if growing with an average interfacial energy; akin to a grains growing from a liquid. It is proposed to be due to impurities at the boundary, mainly sulphur, changing the boundary energy, making it more anodic and thus weaken the nickel-nickel bonds [5]. High resolution STEM-EDS of microanalysis of thin sections has confirmed the increased concentration of sulphur present at the anomalous grain boundary [6]. This will encourage growth of the more energetically favourable larger grain. In extreme cases where the sulphur concentration is sufficiently high a wetting transition has been proposed to accelerate grain boundary movement [7, 8]. In some cases where there is a very low angle boundary, for example a $\Sigma 3$ boundary, the advancing grain boundary sweeps round and engulfs the smaller grain, figure 2.

5. Acknowledgments

The authors acknowledge support from EPSRC.

Journal of Physics: Conference Series **371** (2012) 012076 doi:10.1088/1742-6596/371/1/012076

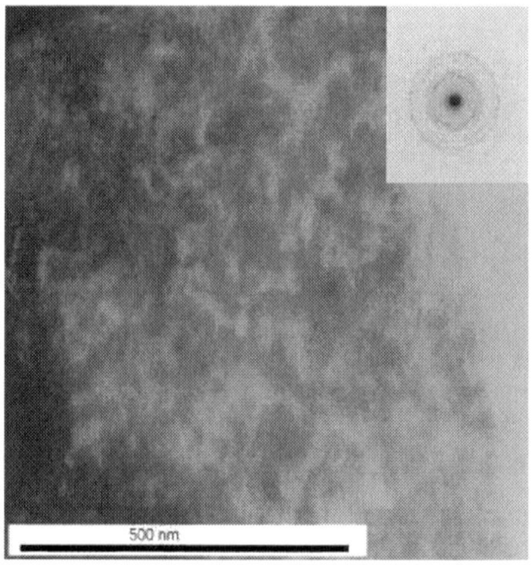

Figure 1: Bright field TEM image of the as received polynanocrystalline nickel, insert- diffraction pattern.

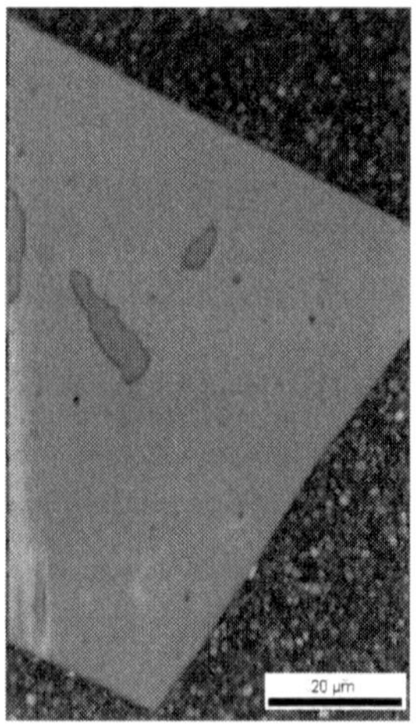

Figure 2: EBSD image of the nickel sample displaying a bi-modal grain size distribution after a heat treatment of 500°C for 3 hours.

Figure 3: Image of the FIB cut used to reveal the angle between planes.

Figure 4: EBSD image showing up to 5 degrees miss-orientation within an anomalous grain.

6. References

[1] H. Gleiter. Nanocrystalline materials. *Progress in Materials Science*, 33:223–315, 1989.
[2] Y. Wang, M. Chen, F. Zhou, and E. Ma. High tensile ductility in a nanostructured metal. *Nature*, 419:912–915, 2002.
[3] L. Cheng and G.D. Hibbard. Abnormal grain growth via the migration of planar growth interfaces. *Materials Science and Engineering A*, 492:128–133, 2008.
[4] J E Darnbrough, B Roebuck, and P E J Flewitt. Investigations into changes of resistivity while altering the microstructure of polynanocrsytalline metals. 2012. to be published.
[5] C.L. Briant and R.P. Messmer. Electronic effects of sulphur in nickel a model for grain boundary embrittlement. *Philosophical Magazine B*, 42(4):569–576, 1980.
[6] J E Darnbrough, S Mahalingam, and P E J Flewitt. Energy dispersive spectroscopy of impurity levels in grain boundaries of a metal with a bi-modal grain size. 2012. to be published.
[7] G. D. Hibbard, J.L. McCrea, G.Palumbo, K.T. Aust, and U. Erb. An initial analysis of mechanisms leading to late stage abnormal grain growth in nanocrystalline ni. *Scripta Materialia*, 47:83–87, 2002.
[8] L M Brown, G Allen, and P E J Flewitt. Nanochemistry of grain boundaries. *Physics World*, page 45, 1997.

Size effects in nanoscale dielectric materials

E Liberti[1], C M McGilvery[1], R Menzel[2], M S P Shaffer[2] and D W McComb[1]

[1] Department of Materials, Imperial College London, London SW7 2AZ, UK
[2] Department of Chemistry, Imperial College London, London SW7 2AZ, UK

E-mail: e.liberti09@imperial.ac.uk

Abstract. The electronic properties of materials change substantially when size and shape are reduced to the nanometre scale. In this study we focus on the size dependence of the dielectric function in novel TiO_2 anatase nanoplatelets. Measurements of the dielectric function are obtained by electron energy-loss spectroscopy (EELS) in the scanning transmission electron microscope (STEM). We present data collected from platelets along the [002] and [200] zone axes and observe that interpretation of the size effects requires a full understanding of the anisotropic effects on the energy-loss function. First principles DFT calculations of the bulk energy-loss function are used as an aid to the interpretation of crystal orientation effects on the EEL spectrum.

1. Introduction

One-dimensional (1D) anatase TiO_2 nanostructures have attracted much interest due to numerous potential applications such as photocatalysis, dye-sensitized solar cells and lithium-ion batteries [1], [2]. This is due to their novel intrinsic properties which strongly depend on shape and size, especially of the exposed facets. For device development it is of particular importance to characterize the dielectric function of these structures, which differs from that of the bulk system. We have investigated the dielectric behaviour of anatase nanoplatelets by combining scanning transmission electron microscopy (STEM) with electron energy-loss spectroscopy (EELS). In the case of uniaxial materials such as TiO_2 the electron energy-loss spectrum is affected by the crystal anisotropy. Therefore, when studying size effects in uniaxial systems, anisotropic effects need to be considered. By varying the convergence and collection angles in EELS it is possible to probe the anisotropy of the inelastic scattering [6] and decouple its effects from the study of the size dependence. In this contribution, challenges to measure the dielectric function of anisotropic anatase nanostructures are discussed. Experimental results are presented together with density functional theory (DFT) calculations of the energy-loss function for the bulk case.

2. Experimental methods

TiO_2 anatase nanoplatelets were synthesized via hydrothermal reaction of titanium (IV) isopropoxide and hydrofluoric acid as described elsewhere [2]. TEM samples were prepared by suspending the platelets in water and depositing the solution on a standard holey carbon film grid. TEM sample preparation of bulk anatase was obtained using a Ga focused ion beam system (FEI Helios NanoLab™ DualBeam™) with different probe sizes. The milling areas were gradually reduced to obtain 100 nm thin samples in the [200] direction. Final thinning of the samples was obtained with an accelerating voltage of 2 kV and 28 pA current. Prior to the insertion of the samples into the TEM, they were plasma cleaned using H_2 and O_2 as the processing gases for 20 seconds. Initial TEM morphology and crystallography studies were carried on a JEOL 2010 microscope operating at 200 kV. STEM-EELS and HRTEM experiments were performed on an FEI Titan microscope operating at 300 kV. The high angle annular dark field (HAADF) inner and outer collection angles were of 12.8 mrad and 63.7 mrad respectively. The EELS spectra were recorded with a Gatan Tridiem 865 EEL spectrometer, with an energy dispersion of 0.05 eV/channel. The convergence and collection semi angles were of 10 mrad and 11.3 mrad respectively. The angles were experimentally calibrated using a known diffraction pattern. The energy resolution, determined as the full width at half maximum (FWHM) of the zero loss peak (ZLP), was 0.75 eV. EELS data were acquired using the Gatan Digiscan software and corrected for the dark current and the gain variation of the CCD detector. A spatial drift correction was applied every 3 pixels to minimize the effects of sample drifting during acquisition.

First principles calculations of the anatase energy-loss function were performed using the full potential linearized augmented plane wave WIEN2k program [3], in the generalized gradient approximation (GGA). The cutoff of the plane-wave expansion was set as $R_{MT} \times K_{max} = 7$. For the calculation of the energy-loss function, the sampling of the Brillouin zone was of 1000 total **k** points. The GGA eigenvalues were computed up to 80 eV above the Fermi level. A "scissor operator" was introduced to displace the valence and the conduction bands relative to each other by a rigid shift of 1.2 eV. Finally, a Lorentzian broadening of 0.6 eV was applied to the theoretical results to account for the energy resolution of the EELS experiments.

3. Results and discussions

Figure 1 (a) shows a bright field TEM image of the platelets. The structures were generally found to be rectangular with an edge length of 20 – 60 nm. The thickness of the platelets was measured in HRTEM to be about 5 nm in the [200] direction (figure 2 (b)). The crystallographic structure of TiO_2 anatase was confirmed by selected area electron diffraction (SAED) (inset in figure 1(a)) and HRTEM. Figure 1 (b), (c) show the measured lattice spacing of 0.18 and 0.4 nm, corresponding to the (200) and (002) diffracting planes of the anatase phase.

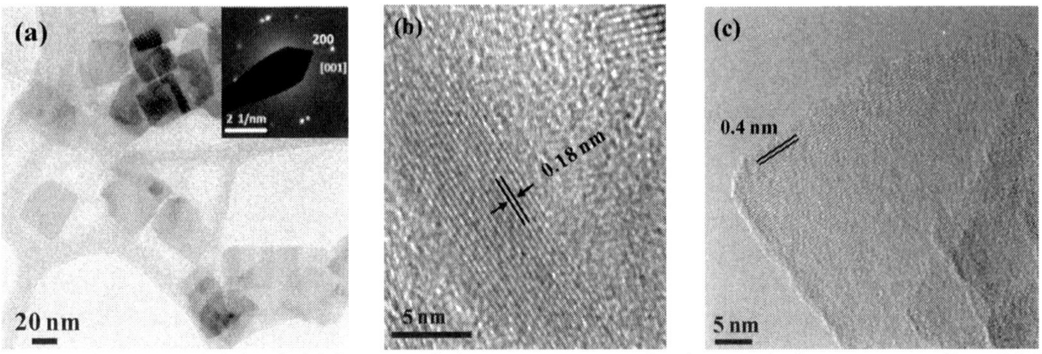

Figure 1. (a) TEM bright field image of the anatase nanoplatelets and the corresponding SAED pattern (inset). (b), (c) HRTEM images of the platelets in the [200] and [002] zone axis, respectively.

EELS data were obtained for both [002] and [200] oriented structures along a line spectrum as shown in figure 2. The illustrated spectra were selected for different positions of the STEM probe along this line. All regions of interest were chosen to be overhanging a hole in the carbon support. As the energy-loss suffered by the incoming electrons is directly proportional to the thickness of the sample, the intensity of the inelastic signal is at the noise level. To get an intense inelastic signal, the spectra were recorded by applying an appropriate drift-tube offset, with an acquisition time of 1 and 0.5 seconds per pixel for the [002] and [200] orientation respectively. The energy scale of the spectra was calibrated using the mean position of the ZLP. The intensity of the signals was normalized to the intensity of the ZLP at 0 eV. Significant differences in the intensities, number and position of the peaks in the low loss spectra were found between the two orientations below the titanium $M_{2,3}$ edge at 50 eV, in both the penetrating and aloof geometry (electrons travelling at short distance from the surface). These differences are mainly due to surface excitations that dominate in the case of figure 1(c). The energies at which surface electronic transitions occur are currently object of investigation. To unambiguously identify the nature (surface and volume plasmons, interband transitions and relativistic losses) of all peaks in figure (b) and (c), the effects of the crystallographic orientation on the energy-loss function must firstly be understood.

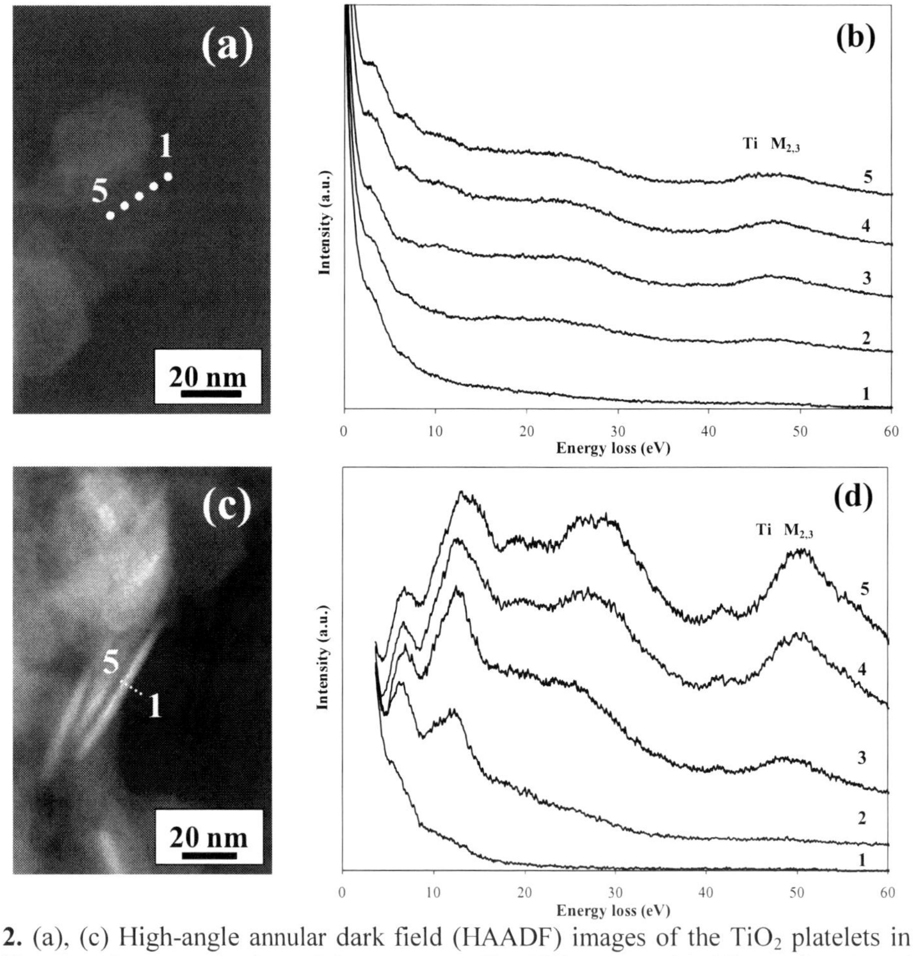

Figure 2. (a), (c) High-angle annular dark field (HAADF) images of the TiO_2 platelets in the [200] and [002] zone axis, respectively, and the corresponding EEL spectra (c), (d) as a function of the probe position (white dots in figures (a) and (c)).

Typically for anisotropic materials such as TiO_2, the structure of the EEL spectrum depends on the orientation of the sample. This is because particular electronic transitions may occur for a specific direction of the momentum transfer with respect to the crystal orientation. Due to the conservation of momentum in any inelastic scattering event, the direction of the momentum transfer vector **q** varies, for a given energy-loss, with the scattering angle θ, changing from parallel to perpendicular to the electron beam as θ increases. Usually, in a STEM/EELS experiment the collection angle of the scattered electrons is such that the momentum transfer vector is given by a combination of the parallel and perpendicular components to the electron beam. Thus, varying the collection geometry in the electron microscope allows selecting a specific direction, or angular range, of the momentum transfer vector. The experimental study of the orientation dependent EEL of bulk TiO_2 thin samples is currently under investigation. Very few studies can be found on the experimental **q** dependence of the low energy-loss region [4]. This is because the crossover between the components of the momentum transfer vector occurs near the characteristic energy loss angle $2\theta_E$ [5] which, for 300 kV incident electrons and energy losses of few electronvolts, is less than 1 mrad. In this condition, the convergence angle is as large as or larger than the collection angle, two effects that can mislead the interpretation [6].

As a first approximation, the angularly integrated energy-loss function can be expressed as a linear combination of the principal components of the dielectric function $\varepsilon_{xx}(E)$ and $\varepsilon_{zz}(E)$ for tetragonal TiO_2, where x, y and z directions coincide with the crystal Cartesian axes. Figure 3 compares the calculated energy-loss function for different linear combinations of $\varepsilon_{xx}(E)$ and $\varepsilon_{zz}(E)$ with the experimental energy-loss function measured for the [200] bulk anatase sample. The EEL spectrum was obtained summing all spectra at each pixel of the scanning area. The acquisition time per pixel was of 0.5 seconds. All spectra were deconvoluted by the ZLP and multiple scattering using a reflected tail method as implemented in DigitalMicrograph^{TM}. It was found that the experimental spectrum is indeed given by a combination of $\varepsilon_{xx}(E)$ and $\varepsilon_{zz}(E)$. However a mismatch in the absolute intensities suggests that this combination is not linear [7]. Also, the introduction of various effects, such as the convergence of the probe, the thickness of the sample and relativistic effects, into the simulation of the energy-loss function have been proven to be important for a satisfying agreement of the peak intensities [8].

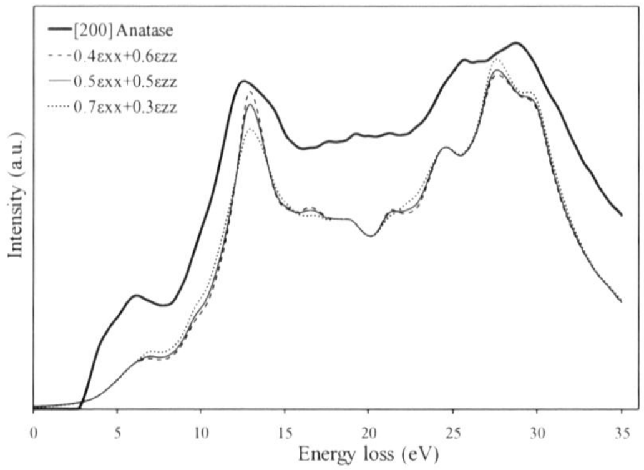

Figure 3. Experimental and theoretical energy-loss functions of anatase. Thick line: EEL experimental spectrum of the [200] anatase FIB section. Dashed, thin and dotted lines: theoretical energy-loss function of bulk anatase obtained with different linear combinations of $\varepsilon_{xx}(E)$ and $\varepsilon_{zz}(E)$.

4. Conclusions

The dielectric behaviour of novel anatase nanoplatelets was investigated at the atomic scale using the STEM-EELS method. Significant differences in terms of number, energy position and intensity of the peaks were found between platelets oriented in the [002] and [200] zone axes, suggesting that both crystal orientation and size affect the energy-loss spectrum. The present work motivates to further investigate the effects of crystal anisotropy on the energy-loss function and to determine the correlation with the size effects in nanoscale anatase systems.

5. References

[1] Han X, Kuang Q, Jin M, Xie Z and Zheng L 2009 *J. Am. Chem. Soc.* **131** 3152
[2] Chen J S and Lou X W 2009 *Electrochem. Commun.* **11** 2332
[3] Blaha P, Schwarz K, Madsen G K H, Kvasnicka D and Luitz J, WIEN2k, An Augmented Plane Wave + Local Orbitals Program for Calculating Crystal Properties (Karlheinz Schwarz, Techn. Universität Wien, Austria) 2001
[4] Hébert C, Schöne W -D and Su D S 2006 *Ultramicroscopy* **106** 1115
[5] Jouffrey B, Schattschneider P and Hébert C, 2004 *Ultramicroscopy* **102** 61
[6] Hébert C, Schattschneider P, Franco H and Jouffrey B 2006 *Ultramicroscopy* **106** 1139
[7] Wessjohann H G 1976 *Phys. Stat. Sol. (b)* **77** 535
[8] Moreau P and Cheynet M C 2003 *Ultramicroscopy* **94** 293.

Characterization of plasmonic nanostructures by analytical TEM

A Suvorova[1], M Saunders[1], P Werner[2], N Zakharov[2], A Berger[2], E Pippel[2], M Losurdo[3], G Bruno[3]

[1]The University of Western Australia, Crawley, WA 6009, Australia

[2]Max Planck Institute of Microstructure Physics, Halle, Germany

[3]Institute of Inorganic Methodologies and Plasmas at National Council of Research (CNR), Bari, Italy

E-mail: alexandra.suvorova@uwa.edu.au

Abstract. A combination of electron microscopy and spectroscopy techniques has been applied to investigate the properties of metallic nanoparticles for applications as plasmonic materials. Gold nanoparticles have been deposited on Si and SiC semiconductor substrates to evaluate their structural and chemical properties at atomic scale resolution. High resolution imaging has been used to study the atomic structure of the nanoparticles and interfaces. The chemistry of the nanostructures has been analysed using energy-filtered TEM, electron energy-loss spectroscopy (EELS) and energy dispersive X-ray spectroscopy (EDS).

1. Introduction

Metallic nanoparticles (NPs) exhibit surface plasmon resonance (SPR) that occurs at a metal/dielectric interface and provides the opportunity to confine light to very small dimensions. Plasmonic NPs are of considerable interest due to plasmon tunability and potential applications as biosensors, photonic, optoelectronic and photovoltaic devices [1]. These applications rely on the fabrication of Au NPs on technologically important substrates and on the possibility to control the SPR properties. A number of experimental and theoretical investigations of the optical and structural properties of Au NPs have been carried out [2-7]. Au NPs have been found to exhibit SPR and it has been found that their optical properties are related to the size and shape of the Au NPs-based nanosystems. In the case of Au NPs on a silicon substrate, it has been shown that interface characteristics such as interfacial SiO_2 and gold silicide formation also affect SPR properties. However, the correlation between the SPR properties and interfaces in Au nanosystems is not clearly understood.

Au NPs have been deposited by Ar sputtering onto Si (111) and SiC (0001) single crystal substrates. The deposition time and the substrate temperature were varied to synthesize NPs of different sizes (10-60 nm). The samples reported in the paper are listed in Table I. Scanning electron microscopy (SEM) imaging has been applied to study the morphology and growth dependent modifications of the Au NPs. Transmission electron microscopy (TEM) and associated analytical tools have been used to determine the structural and compositional properties of the nanostructures at a subnanometer scale. TEM samples were prepared using standard polishing techniques and Ar ion milling at 3kV.

2. Results and Discussion

2.1. SEM imaging

We have used the in-lens secondary electron detector of a Zeiss 1555 VPSEM, operating at low voltage. The SEM images of Au NPs grown on Si (111) at 60 °C show spherical NPs around 20 nm in size. Annealing at 600 °C maintains the surface morphology of the NPs grown at low temperature as revealed in figure 1a. As the growth temperature increases, the NPs begin to coalesce and different morphologies with larger, elongated NPs are observed at 400 °C (figure 1b). The deposition of Au NPs onto a different substrate also changes the morphology of the NPs. Au NPs deposited on silicon carbide exhibit lower surface Au coverage and variation in the Au NPs' shape, including nanoparticles with different aspect ratios. This is possibly due to the different surface properties of the various substrates acting as a template for synthesis of the NPs.

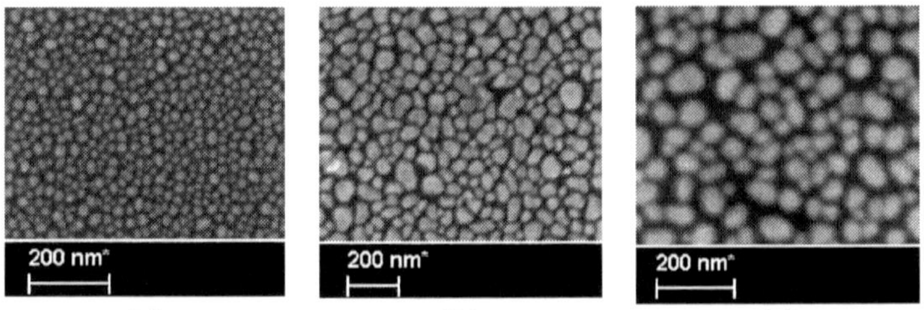

(a) (b) (c)

Figure 1. SEM images of Au NPs grown at low temperature and annealed at 600 °C (a), grown at high temperature (400 °C) (b); Au NPs on SiC grown at high temperature (600 °C) (c).

2.2. High resolution imaging

High resolution imaging (HREM) has been used to study the atomic structure of the nanoparticles and interfaces. The HREM study has been carried out using a JEOL 4010 TEM. The studies have shown that the nanoparticles are polycrystalline, with NP sizes in the range from 10-60 nm depending on growth conditions. For particles grown on the Si substrate, the interface between the NPs and the substrate was found to vary in thickness and composition depending on the deposition and annealing temperatures. The measurements of NPs and the NPs/substrate interfaces are summarized in table 1. A HREM image of a representative sample (grown at 60 °C and annealed at 600 °C) is shown in figure 2a. The HREM image shows the formation of an amorphous interface layer (confirmed by electron energy-loss spectroscopy (EELS) to be SiO_2, see below) 3 nm thick.

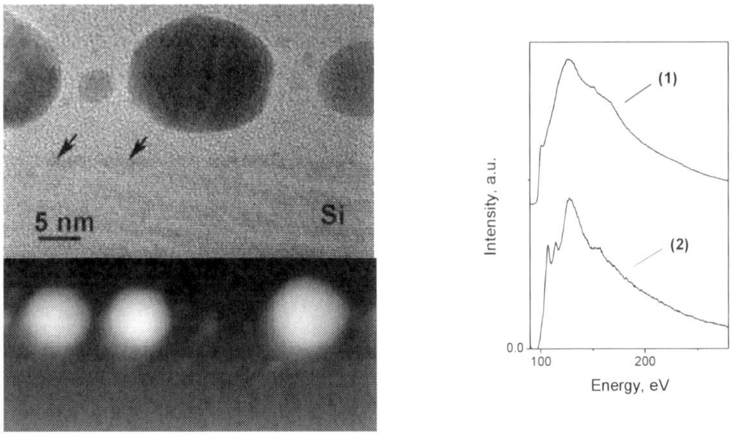

(a) (b)

Figure 2. (a) HREM and HAADF-STEM images of Au/Si interface along [112] Si. **(b).** STEM-EELS: (1)- Si substrate, (2)- interface layer (SiO_2).

The interface layer may form due to air exposure and post-deposition oxidation of the Au/Si interface, since Au stimulates the oxidation of Si [8]. Figure 2a also demonstrates regions of considerably darker contrast along the Si/SiO_2 interface (marked by arrows). This could suggest the presence of Au or a metastable Au_4Si (eutectic) phase at the Si surface. High-angle annular dark field scanning TEM imaging (HAADF-STEM) also revealed regions containing a high atomic number element present at the Si surface. This type of contrast was not observed in the samples as-grown at low temperatures. The formation of the SiO_2 layer is in agreement with extensive work on Au/Si interface formation that revealed Au-catalyzed silicon oxide formation by Au-induced Si bond breaking. The presence of Au at the silicon surface could result from out-diffusion of Au to the Si surface at high temperatures [9].

2.3. Analytical TEM

The chemistry of the nanostructures has been analysed using energy-filtered TEM (EFTEM), EELS and energy dispersive X-ray spectroscopy (EDS). The study has been carried out using a JEOL 3000F FEGTEM and an aberration corrected FEI Titan 80–300 TEM. EELS measurements across the interface confirmed the presence of the SiO_2 layer between the NPs and the substrate (figure 2b). To observe the elemental distribution within a Au/interface/Si stack, EFTEM images were acquired for the silicon $L_{2,3}$ edge, oxygen K edge, and gold $O_{2,3}$ edge and the optimum elemental mapping parameters calculated by use of a reference spectrum. The thickness of the oxide layers was estimated from a line profile across the layers using the full width at half maximum of the profile peak and was compared with the HREM measurements (see table 1). Figures 3a-d show an elastic image and EFTEM maps of the constituent elements. The EFTEM maps indicate that the oxygen is localized at the NPs/Si interface and demonstrate high surface roughness for the sample annealed at 600 °C. The interface thicknesses measured from HREM images and EFTEM data are shown in table 1. It was found that the interface thickness increased with deposition time and surface temperature. However, annealing of low temperature grown samples did not significantly affect the thickness of the interface layer. EFTEM profiles across the Au NPs/Si interface shown in figure 4a were correlated with STEM-EDS profiles shown in figure 4b. The interfaces between Au NPs and a substrate were found to be more stable at high temperatures when deposited onto SiC. TEM showed that synthesis of Au/SiC structures resulted in the formation of a reduced SiO_2 interfacial layer. It has to be noted that the SiO_2 layer was not present before NPs deposition as confirmed by in-situ ellipsometry. In all experiments the native oxide layer has been removed by chemical etching of the substrates before loading into the reactor. It could result from the post-deposition oxidation of the gold silicide interface, since gold silicides quickly oxidize.

Table 1. Summary of HREM/EFTEM analysis of Au NPs and Au/substrate interfaces for 4 representative samples

	Au/Si (111) @60 °C	Au/Si (111) @400 °C	Au/Si (111) Annealed @600 °C	Au/SiC @600 °C
NPs average size (lateral size/height)	20/18 nm	60/44 nm	20/16 nm	36/25 nm
NPs morphology	Spherical NPs	Elongated NPs	Spherical NPs	Elongated NPs
Interfacial SiO_2 :				
thickness from HREM	2.3 ± 0.1 nm	6.1 ± 0.1nm	2.4 ± 0.1 nm	2 ± 0.1 nm
thickness from EFTEM	3.5 ± 0.5 nm	6.9 ± 0.5 nm	3.5 ± 0.5 nm	2.9 ± 0.5 nm

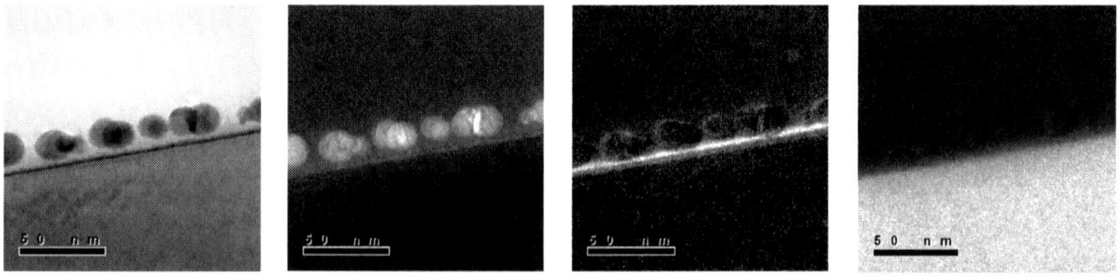

(a) (b) (c) (d)

Figure 3. EFTEM data: (a)- elastic image, (b)- gold $O_{2,3}$ jump-ratio image, (c) - oxygen K map, (d)- silicon $L_{2,3}$ map.

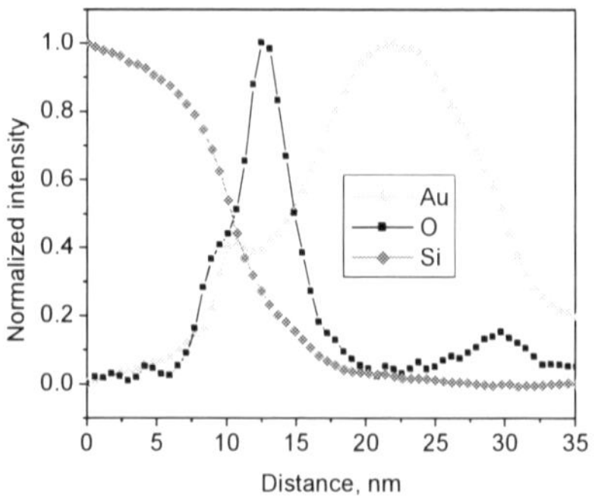

Figure 4a. EFTEM profiles across the Au/SiO_2/Si structure.

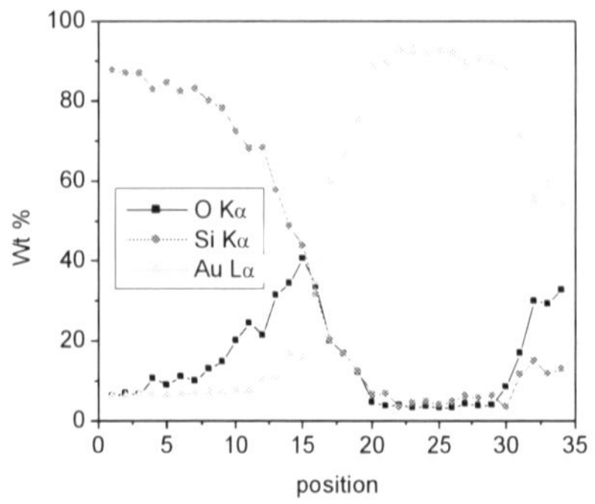

Figure 4b. STEM-EDS measurements across the same interface.

3. Conclusions

SEM analysis has shown that the Au NPs' size and morphology varies as a function of growth conditions (temperature and substrate). Analytical TEM studies have revealed the formation of a SiO_2 interfacial layer in Au/Si and Au/SiC structures. An additional phase was observed at the Si surface in the samples grown/annealed at high temperatures that could be associated with the formation of a gold silicide phase.

References

[1] Ozbay E 2006 *Science* **311** 189

[2] Vernon K C, Funston A M, Novo C, Gomez D E, Mulvaney P, Davis TJ 2010 *Nano Lett.* **10** 2080

[3] Tréguer-Delapierre M, Majimel J, Mornet S, Duguet E, Ravaine S 2008 *Gold Bulletin* **41**/2 195-207

[4] Bruno G, Bianco G V, Giangregorio M M, Sacchetti A, Capezzuto P, Losurdo M 2010 *Appl.Phys.Lett* **96** 043104

[5] Losurdo M Giangregorio M M, Bianco G V, Suvorova A A, Kong C, Rubanov S, Capezzuto P, Humlicek J, Bruno G 2010 *Phys.Rev B* **82** 155451-9

[6] Schaffer B, Grogger W, Kothleitner G, Hofer F 2010 *Ultramicroscopy* **110** 1087–1093

[7] Kuo C L, Clancy P 2004 *Surface Science* **551** 39–58

[8] Pinardi A, Leake S J, Felici R, Robinson I K 2009 *Phys.Rev. B* **79** 045416

[9] Ma Z and Allen L H 1993 *Phys. Rev. B* **48** No 20 15487

Electron Microscopy and Analysis Group Conference 2011 (EMAG 2011) IOP Publishing
Journal of Physics: Conference Series **371** (2012) 012079 doi:10.1088/1742-6596/371/1/012079

Characterisation of Synthetic Two-line Ferrihydrite by Electron Energy Loss Spectroscopy

G. Vaughan[1,2], R Brydson[1], and A. Brown[1]

[1]Institute for Materials Research, SPEME, University of Leeds, Leeds, LS2 9JT, UK.
[2]Johnson Matthey Tech. Centre, Blount's Court, S. Common, Reading, RG4 9NH

pmgmv@leeds.ac.uk / vaughgm@matthey.com

Abstract. The precise atomic structure of the nanocrystalline iron oxyhydroxide mineral, ferrihydrite is currently under intense debate concerning two proposed models. We have synthesized two and six line ferrihydrite and concentrated on the characterisation of the two line form. We show that X-ray scattering data for the two-line form are consistent with the most recent reports and that assessment of the octahedral iron content by electron energy loss spectroscopy in the transmission electron microscope suggests some 90% of the iron is ferric with octahedral co-ordination. We also identify nitrate contamination that has not apparently affected the X-ray scattering and so we infer that it is surface adsorbed.

1. Introduction

Ferrihydrite (Fh) is a highly disordered iron (oxy)hydroxide which exists exclusively in nanocrystalline form (ranging from 2-6 nm in crystallite size). It occurs widely in natural waters, sediments, is a constituent of a wide variety of soils and is similar in form to the iron-rich mineral core of the iron-storage protein, ferritin [1]. Generally formed by rapid hydrolysis of dissolved iron in solution, Fh is often found co-present with other chemisorbed species such as arsenic, lead, phosphate and some organic molecules. The high sorption-capacity makes it an ideal candidate for use in water purification since it is capable of regulating the concentration and distribution of a host of impurities. First described in 1967 [2], there still exists considerable controversy over the precise atomic structure of Fh since its size and defective nature make a definitive characterization challenging. One structural model, based principally on simulation of powder XRD patterns, describes Fh as three intermixed phases for which iron occupies octahedral interstices only [3]. Recently, a new model challenging the standard view has been proposed, based predominantly on structural refinement of atomic pair-distributions measured in the synchrotron using total X-ray scattering methods [4,5]. In this model Fh is described as a single phase in which iron occupies both tetrahedral ($<=$ 20%) and octahedral interstices. This new model has received some criticism [6,7] however the potential identification of 'under saturated' iron coordination sites has re-invigorated the structural debate.

In this study we have produced 2- and 6-line ferrihydrite (2LFh and 6LFh) and concentrated on the characterization of the 2LFh. In particular we present transmission electron microscope (TEM) based electron energy loss spectroscopic (EELS) characterization since this permits the assessment of iron valence and co-ordination at the nanometre scale [8].

2. Experimental

The Fhs used in this study were prepared following standard methods [9]. Briefly, for the synthesis of 2LFh, 10g of $Fe(NO_3)_3.9H_2O$ was added to 125 ml of distilled water and the pH of the resulting solution was raised to 7.2 by the addition of a 1 M solution of NaOH. Washing of the product was achieved by first centrifuging the solution, in order to separate the precipitate from the supernatant, and then re-dispersing the sediment in distilled water. This process was repeated 5 times before the product was freeze-dried and stored. For the preparation of 6LFh, 10 g of $Fe(NO_3)_3.9H_2O$ were added to 1 L of distilled water which had been preheated to a temperature of 70 °C. The resulting solution was held at this temperature for 10 minutes before being quenched in an ice bath. Washing was done by dialysis over several days. The cleaned product was then freeze-dried. Both Fhs have been stored as dry powders at 4 °C.

Powder X-ray Diffraction was carried out using a Philips X´pert (PW 3040/60) diffractometer, high angle annular dark field scanning TEM (HAADF-STEM) imaging was undertaken using the VG HB501 STEM retro-fitted with a NION spherical aberration corrector (SuperSTEM 1, Daresbury, UK). Pair distribution functions (PDFs) were kindly produced by Marc Michel (Stanford University) who acquired the total X-ray scattering data for the 2LFh at the Advanced Photon Source (Argonne, USA) following a similar procedure to that reported in [4,5]. Fourier Transform Infrared Spectroscopy (FTIR) was conducted using a Thermoscientific Nicolet iS10 spectrometer run in 'SMART' attenuated total reflectance (ATR) mode with a diamond disc. Each spectrum is a background subtracted sum of 36 scans at 1.92 cm^{-1} data spacing.

TEM investigations were carried out with a FEI/Philips CM200 FEG TEM operating at 197 kV and fitted with a Gatan Imaging filter (GIF 200). EELS Fe-$L_{2,3}$ edges were acquired in diffraction mode (image-coupled) on the microscope using the procedure described in [8]. We use here the linear least-squares fitting (LLS) of four well characterized reference minerals to our experimental Fe-L_3 edges to estimate the fractional compositions of octahedrally and tetrahedrally coordinated ferric and ferrous iron. The method and reference standards are detailed in [8].

3. Results and discussion

Powder X-ray diffraction (Figure 1a) and STEM-HAADF imaging (Figure 1b) of the 2LFh confirm the phase purity and nanocrystalline nature of the Fh. The XRD patterns for both 6LFh and 2LFh are consistent with those reported by Drits *et al.*, [3]

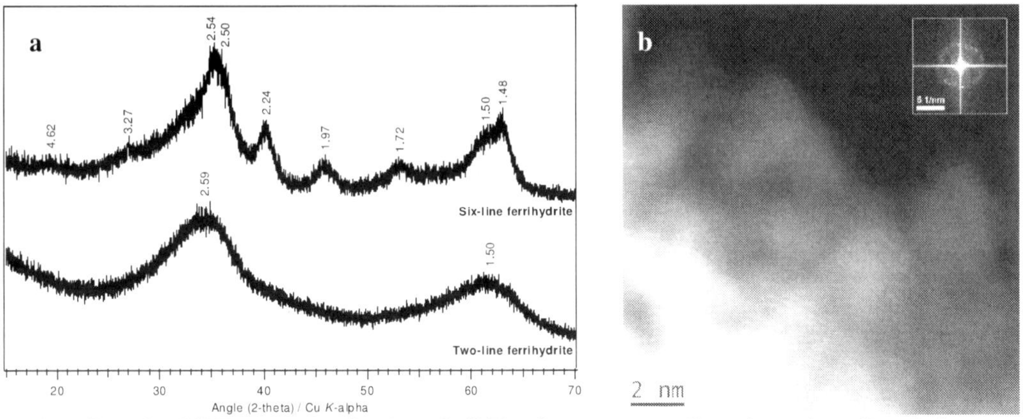

Figure 1 a) Powder XRD patterns for 2 and 6LFh, the corresponding *d*-spacings listed above each peak are consistent with [3]. b) STEM-HAADF image of the 2LFh confirming the nanocrystalline nature of the powder. Inset is a power spectrum of the image demonstrating that the imaged region is polycrystalline.

The experimental pair distribution function (PDF) for the 2LFh can be compared with that of an equivalent Fh (Fhdy2) synthesized by the same protocol used here and reported in [4] (Figure 2).

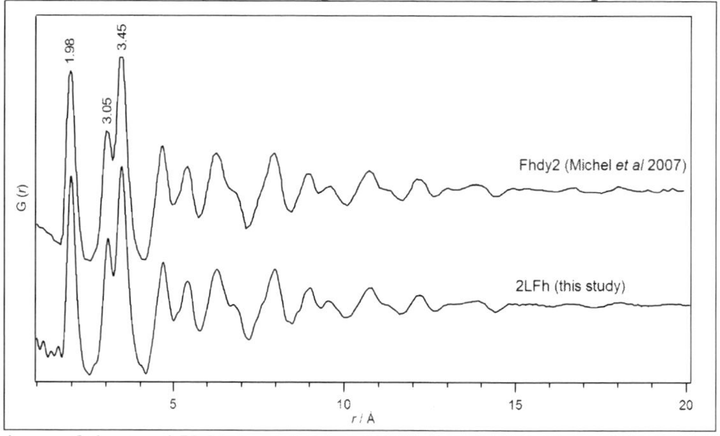

Figure 2. Comparison of the total X-Ray scattering PDF data for 2LFh and Fhdy2 from [4]. The atom-atom separations of the first three correlations, are labeled in units of Å. The profiles of the two samples are very similar.

Besides the high-frequency noise in the region $r < 1.7$ Å, which is an artifact resulting from the Fourier transformation and the finite Q-range used to obtain the PDF, the PDFs of the two samples are essentially identical. The positions and relative intensities of the correlation peaks, which extend out to approximately $r \sim 20$ Å, when overlaid are practically indistinguishable. The first correlation peak at $r = 1.98$ Å is typically assigned to the Fe-(O, OH) distances, the next two at r = at 3.05 Å and 3.45 Å can be assigned to Fe-Fe distances and, as alluded to in the Introduction, the precise structure that describes this PDF is currently under intense debate [5, 7].

FTIR spectra for 2 and 6LFh are broadly similar to those previously published [10,11] (Figure 3).

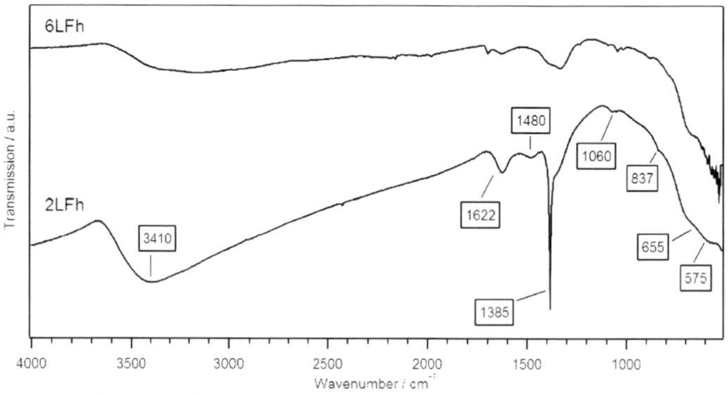

Figure 3. Infrared spectra obtained from 2 and 6LFh. The 2LFh has a steeper background and a sharp peak at 1385 cm^{-1} that can be attributed to nitrate.

The background in the range 2,000 to 4,000 cm^{-1} can be attributed to the strong scattering of the IR radiation which results from the "small sized" particles and aggregates of Fh. The background in this region is more pronounced for the 2LFh probably reflecting the smaller average crystallite sizes; however, this also may be in part due to the extent of hydration of this sample. This last point is supported by the increased intensity of the band around 1622 cm^{-1} (OH bending) for the 2LFh. This band, along with one at 3450 cm^{-1} (OH stretching), was noted by Towe and Bradley [2] to be associated with molecular water in "hydrogen-bonded association". Cornell and Schwertmann [11] attribute bands at 3615 cm^{-1}, 3430 cm^{-1} and 650 cm^{-1} to be due to free surface OH, bulk OH stretch and bulk OH deformations respectively.

The sharp spike at 1385 cm^{-1} observed for the 2LFh sample can be attributed to NO_3^- which is due to the ferric nitrate reagent used during synthesis [11]. The absence of this feature in the 6LFh spectrum highlights the efficacy of the dialysis method of washing over that used for the 2LFh.

The amount of iron in ferrihydrite that is ferric and in octahedral co-ordination was assessed by Fe L_3 ELNES fitting and it decreases as a function of accumulated electron-fluence for 2LFh (Figure 4). Error bars shown have been derived from the reported errors in the fitting coefficient values provided by the fitting routine.

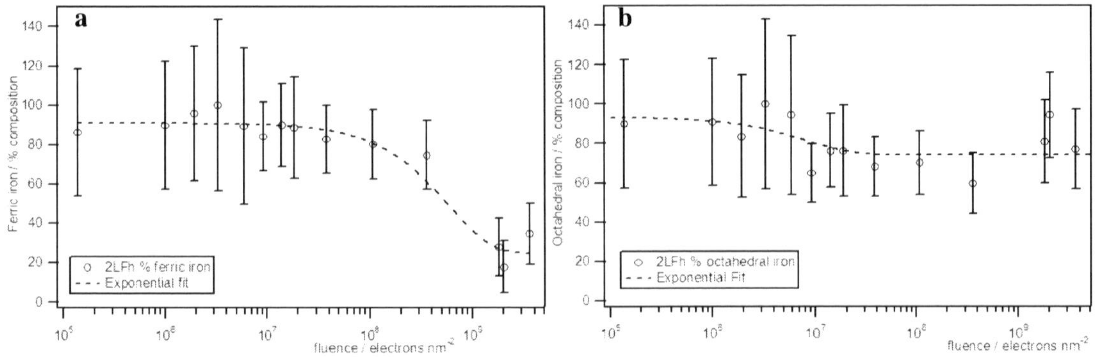

Figure 4. Fe-L_3 ELNES analysis of the ferric (a) and octahedral (b) iron content of 2LFh as a function of cumulative electron fluence. The extrapolated ferric and octahedral content at very low fluence are 91 and 93% respectively.

There is a slight decrease in the amount of ferric and octahedrally coordinated iron that can be fitted by a characteristic exponential 'dose-response' function. When extrapolated back to the zero fluence limit the exponential function estimates 91 and 93% of the total iron is ferric and in octahedral coordination. This is consistent with the values measured by the same method for 6LFh [8]

Overall, the results show that probes of the 'bulk' structure of 2LFh are insensitive to nitrate contamination. This indicates that the nitrate is surface adsorbed rather than co-precipitated and this has implications for how other contaminants such as phosphate and arsenate bind to Fh. The TEM-EELS results confirm the beam sensitive nature of Fh and suggest that 2LFh contains some 10% less tetrahedral iron than proposed by the revised Michel model (which suggests 20% tetrahedral iron) [4].

4. Conclusions

X-ray scattering profiles of synthetic 2LFh are consistent with other reports however the measurement of ~93% octahedrally co-ordinated ferric iron by TEM-EELS is not in complete agreement with the most recently proposed structural model. Nitrate has also been detected and identified as a surface contaminant because its presence does not apparently affect the bulk structure.

References

[1] Jambor J L and Dutrizac J E 1998 Chem. Rev. **98** 2549-2585.
[2] Towe K and Bradley W 1967 J. Colloid Interface Science **24** 384-392
[3] Drits V, Sakharov B, Salyn A, and Manceau A 1993 Clay Minerals **28** 185-207.
[4] Michel F M, Ehm L, *et al.* 2007 Science **316** 1726-1729.
[5] Michel F M, Barrón V, *et al.* 2010. PNAS **107**(7): 2787-2792.
[6] Manceau A 2009 Clay Minerals **44**(1) 19-34.
[7] Manceau A 2011 Am. Min. **96** 521-533.
[8] Pan Y H, Vaughan G, et al. 2010 Ultramicroscopy **110** 1020-1032.
[9] Schwertmann U and Cornell R M, 1991. The Iron Oxides in the Laboratory, VCH.
[10] Russell J D 1979 Clay Minerals **14**(2) 109-114.
[11] Cornell R M and Schwertmann U 2003 The Iron Oxides: Structure, Properties, Reactions.. Wiley-VCH.

Characterisation of ZnO nanoparticle suspensions for toxicological applications

R. Wallace[1], A.P. Brown[1], R. Brydson[1], S.J.Milne[1], N. Hondow[1], P.Wang[2]
[1] Institute for Materials Research, University of Leeds, Leeds LS2 9JT
[2] Institute of Molecular and Cellular Biology, University of Leeds, LS2 9JT

R.Wallace06@leeds.ac.uk

Abstract. ZnO nanoparticles are used in a variety of commercial products including sunscreens. The potential for accumulation in the aquatic environment has driven investigation into the toxicity of nanoparticulate ZnO. Thorough characterisation of the physicochemical properties of nanoparticulates is required in order to gain insight into the key factors which result in a nanoparticle type exhibiting a toxic response. This includes measuring the agglomeration state of the nanoparticles when dispersed in solution. This work focuses on the synthesis and characterisation of dietheylene glycol coated ZnO nanoparticles (average particle size ~ 40 nm) suspended in water. Agglomeration measurements by light scattering techniques are compared to that by TEM imaging of nanoparticle dispersions plunge frozen in liquid ethane. The two techniques produce similar size distributions suggesting that the ZnO is polydispersed in the solution with an average agglomerate size of 50-70 nm.

1. Introduction

The increased use of nanomaterials in commercial applications such as cosmetics and pharmaceuticals has led to concerns surrounding their fate and potential toxicity in the aquatic environment [1][2]. ENNSATOX is a European NMP FP7 small collaborative project that aims to relate the structure and size of a set of well characterised engineered nanoparticles to their biological activity in environmental aquatic systems (www.ennsatox.eu). The interaction of the nanoparticles with a series of biological models is being investigated including with supported phospholipid membranes, in vitro models of cell and tissue cultures and in vivo models of several different species of key indicator organisms. Zinc Oxide is one set of nanoparticles that are being investigated.

Zinc Oxide (ZnO) is a wide band gap semiconductor that has excellent ultraviolet (UV-) light absorbance. ZnO nanoparticles <100nm are being increasingly used in the manufacture of sunscreens due to their increased absorbance in the UV-A, UV-B and even into the UV-C range [3]. As well as the benefits to the safety of the sun screen user, there are also those associated with aesthetics because ZnO particles smaller than 200nm are virtually transparent in thin film coatings that are applied as a sunscreen [4].

For toxicological testing, ZnO NPs are dispersed in a range of different media and characteristics, such as degree of agglomeration and solubility need to be understood fully in each of the different media in order to ascertain whether any response is induced by the primary particles or secondary agglomerates, or whether they are a result of released free zinc ions which are already known to be toxic to aquatic organisms [5] [6].

Dynamic Light Scattering (DLS) is the technique most commonly employed for investigating the agglomeration state of a dispersion of particles in solution. However; there are intrinsic limitations to the technique. Firstly, DLS calculates the size of suspended particles by measuring Brownian motion and calculating the hydrodynamic diameter associated with the translational diffusion coefficient [7]. DLS therefore works on the assumption that all of the particles suspended in the liquid are spherical. As well as this, it is not capable of distinguishing between two different particles in a medium and it also relies on knowledge of the refractive index of the suspended particle as well as the viscosity and concentration of the suspension medium [8]. DLS is more sensitive to the larger size fraction of a polydispersed suspension due to the intensity of scattered light being proportional to d^6 (where d is the diameter of the scattering 'particle')[7]. Finally, if the dispersing media contains a solid fraction of un-dissolved macromolecules, there is a chance that the DLS will measure these as suspended particles if they are significantly more concentrated than the particles of interest.

Transmission electron microscopy (TEM) is often employed to look at the primary particles in dispersions however, just dropping the solution onto a support film and air-drying before imaging in the TEM (drop-casting) does not give a representative view of the particles in suspension because of drying effects such as nanoparticle coalescence.

With this in mind, an alternative sample preparation technique has been investigated for the TEM [9]. This preparation route involves the blotting of the suspension on the TEM support film, immediately followed by plunging it into liquid ethane cooled by liquid nitrogen. This produces an electron transparent thin film of the vitrified dispersant with the nanoparticle dispersion trapped in the film. We have shown that when the specimens are warmed under the vacuum conditions of the TEM the dispersant devitrifies and sublimes leaving the dispersion of nanoparticles on the carbon film unaltered [10]. Consequently for this work, thin film ZnO-suspensions prepared for TEM by plunge freezing have been warmed under rotary vacuum conditions and imaged in the TEM at room temperature. Quantitative analysis of the ZnO agglomerate sizes visible in the TEM images is compared with DLS results from the same sample.

2. Experimental

Colloidal ZnO nanoparticles were prepared based on the polyol method [11] [12]. 5g of Zinc Acetate dihydrate (ZnAc.2(H_2O)) (Sigma Aldrich Reagent > 99%) is added to 50 ml of diethylene glycol (DEG) (Sigma Aldrich ACS Reagent 99%) and 1ml of triple distilled milliQ water. The mixture is then heated under reflux at 140°C for 1 hour and for a further 2 hours at 180°C, after which a milky white suspension is observed. The suspension is then diluted with milliQ water to a concentration where the ZnO NPs are 0.1% w/v with respect to water. The DEG concentration is 1.25 % v/v with respect to water. The suspension is then sonicated for 20 minutes. Samples for TEM were prepared by two methods:

(i) Plunge frozen TEM specimens were prepared by placing a 3.5 μL droplet on a glow discharge treated carbon film, the grid is then blotted and immediately plunge frozen in liquid ethane [9]. The grid was then allowed to warm to room temperature under rotary vacuum conditions.

(ii) Drop cast specimens were prepared by placing a 3.5 μL droplet on a holey carbon support film and allowed to dry in air.

TEM was then performed on an FEI Tecnai F20 FEG-TEM operated at 200 kV and equipped with a Gatan Orius SC600A CCD camera. DLS of the suspension was performed on a Malvern Zetasizer Nano ZS with the data manipulation performed by the DTS Nano software.

3. Results and Discussion

TEM micrographs were obtained for the grids prepared by both drying techniques. The average primary particle size (38 ± 10 nm) was calculated from measurement of 200 particles over a set of mid-range magnification TEM images (e.g. Figure 1 a). The majority of the nanoparticles are faceted with an aspect ratio of approximately 1:1, however a number of the particles have grown along the c-axis to produce elongated nanoparticles with bullet like edges. The average aspect ratio for the elongated particles is approximately 2.5:1. Ignoring the length and measuring only the width of the elongated particles and the diameter of the faceted particles, the sample has an average size distribution of 30 nm with a standard deviation of less than 5 nm. In comparison to commercially available samples of ZnO nanoparticles, the particles synthesized have a relatively narrow size and shape distribution.

Low magnification images of the two different sample preparation methods are shown in Figure 1 (b) and (c). From the images of the drop cast sample, 100 agglomerates were measured and the average agglomerate size was calculated to be 2.4 ± 2.1 μm however there were agglomerates of up to 10 μm and also some individual particles on the film (Figure 1 (b)). In stark contrast, the particles are dispersed to a far greater degree in the plunge frozen preparation method (Figure 1(c)). There are very few agglomerates larger than 400 nm, and there is also a significant fraction (~50%) of individual particles as well as very small clusters of 2 or 3 particles on the film. The agglomerate size distribution in this specimen shows a distribution centred around ~ 50 nm (Figure 1(d)) which implies that in the drop casting method most of the small aggregates are swept into the big agglomerates in the drying process (Figure 1(b)).

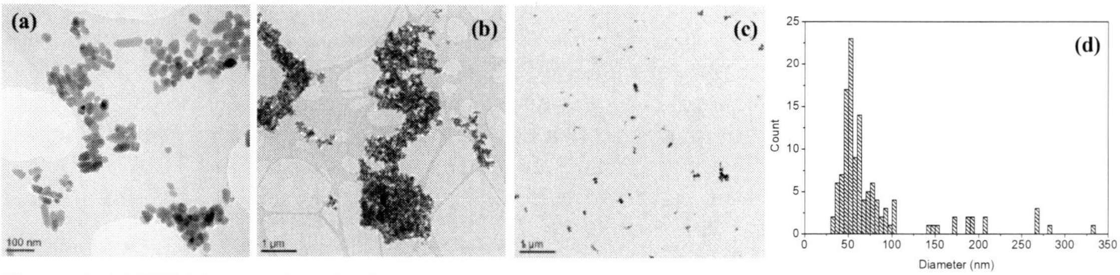

Figure 1: (a) TEM image of the ZnO suspension drop cast onto holey carbon support film. Primary particles of 30 – 50 nm are clearly visible. Low magnification TEM images of samples prepared by (b) the drop-casting method and (c) by the plunge-freezing method. The maximum agglomerate size in (b) is orders of magnitude bigger than in (c). (d) Histogram showing agglomerate size (nm) vs. frequency obtained by quantifying several TEM images of the plunge frozen specimen seen in (c).

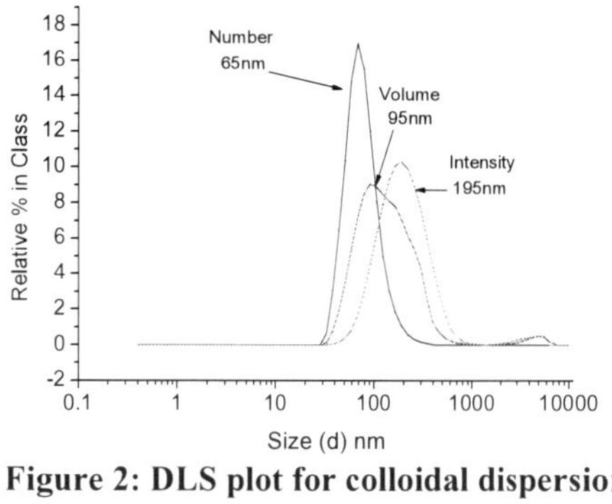

Figure 2: DLS plot for colloidal dispersion of ZnO nanoparticles showing (a) number (b) volume and (c) intensity plots

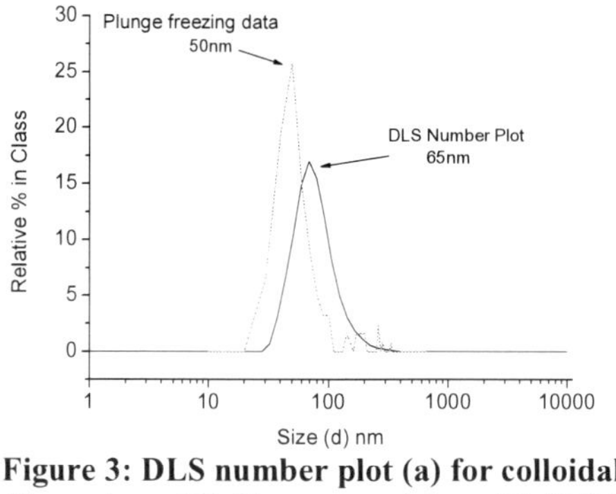

Figure 3: DLS number plot (a) for colloidal dispersion of ZnO nanoparticles with TEM plunge freezing data (b) overlaid

The light scattering profiles of the suspension obtained by DLS are converted into plots of Size (nm) vs Intensity/Volume/Number and all have a principle scattering peak with a maximum at a hydrodynamic diameter that drops from 200 to 70 nm respectively (figure 2). The intensity plot shows the relative intensity of light scattered by each of the size groups. This can then be converted into a volume plot using Mie theory which takes into account the greater light scattering of the larger particles compared to small. This explains why the volume peak is at a lower size than the intensity peak and consequently why the number plot has an even lower average particle size. Both the volume and intensity plots both show a secondary maximum in the 5-10 μm region, this may be a result of scattering from a very large agglomerate or from dust or contaminants in the suspension. Comparing the number plots from DLS and the TEM plunge freezing data, it can be seen in (figure 3) that the shapes of the curve and hence the overall size distribution is in good agreement even if however, the TEM data peak is at a smaller particle size than for DLS (50 nm vs. 65 nm). This could be due to a number of different reasons.

Firstly, there are a series of assumptions that take place in acquiring data through DLS, the Mie theory approximation as well as the assumption that the particles are spherical will undoubtedly lead to inaccuracies in particle size measurements because of the presence of nanoparticles with elongated morphologies [7]. Secondly, following the synthesis route and even after dilution with water there will be a coating of DEG on the surface of the ZnO NPs that is likely to be fully extended in solution due to its miscibility in water. DLS will measure the hydrodynamic diameter of both the particle and the coating on the surface and so this may account for a small discrepancy in size [13]. Thirdly there will be uncertainty in the TEM measurement due to potentially non representative and insufficient sampling which we are trying to reduce with further work. Either way, the two results are sufficiently similar to suggest that although the current ZnO suspension is polydisperse the average agglomerate length is no more than twice the primary particle length.

4. Conclusions

A highly dispersed sample of ZnO NPs with a reasonably tight size distribution has been produced using the polyol route. The dispersion of DEG-coated ZnO nanoparticles in water has been investigated by DLS and a plunge-freeze TEM method. Both techniques indicate a polydispersion of ZnO nanoparticles with an number average agglomerate length no more than twice the primary ZnO length and with 90% of the agglomerates (in a number based size distribution) having lengths less than 150 nm. Accurate measurement of ZnO agglomeration in solution will assist the interpretation of Zn solubility and toxicity measurements.

References

[1] Colvin V L 2003 *Nature biotechnology* **21** 1166
[2] Crane M, Handy R, Garrod J, Owen R 2008 *Ecotoxicology* **17** 421
[3] Wong S W Y, Leung P T Y, Djurisić A B, Leung K M Y 2010 *Analytical and bioanalytical chemistry* **396** 609
[4] Mitchnick M, Fairhurst D, Pinell S 1999 *Journal of the American Academy of Dermatology* **40** 85
[5] Van Straalen N, 2002 *Environmental Toxicology and Pharmacology* **11** 167
[6] Santore R C, Rooni M, Paquin P R, DiToro D 2002 *Comparative biochemistry and physiology. Toxicology & pharmacology* **133** 271
[7] *Zetasizer Nano Series Technical Note*. Dynamic Light Scattering : An Introduction in 30 Minutes
[8] Berne B J, Pecora R 1990 *Dynamic light scattering: with applications to chemistry, biology, and physics*. (New York, Dover Publication Inc.)
[9] White, H D, M L Walker, and J Trinick 1998 *Journal of structural biology* **121** 306
[10] Hondow N, 2011 Submitted to EMAG Conference
[11] Jézéquel D, Guenot J, Jouini N, Fiévet F 1995 *Journal of Materials Research* **10** 77
[12] Feldmann C, Merikhi J 2000 *Journal of colloid and interface science* **223** 229
[13] Zheng J, Stevenson M S, Hikida R S, Van Patten G P, 2002 *The Journal of Physical Chemistry* **106** 1252

Journal of Physics: Conference Series **371** (2012) 012081 doi:10.1088/1742-6596/371/1/012081

Electron microscopy of electrochemically deposited copper nanowires

X Zhao, Y L Chiu and I P Jones

School of Metallurgy and Materials, University of Birmingham, Birmingham B15 2TT, UK

Email: xxz928@bham.ac.uk

Abstract. Copper nanowires have been synthesized using a direct-current electrochemical deposition method with current densities down to 1 mA/cm^2 on a honeycomb anodic aluminium oxide template. A high pore-filling rate was achieved on the template and uniform arrays of nanowires with diameter of 60-80 nm were obtained. This paper reports SEM observations of the morphology of the deposits and cross-sectional TEM observations of the template and deposited nanowires. The template formation and the nanowire growth mechanism will be discussed in the light of the microscopy observations.

1. Introduction

In recent years, many efforts have been made to synthesize nanowires by electrochemical deposition on ordered porous anodic aluminium oxide (AAO) templates [1]. AAO templates have unique characteristics such as uniform and well-aligned pores, high temperature resistance and a large range of pore diameters. Compared to other methods, AAO-based nanowire fabrication is simple and of relatively low cost. Copper nanowires are of great interest due to their various applications especially in the field of nanoelectronics, in which they can serve as interconnections and building blocks [2]. An understanding of the microstructure and mechanical behaviour of the nanowire, is essential to realise reliable and durable applications. Here we report electron microscope observations of copper nanowires deposited on AAO templates.

2. Experiment

The well established two-step anodization technique[3] to fabricate the AAO membrane was adopted in this work. An aluminium sheet was first anodized at 0-10°C and 40 V in 0.3 mol/L oxalic acid for 1 hour. The alumina layer formed was then removed by immersing in a mixture of phosphoric acid (6 wt%) and chromic acid (1.8 wt%) at 60°C for 1 hour in order to achieve a concave metal substrate on which ordered porous aluminium oxide can be produced by repeating the anodization process for a further 3 hours. After removing the bottom barrier layer, the AAO formed was used as the template for the following electrochemical deposition. A gold layer approximately 200 nm thick was sputtered on one side of the AAO before the deposition process to serve as the cathode. An electrolyte containing 0.2 mol/L Cu_2SO_4 was used for the electrochemical deposition and its pH value maintained at 2.5 by adding sulphuric acid. Direct current electrochemical deposition with a current density of 1 mA/cm^2 was carried out at different temperatures ranging from 0 to 60 °C.

The microstructures of the AAO template and the copper nanowires are characterized using scanning electron microscope (SEM) and transmission electron microscopy (TEM). For SEM characterisation of nanowires, the AAO template was partially dissolved after the electrochemical deposition. To prepare freestanding nanowires for TEM observations, the AAO template was completely dissolved away and the nanowires ultrasonically dispersed in ethanol. A small droplet of the ethanol-nanowire suspension was then placed onto a carbon coated TEM grid. The AAO template and the cross sectional sample for TEM study were prepared using Focused Ion Beam milling.

3. Results and Discussion

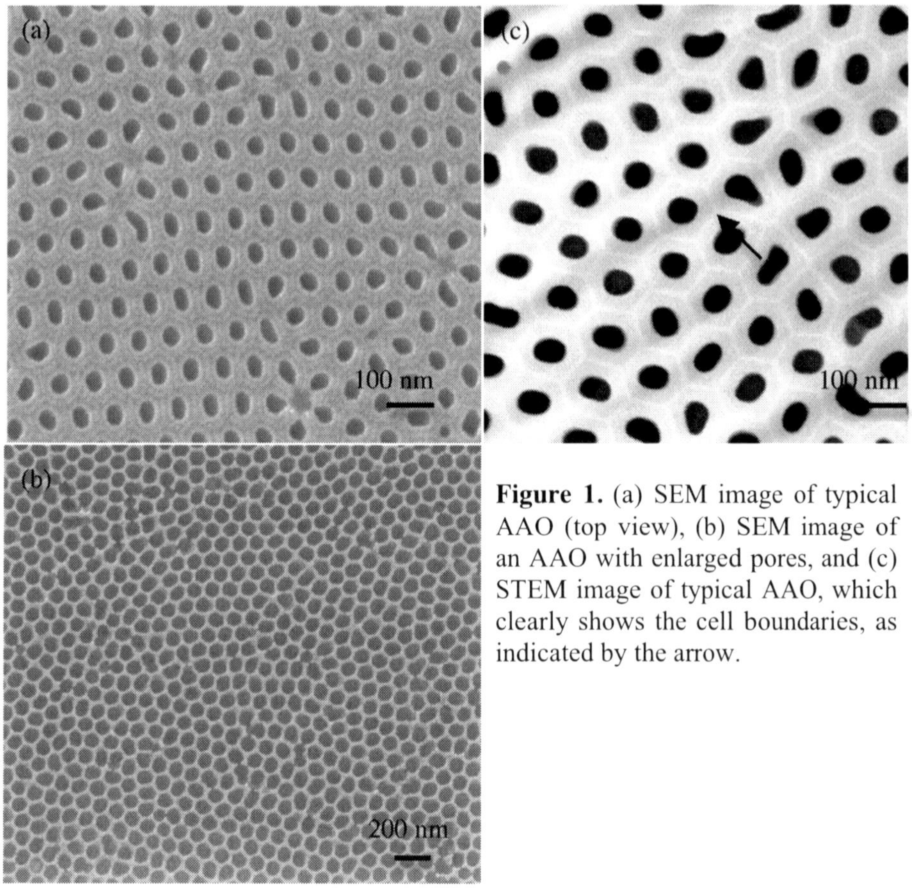

Figure 1. (a) SEM image of typical AAO (top view), (b) SEM image of an AAO with enlarged pores, and (c) STEM image of typical AAO, which clearly shows the cell boundaries, as indicated by the arrow.

Figure 1 (a) shows an SEM image of a typical AAO template obtained in this study. The pore diameter and inter-pore distance measured from this SEM image are 60-70 nm and 100 nm respectively. The pore diameter can be further enlarged by chemical etching, as indicated by (b). Figure 1 (c) shows the corresponding STEM image of the template. Although the whole AAO template is amorphous, a duplex oxide layer is clearly indicated by the different contrast in the image. The inner oxide layer, which is also called the cell boundary band, is known to be composed of relatively pure alumina; while the outer oxide layer, which was adjacent to the electrolyte during the anodization, is thought to have acid anion incorporation[4]. Details of this AAO microstructure are quite similar to those reported by Thompson *et al.*[5]. They performed STEM/EDX analysis on the AAO formed in phosphoric acid and found significant phosphorus in the EDX spectrum for the outer oxide layer. This duplex oxide layer indicates the simultaneous existence of two different growth mechanisms for the

AAO template: electrochemical reaction based on direct ion transfer across the pure alumina region [6] and deposition of the outer layer from the colloidal hydrous alumina [7].

Figure 2 shows the morphology of the as-deposited copper nanowires at different magnifications. As shown in figure 2(a), a top view of the nanowires embedded in the AAO template, the nanowires are very densely packed suggesting that a high pore-filling rate has been achieved. After dissolving the template, the nanowires are more recognizable (b). A cross-section view of the nanowires/AAO system was observed using TEM, as shown in (c). Note that the nanowires have uniform diameters of approximately 80 nm along their length, while the length shown in this image is not accurate due to truncation during the TEM sample preparation via FIB milling.

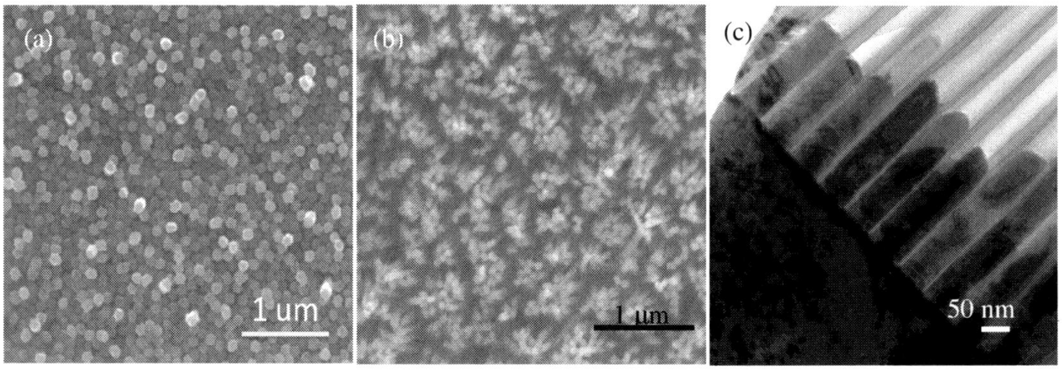

Figure 2. (a) SEM image of NWs grown in the AAO template from top. (b) SEM image of free-standing NWs after dissolving the AAO template completely and (c) cross-section TEM image of NWs/AAO composite.

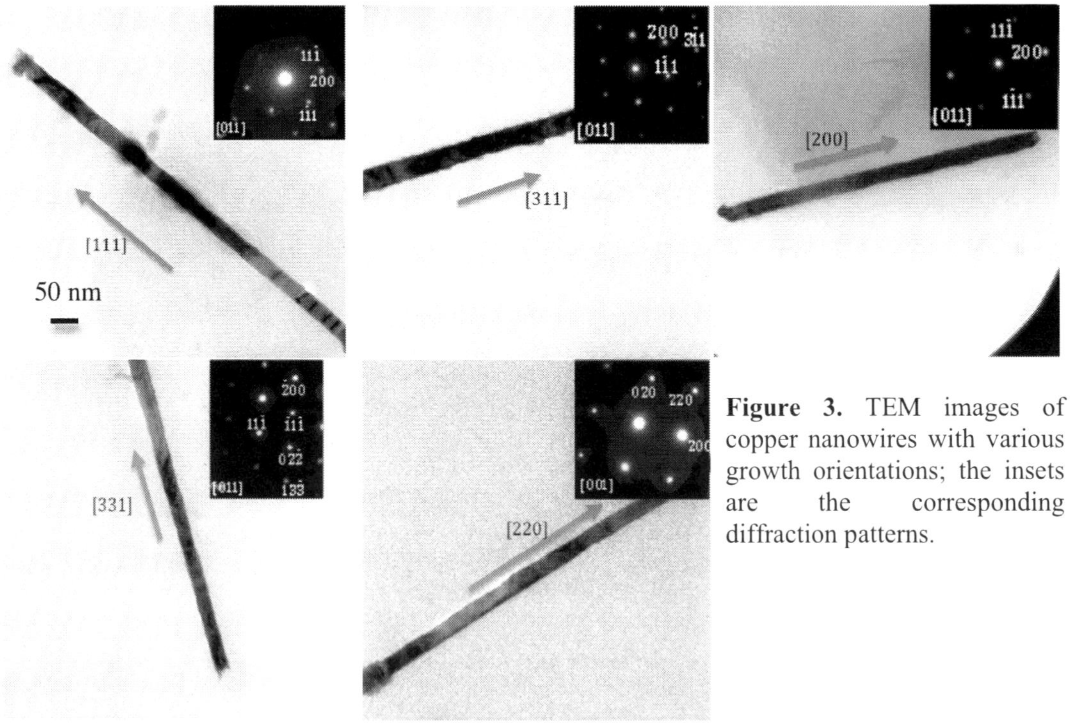

Figure 3. TEM images of copper nanowires with various growth orientations; the insets are the corresponding diffraction patterns.

During the TEM observation, the diffraction pattern remains the same along the length of a single nanowire, which indicates their single crystalline nature. Moreover, various growth orientations have been observed, as demonstrated in Figure 3. The bend contours shown here may be created by the deformation resulted from ultrasonication during sample preparation. Tian *et al.* [8] proposed that single-crystal nanowires have a two-dimensional nucleation/growth mechanism while polycrystalline nanowires results from a three-dimensional nucleation/growth mechanism. The critical dimension N_c for a 2D-like nucleus can be expressed as $N_c = bs\varepsilon^2/(Ze\eta)^2$, where s is the area occupied by one atom on the surface of the nucleus, ε is the edge energy, Z is the effective electron number, η is the over potential and b is a constant. During electrodeposition, if the size of the initial cluster exceeds the critical value N_c, new grains will grow and thus a polycrystalline structure will form. According to the above equation, a lower over potential secures a larger N_c and the chance of obtaining single crystals. It is possible that in the current study, the low current density resulted in a low over-potential, leading to 2D nucleation and growth of the nanowires. The different growth orientations may arise from the polycrystalline nature of the sputtered gold layer, which serves as the deposition substrate.

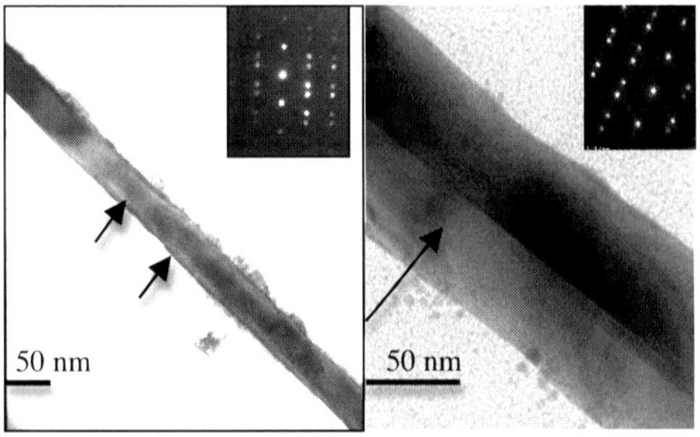

Figure 4. TEM images of twins (indicated by the arrows) in single-crystal copper nanowires; the insets are the corresponding diffraction patterns.

Twinning can also be observed in some of the nanowires, the corresponding diffraction pattern is characteristic of an fcc material twinned on {111}, as shown in Figure 4. Twin boundaries across the width and along the growth orientation of the wires are both present here.

4. Conclusions

Morphologies and microstructures of the copper nanowires as well as the AAO template were studied using electron microscopy. The results indicate that the AAO template consists of duplex oxide layers, that the copper nanowires have single crystalline structures with different growth orientations and that twin boundaries appear in some of the wires.

References

[1] Gao T, Meng G W, Wang Y W, Sun S H and Zhang L 2002 *J. Phys-Condens. Mat.* **14**(3) 355
[2] Riveros G, Gomez H, Cortes A, Marotti R E and Dalchiele E A 2005 *Appl. Phys. A-Mater.* **81**(1) 17
[3] Wang Z and Brust M 2007 *Nano. Scale. Res. Lett.* **2**(1) 34
[4] Thompson G E and Wood G C 1981 *Nature.* **290**(5803) 230
[5] Thompson G E, Furneaux R C, Wood G C and Hutchings R 1978 *J. Electronchem. Soc.* **125**(9) 1480
[6] Jessensky O, Muller F and Gosele U 1998 *Appl. Phys. Lett.* **72**(10) 1173
[7] Thompson G E, Furneaux R C, Wood G C, Richardson J A and Goode J S 1978 *Nature.* **272**(5652) 433
[8] Tian M L, Wang J U, Kurtz J, Mallouk T E and Chan M H W 2003 *Nano. Lett.* **3**(7) 919

Characterization and structure of precipitates in 6xxx Aluminium Alloys

Randi Holmestad[1], Ruben Bjørge[1], Flemming J H Ehlers[1], Malin Torsæter[1], Calin D Marioara[2] and Sigmund J Andersen[2]
[1]Department of Physics, Norwegian University of Science and Technology (NTNU), 7491 Trondheim, Norway
[2]Department of Synthesis and Properties, SINTEF Materials and Chemistry, 7465 Trondheim, Norway

E-mail: randi.holmestad@ntnu.no

Abstract. Solute atom nanoscale precipitates are responsible for the favourable mechanical properties of heat treatable aluminium alloys such as Al-Mg-Si (6xxx). The shape, structure and strengthening properties of age-hardening precipitates depend on alloy composition and thermo-mechanical history. We seek an improved understanding of the physics related to nucleation and precipitation on the atomistic level in these alloys. Once these mechanisms are sufficiently well described and understood, the hope is that 'alloy design' simulations can assist tailoring of materials with desired properties. In pure Al-Mg-Si we have determined the structure of nearly all the known metastable precipitate phases, by combining advanced TEM techniques (such as high resolution TEM and nano-beam diffraction) with atom probe tomography and density functional theory. We are now studying effects of additions /substitutions of Cu, Ag and/or Ge that promote formation of more disordered precipitates, employing aberration corrected high angle annular dark field scanning TEM. We find that all metastable precipitates contain variations of a widely spaced 'Si/Ge network'. In spite of disorder or defects, this network is surprisingly well ordered, with hexagonal projected sub-cell dimensions $a = b \approx 0.4$ nm and c (along the fully coherent precipitate main growth direction) equal to 0.405 nm or a multiple of it.

1. Introduction

The Al-Mg-Si(-Cu) alloys are an important group of age-hardenable materials. Properties like high strength/weight ratio, good formability and weldability, combined with an excellent corrosion resistance, rank them among the most attractive alloys and have made them widely used by the industry. 6xxx alloys find applications in a wide range of areas, such as in building constructions, automobiles and marine applications. The significant increase in hardness during the ageing heat treatment is caused by the formation of a high density of metastable (semi-)coherent precipitates in the Al matrix. There is an increasing demand for improved control of properties ('alloy design'). This in turn requires knowledge of the finest details of the precipitation sequence, as some precipitates have considerably stronger hardening effect than others. The task is manifold, including diffusion and solute clustering, nucleation, growth and phase transformations of the precipitates. These topics are inherently atomistic, meaning that knowledge and understanding of the atomic structures of the precipitates and their interfaces with Al is paramount. The quality of modern transmission electron microscopes (TEM), the use of recently developed quantitative analysis techniques and the availability of computer power and programs to perform first principles calculations have enabled the crystal structures of most precipitates in the ternary Al-Mg-Si alloys to be solved. Disregarding the actual age temperature and time, for this system the following precipitation sequence can be given (with the structures largely determined within our group):

$$\text{SSSS} \rightarrow \text{Clusters} \rightarrow \text{GP (pre-}\beta\text{''[1,2])} \rightarrow \beta\text{''[3]} \rightarrow \beta\text{'[4], U1[5,6], U2[5,7], B'[8,9]} \rightarrow \beta\text{, Si}$$

Here, SSSS denotes the supersaturated solid solution. The U1, U2 and B' phases are also called Type A, Type B and Type C, respectively [10]. With exception of the equilibrium phase β (Mg_2Si) and Si, all phases are assumed metastable. Every stage of the sequence is complex and is a function of alloy composition, heat treatment (including cooling and heating rates) and storage time at room temperature prior to ageing. Cu additions to the ternary Al-Mg-Si system change the precipitation sequence and produce new phases [11-14]:

$$\text{SSSS} \rightarrow \text{Clusters} \rightarrow \text{GP} \rightarrow \beta\text{''} + \text{L} + \text{S} + \text{QC} \rightarrow \beta\text{'} + \text{Q'} \rightarrow \text{Q}$$

This paper firstly reviews the methodology used in our studies [1,3-7,15] of precipitates in 6xxx alloys. We then show some examples of how advanced TEM methods are used to acquire information on precipitate atomic structures at different stages, including quantitative diffraction and probe corrected annular dark field scanning TEM (ADF-STEM).

2. Methodology

Figure 1 shows the length scales involved in aluminium alloy development. Clustering, nucleation and precipitation are phenomena that must be studied near and at the atomic scale, as demonstrated on the right hand side of the figure. Usually, before attempting such detailed studies, in order to correlate the information obtained at the smallest scale, it is normal to characterize the materials properties and investigate at lower magnifications using more conventional methods, like optical and scanning electron microscopy. The length scales are demonstrated on the images on the left hand side of Figure 1.

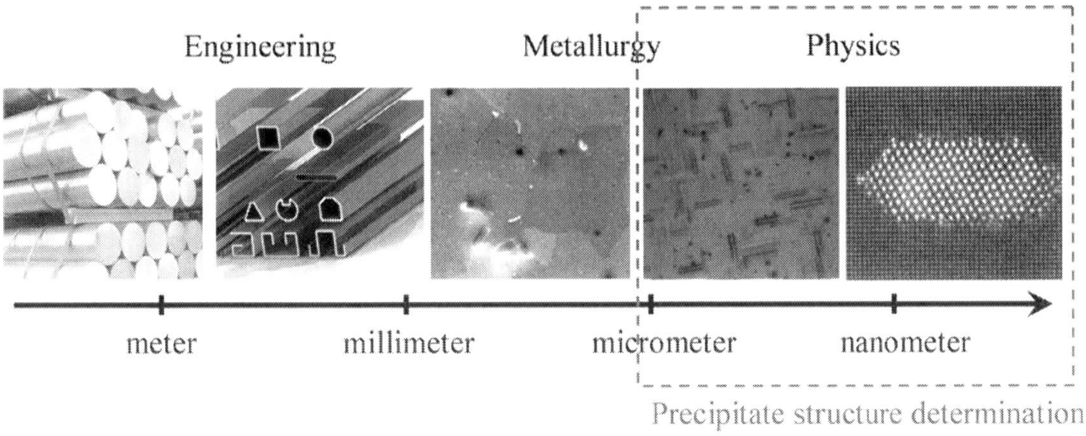

Figure 1. The different length scales investigated in aluminium alloys. Tools for studying the nano/micro-structure as shown in the stippled box would include high resolution TEM and annular dark field scanning TEM, nano-beam diffraction, atom probe tomography and density functional theory calculations.

2.1. Experimental methods

It is well established that age hardening strongly depends on the number density of precipitates as well as their type. Because all metastable precipitates in Al-Mg-Si(-Cu) have needle/lath/rod/plate morphologies with main growth direction along $<001>_{Al}$, for most purposes the TEM specimens should be oriented in this zone axis. Images recorded in bright field TEM are used for finding precipitate number density and measuring the average precipitate length and cross-section that enables calculation of corresponding precipitate volume fractions. Early on, our group developed a methodology for accurate microstructure determination by taking into account errors originating from underestimation of average precipitate needle/lath/rod/plate length when some precipitates close to the specimen surface are cut during sample preparation [16]. Thicknesses of the samples are needed here, and are measured using electron energy loss spectroscopy. Often, precipitates can be distinguished based on morphology and orientation, see e.g. [17].

Atomic models of precipitate structures are constructed by analyzing high resolution TEM (HRTEM) images that give input on atomic column positions and crystal symmetry. Since these images can only be acquired in main growth axis projection (with the precipitate occupying the entire region across the sample thickness direction), they do not contain full information about atomic heights or atomic column composition (type).The heights can be inferred from projected inter-column distances, and approximate precipitate compositions can be obtained by Energy Dispersive X-ray Spectroscopy (EDS). With heavier atoms like Cu, Ag or Ge present in the precipitates, annular dark field scanning TEM (ADF-STEM) is a superior technique due to its Z-contrast information. In addition, ADF-STEM images are more directly interpretable than HRTEM images which rely on phase contrast [18]. The ADF-STEM technique has been greatly enhanced recently by the development of probe Cs-corrected STEM machines that are capable of achieving sub-Ångström resolutions at superior signal to noise ratios [19]. Consequently, ADF-STEM is increasingly replacing HRTEM. One drawback of the ADF-STEM technique is the presence of distortions in the images due to specimen drift during the electron scan.

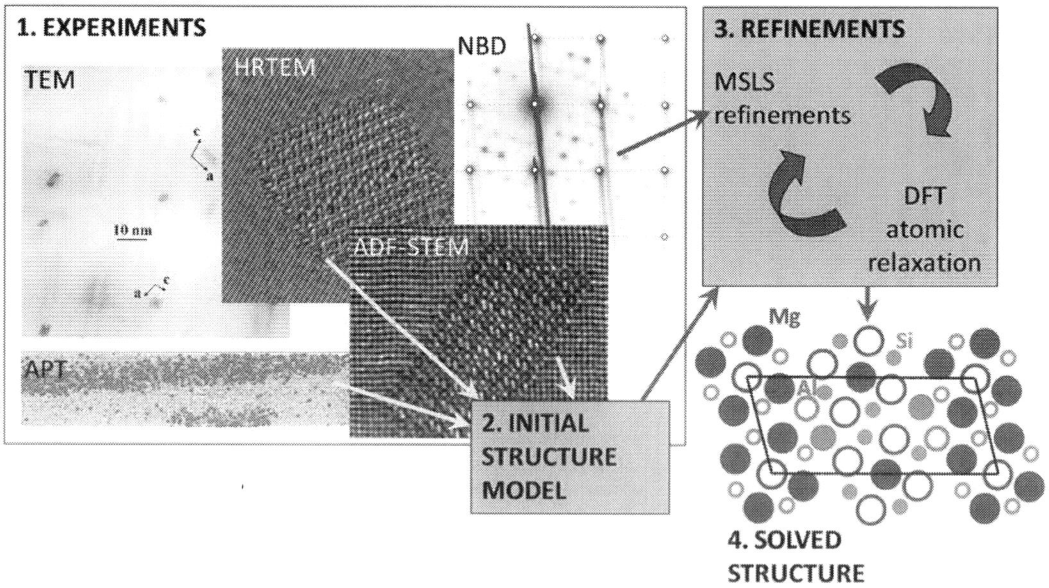

Figure 2. Overview of the methodology used to determine precipitate structures. Examples are shown for the main hardening phase β". Experiments (1) include medium magnification bright field TEM, HRTEM, APT, ADF-STEM and nano-beam diffraction (NBD). Theses are used to construct an initial structure model (2) as input to the quantitative refinements (3) where NBD intensities are used in a Multi Slice Lest Squares (MSLS) program and density functional theory (DFT) calculations are done in a self-consistent loop to solve the structure (4). Abbreviations not explained are given in the text.

In the early stages of clustering, when randomly distributed clusters have small sizes and are highly coherent with the face centred cubic (fcc) Al matrix, TEM based techniques have proven difficult. For Al-Mg-Si, even the exact pre-β'' structure is still a matter of debate [1, 2]. Atom probe tomography (APT) [20], is a unique method enabling real-space mapping of atoms with near-atomic resolution, and it is superior in the characterization of small clusters and precipitates in metallic alloys [21]. In addition to providing particle number density and morphology, APT gives information complementary to that of the TEM. For the case of well-developed precipitates APT can be used to accurately measure the particle composition. If precipitate compositional disorder is present, this will be most clearly revealed with APT [20]. Figure 2 shows an overview of the methodology.

2.2 *Quantitative refinements*

By combining some of the above-mentioned experimental techniques, good initial models of the precipitate atomic structures can be constructed. When possible, verification of these models is achieved by using quantitative electron diffraction. For this purpose, relative intensities of diffraction spots in nano-beam diffraction (NBD) patterns recorded from individual particles are compared with the corresponding relative intensities generated from the proposed atomic model, taking into account dynamical diffraction. This is acquired by use of a multi slice least squares (MSLS) method [22]. After the experimental and theoretical intensities are compared, the atomic model is adjusted to optimize the fit with experimental data. For 3D refinement of atomic coordinates, the MSLS program is dependent on input from a collection of electron diffraction patterns originating from different zone axes. In addition to atomic coordinate refinement, MSLS can also refine cell parameters, atomic occupancies and temperature factors. MSLS is presently an integrated part of our methodology [3-7,15]. However, since the MSLS program only optimizes a structure to fit experimental data, it needs to be verified whether or not the refined atomic structure has a favourable formation enthalpy. Also, precipitate compositional disorder can prove challenging, given the highly similar scattering cross sections of Al, Mg and Si.

First principles density functional theory (DFT) [23,24] based total energy calculations represent the selected complementary theoretical tool to MSLS for clarifying the structure and composition of the experimentally observed precipitates [3-7,15]. We generally employ the Vienna ab initio simulation package (VASP) [25], a plane wave based benchmark tool for condensed matter theoretical studies. A combined MSLS and VASP analysis is highly advantageous: MSLS makes direct reference to experimental observations, but can leave doubts with regards to the composition [15]. DFT studies, on the other hand, involve well defined configurations, but within user defined restrictions (choice of unit cell size e.g.) of the analysis. When fed with input from a preliminary structural analysis, total energy calculations can provide a set of candidates for MSLS to analyze. In turn, the MSLS optimized structures may provide new information (such as the need of a larger unit cell) for VASP to address. In principle, this 'self-consistency loop', as shown in figure 2, can fully clarify the precipitate structure. Inclusion of APT investigations in the above described loop promises increased reliability of the results.

When used for structural determination, theory conventionally ignores the precipitate interfaces with Al for reasons of computational efficiency. Evidently, a proper description of the precipitate interactions with Al requires interface studies [26-28].

As expected from the comparative dominance of bulk over interface energy contributions, bulk calculations do tend to get the relative stabilities of the phases qualitatively right [8]. However, we emphasize the crucial importance of experimental input for the theory: For Al-Mg-Ge, only some of the Al-Mg-Si precipitates remain stable according to experiment [29], whereas bulk calculations reveal no exclusion of any of these precipitates.

3. ADF-STEM on Cu-containing precipitates and the Si network

The advantage of the ADF-STEM technique when investigating metastable precipitates in the 6xxx system was first realized in the Al-Mg-Si-Cu system [14]. Here, the distribution of Cu-columns along the needles' directions demonstrated that many of these precipitates had disordered structures. This, corroborated with the additional information provided by HRTEM images and NBD patterns enabled us to conclude that, as for the Al-Mg-Si system, the precipitates can be basically described as different arrangements of Al, Mg, Cu on a Si-network with projected hexagonal dimensions a = b ≈ 0.4 nm when viewed along the precipitates longest directions. A direct confirmation of this has been provided by aberration corrected ADF-STEM, where the Si atomic columns are clearly visible in Z-contrast. Figure 3 shows as example a partly disordered Q' precipitate image in aberration corrected ADF-STEM mode, with the Cu and Si atomic columns identified solely based on the Z-contrast information. The importance of ADF-STEM is of equal importance when studying alloys where Si is replaced with Ge [29].

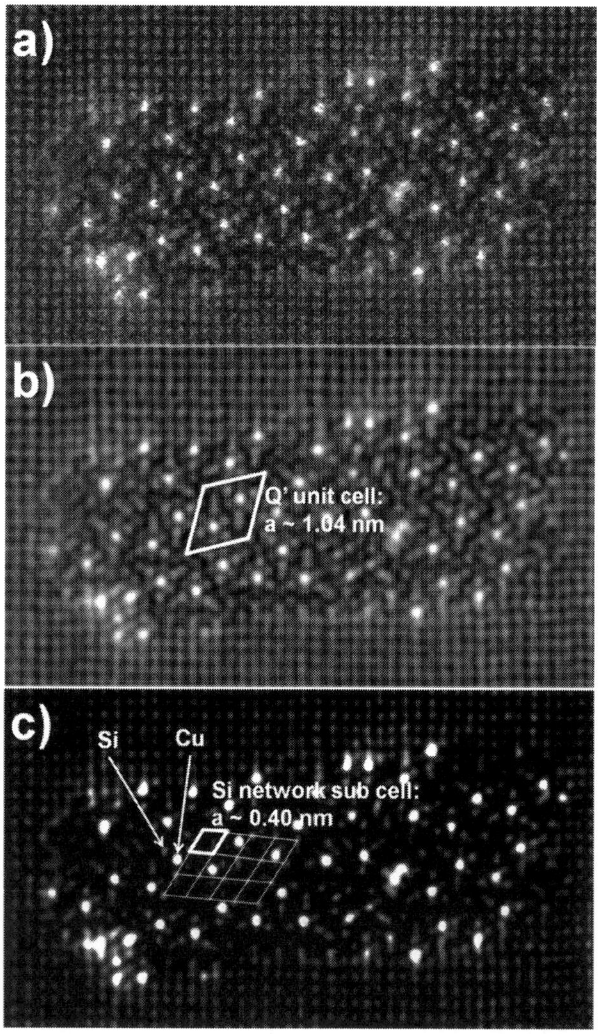

Figure 3. a) Unprocessed ADF-STEM image recorded along <001>Al of a partly disordered Q' precipitate viewed along its lath direction; b) Noise was removed by applying a band pass mask that removed all distances shorter than 0.17 nm. A Q' unit cell is marked; c) By enhancing contrast and reducing brightness of the image in b), the Cu and Si atomic columns inside the precipitate become visible, see arrows. The Si-network with projected hexagonal a = b ~ 0.4 nm is indicated.

4. Conclusions

Advances on optimizing alloy properties for specific purposes will increasingly be relying on the fundamental understanding of precipitates as exemplified in this paper. We have developed a methodology combining several experimental techniques to study precipitate structures in Al-Mg-Si/Ge(-Cu) alloys. This has revealed the existence of structural similarities between all metastable precipitates through a common Si network. ADF-STEM and NBD are central techniques in this methodology, and probe Cs correction has proven extremely useful. Linking alloy composition and heat treatment with these results, 'alloy design' can be used to tailor materials with desired properties.

Acknowledgments
This work was financially supported by The Research Council of Norway via two projects "Fundamental investigations of solute clustering and precipitation" and "Nucleation Control", the last which is supported by Norwegian industry: Hydro Aluminium AS, Steertec Raufoss AS. Simulations are performed through access to the NOTUR facilities. Collaborations with Rouen University (Williams Lefebvre), Monash University (Joanne Etheridge) and Denmark Technical University (Rafal Dunin-Borkowski) are highly acknowledged.

References
[1] Marioara C D, Andersen S J, Jansen J and Zandbergen H W 2001 *Acta Mater.* **49** 321
[2] Chen J H, Costan E, van Huis M A, Xu Q and Zandbergen H W 2006 *Science* **312** 416
[3] Andersen S J, Zandbergen H W, Jansen J, Træholt C, Tundal U and Reiso O 1998 *Acta Mater.* **46** 3283
[4] Vissers R, van Huis M A, Jansen J, Zandbergen H W, Marioara C D and Andersen S J 2007 *Acta Mater.* **55** 3815
[5] Frøseth A, Høier R, Derlet P M, Andersen S J and Marioara C D 2003 *Phys. Rev. B* **67** 2241061
[6] Andersen S J, Marioara C D, Vissers R, Frøseth A and Zandbergen H W 2007 *Mater. Sci. Eng.* A **444** 157
[7] Andersen S J, Marioara C D, Frøseth A, Vissers R and Zandbergen H W 2005 *Mater. Sci. Eng.* A **390** 127
[8] Ravi C and Wolverton C 2004 *Acta Mater.* **52** 4213
[9] Wolverton C 2001 *Acta Mater.* **49** 3129
[10] Matsuda K, Sakaguchi Y, Miyata Y, Uteni Y, Sato T, Kamio A and Ikeno S 2000 *J. Mater. Sci.* **35** 179
[11] Matsuda K, Ikeno S, Uetani Y and Sato T 2001 *Metall. Mater. Trans. A* **32** 1293
[12] Chakrabarti D J and Laughlin D E 2004 *Progr. Mater. Sci.* **49** 389
[13] Cayron C, Sagalowicz L, Beffort O and Buffat P A 1999 *Phil. Mag.* A **79** 2833
[14] Marioara C D, Andersen S J, Stene T N, Hasting H, Walmsley J C, van Helvoort A T J and Holmestad R 2007 *Phil. Mag.* **87** 3385
[15] Hasting H S, Frøseth A G, Andersen S J, Vissers R, Walmsley J C, Marioara C D, Danoix F, Lefebvre W and Holmestad R 2009 *J. Appl. Phys.* **106** 123527
[16] Andersen S J 1995 *Metall. Mater. Trans.* A **26** 1931
[17] Torsæter M, Lefebvre W, Marioara C D, Andersen S J, Walmsley J C and Holmestad R 2011 *Scripta Mater.* **64** 817
[18] Nellist P D and Pennycook S J 1999 *Ultramicr.* **78** 111
[19] Krivanek O L, Chisholm M F, Nicolosi V, Pennycook T J, Corbin G J, Dellby N, Murfitt M F, Own C S, Szilagyi Z S, Oxley M P, Pantelides S T and Pennycook S J 2010 *Nature* **464** 571
[20] Miller M K 2000 *Atom Probe Tomography: analysis at the atomic level* (New York: Plenum)
[21] Marquis E A, Hyde J M 2010 *Mater. Sci. and Eng.* **69** 37
[22] Jansen J, Tang D, Zandbergen H W and Schenk H 1998 *Acta Cryst.* A **54** 91
[23] Kohn W and Sham L 1965 *Phys. Rev.* **140** A1133
[24] Martin R M 2004 in *Electronic Structure: Basic Theory and Practical Methods* (Cambridge University Press) 119.
[25] Kresse G and Hafner J 1993 *Phys. Rev.* B **47** 558.
[26] Mishin Y, Asta M and Li J 2010 *Acta Mater.* **58** 1117
[27] Wang Y, Liu Z K, Chen L Q and Wolverton C 2007 *Acta Mater.* **55** 5934
[28] Ehlers F J H and Holmestad R 2010 *Mater. Res. Soc. Symp. Proc* **1229** LL02
[29] Bjørge R, Marioara C D, Andersen S J and Holmestad R 2010 *Metall. Mat. Trans.* A **41** 1907

A study on microstructure and strain-hardening rate of friction stir welded Al-Mg-Si alloys using a weak beam technique

E. Sukedai and T. Yokoyama

Okayama University of Science, Okayama 700-0005, Japan

E-mail: sukedai@mech.ous.ac.jp

Abstract. Mechanical properties of a friction stir welded Al-Mg-Si (6061-T6Al) alloy are evaluated by a tensile test. It is found that the strain-hardening rate is higher than that of a base material. In order to investigate the origin, TEM observations have been performed about 4 kinds of materials; base- and friction stir welded-materials, and both materials deformed to 5 % strain by tension. There are not so large differences about dislocation density, size and density of precipitates and crystal defects between the base material and the friction stir welded-materials, but a significant decrease of grain-size in the friction stir welded-materials is recognized. These results suggest a dynamic recovery occurs during FSW process, and it is speculated that the recovery leads to the differences of yield stress and strain-hardening rate between both materials.

...

1. Introduction

Friction Stair Welding (FSW) method is relatively a simple method, and it has been widely applied to weld light metals such as Al, Mg and Ti alloys. Therefore, this method has been widely used in many industrial fields such as automobile, railway carriage and air-plane companies. In order to redound the reliability of this method, many kinds of mechanical properties of friction stair welded materials (FS welds) have been studied [1-3)]. It is reported on tensile properties of FS welded Al-Mg-Si (6061-T6Al) alloys that strain-hardening rates of their stress-strain curves are higher than those of the base materials [4, 5)]. Instinctively, the difference of dislocation density between FS welds and base materials is visualized. Therefore, the dislocation density of a FS welded 6061-T6Al alloy has been measured by X-ray peak profile analysis method [4)]. However, before FSW process, the 6061-T6Al alloy sheets were thermal refined, i.e. solution treating and aging. Therefore, changes of not only dislocation density, but also microstructure due to the FSW process are considered as important factors to decide mechanical properties, including the phenomenon of strain-hardening rates.

In the present study, to elucidate the origin of the increase in strain-hardening rate of a FS welded 6061-T6Al alloy, microstructures of (1) the base-material and a FS welds, and (2) both materials deformed to 5 % strain by tension have been investigated using a transmission electron microscope (TEM) observation method, especially a weak beam technique, which is probably the best technique to observe dislocations and crystal-lattice defects [6)].

2. Experimental procedures

A material used in the present study is a 6061-T6Al alloy sheet and the chemical composition was 0.4-0.8 % Si, 0.8-1.2 % Mg, Fe, Cu, Mn etc, and balance of Al (mass%). The sheet was thermal refined by a solution-treating at 788-823 K, followed by aging at 428-438 K for 18 hrs. FSW conditions were as follows; the rotation speed of the rotating tool was 2000 rpm and welding speeds of specimens A, B and C were 600, 900 and1200 mm/min, respectively [5)]. The FSW process was carried out along the rolling direction. Hardness testing on a cross-section perpendicular to the FSW direction was performed. Tensile testing was carried out at a strain rate of 6.2×10^{-4}/ s. Tensile tests of FS welds were performed to obtain data from FS welds part (weld nugget = WN) as follows; as the width of WN is approximately 25 mm [5)], an extensometer with a gauge length of 25 mm was used. Stress-strain curves obtained prove almost tensile properties of WN parts. For TEM observations, thin-foil specimens of the base material, a WN part of FS welds A, and both of them deformed to 5 % strain were prepared. A JEM 2010F electron microscope operated at 200 kV was used for the observations.

3. Results and discussion

3.1. Stress-strain curves and grain sizes of base material and FS welds

Figure 1 indicates tensile stress-strain curves for the base material and the three FS welds. The tensile strength and ductility of the base material are greatly reduced by FSW process. It is obvious that the yield and tensile strengths of the FS welds A, B and C remain almost unchanged, and they are lower than those of the base material. The elongation of a FS weld A decreases compared with that of the base material. In Fig. 1, it is also clear that strain-hardening rates of FS welds are higher than that of the base material; that is approximately 2 times higher at 3 % and 5 % strains [5]. In order to elucidate the origin on this phenomenon, the TEM observation was carried out and the results will be discussed in the following section.

It is also known that grain sizes affect the yield and tensile strengths of many kinds of materials. So, grain sizes of a base material and a FS weld A were measured. While the average grain size of the base material, measured by an optical micrograph is approximately 150 μm [5], the size of a FS weld A is 3-7 μm, estimated by a TEM figure shown in Fig. 2. A significant decrease in grain-size due to FSW process was recognized. The decrease of grain-size in Al alloys due to a FSW process has been reported [2, 3].

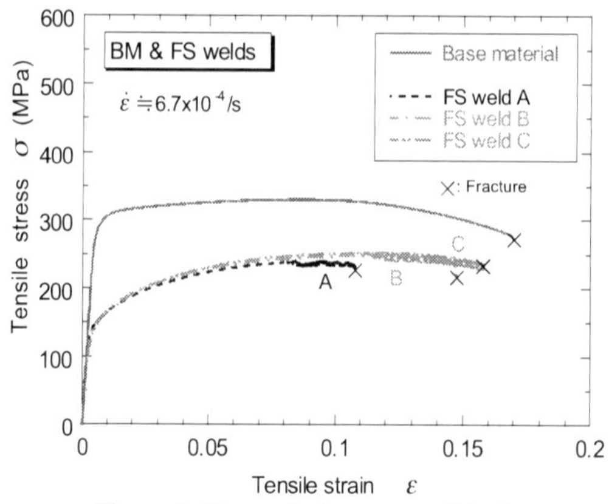

Figure 1 Stress-strain curves of the base material and FS welds A, B and C.

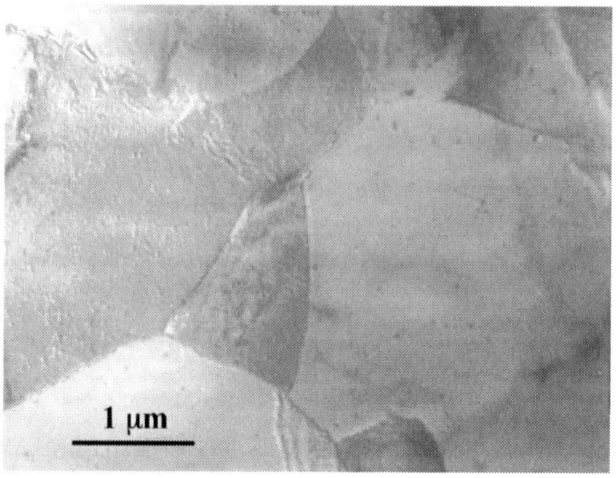

Figure 2 TEM bright field image of FS weld A.

3.2. TEM observation results of base material and FS weld A

Figure 3 (a) and (b) show results of the base material and the FS weld A, respectively. In Fig. 3(a), dislocation segments (see an arrow d), a large number of small defects in 20-40 nm length and bright dots are visible. The small defects are parallel to two directions, and as their visibility depends on reflection conditions, they might be dislocation segments. High resolution EM images indicate many stacking faults parallel to {111} slip planes. Since two directions of small defects are not consistent with the traces of {111} planes, small defects are not stacking faults. In addition, it seems that metastable phases are not observed because of the thermal refining. Bright field images indicate rod- and round-shape precipitates in 100-150 nm in length and 30-50 nm in diameter. The microstructure shown above results from a rolling, followed by the thermal refining. In Fig. 3(b), a few dislocation segments are visible. A large numbers of small defects in 20-80 nm in length also appear parallel to two directions, which are not consistent with slip-plane traces. It is found that the density of small defects is lower than that of the base material. The decrease in the density is considered as a result

from heating due to the FSW process. An arrow indicates a round shape precipitate in 150 nm in diameter.
Figure 4(a) and (b) indicate results of a 5 % strained base material and a 5 % strained FS weld A, respectively. In Fig. 4(a), it is recognized that dislocation density increases and small defect density decreases compared with the base material (Fig. 3(a)). In Fig. 4(b), dislocation lines can be seen clearer than those of the 5 % strained base material, and in the matrix, few defects are visible. It seems considered that many dislocations swept out them due to their interactions.

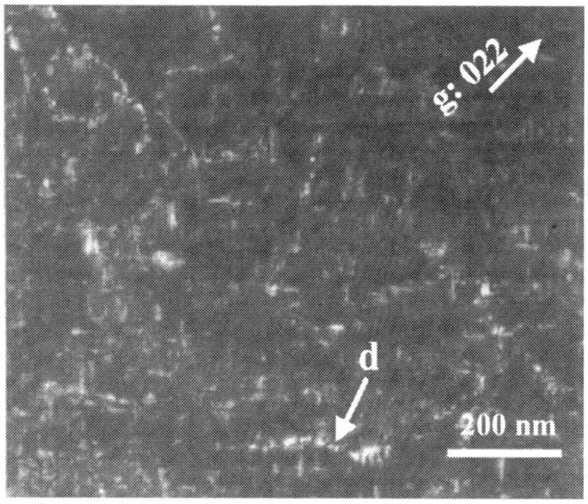

Figure 3 (a) Weak beam dark field (WBDF) image of the base material. d: dislocation

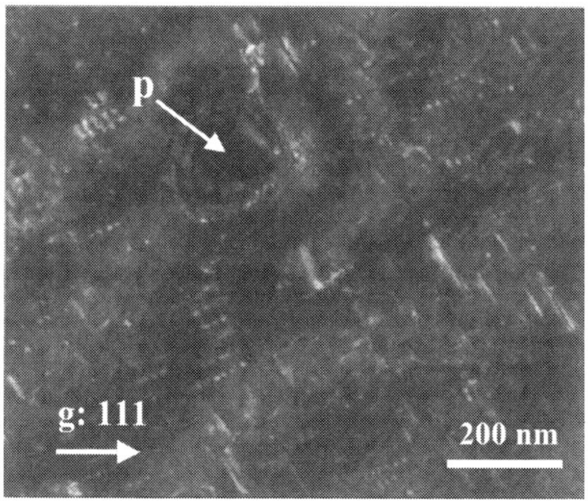

Figure 3 (b) WBDF image of the welded part of FS weld A. p : precipitate

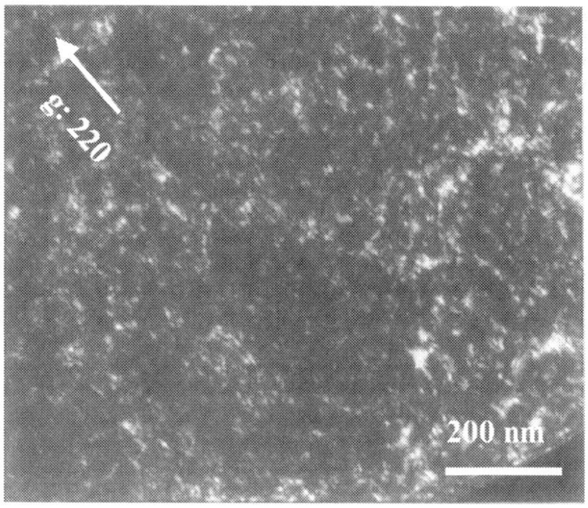

Figure 4 (a) WBDF image of 5 % strained base material.

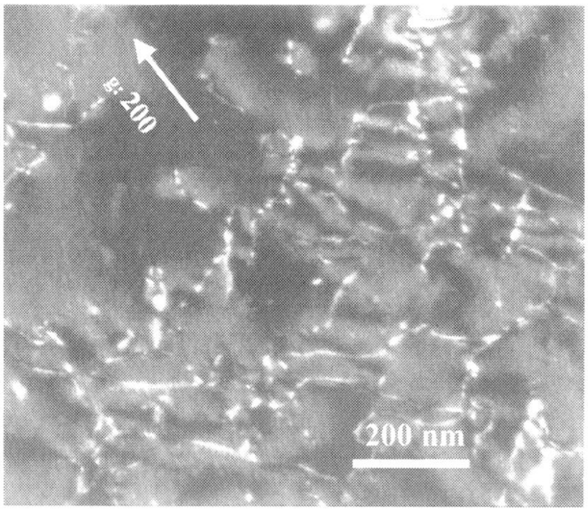

Figure 4 (b) WBDF image of the welded part of 5 % strained FS weld A.

Figure 5 shows an image of dislocation-lines on a slip plane {111} of a 5 % strained FS weld A. Long and uniformly distributed dislocation lines are visible, and small defects and bright dots are hardly observed. A comparison of Fig. 5 and Fig. 4(a) makes us speculate that dislocation motions in FS weld A are easier than those in the base material.
From the results mentioned above, it is found that a significant decrease in grain-size, and also a few decrease in densities of dislocations and small defect occurred due to FSW process. The decrease in grain-size suggests that a dynamical recovery occurred and consequently, a relaxation of internal stress and a softening in the base material occurred. Therefore, it is considered that one of the origins for the

decrease in yield stress of FS weld A is the dynamic recovery. About the increase in strain-hardening rate of FS weld A, two possibilities are speculated on a prediction that strain hardening results from interactions between dislocations and obstacles such as dislocations, precipitates and grain boundaries: 1) Figures 4(b) and 5 suggest the moving distance of dislocations in 5 % strained FS welds is longer than that of 5 % strained base materials. As the result, it is considered that the probability of the interactions in FS welds is higher than that in base materials. This leads to increase the strain-hardening rate of FS welds. 2) The flow stress of base materials is higher than that of FS welds. This higher flow stress might help to enhance for dislocations to surmount the obstacles due to the interaction. This leads to decrease the strain-hardening rate of the base materials. In addition, the problem about elongation of FS welds will be a future one because of lack of TEM observation results.

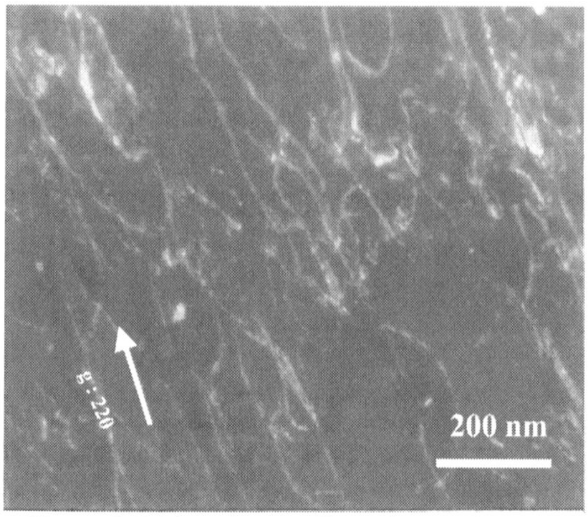

Figure 5 Dislocation configuration on {111} of a welded part of 5% strained FS weld A.

4. Summary

1) A significant difference of microstructure between the base material and a FS weld A due to FSW process is a decrease in grain size. The decrease was caused by a dynamic recovery during the process.
2) The dynamic recovery occurs a softening of FS weld A and a decrease in the yield stress.
3) Dislocation density in the 5 % strained base material is much higher than that of a 5 % strained FS weld A. This result suggests that a strain-hardening rate of base material is saturated. Observed long and isolated dislocations in the 5 % strained FS weld A suggest that these dislocations lead to increase the strain-hardening rate.

Acknowledgments
Authors wish to thank Messrs. Hinoue R, Kashiwa Yand Higashiyama A, graduate students of Okayama University of Science for preparing TEM specimens.

5. References

[1] Z. Y. Ma 2008 *Metallurgical and Materials Transactions A* vol 39A pp 642-658
[2] Murr L E, Liu G and McClure J C 1998 *Journal of Materials Science* vol 33 pp 1243-1251
[3] Reddy G M, Mastanaiah P, Prasad K S and Mohandas T 2009 Transactions of the Indian Institute of Metals vol 62 pp 49-58
[4] Woo W, Balogh L, Ungár T, Choo H and Feng Z 2008 *Materials Science and Engineering A* vol 498 pp. 308-313
[5] Yokoyama T, Nakai K, Sukedai E, Journal of JSEM, Special issue (2011) SS210-SS215
[6] Cockayne D J H, Ray I L F and Whelan M J 1969 Philos. Mag. vol 20 1265

Microstructure of Ti-6Al-4V produced by selective laser melting

M Simonelli[1]**, Y Y Tse**[1] **and C Tuck**[2]
[1] Department of Materials, Loughborough University, Loughborough, LE11 3T, UK.
[2] School of Mechanical and Manufacturing Engineering, Wolfson Building, Loughborough University, Loughborough LE11 3TU, UK

E-mail: M.Simonelli@lboro.ac.uk

Abstract. Ti-6Al-4V is the most widely used titanium alloy. Manufacturing of Ti-6Al-4V components using novel additive processing techniques such as selective laser melting is of great interest. This study focuses on the microstructure characterisation of Ti-6Al-4V components produced by selective laser melting (SLM) with full (Ti-6Al-4V base plate) and partial (Ti-6Al-4V needle-shaped bed) support. The starting material, a plasma atomised powder, and the component products are studied using various microscopy techniques including optical, scanning electron and transmission electron microscopy and electron backscattered diffraction (EBSD). Powder particles are fully dense, possess a spherical shape and are composed of acicular α phase. The as-built material shows oriented acicular martensitic phase with well defined columnar grains. The morphology of martensitic phase and microstructural evolution will be discussed in relation to the SLM processing parameters employed and the different cooling rates experienced by the components.

Introduction

Recently, selective laser melting (SLM) of powder materials has been proposed as an alternative manufacturing route to produce Ti-6Al-4V (Ti64) components, as this process, similar to other rapid prototyping techniques, allows a production of near-net-shape parts with high material utilization and freedom of design[1]. SLM products do not require further downstream processes and therefore this technology is suggested to reduce costs and lead time on production. One of the most attractive advantages offered by the SLM manufacturing technique is that it allows tailoring, to some extent, of the build microstructure by tuning the process parameters in a strategic manner [2]. In particular, it has been demonstrated that the quantity of energy delivered to the metallic powders (i.e. energy density) plays a crucial role on the microstructure of the SLM products [1,2]. When it comes to Ti-6Al-4V, the fundamental question is whether the SLM Ti64 products could have mechanical properties comparable to the wrought or cast Ti64. The intrinsic characteristic of the SLM technology is the high solidification rate experienced by the powder material which determines, in turn, a characteristic fine microstructure with low ductility [3].This study is aimed at understanding the origin and the evolution of the microstructure of Ti-6Al-4V parts produced by SLM. Attention was focused on the stability of the process and the role of the supporting structures on the microstructural evolution of the built components.

Materials and methods

The powder material used for this study was Ti-6Al-4V grade 23 type 5 with nominal chemical composition as shown in Table 1. All the SLM components were built using a MTT SLM 250 machine; the adopted process parameters are listed in Table 2. The volumetric energy density E was calculated as:

$$E=\frac{P}{v\,d\,h}$$

where P is the laser power (W), v is the scanning velocity (μm/s) given by the ratio of the point distance (μm) to the exposure time (μs), h is the hatch spacing (μm) and d is the layer thickness (μm). The parts were built in a controlled Ar atmosphere to avoid any possible oxygen contamination. The build plate under the powder bed was kept at a temperature of 65°C, in order to reduce the residual stresses that might develop during the process. Simple cubic specimens with volume of 1 cm^3 were built. SLM of Ti-6Al-4V powders was carried out directly on top of a Ti-6Al-4V build plate and onto Ti-6Al-4V needle-shaped supports respectively using the same scanning strategy. These latter structures supported the parts along their entire contour and only a part of their bottom surface through thin diagonal walls. Different supports provided different contact area and hence different thermal histories to the built component.The microstructural evolution of the SLM built components on different heat sink were studied in three orthogonal planes using optical microscopy, scanning and transmission electron microscopy (SEM and TEM). Backscattered electron diffraction (EBSD) was used to study the phase and orientation distribution of Ti64 powder and built parts. The samples were ground on SiC papers and polished with a diamond paste of 9 μm first, and a SiO_2-H_2O_2 solution after. Kroll's reagent was used as etching agent. The 3-mm diameter thin TEM foil were prepared using a twin-jet electropolisher in an electrolyte of perchloric acid (10%), ethanol and distilled water at -20°C. TEM was carried out using a Jeol 2000FX microscope operated at 200kV.

Table 1 Chemical composition of Ti-6Al-4V powders

Element	N	C	H	Fe	O	Al	V	Ti
wt%	0,03	0,08	0,01	0,02	0,13	6,50	4,50	Bal.

Table 2 Process parameters

Laser power [W]	200
Hatch spacing [μm]	180
Layer thickness [μm]	50
Point distance [μm]	50
Exposure time [μs]	251
Energy density [J mm^{-3}]	11.1 * 10^{-5}

Results and discussion

The Ti64 powder were produced by a plasma atomization process and therefore appear spherical, smooth, fully dense and with few satellites as shown in Figure 1a.The measured particle size distribution was within 15 and 70 μm, however, about 75% of the examined particle size ranged between 25 to 50 μm. Preliminary EBSD results show that the particles are randomly oriented and contain lamellar α phase (Figure 1b and c). Due to the line and layer-wise building strategy, the microstructure of the SLM samples shows different morphology when viewing from different planes. A commonly used coordinate system is the scanning plane *x-y* and the build direction *z*. The microstructure of the top view (*x-y* plane) consists of features that are correlated to the scan strategy and in particular to the heat conducting direction. Coherently to what has been reported by Thijs et al. [1], individual laser scan tracks can be identified on Figure 2 which shows the top surface of the specimens built on needle supports. Each laser scan track interacts with a portion of material about 235±15 μm wide. When the top surface is polished and etched it reveals a fully dense microstructure (no pores are observed) with relatively coarse α phase were needles of martensitic α' emerge as dark contrast. It should be noted that the top surface was scanned twice in order to improve the overall geometrical accuracy and surface finish of the component. This morphology is similar to what Ahmed et al. have reported [4].

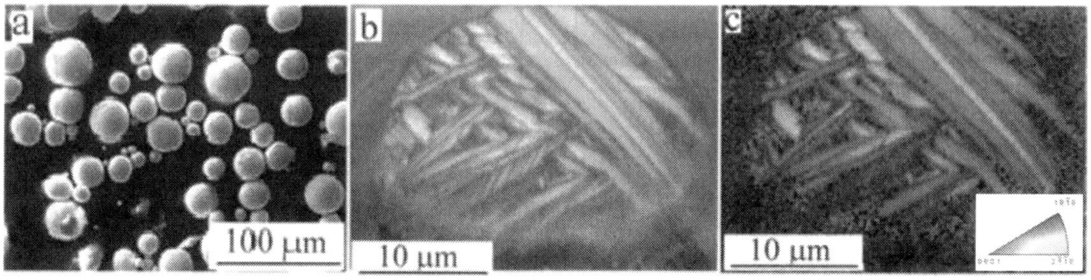

Figure 1. a) SEM micrograph of Ti-6Al-4V powder particles, b) EBSD image quality map of a randomly selected particle, c) Orientation map of a particle and inset showing the inverse pole figure colour scheme

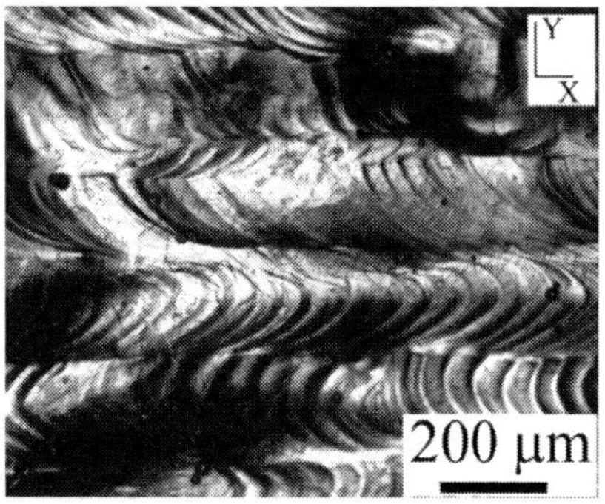

Figure 2. Top surface of the as-built component

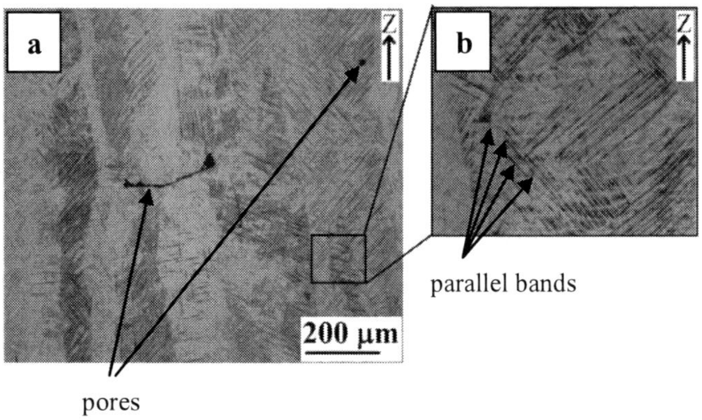

Figure 3. a) Frontal cross section exhibiting columnar grains and pores, b) parallel bands as marked by arrows

β phase was not detected by X-ray diffraction and EDS analysis as a result of the high cooling rate that occurs during SLM. The microstructure of the bottom surface (*x-y* plane) is slightly different from that observed on the top surface: spherical pores can be noted and the volumetric fraction of acicular martensite increases. The microstructure of the front and side cross-sections (*y-z* and *x-z* plane, respectively) reveals information regarding to the thermal history that the layers have experienced. The observed porosity has to characteristic morphologies: pores are flat and oriented in the *x-y* plane or spherical (pin-hole morphology) as shown in Figure 3a. As re-heating of the previously solidified layers gives rise to re-growth, morphology and microstructure of the components differ along the build height. Columnar grains are observed in cross sections (Figure 3a). According to the present results, two different regions can be identified: 1) a region that goes from the middle to the top portion of the specimen and includes the external surfaces (wall surfaces) 2) a region representative for the first layers in contact with the supportive needles (bottom region). Near the external surfaces (wall surfaces) and from the middle to the top region, the morphology exhibits a colony pattern with parallel acicular needles (Figure 4a). These etch resistant needles are considered martensitic α'regions between alpha lamellae [5]. At higher magnification it is possible to distinguish smaller needles that can be considered as secondary α structures, probably grown because of the subsequent heating and cooling that occur during the SLM process. Most of the acicular needles are arranged at ± 45° to the z-axis in accord with the preferential planes of crystallographic growth (Figures 3a and 4a). Further TEM investigations and XRD are required to confirm the crystal structure and orientation of the martensitic needles. Beside the slight crystallographic difference, it has been reported that α' structures are generally contain a high dislocation density, sometimes twins and higher content in V with respect to HCP α lamellae as shown in Figures 5 a and b [5]. In the bottom region, the morphology of the martensitic α' needles shows a basket-weave pattern (Figure 4b); as it can be noticed, some of the acicular needles appear entangled as a result of competitive growth. This morphology is coherent to that described by Baufeld et al. in wire deposited Ti-6Al-4V specimens [6].

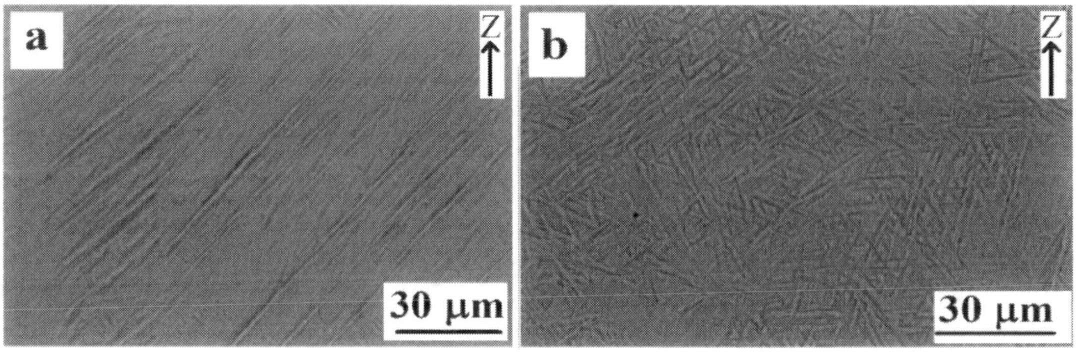

Figure 4. a) Typical wall surface and top region microstructural morphology, b) baske-weave morphology found prevalently in the bottom portion of the specimens

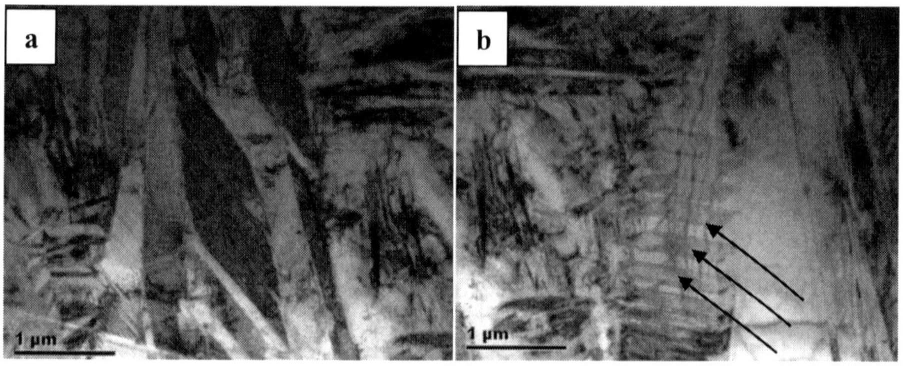

Figure 5. Bright field TEM micrograph showing a) probable α' lamellae, b) stacking faults (arrows) within the lamellae

In general, oriented structures were observed at wall surfaces while basket-weave morphology was present from the bottom to the middle region of the specimens. It is noteworthy that basket-weave morphology is not observed in correspondence of the needle supports but just in the zone that where in contact with the loose powder. Since the heat conduction is lower in the zones without needle supports, the basket-weave structure might be associated with a slower cooling rate and different thermal cycling. In the middle portion, the processed components show a layered structure and elongated columnar grains which grow across the layer (Figure 3a). The columnar grains are a result of epitaxial growth of the previous layer upon successive laser scans [7].Contrary to the result obtained by Thijs et al. [1], the observed columnar grains do not possess an inclination with respect to z direction (Figure 3a): this is probably due to the different adopted scanning strategy. The average width of the columnar grain is of 210±50 μm. Interestingly, the ratio of the scan track width to the columnar grain width is approximately unitary. Finally, the middle portion of specimens displays periodic bands parallel to the base substrate (Figure 3b): these parallel bands, tend to disappear near the top surface of the components. Focused ion beam assisted TEM and EDS analysis are on-going to understand whether these structure could be intermetallic compounds (i.e. where micro-segregation has occurred) or areas with a different specific crystallographic orientation. The microstructures of the samples built on the Ti64 plate show similar feature as the ones built on needle supports; however, smaller pores are observed.

Conclusion

Selective laser melting has proved to be a new and promising method to directly fabricate metal parts. In this work, Ti-6Al-4V cubic blocks were built and characterized; the resulting microstructure is different from that observed in cast or wrought materials but given its complexity further investigations are required. The microstructural features in the as built SLM component include fine martensitic and secondary alpha phase, oriented columnar grains (that probably imply a degree of anisotropy in the build direction), flat defects (pores) and periodic band structures.

References

[1] Thijs L, Verhaeghe F, Craeghs T, Humbeeck JV, Kruth JP. 2010 *Acta Materialia* **58** 3303-3312.
[2] Yadroitsev I, Thivillon L, Bertrand P, Smurov I.. 2007 *Applied Surface Science* **254** 980-983.
[3] Facchini L, Magalini E, Robotti P, Molinari A, Hoeges S, Wissenbach K. 2010 *Acta Materialia* **58** 3303-3312.
[4] Ahmed T, Rack H. *Materials Science and Engineering A* 1998 **243** 206-211.
[5] Qazi J, Rahim J, Fores F, Senkov O, Genc A. 2001 *Metallurgical and Materials Transactions A* **32** 2453-2463.
[6] Baufeld B, Brandl E, van der Biest O. *Journal of Materials Processing Technology* 2011 **211** 1146-1158.
[7] Kelly S, Kampe S. 2004 *Metallurgical and Materials Transactions A* **35** 1861-1867

Journal of Physics: Conference Series **371** (2012) 012085 doi:10.1088/1742-6596/371/1/012085

An electron microscopy study of the effect of Ce on plasma sprayed bronze coatings

Li Wensheng[1], S.C. Wang[2], Chao Ma[2], Wang Zhiping[1]

1. State Key Laboratory of Advanced Nonferrous Materials, Lanzhou University of Technology, Lanzhou, 730050, China
2. National Centre for Advanced Tribology at Southampton, School of Engineering Sciences, University of Southampton, SO17 1BJ, United Kingdom

Abstract: The Cu-Al eutectoid alloy is an excellent material for mould due to its superior low friction. The conventional sand casting technique, however, is not feasible to fabricate high Al bronze because of high hardness and brittleness. Plasma arc spray has been used to produce high Al/Fe bronze coatings for mould. The inherent impurities such as H, O, N, S during the spray, however, may affect the coating's mechanical strength. One approach is to utilise the active rare earth Ce to clean up these impurities. The study is to investigate the effect of Ce on the microstructure, which has few reported in the literature.

1. Introduction

The copper based aluminium (5-12 wt.%) bronzes possess low wear rate and coefficient of friction, and have been widely used for tool, gears, bearings, valves, propellers especially on drawing and rolling die industry. Their hardness, which is between 190-240 HB, is not high enough to make a more precision die finish. Therefore, new interest has been raised on aluminum bronze with Al contents over 12 wt.% with the hardness at 400 HB [1,2]. Fe was also added in aluminum bronzes to enhance the corrosion resistance as long as the iron remains in solution and not precipitated as pure iron [3]. Since the conventional sand casting techniques are not feasible to fabricate Al/Fe bronze over the equilibrium solubilities (9.4 wt.% Al and 3.5 wt% Fe), thermally plasma spray has been attempted to make a saturated Cu-14Al-4.5Fe (wt.%) copper–aluminum–iron coating. As the unavoidable impurities such as H, O, N, S will affect the coating's mechanical strength especially the interface adhesion strength, the rare earth Ce has been added to clean up these impurities. The paper aims to study the Ce effect on the coating microstructure. Our previous study [4] showed the possible intermetallics, as listed in Table 1. Due to the lattice similarity among these phases, X-ray diffraction was found difficult to identify the structures. Transmission electron microscopy (TEM) is therefore used to investigate the microstructure with and without Ce, in order to understand the effect of Ce on the microstructure and properties.

Table 1 Characteristics of the equilibrium phases in aluminum bronze

Phases	Crystal Structure	Lattice Spacing	Micro-Hardness (HV)
α (Cu)	A1 ($Fm\bar{3}m$)	0.364 nm	200-270
γ_2 (Cu_9Al_4)	$D8_3$ ($P\bar{4}3m$)	0.869 nm	360-570
β' (Cu_3Al)	L12 ($Pm\bar{3}m$)	0.353 nm	290-407
κ_1 (Fe_3Al)	DO_3 ($Fm\bar{3}m$)	0.571 nm	>700
κ_2 (FeAl)	B2 ($Pm\bar{3}m$)	0.29 nm	>650

2. Experimental

The Cu-Al-Fe alloys with free Ce and 0.6 wt.% Ce respectively were melted at 1200°C ~ 1260°C, followed by water cooling atomization to form the alloys powders with the compositions listed in Table 2. These powders were then plasma sprayed thermally on mild steel. The detail was reported in ref [4]. Rockwell hardness was tested on the HD1-187.5 sclera-meter with a load of 150 Kg. The specimens were etched using a solution consisting of 5 g $FeCl_3$, 10 ml HCl and 100 ml distilled water. The image observation and element mapping were performed in scanning electron microscope (SEM) JSM6500F and EPMA-1610 Electronic Probe Microanalysis equipped with an OXFORD ISIS 300 EDS analyzer. The TEM specimens were prepared using electro-polishing solution of 20% HNO_3, 15% 2-butoxyethanol and 65% methanol in volume, and JEM3010 at 300 kV was used for analysis.

Table 2 Chemical composition of Cu-14Al-4.5Fe alloy

Ingredients	Cu	Al	Fe	Co	Ni	Others
Weight percent (%)	75-80	13-16	2-4.5	0.5-0.8	0.4-0.6	1.0-2.6

3. Results and Discussion

The Ce-free coating consists of inhomogenous intermetallics ranged from a few to tens of microns, as shown by backscattered electron (BSE) image in Fig.1. These intermetallics show dark contrast in the element map of Cu, the bright contrast in the element map of Fe and grey colour in the element map of Al. It indicates that they are most likely κ_2 (FeAl). It is also interesting that the intermetallics in the element map of Fe appear larger than those in the element map of Cu, as compared in the same area (e.g. highlighted circle). The high magnified color image in the insert of Cu map shows clearly that a thin layer surrounds the κ_2 phase. The composition is close to γ_2 (Cu_9Al_4) with Iron rich.

In contrast, the Ce added coating has more homogenous intermetallics at roughly 10 μm as shown in Fig. 2. Again the intermetallics in the element map of Fe are larger than those in the element map of Cu. These intermetallics have the dominating κ_1 (Fe_3Al) and γ_2 (Cu_9Al_4) with Fe rich. Figs. 3 & 4 show the magnified secondary electron (SE) images in the rectangular areas of Figs. 1 & 2, respectively. Two coatings have shown the different microstructures: for Ce-free coating, EDS confirm that there exist large γ_2 intermetallics (position 1 in Fig. 3), and a number of small dispersoid particles κ_2 (position 2, in a diameter of a few hundred nanometre). The high concentration of Cu in κ_2 phase is due to the contribution from the matrix β'(Cu_3Al). For Ce-added coating, the matrix β' zones are precipitates free (Fig. 4), and the κ_1 and γ_2 particles seem to coexist in "boat" shaped areas (Position 2 & 3).

Journal of Physics: Conference Series **371** (2012) 012085 doi:10.1088/1742-6596/371/1/012085

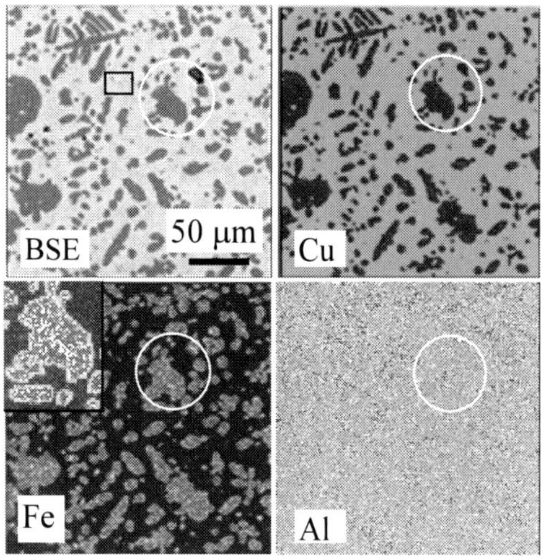

Fig.1 BSE image and element maps of Cu, Fe & Al of the Ce-free coating

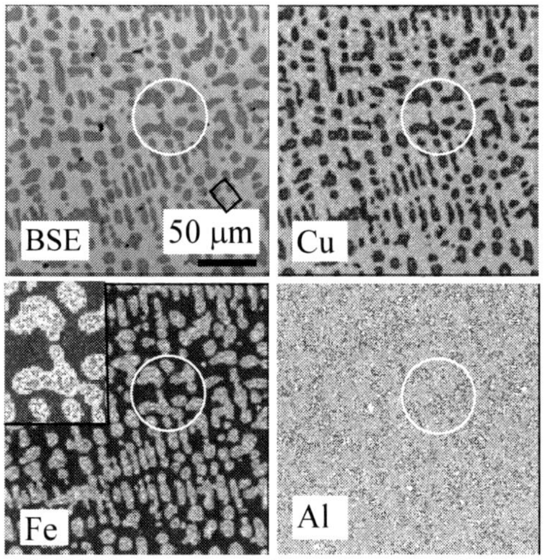

Fig.2 BSE image and element maps of Cu, Fe & Al of the Ce-added coating

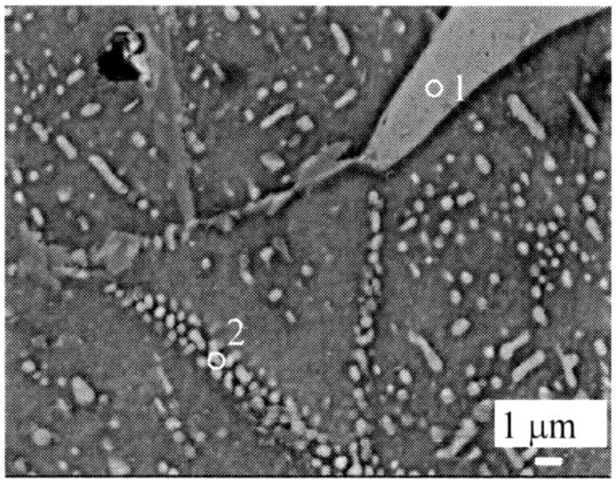

Fig.3 Magnified SE image on the Ce-free coating

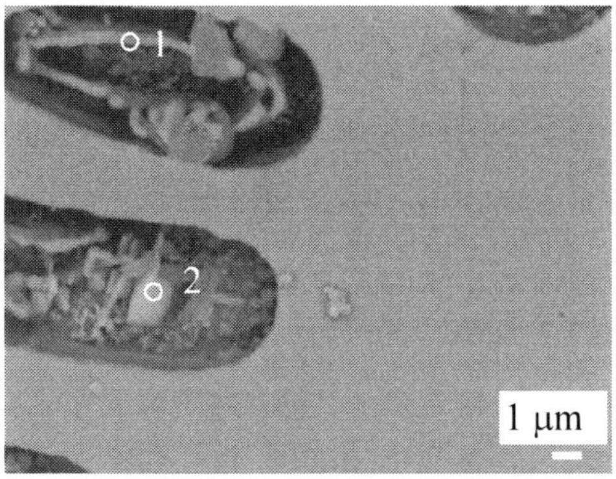

Fig.4 Magnified SE image on the Ce-added coating

Position	Al(at.%)	Fe(at.%)	Cu(at.%)	Phase	Position	Al(at.%)	Fe(at.%)	Cu(at.%)	Phase
1	16.3	6.8	76.9	$\gamma_2(Cu_9Al_4)$	1	17.5	6.8	75.7	$\gamma_2(Cu_9Al_4)$
2	16.7	22.6	60.7	κ_2 (FeAl)	2	20.0	63.5	16.5	$\kappa_1(Fe_3Al)$

EDS in SEM could be used for the preliminary work on phase identification via composition, but it could be uncertain without diffraction analysis. TEM has therefore been used for the structure determination. The bright field (BF) of the Ce-free coating in Fig. 5a shows the round dispersoid at a hundred nanometres. The corresponding diffraction pattern in Fig. 5b to position 1 in Fig. 5a confirms the matrix indeed as $\beta'(Cu_3Al)$. The very fine grey particles (~ten nanometers) are the artifact caused by the eletropolish. The dispersoid (position 2 in Fig. 5a) has been determined as κ_2 (FeAl) as indexed in Fig. 5c (some extra spots come from the matrix). For Ce-added coating, the matrix has again $\beta'(Cu_3Al)$ indexed in Fig. 6b for Position 1 of Fig. 6a. The micron-sized κ_1 (Fe_3Al) phase has been confirmed in Fig. 6c for Position 2 of Fig. 6a. The different structure of the Fe-containing intermetallics could result in different hardness on coatings (Table 1). In addition, stacking faults have been observed with Ce coating as shown in Fig. 6d. It indicates that the addition of Ce has reduced the stacking fault energy (SFE). It has been reported that the wear rate would be decreased with the decrease of SFE in Cu-Al alloy [5]. This observation is consistent to our tribological results that the coatings with Ce has reduced the wear rate by 10%, and enhanced the strengthening by 10% [4].

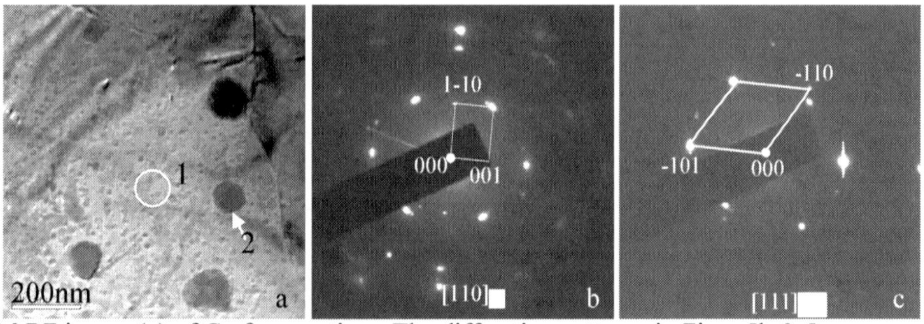

Fig. 5 TEM BF image (a) of Ce-free coating. The diffraction patterns in Figs. 5b & 5c corresponding to Positions 1 & 2, respectively.

Position	Al(at.%)	Fe(at.%)	Cu(at.%)	Mn(at.%)	Co(at.%)	Ni(at.%)	Phase
1	19.3	2.8	77.9	-	-	-	β'(Cu_3Al)
2	24.1	15.2	44.4	1.1	3.1	12.0	κ_2 (FeAl)

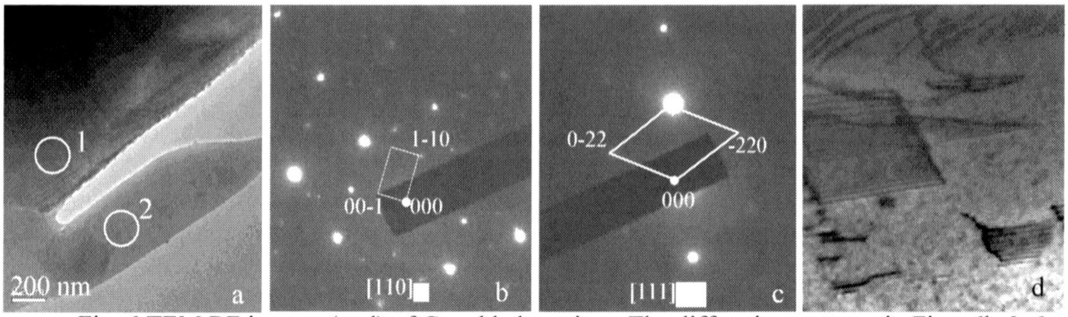

Fig. 6 TEM BF images (a, d) of Ce-added coating. The diffraction patterns in Figs. 6b & 6c corresponding to Positions 1 & 2, respectively.

Position	Al(at.%)	Fe(at.%)	Cu(at.%)	Mn(at.%)	Phase
1	23.4	1.9	73.8	0.83	β' (Cu_3Al)
2	21.3	69.2	8.2	1.3	κ_1 (Fe_3Al)

4. Conclusions

The effect of Ce addition on the microstructure of thermally plasma spray coatings was studied by electron microscopy. For Ce-free coating, the intermetallics were κ_2 (FeAl) surrounding by γ_2 (Cu_9Al_4) with Fe rich phase. In contrast, κ_1 (Fe_3Al) were dominating intermetallics in Ce-added coating. This indicates that Ce stimulates the transition of the Fe-containing intermetiallics from FeAl to Fe_3Al. Stacking faults were also observed in the coating with Ce addition, which is responsible for the reduced wear rate and increase of the strengthening.

References

[1] Heide E, Stam ED, Girand H, Lovato G, Akdut N, et al, 2006 Wear **261** 68-73
[2] Li WS, Wang ZP, Lu Y, Jin YH, Yuan LH, Wang F 2006 Wear **261** 155-163
[3] Moreton BB 1985 Corr. Prev. Control **6** 122-126
[4] Li WS, Lu Y, Yuan K, and Yuan C 2011 Journal of Rare Earths **29** 363
[5] Wert JJ and William MC 1988 Wear **123** 171

Understanding surface oxidation in stainless steels through 3D FIB sequential sectioning

Sergio Lozano-Perez[1], Karen Kruska[1], Ilya Iyengar[2], Takumi Terachi[3] and Takuyo Yamada[3]

[1]Department of Materials, University of Oxford, Parks Rd, Oxford, OX1 3PH, UK
[2]Winchester College, College Street, Winchester, SO23 9LX, UK
[3]Institute of Nuclear Safety System (INSS), 64 Sata, Mihama-cho, Mikata-gun, Fukui 919-1205, Japan
Sergio.lozano-perez@materials.ox.ac.uk

Abstract. Focused Ion Beam (FIB) 3D sequential sectioning has been used to reconstruct the morphology of surface oxides in 304 stainless steels in nuclear reactors in order to characterize environmental degradation. Several coupon specimens with prior cold work and oxidized under applied stress have been studied in detail. A novel statistical approach based on more than 250k individual measurements will be presented as an alternative to conventional methods.

1. Introduction

The characterization of oxide layers is a well-studied topic and in the literature, where we can find studies based on surface techniques, TEM or atom-probe tomography (APT). There is, however, a gap between the highest resolution atomic characterization with APT (which provides information on just a few nm3) and a more macroscopic approach such as SIMS or Auger depth profiling (which provides averaged information on areas of a few mm2). In this work, we demonstrate that 3D FIB sequential slicing can provide a high spatial resolution over volumes of several micrometres with minimum sample preparation.

Low potential stress corrosion cracking (LPSCC) of stainless has been identified as a potential problem for pressurized water reactors (PWRs). Although initially thought to be immune, autoclave testing under controlled environments recently confirmed that highly cold-worked stainless steels are susceptible to SCC even in PWR primary water conditions [1]. Not much later, SCC was reported in an operating PWR plant [2]. A necessary step towards understanding the mechanisms controlling SCC and the influence of cold-work is the development of better methods for characterizing oxidation at the nanoscale. It is well known that when 304 stainless steel is oxidized under PWR primary water conditions it develops a dual oxide layer with an inner layer of Cr-rich spinel and an outer layer of Fe-rich spinel [3]. If the alloy has been previously cold-worked, the oxidation process is much more complex and enhanced oxidation of twin deformation bands and high-dislocation-density regions has been observed [4]. In this work, we have focused on the inner Cr-rich oxide layer and will demonstrate that oxidation can be better described statistically, instead of by a compositional line profile or by a single diffusion coefficient.

2. Materials and methods

A nuclear grade 304 stainless steel (SUS304) was solution treated at 1060 °C for 100 min and water-quenched. The material was then unidirectionally cold-rolled to a 20% reduction in thickness (referred to as 20% CW). The surface was polished down to mirror finish with colloidal silica. Oxidation tests were performed at the INSS laboratories (Japan) in simulated PWR primary water (hydrogenated water, containing 500ppm B, 2ppm Li and 30cc STP Hd/kg H_2O dissolved H) in an autoclave for 1500 hrs at 360°C. The coupon specimens were oxidized stress-free and the tensile test specimens under a load below the yield stress of the material

FIB sectioning was carried out on a Zeiss NVision 40 dual beam FIB instrument (2-30kV Ga+ incident beam energy with currents of 80-1000 pA). Samples were orientated such that the focused Ga+ ion beam is at normal incidence to the cross-sectional surface (x-y plane) and SEM secondary electron (SE) images were taken of each sequential 2D slice (in the x-y plane) at a tilt angle of 54. The image stretching caused by the tilt was automatically corrected by the microscope software and, therefore, all SEM images shown in this work have been tilt corrected. A beam current of 700 pA was used to carry out the sectioning, producing slices every 40nm. The 2D SE images were aligned and corrected for drift using cross-correlation of static features on the sample original surface following the method described in [5]. Then, an area containing the region of interest and with its top side coinciding with the sample's top surface is extracted from each image and arranged as a 3D dataset.

The 3D reconstruction was carried out using Mercury Software's Avizo .6.1. The different regions of interest appearing in the 2D images (e.g. metal, oxides) were automatically selected (using intensity thresholding methods) or manually drawn when no automated segmentation method was good enough to avoid the channelling contrast present in some SE images. The dimensions of the reconstructed volumes are typically ~20x20x5μm with a voxel size of ~11x11x40nm. The Cr-spinel segmentation was performed in a semi-automatic way, relaying on intensity thresholding to separate the Cr-rich oxide from the other phases and drawing the boundaries manually when this was not satisfactory (See Figure 1).

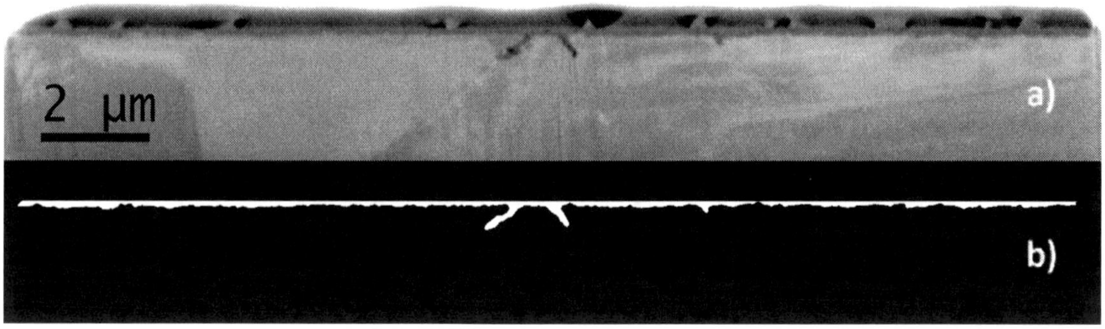

Figure 1: (a) SEM SE image showing the cross-sectional view used to reconstruct the surface oxides. Deposited carbon appears darker at the top, while the Cr-rich oxide appears covering the top surface with a deeper oxidised region along the grain boundaries; (b) Binary mask representing the Cr-rich spinel oxide after segmentation.

3. Results

For all the four specimens characterized, a 3D model of the inner Cr-rich oxide was created (See Figure 2). The depth of oxidation was measured in over 500000 points on average per reconstructed Cr-rich spinel models and subsequently statistically analysed. The nature of the data allowed the individual analysis of features such as grains, grain boundaries (gbs) or twin deformation bands (tbds), as will be shown next.

Journal of Physics: Conference Series **371** (2012) 012086 doi:10.1088/1742-6596/371/1/012086

A Cr-spinel 3D model for the sample oxidized without applied stress is shown in Figure 2a. This region includes two gbs and three grains. A similar model for the sample oxidized under applied stress is shown in Figure 2b, including three gbs and three grains. An individual histogram for these gbs can be found in Figure 3a, as an example of the kind of detailed data that can be extracted. A comparison between the oxidation behavior of the samples oxidized with and without applied stress can be found in Figure 3b. Oxidation data for each of the grains and gbs can be found in table 1.

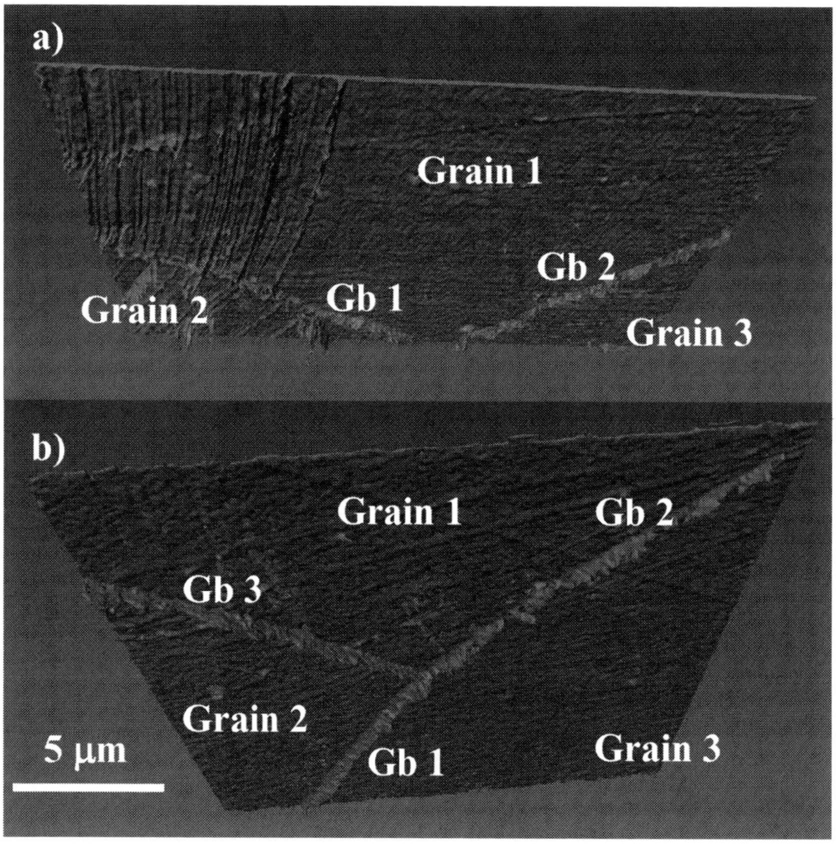

Figure 2: 3D model of the Cr-rich spinel oxide for the sample with 20%CW oxidized without applied stress (a) and under applied stress (b). The longer parallel bands in the top figure are oxidized deformation bands

	No stress - CW		**Stress and CW**	
	Average oxidation	σ	Average oxidation	σ
Grain 1	132	53	177	48
Grain 2	153	93	224	52
Grain 3	105	23	160	42
Grain boundary 1	203	107	394	151
Grain boundary 2	210	123	381	133
Grain boundary 3			426	282

Table 1: Oxidation depths in nm for each of the grains and grain boundaries in Figure 2

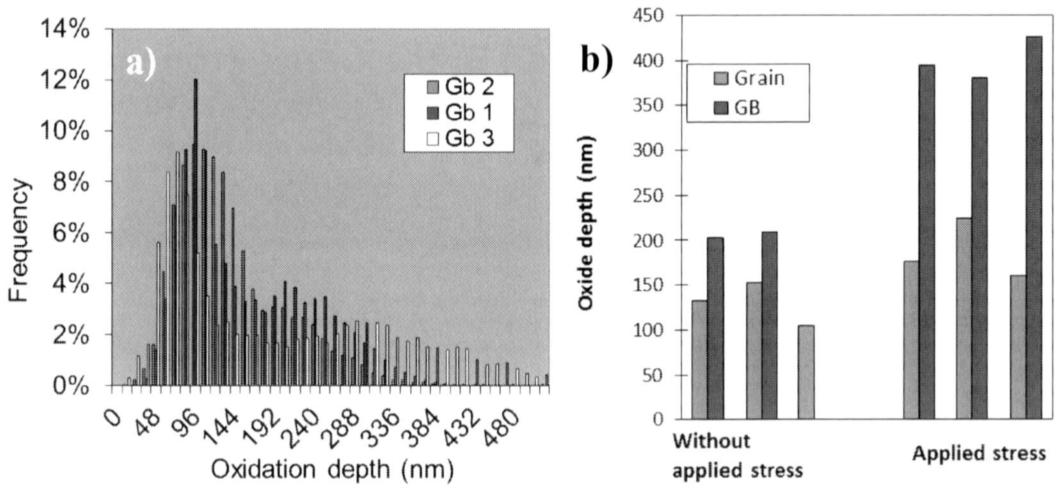

Figure 3: (a) Histogram showing the different oxidation behaviour of the three grain boundaries in the model in Figure 2b. (b) Plot showing effect of applied stress in grain boundary and grain oxidation.

4. Conclusions

Surface oxidation in a matrix of 304 stainless steel samples oxidized under simulated PWR primary water conditions has been characterized by 3D sequential FIB sectioning. The individual effect of applied stress and prior cold work has been studied. It was found that the applied stress greatly increases the oxidation depth, an effect which is magnified at grain boundaries. The technique has allowed the reconstruction of volumes containing several grains and grain boundaries with nm resolution. This remarkable spatial resolution enabled the visualization of preferential oxidation down microstructural features such as grain boundaries and deformation bands. It also produced results of high statistical significance, where a histogram of oxide depths is used to characterize surface oxidation.

Acknowledgements

INSS (Japan) is acknowledged for sponsoring this project and continuous support.

References

[1] Arioka K, Yamada T, Terachi T, Staehle RW. Corrosion **62** 74.

[2] Couvant, T, Legras, L, Pokor, C, Vaillant, F, Brechet, Y, Boursier, JM et al. Canadian Nuclear Society - *Proceedings of the13th International Conference on Environmental Degradation of Materials in Nuclear Power Systems* 2007 p499.

[3] Robertson, J. *Proceedings of the Water Chemistry of Nuclear Reactor Systems* **5** 1989 p254.

[4] Lozano-Perez S, Schröder M, Yamada T, Terachi T, English CA, Grovenor CRM. App Surf Sci **255** 1541.

[5] Schaffer B, Grogger W, Kothleitner G. Ultramicroscopy 2004 **102** 27.

The Influence of Grain Size on Brittle Crack Propagation

S Mahalingam[1] PEJ Flewitt [1, 2]

[1]Interface Analysis Centre, University of Bristol, Bristol BS2 8BS, UK

[2]School of Physics, H H Wills Physics Laboratory, University of Bristol, Bristol BS8 1TL,UK

Email:Sunthar.Mahalingam@bristol.ac.uk

Abstract

Ferritic steels undergo a ductile to brittle transition as the temperature is progressively reduced. The strain annealing method has been developed to produce a microstructure in EN1A ferritic steel that contains a transition from a coarser grain size, ~40 μm, to a finer grain size, ~20 μm. Small pre-notched bend geometry specimens have been tested at -196°C to investigate crack propagation from the coarser to finer grain region and vice-versa. Details of the fracture path have been investigated using a combination of high resolution scanning electron microscopy, electron back scattered diffraction and ion milling and imaging. The results are considered with respect to the transgranular and intergranular modes observed.

Introduction

The fracture of a solid arises from the formation of new surfaces in a body created by thermodynamically irreversible processes. The failure mechanisms including crack initiation and propagation may change with temperature. For example, polycrystalline ferritic steels are ductile at room temperature but when temperature is reduced they undergo brittle failure. The ferritic steels fail by transgranular cleavage at low temperatures on {100} planes, but encounter obstacles such as grain boundaries or sub-boundaries as the crack propagates. Hence there is a need to accommodate grain orientation mismatch as the cleavage crack propagates from one grain to another [1]. Crack arrest at the grain boundaries has been studied by various workers [2]. The recent advancement in high resolution electron and ion beam techniques has allowed the local grain boundary failures as well as general fractographic details to be considered in more detail. In this paper, brittle crack propagation from a coarser to finer and a finer to coarser grain size in ferritic steel is analysed.

Crack Plane Determination

Two surface analysis is a recognised technique for determining the transgranular fracture planes in polycrystalline materials. As described by Davies and Randle [3] this is usually achieved by providing two macro surfaces at right angles to intersect the fracture plane to be evaluated. However, as the grain size decreases it becomes increasingly difficult to maintain the accuracy of the determination using this method of preparation. Moreover, to select the specific feature to be measured often requires significant destruction of the specimen containing the fracture surface. As shown by Hughes et al. [4] focussed ion beam milling provides a tool for interrogating 3-D crack morphology. When combined with electron backscattered diffraction, EBSD, it becomes a powerful tool for undertaking two surface analyses of fracture planes. A difficulty is that the surface to be oriented by EBSD has to be inclined at an angle of 20° to the incident electron beam, 70° to the horizontal, and there has to be an unimpeded path for the diffracted beam to reach the detector. Two approaches can be adopted to achieve these requirements: (i) FIB milling to introduce two surfaces intersecting the crack surface at 90° at a position on a pre-cut edge of the specimen, Figure 1(a) and (ii) Initially FIB milling a single trench parallel to the incident ion beam, Figure 1 (b), stage (i), and then rotating the specimen, Figure 1(b), stage (ii), through 45° and creating a new surface, surface 3, that intersects the fracture surface. The specimen is then rotated by 25° as shown in Figure 1(b), stage (iii) to provide a free path to the detector for the diffracted beam. Hence a pre-selected feature can be evaluated at any position on a fracture surface.

Experimental

A section of a 16 mm square EN1A ferritic steel bar was heat treated at the temperature of 920℃ for 1h and air cooled to room temperature. The chemical analysis, (wt%) for this EN1A steel gives 0.06% C, 0.30% Si, 0.78% Mn, 0.015% P, 0.02% Ni, 0.01% Mo, 0.05% S as main elements plus Fe. These, specimens were compressed using a Zwick tensometer to a strain level of 2.2% and annealed at 690℃ for 3hours. As a result of greater near surface strains, this heat treatment resulted in layered grain size microstructure which is used to propagate cracks from coarser to finer grains and vice-versa; Coarser grain size 40 µm and finer grain size 20 µm (mean linear intercept).

An unique three point bend test has been developed to study the crack propagation in this ferritic steel. Specimens with dimensions of 20 mm × 5 mm × 2.7 mm were extracted from the bar by electric discharge machining (EDM) after the heat treatment. The notches were cut to a depth of ~ one third the width (a/w = 1/3) by electric discharging machining with a wire of diameter 0.1 mm. At a temperature of -196℃, the specimens were loaded with a load rate of $3Ns^{-1}$ and three repeated tests were undertaken. Load - displacement was acquired using the embedded software in the system. Fracture surfaces of the failed specimens were examined using the SEM (Hitachi S-2300) at an operating voltage of 25 kV. The high resolution FEI Dual Beam (Helios) work station was used for milling and imaging; a high beam current (11 500pA) to mill and a low beam current (70 pA) to polish specimens. Cross-sectioning, as described in the above section was used to establish the cleavage planes using two surface analyses to provide unique solutions for the orientation of the fracture facets and cleavage planes.

Results

The load-displacement characteristic for the finer to coarser microstructure is shown in Figure 2a. The curve shows two prominent regions before the final fracture. The similar behaviour was observed for all the samples tested. The load-displacement curves for coarser to finer microstructure showed three discrete regions before the final fracture, Figure 2b. There is linearity up to ~180 N and thereafter an increase in energy corresponding to crack arrest and then a sudden drop in load before proceeding to the final fracture. The work of fracture is calculated from the area under the curve, Table 1.

Figure 3a shows the low magnification image of the fracture surface obtained at -196 ℃ for the coarser to finer region microstructure. The corresponding fracture features seen from the notch root are illustrated schematically, Figure 3b. From the notch root to ~ 0.2 mm, in the coarse grain region the cleavage planes are observed, Figure 3c - region X. At this point there is a crack arrest at the finer grain region consistent with the load-displacement curve and this is a region, Y, of intergranular fracture. Figure 3d reveals a cleavage plane meeting at an interface (region Y) where the intergranular failure dominates. The crack then propagates in a cleavage mode until the final failure, Figure 3e-region Z. The crack propagation from finer to coarser grain show continuous cleavage fracture.

FIB cross-sections were milled at the edge of the specimen for cleavage plane evaluation in the coarser grain region and two surface analysis was undertaken as Figure 1 (a). The surfaces were carefully milled using a low beam current to obtain clear bright EBSD patterns. The pole figure for surface 1 is shown in Figure 4(a). The two milled surfaces lie close to {001} and the cleavage plane was established to be {001}. Figure 4(b) shows the EBSD pattern obtained for the finer grain region using the general preparation method, Figure 1(b). Here a cut was made on the surface at 45°, surface 3, to achieve stage (ii), and then the specimen was transferred to the SEM and aligned as shown in stage (iii). In this way clear EBSD patterns were obtained and the main pattern showing the [100] normal, Figure 4(b). The subsequent two surface analysis showed that the cleavage facets in this finer grain region were {001}.

Concluding Comment

Two preparation methods for two surface analyses have been adopted which provide an accurate evaluation for fracture surfaces at smaller grain sizes than the macro-sectioning method as described by, for example, by Davies and Randle [3]. Although more convenient it is limited since macro-sections have to be prepared close to the fracture surface feature to be analysed. The new method described in Figure 1(b) provides a general approach. In particular a fracture facet of interest can be selected at any position on the surface and by adopting the more complex FIB milling sequence described in Figure 1(b) it is possible to undertake a two surface analyses to provide a unique solution. Indeed this can be applied to fracture facets over a wide range of sizes from about one micrometer up to tens of micrometers. In the present case for both the coarser and finer grain size regions the cleavage fracture is {100}.

Acknowledgement: The authors acknowledge the EPSRC for financial support.

References

1. Crocker, A. G., P. E. J. Flewitt, and G. Smith (2005). International Materials Reviews **50**(2): 99-124.
2. Qiao, Y. and X. Kong (2004). Materials Letters **58**(25): 3156-3160.
3. Davies, P. A. and V. Randle (2001). Journal of Microscopy-Oxford **204**: 29-38.
4. Hughes, G. M., G. Smith, A.G. Crocker and P.E.J. Flewitt (2005). Materials Science and Technology **21**(11): 1268-1274.

Table 1 Average values for work of fracture (Nmm) calculated from area under load-displacement curves

Work of fracture (Nmm)	Fine to coarse	Coarse to fine
Total	91±10	106±10
Coarser region	80	21
Fine region	11	85

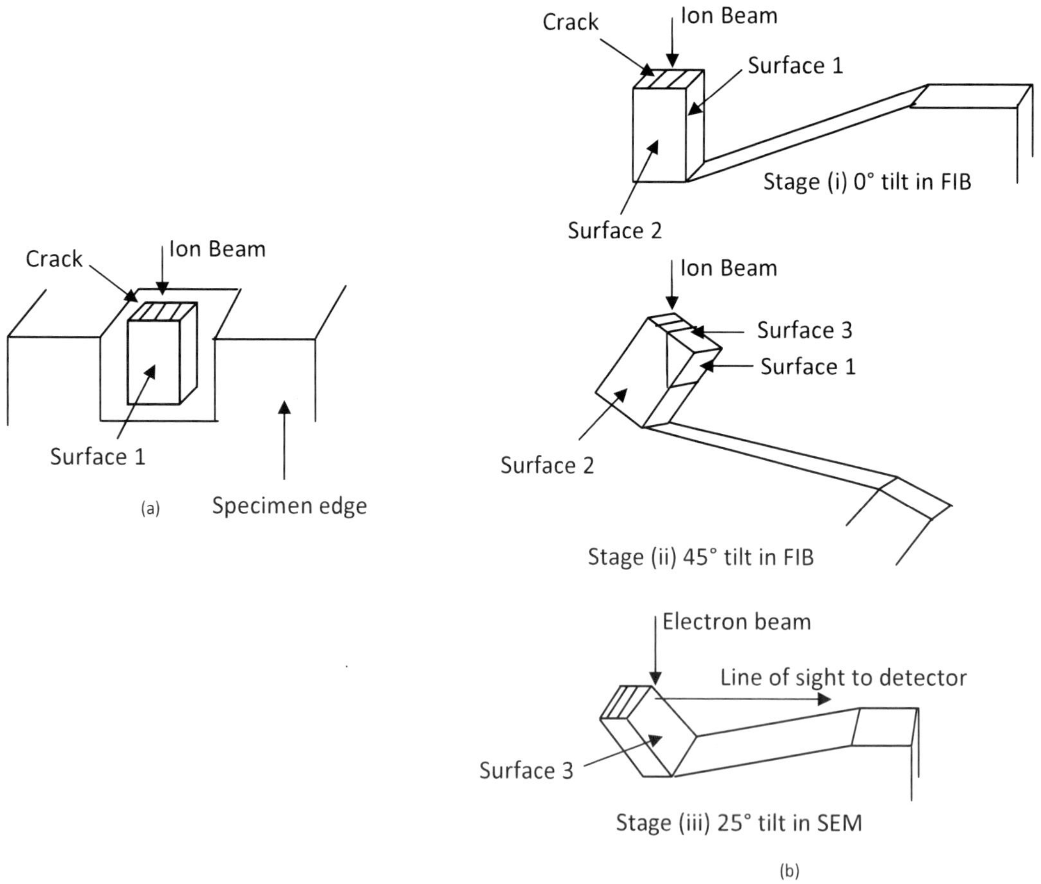

Figure 1 Schematic drawing showing the FIB milled surfaces for EBSD analyses; (a) surface, created at the prior edge of the specimen, (b) the general method, stage (i) surfaces created within the specimen at 90° to each other, stage (ii) showing inclined angle surface 3 prepared at 45°and stage (iii) subsequent rotation to allow EBSD detection

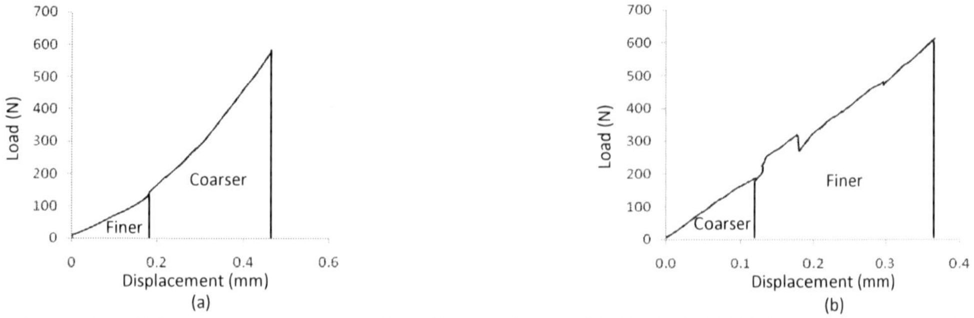

Figure 2 Load-displacement characteristics of three point bend tests at -196 °C (a) finer to coarser grain structure (b) coarser to finer grain structure

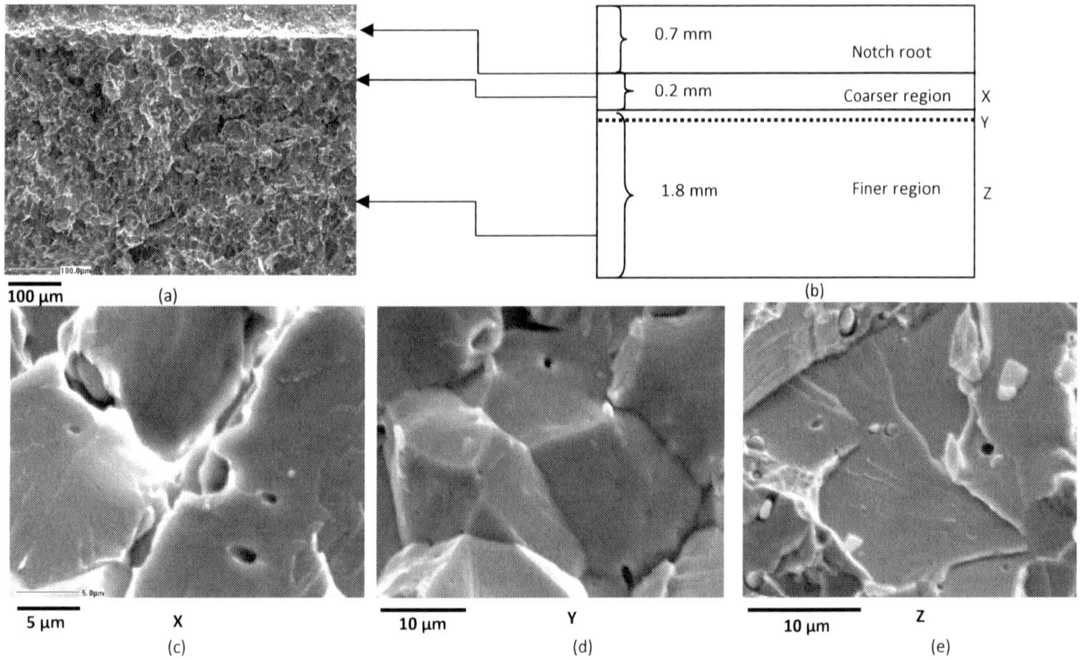

Figure 3 Fractography of coarser to finer grain structure; (a) low magnification SEM image from the notch root, (b) schematic drawing of the microstructural regions from the notch root, (c) transgranular cleavage at region X, (d) intergranular fracture at region Y, and (e) transgranular cleavage at region Z

Figure 4 (a) An Inverse pole figures for the surface1, Figure 1(a); (b) electron back scattered pattern of the surface 3, Figure 1(b) shown [001] normal

AUTHOR INDEX

AUTHOR INDEX

AUTHOR INDEX

Institute of Physics
Dirac House, Temple Back
Bristol BS1 6BE UK

Curran Associates Inc.
proceedings.com